Flavonoids and Their Disease Prevention and Treatment Potential

Flavonoids and Their Disease Prevention and Treatment Potential

Editor

H.P. Vasantha Rupasinghe

MDPI • Basel • Beijing • Wuhan • Barcelona • Belgrade • Manchester • Tokyo • Cluj • Tianjin

Editor
H.P. Vasantha Rupasinghe
Dalhousie University
Canada

Editorial Office
MDPI
St. Alban-Anlage 66
4052 Basel, Switzerland

This is a reprint of articles from the Special Issue published online in the open access journal *Molecules* (ISSN 1420-3049) (available at: https://www.mdpi.com/journal/molecules/special_issues/flavonoids_disease_prevention).

For citation purposes, cite each article independently as indicated on the article page online and as indicated below:

LastName, A.A.; LastName, B.B.; LastName, C.C. Article Title. *Journal Name* **Year**, *Volume Number*, Page Range.

ISBN 978-3-0365-0000-3 (Hbk)
ISBN 978-3-0365-0001-0 (PDF)

Cover image courtesy of H.P. Vasantha Rupasinghe.

Contents

About the Editor

H.P. Vasantha Rupasinghe, Professor and Killam Chair in Functional Foods & Nutraceuticals.

Dr. Rupasinghe is a Full Professor and Killam Chair in Functional Foods & Nutraceuticals at the Department of Plant, Food, and Environmental Sciences of Dalhousie University, Nova Scotia, Canada. He obtained his Bachelor's degree from Sri Lanka and moved to Iowa State University, USA, as a Fulbright Scholar to complete his M.Sc. in Biochemistry. He received his Ph.D. from the University of Guelph, Ontario, Canada. Dr. Rupasinghe's research interests include understanding the health benefits of plant-food flavonoids, as well as exploring novel value-added opportunities for primary agricultural products, especially cool-climate fruits. He has published 197 peer-reviewed articles (author h-index 53 and total citations over 9000), 22 book chapters, and over 250 conference abstracts, and has delivered over 75 invited talks. His contributions were recognized with the Fellow Award of the Canadian Institute of Food Science and Technology (CIFST).

Preface to "Flavonoids and Their Disease Prevention and Treatment Potential"

During the last decade, interest in the use of dietary flavonoids to prevent human diseases and as pharmaceutical leads has been exponentially extended. Flavonoids, a sub-class of plant polyphenols, have exhibited numerous health-promoting physiological benefits in a wide range of investigations, from cell-based assays to randomized, controlled human clinical trials. This e-book consists of an editorial overview, four reviews, and fifteen original research articles published in the Special Issue of *Molecules* titled "Flavonoids and Their Disease Prevention and Treatment". Each of these papers contributes significantly to discovering and validating the beneficial physiological functions and therapeutic potential of dietary flavonoids. Beyond their role as plant food dietary antioxidants, new insights are presented for the potential use of flavonoids—and their derivatives—as effective therapeutics to manage many chronic and metabolic disorders, including certain cancers. Flavonoids also offer promising applications in the management of infectious diseases, including COVID-19.

H.P. Vasantha Rupasinghe
Editor

 *molecules*

Editorial

Special Issue "Flavonoids and Their Disease Prevention and Treatment Potential": Recent Advances and Future Perspectives

H.P. Vasantha Rupasinghe [1,2]

1   Department of Plant, Food, and Environmental Sciences, Faculty of Agriculture, Dalhousie University, Truro, NS B2N 5E3, Canada; vrupasinghe@dal.ca; Tel.: +1-902-893-6623
2   Department of Pathology, Faculty of Medicine, Dalhousie University, Halifax, NS B3H 4R2, Canada

Received: 13 October 2020; Accepted: 15 October 2020; Published: 16 October 2020

In recent years, the interest in flavonoids as dietary bioactives to prevent human diseases, as well as their candidacy as pharmaceutical leads, has exponentially expanded. Flavonoids are a sub-class of plant polyphenols that have been shown to possess numerous health-promoting physiological benefits in a wide range of investigations from cell-based assays to epidemiological and human intervention studies. In this e-book editorial, a brief overview and insights into articles published in the Special Issue of *Molecules* titled "Flavonoids and Their Disease Prevention and Treatment" are provided. It is evident that all these papers contribute toward new knowledge to discover and validate beneficial physiological functions and the therapeutic potential of dietary flavonoids. Beyond their role as biologically active molecules of plant food, new insights are presented for the potential use of flavonoid derivatives as effective therapeutics to manage certain cancers. Flavonoids also offer promising applications in the management of obesity- and inflammation-associated disorders as well as the control of infectious diseases, including COVID-19.

Flavonoids are ubiquitously present in plant-based foods and natural health products. The molecule of flavonoids is characterized by a 15-carbon skeleton of C6-C3-C6, with the different structural configuration of subclasses. The major subclasses of flavonoids with health-promotional properties are the flavanols or catechins (e.g., epigallocatechin 3-gallate from green tea), the flavones (e.g., apigenin from celery), the flavonols (e.g., quercetin glycosides from apples, berries, and onion), the flavanones (e.g., naringenin from citrus), the anthocyanins (e.g., cyanidin-3-*O*-glucoside from berries), and the isoflavones (e.g., genistein from soya beans). Scientific evidence has strongly shown that regular intake of dietary flavonoids in efficacious amounts reduces the risk of oxidative stress- and chronic inflammation-mediated pathogenesis of human diseases such as cardiovascular disease, certain cancers, and neurological disorders [1]. The physiological benefits of dietary flavonoids have been demonstrated to be due to multiple mechanisms of action, including regulating redox homeostasis, epigenetic regulations, activation of survival genes and signaling pathways, regulation of mitochondrial function and bioenergetics, and modulation of inflammation response. The role of flavonoids on gut microbiota and the impact of microbial metabolites of flavonoids on optimal health has begun to unravel. The complex physiological modulations of flavonoid molecules are due to their structural diversity. However, some flavonoids are not absorbed well, and their bioavailability could be enhanced through structural modifications and applications of nanotechnology, such as encapsulation. This Special Issue consists of four review articles on flavonoids and 13 original research articles, which cover the latest findings on the role of dietary flavonoids and their derivatives in disease prevention and treatment.

1. Dietary Flavonoids Exhibit Cancer Preventive Properties

Cancer chemoprevention is defined as the application of safe natural compounds, synthetic molecules, or their combinations to interfere in multistage carcinogenesis. In recent years, interest in plant-based food for reducing the risk of cancer has emerged as a realistic and cost-effective approach [2]. Some of the ancient plant food that has been used as nature's medicine has now been investigated for their disease preventive and treatment properties. Haskap berry, also called blue honeysuckle (*Lonicera caerulea* L.), is a plant food used by Japanese native people for many curative purposes [3]. In this issue, we have made a milestone demonstration that dietary supplementation of haskap berry can suppress the carcinogen-induced lung tumorigenesis in A/JCr mice [4]. Supplementation of haskap berry either before or after the induction of tumors by intraperitoneal injection of tobacco-specific nitrosamine ketone reduced the lung tumor multiplicity, tumor area, and the expression of cancer proliferative biomarkers (proliferating cell nuclear antigen and Ki-67) in lung tissues. Haskap is a unique berry in that 80% of its anthocyanins are contributed by a single molecule, cyanidin-3-*O*-glucoside [5]. Hydroxybenzoic acids are the major degrative products as well as in vivo metabolites of cyanidin-3-*O*-glucoside [6]. The review written by Sankaranarayanan and colleagues discussed the role and mechanism of hydroxybenzoic acids in the inhibition of cancer cell proliferation [7]. The majority of dietary flavonoids are not absorbed in the intestinal lumen but are subject to degradation by colonic microbiome generating hydroxybenzoic acids such as 2,4,6-trihydroxybenzoic acid (2,4,6-THBA), 3,4-DHBA, 3,4,5-THBA, 4-HBA, 2,4-DHBA, 2,6-DHBA. Evidence suggests that these microbial flavonoid metabolites could contribute to the prevention of certain cancers [8]. Natural hydroxybenzoic acids are also found in many medicinal plants known to have anti-cancer properties. Cellular mechanisms of cancer prevention by flavonoid metabolites such as hydroxybenzoic acids are relatively underexplored, and further investigations are warranted to fully explore flavonoids and their microbial metabolites as anti-cancer molecules.

2. Flavonoid Derivatives Exhibit Enhanced Proapoptotic Activity in Cancer Cells

The dietary intervention of flavonoids, such as anthocyanins, as an anti-cancer therapy, has also been quite extensively studied using mouse models [9]. Tutino and colleagues have demonstrated that flavonoids extracted from grape skin possess anti-proliferative and proapoptotic activity in colon cancer cells in vitro [10] and the possible mechanism involved the changes in the membrane polyunsaturated fatty acid profile [11]. Regulation of the expression of 15-lipoxygenase-1 (15-LOX-1) and peroxisome proliferator-activated receptor gamma (PPAR-γ) by grape skin flavonoids has been observed; however, future investigations should aim at understanding the role of flavonoids in membrane lipid synthesis and degradation in relation to proapoptotic activity in cancer cells. Similarly, Ko and colleagues have shown that tangeretin, a citrus peel-derived flavonoid, inhibits breast cancer stem cell formation through the suppression of signal transducer and activator of transcription 3 (Stat3) signaling [12]. Cancer stem cells are responsible for chemoresistance and recurrence of many cancers. Therefore, targeting cancer stem cells by flavonoids has the potential to become a novel cancer therapy. A limitation of the therapeutic application of flavonoids is their poor absorption in intestinal lumen and uptake, rapid metabolism, and poor uptake by targeted tissues. Nevertheless, attention should also be paid to specific flavonoids that could be cancer-promoting through the activation of the Keap1/Nrf2/ARE pathway in cancer cells [13,14].

Recent investigations revealed that the structural modification of flavonoids such as acylation can improve their selective anti-proliferative activity against cancer cells [15] and anti-metastatic activity [16]. Chalcone is the precursor of flavonoids, and Pawlak and colleagues have shown that methoxy derivatives of 2′-hydroxychalcone exhibit anti-proliferative and proapoptotic activity on canine lymphoma and leukemia cells [17]. The two most active derivatives 2′-hydroxy-2″,5″-dimethoxychalcone and 2′-hydroxy-4′,6′-dimethoxychalcone had the greater proapoptotic potential due to the addition of two methoxy groups. The methoxy derivatives of chalcones triggered DNA damage in the cell lines (GL-1 B/T-cell leukemia) resistant to chalcone-induced apoptosis. The findings demonstrate the potential of flavonoids and their analogs to develop as anti-cancer therapeutics.

3. The Relevance of Flavonoids in Obesity Prevention

Obesity is characterized as an excessive accumulation of fat due to an imbalance of high energy intake and less energy expenditure. Dietary phytochemicals such as flavonoids have been shown to modulate lipid metabolism by increasing basal metabolic rate and thermogenesis [18]. Kim and colleagues have shown that flavonoids isolated from *Acer okamotoanum* inhibit adipocyte differentiation and promote lipolysis in the 3T3-L1 adipocytes [19]. The study also showed significant down-regulation of adipogenic transcription factors, such as γ-cytidine-cytidine-adenosine-adenosine-thymidine/enhancer-binding protein-α, -β, and PPAR-γ. In addition, the flavonoids significantly activated 5′-adenosine monophosphate-activated protein kinase (AMPK), leading to inhibition of triacylglyceride accumulation, potentially inactivating the key regulatory enzyme acetyl Co-A carboxylase (ACC). Flavonoids need to be further explored for their ability to regulate adipocyte differentiation, lipolysis, and AMPK signaling in relation to weight management. It is important to understand the efficacious dose, which may be dependent on various factors, including the individual's body mass index, lifestyle, age, and gut microbiome composition and diversity, etc.

Obesity-associated high blood pressure or hypertension is a serious public health concern. Obesity is associated with an increased renin-angiotensin-aldosterone system (RAAS), which regulates blood pressure [20]. Certain dietary flavonoids possess the potential to inhibit angiotensin-converting enzyme (ACE) that is the key regulatory enzyme of the RAAS and therefore regulates the blood pressure [21]. Here, Kamkaew and colleagues have demonstrated vasodilatory effects and mechanisms of action of flavonoids present in a traditional medicinal plant, *Bacopa monnieri*, using intact endothelial rings of mesenteric arteries of rats [22]. Flavonoids (luteolin and apigenin) at 0.1–100 μM caused vasorelaxation in a concentration-dependent manner, potentially by inducing endothelial nitric oxide synthase (eNOS) phosphorylation at Ser1177, leading to nitric oxide (NO) production. It is interesting to note that flavonoids have about twice the potency of saponins as vasodilators. However, the application of dietary flavonoids in the management of high blood pressure needs to be assessed and validated using appropriate human clinical trials.

The biological processes of aging and senescence are accelerated by obesity, linked to metabolic syndrome, and the risk of developing chronic diseases increases with age [23]. The extension of lifespan and healthspan is dependent on behavioral, pharmacologic, and dietary factors, which remain largely unknown [24]. However, non-nutrient dietary antioxidants such as flavonoids are considered potential anti-aging agents or lifespan essentials. The ability of flavonoids to modulate some of the evolutionarily conserved hallmarks of aging, including oxidative damage, inflammation, cell senescence, and autophagy, have been reported. Guo and colleagues have investigated the anti-aging potential of neohesperidin and its synergistic effects with other citrus flavonoids in extending the lifespan of *Saccharomyces cerevisiae* [25]. Authors postulated that autophagy promoted by the decreased target of rapamycin complex 1 (TORC1) signaling is critically important for a long chronological lifespan. This and many similar studies also suggest the important considerations to make for the best combination of flavonoids or what nature has given us in certain plant foods when designing new dietary supplements or selecting functional foods for regular consumption with the aim of optimal healthy aging.

Obesity, an inflammatory disease, has reached epidemic levels worldwide and has become the modern socioeconomic burden of the 21st century. Affecting nearly 30% of the world population, obesity is a risk factor for many chronic diseases including type 2 diabetes mellitus, cardiovascular disorders, certain cancers, hypertension, non-alcoholic fatty liver disease, and steatohepatitis, consequentially impacting the quality of life [26]. Therefore, reducing obesity and associated life-threatening diseases has become an emerging area of health sciences. In this Special Issue, Sudhakaran and Doseff have contributed to a comprehensive review of the impact of dietary flavones on obesity-induced inflammation [27]. The common flavones are luteolin, apigenin, chrysin, baicalin, acacetin, orientin, and apigenin, which could be found in food sources such as citrus fruits, vegetables, herbs, and grains.

Based on the preclinical studies, flavones exhibit a potential role in suppressing adipogenesis, inducing the browning of white adipocytes, modulating immune responses in the adipose tissues, and hindering obesity-induced inflammation, which has been positively correlated with an enhanced cancer incidence. This review also summarizes the crosstalk between adipocytes and macrophages. However, further investigations are required to further understand the molecular mechanisms responsible for the anti-obesogenic activity of flavones.

Similarly, Ferraz and colleagues reviewed the therapeutic potential of flavonoids in pain and inflammation [28]. Pathological pain results from inflammation as well as peripheral nerve injury. In this review, the authors discussed the preclinical and clinical evidence on the analgesic and anti-inflammatory proprieties of flavonoids. Flavonoids can suppress the expression and activation of many inflammatory mediators such as interleukin (IL)-1β, tumor necrosis factor α (TNF-α), NO, cyclooxygenase-2 (COX-2), vascular endothelial growth factor (VEGF), and intercellular adhesion molecule-1 (ICAM-1). Interestingly, Zaragozá and colleagues showed the anti-inflammatory as well as immunomodulatory activities of dihydroflavones, flavones, and flavonols [29]. Based on the lipopolysaccharide (LPS)-stimulated whole-blood experimental model, the authors suggested that quercetin, naringenin, naringin, and diosmetin could have a potential therapeutic effect in the inflammatory process of cardiovascular disease. With this emerging intense scientific evidence for ameliorating inflammatory conditions by flavonoids and due to their safe and cost-effective attributes, novel flavonoid-inspired nutraceuticals and therapeutics have begun to enter into the medical shelves to improve human health.

4. Potential Anti-COVID-19, Anti-Infectious, Anti-Inflammatory, Immunity-Enhancing Properties of Flavonoids

The epidemic of the infection COVID-19 by an emerging coronavirus SARS-CoV-2 in December 2019 poses significant threats to global health security and the economy [30]. Currently, there is no registered treatment or effective COVID-19-related vaccine. Therefore, investigators across the world are exploring alternative methods to manage this novel coronavirus through prevention and control of the propagation and transmission of COVID-19, antiviral and anti-infectious treatments, natural and effective inhibitors of COVID-19 entry proteins such as ACE2, approaches to enhance host immune response against RNA viral infection, and passive immunotherapy among many others.

In this Special Issue, Ngwa and colleagues communicate the potential of flavonoid-inspired prophylactics or therapeutics against COVID-19 [31]. Based on in silico studies, the authors show that flavonoids such as caflanone, hesperetin, myricetin, and flavonoid derivatives such as Equivir can bind with high affinity to the spike protein, helicase, and protease sites on the ACE2 receptor. Interestingly, caflanone, a unique prenylated flavonoid of *Cannabis*, inhibits virus entry factors including tyrosine-protein kinase ABL-2, cathepsin L, cytokines (IL-1β, IL-6, IL-8, macrophage inflammatory protein 1α (Mip-1α), TNF-α), and lipid kinase PI4Kiiiβ, as well as tyrosine-protein kinase receptor AXL-2, which facilitates mother-to-fetus transmission of coronavirus.

Uncontrolled microbial infection can lead to inflammation as a response to the activation of the innate immune system. The persistence of chronic inflammation can cause fatal diseases such as sepsis. The severe conditions of sepsis due to cytokine storm may result in multiple organ failure, including the lungs. The activation of toll-like receptors (TLR) by microbes or microbial peptides is the critical initial step in developing sepsis [32]. *Escherichia coli*-induced sepsis in rodents has been used as an effective experimental model to assess potential anti-inflammatory compounds. Using this approach, Chauhan and colleagues found that isorhamnetin has the potential to prevent Gram-negative sepsis. Isorhamnetin treatment has significantly enhanced survival and reduced proinflammatory cytokines in the serum and lung tissue of *E. coli*-infected mice. Using docking studies, the authors demonstrated that isorhamnetin can interact directly with the TLR4/myeloid differentiation factor 2 (MD-2) complex that recognizes LPS on Gram-negative bacteria; therefore, it has the potential to inhibit the TLR4 cascade, and thus systemic inflammation and cytokine storm-mediated organ injury.

Urinary tract infections (UTIs) are the second most common type of infection worldwide. The protective effects of flavonoids and phenolic acids present in cranberry (*Vaccinium macrocarpon*) against UTIs are reviewed in this Special Issue [33]. The review reveals that besides uropathogenic *E. coli*, other bacteria such as *Klebsiella pneumoniae* or the Gram-positive bacteria of Enterococcus and Staphylococcus genera seem to be widely involved in UTIs. The reported clinical trials provide substantial evidence of cranberry as total or partial therapeutic alternatives to antibiotics in UTIs. However, individual and/or case-dependent variations of effectiveness have been seen. A-type proanthocyanidins, the oligomeric and polymeric flavonoids, are reported to be responsible for these preventive effects against UTIs. Unabsorbed proanthocytanidins are metabolized by the colon microbiota to generate many low-molecular weight microbial metabolites that can be further absorbed [34]. Future studies need to be focused on understanding the antimicrobial activities of proanthocyanidin-microbial metabolites and metabotypes in relation to UTIs.

In this Special Issue, Quintanilla-Licea and colleagues also discuss the activity of flavonoids isolated from *Lippia graveolens* Kunth (Mexican oregano) as an anti-protozoal agent against *Entamoeba histolytica*, which causes amebiasis, a serious public health problem in developing countries. The major flavonoids of the extracts were pinocembrin, sakuranetin, cirsimaritin, and naringenin, which showed IC_{50} ranging from 28 to 154 µg/mL. These findings provide the basis for the development of flavonoid-based therapeutics against infectious diseases.

5. Future Perspectives of Flavonoid Research

In conclusion, regardless of their broad and multi-potent pharmacological properties, flavonoids are present in low amounts in many dietary sources, most of them possess low water solubility, some of them are unstable under certain conditions, have low intestinal absorption and bioavailability, and possibly need high doses to show efficacy in human studies. Most of the reported in vitro investigations have used metabolically unrealistic high concentrations of flavonoids to demonstrate the targeted efficacy and mode of action; however, these findings need to be validated using standardized in vivo experimental models. Investigations are still lacking in demonstrating the therapeutic efficacy of standardized flavonoid products isolated from plant-sources using prospective human studies. Scale-up, consumer- and environment-friendly green technologies are required for producing cost-effective flavonoid-based natural health products. Flavonoid supplementation to cancer patients should be done cautiously since they could interfere with radiotherapy and various chemotherapies. Multidisciplinary research collaborations are required in investigating enhanced and safe delivery systems of flavonoids to overcome bioavailability limitations, targeted delivery, and improve the therapeutic efficacy of certain flavonoids. Phase 2 metabolism and pharmacokinetics of most of the major flavonoids have been reported. However, the interaction of unabsorbed flavonoids with colon microbiota and resulting metabolites and their role in disease prevention and treatment still need to be understood. Thus, flavonoid nanotechnology, flavonoid-microbiome pharmacology, and flavonoid-inspired therapeutics remain an emerging discipline in life science. We welcome such novel investigations to present through MDPI *Molecules* special issue **Flavonoids and Their Disease Prevention and Treatment—2021.**

Funding: This research received no external funding.

Acknowledgments: The Guest Editor would like to thank all the authors for their contributions to this Special Issue as well as all the reviewers for their voluntary help to assess and improve the submissions.

Conflicts of Interest: The authors declare no conflict of interest.

References

1. Williamson, G.; Kay, C.D.; Crozier, A. The Bioavailability, Transport, and Bioactivity of Dietary Flavonoids: A Review from a Historical Perspective. *Compr. Rev. Food Sci. Food Saf.* **2018**, *17*, 1054–1112. [CrossRef]
2. George, V.C.; Dellaire, G.; Rupasinghe, H.P.V. Plant Flavonoids in Cancer Chemoprevention: Role in Genome Stability. *J. Nutr. Biochem.* **2017**, *45*, 1–14. [CrossRef] [PubMed]
3. Rupasinghe, H.P.V.; Arumuggam, N.; Amararathna, M.; De Silva, A.B.K.H. The Potential Health Benefits of Haskap (*Lonicera caerulea* L.): Role of Cyanidin-3-*O*-glucoside. *J. Funct. Foods* **2018**, *44*, 24–39. [CrossRef]
4. Amararathna, M.; Hoskin, D.W.; Rupasinghe, H.P.V. Cyanidin-3-*O*-Glucoside-Rich Haskap Berry Administration Suppresses Carcinogen-Induced Lung Tumorigenesis in A/JCr Mice. *Molecules* **2020**, *25*, 3823. [CrossRef]
5. Amararathna, M.; Hoskin, D.W.; Rupasinghe, H.P.V. Anthocyanin-rich Haskap (*Lonicera caerulea* L.) Berry Extracts Reduce Nitrosamine-induced DNA Damage in Human Normal Lung Epithelial Cells In Vitro. *Food Chem. Toxicol.* **2020**, *141*, 111404. [CrossRef] [PubMed]
6. Pace, E.; Jiang, Y.; Clemens, A.; Crossman, T.; Rupasinghe, H.P.V. Impact of Thermal Degradation of Cyanidin-3-*O*-Glucoside of Haskap Berry on Cytotoxicity of Hepatocellular Carcinoma HepG2 and Breast Cancer MDA-MB-231 Cells. *Antioxidants* **2018**, *7*, 24. [CrossRef]
7. Sankaranarayanan, R.; Kumar, D.R.; Patel, J.; Bhat, G.J. Do Aspirin and Flavonoids Prevent Cancer through a Common Mechanism Involving Hydroxybenzoic Acids?—The Metabolite Hypothesis. *Molecules* **2020**, *25*, 2243. [CrossRef]
8. Thilakarathna, W.P.D.W.; Rupasinghe, H.P.V. Microbial Metabolites of Proanthocyanidins Reduce Chemical Carcinogen-induced DNA Damage in Human Lung Epithelial and Fetal Hepatic Cells In Vitro. *Food Chem. Toxicol.* **2019**, *125*, 479–493. [CrossRef]
9. Rupasinghe, H.P.V.; Arumuggam, N. Chapter 5: Health Benefits of Anthocyanins. In *Anthocyanins from Natural Sources*; Springer: Berlin/Heidelberg, Germany, 2019; pp. 121–158.
10. Tutino, V.; Gigante, I.; Scavo, M.P.; Refolo, M.G.; Nunzio, V.D.; Milella, R.A.; Caruso, M.G.; Notarnicola, M. Stearoyl-CoA Desaturase-1 Enzyme Inhibition by Grape Skin Extracts Affects Membrane Fluidity in Human Colon Cancer Cell Lines. *Nutrients* **2020**, *12*, 693. [CrossRef]
11. Tutino, V.; Gigante, I.; Milella, R.A.; De Nunzio, V.; Flamini, R.; De Rosso, M.; Scavo, M.P.; Depalo, N.; Fanizza, E.; Caruso, M.G.; et al. Flavonoid and Non-Flavonoid Compounds of Autumn Royal and Egnatia Grape Skin Extracts Affect Membrane PUFA's Profile and Cell Morphology in Human Colon Cancer Cell Lines. *Molecules* **2020**, *25*, 3352. [CrossRef]
12. Ko, Y.-C.; Choi, H.S.; Liu, R.; Kim, J.-H.; Kim, S.-L.; Yun, B.-S.; Lee, D.-S. Inhibitory Effects of Tangeretin, a Citrus Peel-Derived Flavonoid, on Breast Cancer Stem Cell Formation through Suppression of Stat3 Signaling. *Molecules* **2020**, *25*, 2599. [CrossRef] [PubMed]
13. Fernando, W.; Rupasinghe, H.P.V.; Hoskin, D.W. Dietary Phytochemicals with Anti-oxidant and Pro-oxidant Activities: A Double-edged Sword in Relation to Adjuvant Chemotherapy and Radiotherapy? *Cancer Lett.* **2019**, *452*, 168–177. [CrossRef] [PubMed]
14. Suraweera, T.L.; Rupasinghe, H.P.V.; Dellaire, G.; Xu, Z. Regulation of Nrf2/ARE Pathway by Dietary Flavonoids: A Friend or Foe for Cancer Management? *Antioxidants* **2020**, *9*, 973. [CrossRef]
15. Fernando, W.; Coombs, M.R.P.; Hoskin, D.W.; Rupasinghe, H.P.V. Docosahexaenoic Acid-acylated Phloridzin, a Novel Polyphenol Fatty Acid Ester Derivative, is Cytotoxic to Breast Cancer Cells. *Carcinogenesis* **2016**, *37*, 1004–1013. [CrossRef]
16. Fernando, W.; Coyle, K.; Marcato, P.; Rupasinghe, H.P.V.; Hoskin, D.W. Phloridzin Docosahexaenoate, a Novel Fatty Acid Ester of a Plant Polyphenol, Inhibits Mammary Carcinoma Cell Metastasis. *Cancer Lett.* **2019**, *465*, 68–81. [CrossRef]
17. Pawlak, A.; Henklewska, M.; Hernández Suárez, B.; Łużny, M.; Kozłowska, E.; Obmińska-Mrukowicz, B.; Janeczko, T. Chalcone Methoxy Derivatives Exhibit Antiproliferative and Proapoptotic Activity on Canine Lymphoma and Leukemia Cells. *Molecules* **2020**, *25*, 4362. [CrossRef]
18. Rupasinghe, H.P.V.; Sekhon-Loodu, S.; Mantso, T.; Panayiotidis, M.I. Phytochemicals in Regulating Fatty acid β-Oxidation: Potential Underlying Mechanisms and Their Involvement in Obesity and Weight Loss. *Pharmacol. Ther.* **2016**, *165*, 153–163. [CrossRef] [PubMed]

19. Kim, J.H.; Lee, S.; Cho, E.J. Flavonoids from *Acer okamotoanum* Inhibit Adipocyte Differentiation and Promote Lipolysis in the 3T3-L1 Cells. *Molecules* **2020**, *25*, 1920. [CrossRef]

20. Tanaka, M. Improving Obesity and Blood Pressure. *Hypertens. Res.* **2020**, *43*, 79–89. [CrossRef]

21. Balasuriya, B.W.N.; Rupasinghe, H.P.V. Plant Flavonoids as Angiotensin Converting Enzyme Inhibitors in Regulation of Hypertension. *Funct. Foods Health Dis.* **2011**, *1*, 172–188. [CrossRef]

22. Kamkaew, N.; Paracha, T.U.; Ingkaninan, K.; Waranuch, N.; Chootip, K. Vasodilatory Effects and Mechanisms of Action of *Bacopa monnieri* Active Compounds on Rat Mesenteric Arteries. *Molecules* **2019**, *24*, 2243. [CrossRef] [PubMed]

23. Bonomini, F.; Rodella, L.F.; Rezzani, R. Metabolic Syndrome, Aging and Involvement of Oxidative Stress. *Aging Dis.* **2015**, *6*, 109–120. [CrossRef] [PubMed]

24. Russo, G.L.; Spagnuolo, C.; Russo, M.; Tedesco, I.; Moccia, S.; Cervellera, C. Mechanisms of Aging and Potential Role of Selected Polyphenols in Extending Healthspan. *Biochem. Pharmacol.* **2020**, *173*, 113719. [CrossRef]

25. Guo, C.; Zhang, H.; Guan, X.; Zhou, Z. The Anti-Aging Potential of Neohesperidin and Its Synergistic Effects with Other Citrus Flavonoids in Extending Chronological Lifespan of *Saccharomyces Cerevisiae* BY4742. *Molecules* **2019**, *24*, 4093. [CrossRef] [PubMed]

26. Blüher, M. Obesity: Global Epidemiology and Pathogenesis. *Nat. Rev. Endocrinol.* **2019**, *15*, 288–298. [CrossRef]

27. Sudhakaran, M.; Doseff, A.I. The Targeted Impact of Flavones on Obesity-Induced Inflammation and the Potential Synergistic Role in Cancer and the Gut Microbiota. *Molecules* **2020**, *25*, 2477. [CrossRef]

28. Ferraz, C.R.; Carvalho, T.T.; Manchope, M.F.; Artero, N.A.; Rasquel-Oliveira, F.S.; Fattori, V.; Casagrande, R.; Verri, W.A. Therapeutic Potential of Flavonoids in Pain and Inflammation: Mechanisms of Action, Pre-Clinical and Clinical Data, and Pharmaceutical Development. *Molecules* **2020**, *25*, 762. [CrossRef]

29. Zaragozá, C.; Villaescusa, L.; Monserrat, J.; Zaragozá, F.; Álvarez-Mon, M. Potential Therapeutic Anti-Inflammatory and Immunomodulatory Effects of Dihydroflavones, Flavones, and Flavonols. *Molecules* **2020**, *25*, 1017. [CrossRef] [PubMed]

30. Zhang, L.; Liu, Y. Potential Interventions for Novel Coronavirus in China: A Systematic Review. *J. Med. Virol.* **2020**, *92*, 479–490. [CrossRef]

31. Ngwa, W.; Kumar, R.; Thompson, D.; Lyerly, W.; Moore, R.; Reid, T.-E.; Lowe, H.; Toyang, N. Potential of Flavonoid-Inspired Phytomedicines against COVID-19. *Molecules* **2020**, *25*, 2707. [CrossRef]

32. Hotchkiss, R.S.; Moldawer, L.L.; Opal, S.M.; Reinhart, K.; Turnbull, I.R.; Vincent, J.-L. Sepsis and Septic Shock. *Nat. Rev. Dis. Primers* **2016**, *2*, 16045. [CrossRef] [PubMed]

33. González de Llano, D.; Moreno-Arribas, M.V.; Bartolomé, B. Cranberry Polyphenols and Prevention against Urinary Tract Infections: Relevant Considerations. *Molecules* **2020**, *25*, 3523. [CrossRef] [PubMed]

34. Rupasinghe, H.P.V.; Parmar, I.; Neir, S.V. Biotransformation of Cranberry Proanthocyanidins to Probiotic Metabolites by *Lactobacillus rhamnosus* Enhances Their Anticancer Activity in HepG2 Cells In Vitro. *Oxidative Med. Cell. Longev.* **2019**, *2019*, 1–14. [CrossRef] [PubMed]

Sample Availability: Samples of the compounds are not available from the authors.

Publisher's Note: MDPI stays neutral with regard to jurisdictional claims in published maps and institutional affiliations.

 *molecules*

Article

Cyanidin-3-*O*-Glucoside-Rich Haskap Berry Administration Suppresses Carcinogen-Induced Lung Tumorigenesis in A/JCr Mice

Madumani Amararathna [1], David W. Hoskin [2,3] and H. P. Vasantha Rupasinghe [1,2,*]

[1]   Department of Plant, Food, and Environmental Sciences, Faculty of Agriculture, Dalhousie University, 50 Pictou Road, Truro, NS B2N 5E3, Canada; madu.ama@dal.ca
[2]   Department of Pathology, Dalhousie University, Halifax, NS B3H 4R2, Canada; d.w.hoskin@dal.ca
[3]   Department of Microbiology and Immunology, Dalhousie University, Halifax, NS B3H 4R2, Canada
*   Correspondence: vrupasinghe@dal.ca; Tel.: +1-902-893-6623

Academic Editor: Fabio Sonvico
Received: 26 July 2020; Accepted: 21 August 2020; Published: 22 August 2020

Abstract: In our previous study, we demonstrated that cyanidin-3-*O*-glucoside (C3G)-rich haskap (*Lonicera caerulea* L.) berry extracts can attenuate the carcinogen-induced DNA damage in normal lung epithelial cells *in vitro*. Here, the efficacy of lyophilized powder of whole haskap berry (C3G-HB) in lowering tobacco-specific nitrosamine, 4-(methylnitrosamino)-1-(3-pyridyl)-1-butanone, (NNK)-induced lung tumorigenesis in A/JCr mice was investigated. Three weeks after daily oral administration of C3G-HB (6 mg of C3G in 0.2 g of C3G-HB/mouse/day), lung tumors were initiated by a single intraperitoneal injection of NNK. Dietary C3G-HB supplementation was continued, and 22 weeks later, mice were euthanized. Lung tumors were visualized through positron emission tomography (PET) and magnetic resonance imaging (MRI) 19 weeks after NNK injection. Dietary supplementation of C3G-HB significantly reduced the NNK-induced lung tumor multiplicity and tumor area but did not affect tumor incidence. Immunohistochemical analysis showed reduced expression of proliferative cell nuclear antigen (PCNA) and Ki-67 in lung tissues. Therefore, C3G-HB has the potential to reduce the lung tumorigenesis, and to be used as a source for developing dietary supplements or nutraceuticals for reducing the risk of lung cancer among high-risk populations.

Keywords: anthocyanin; tobacco-specific nitrosamine; carcinogenesis; cell proliferation; cancer chemoprevention; lung cancer

1. Introduction

Lung cancer is the most commonly diagnosed cancer (11.6% of all the cancers), and the leading cause of cancer deaths (18.4% of all cancer deaths) among both men and women worldwide. Among western populations, over 80% of lung cancer incidence is attributed to tobacco smoking [1]. Tobacco-specific nitrosamine, 4-(methylnitrosamino)-1-(3-pyridyl)-1-butanone (NNK), is a known lung carcinogen that causes lung tumors in laboratory animals and is likely to cause lung cancer in humans [2,3]. In the lungs, NNK is converted into reactive electrophilic metabolites, which cause point mutations in critical genes that involve cellular functions. Hence, NNK can deregulate the cell cycle, apoptosis, and DNA damage repair [4–6]. NNK also activates cell growth and proliferation signaling cascades, such as extracellular signal-regulated kinase 1/2 (ERK1/2), mitogen-activated protein kinase (MAPK) [5], and phosphatidylinositol 3-kinase (PI3K)/protein kinase B (AKT) [7], resulting in tumorigenesis.

Numerous *in vitro, in vivo,* and epidemiological studies, have reported the benefits of flavonoids and flavonoid-rich plant extracts in preventing or curing cancer, including lung cancer [8,9]. For example, oral administration of anthocyanin-rich pomegranate juice reduces lung tumorigenesis in mice by inhibiting ERK1/2 and PI3K/AKT [10]. Anthocyanin is also able to suppress MAPK, Wnt/β-catenin signaling, and induce apoptosis [11–13]. Haskap (*Lonicera caerulea* L.), also known as blue honeysuckle, is a berry fruit with abundant anthocyanin, particularly cyanidin-3-*O*-glucoside (C3G). Haskap berry has a higher antioxidant capacity than other common fruits [14,15]. Recent studies have demonstrated anti-inflammatory [16,17], antiarthritis [18], antiobesity [19], and antidiabetic [20] properties of haskap berry. We have demonstrated that C3G-rich haskap berry extract can reduce NNK acetate-induced DNA double-strand breaks and oxidative stress in healthy human bronchial epithelial (BEAS-2B) cells *in vitro* [21]. The objective of this study was to investigate the chemopreventive ability of lyophilized C3G-rich whole haskap berry powder (C3G-HB) against NNK-induced lung tumorigenesis in A/JCr mice. The number of tumors, tumor area, and proliferative markers were used as parameters to detect the chemopreventive ability of C3G-HB. We found that C3G-HB can suppress NNK-induced lung tumorigenesis *in vivo*.

2. Results

Two mice from groups C3G-HB supplemented diet continuously before and after NNK-injection (conti.-C3G-HB) and C3G-HB supplemented diet only after NNK-injection (post-C3G-HB) were euthanized due to weight loss and eliminated from the study. Observations for symptoms of stress, i.e., changes in fur color or texture, food consumption, and behavioral abnormalities such as hunched posture, fast movements, and vocalization, were performed daily.

2.1. The Composition of C3G-HB

The nutritional composition of the C3G-HB cv. Tundra is presented in Table 1. C3G-HB is rich with proteins (68%), fat (3.3%), and fiber (8%). The C3G content of the studied haskap berry sample was 3.4%. Additionally, C3G-HB was rich in minerals, particularly manganese, magnesium, zinc, and copper (Table 1).

Table 1. The nutritional composition of lyophilized powder of haskap berry cv. Tundra.

Nutrient	%	Mineral Content	%
Dry matter	91.64	Potassium	1.175
Protein digestibility	67.76	Magnesium	0.056
Crude protein	4.86	Phosphorous	0.176
Bound protein	5.10	Calcium	0.105
ADIN	0.25	Sodium	0.016
Crude fat	3.33	Copper (mg/kg)	6.34
Acid detergent fiber	3.64	Manganese (mg/kg)	<10.00
Neutral detergent fiber	4.28	Zinc (mg/kg)	7.74
Ash	2.57	Cyanidin-3-*O*-glucoside	3.4

All the parameters are presented in percentages unless the unit is indicated in front of the parameter in the table. ADIN, acid detergent insoluble nitrogen.

2.2. The Response of Mice to C3G-HB Supplementation and NNK-Injection

Dietary supplementation of C3G-HB and NNK injection affects the body weight of mice (Figure 1A). Body weight of naive mice was significantly higher (paired *t*-test, $p < 0.0001$) in comparison to the control and no-C3G-HB groups. For instance, at the termination, the body weight of mice given C3G-HB supplement and NNK was reduced by 3.5% (naive vs. control) and 7.6%, respectively (naive vs. no-C3G-HB). Conversely, long-term C3G-HB supplementation significantly ($p < 0.001$) increased the body weight of NNK-injected mice by 2% (no-C3G-HB vs. conti.-C3G-HB), and 3.2% (no-C3G-HB vs. post-C3G-HB), respectively (Figure 1A). In fact, the weight loss in NNK-injected, no-C3G-HB mice could be linked with their dietary intake as no-C3G-HB group had significantly lower feed intake compared to the control (paired *t*-test, $p < 0.0035$) (Figure 1B).

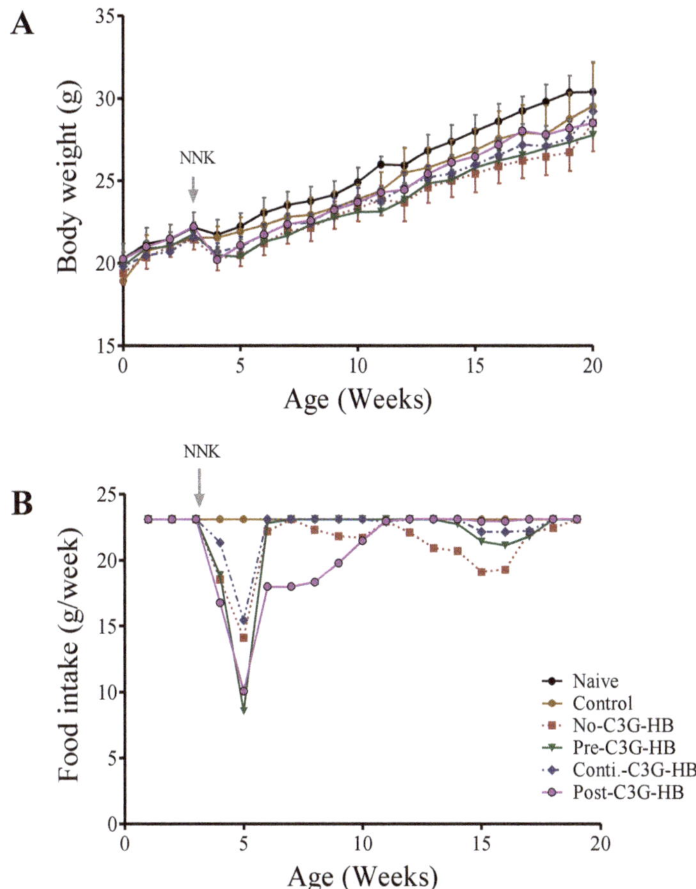

Figure 1. The effect of cyanidin-3-*O*-glucoside (C3G)-HB and 4-(methylnitrosamino)-1-(3-pyridyl)-1-butanone (NNK) injection on the body weight of A/JCr mice. Naive mice ($n = 5$) and NNK-injected mice in group no-C3G-HB ($n = 10$) were fed a regular mouse diet. Mice in the control ($n = 5$) and NNK-injected (pre-C3G-HB, conti.-C3G-HB, and post-C3G-HB) ($n = 10$) groups were fed with the C3G-HB supplemented diet as presented in Figure 5. (**A**) Average body weight of the mice and (**B**) Average food intake over the experimental period. The effect of C3G-HB dietary supplement and NNK carcinogen injection on the body weight of mice was determined by paired *t*-test at $\alpha = 0.05$. No-C3G-HB, not given C3G-HB supplemented diet; Pre-C3G-HB, C3G-HB supplemented diet only before NNK injection; Conti.-C3G-HB, C3G-HB supplemented diet continuously before and after NNK injection; Post-C3G-HB, C3G-HB supplemented diet only after NNK injection.

2.3. Lung Tumorigenesis and Tumor Incidence

PET/MRI images confirmed the presence of tumors in the lungs of NNK-injected mice (Figure 2A). The effect of C3G-HB dietary supplementation on lung tumorigenesis was determined by the number of peripheral lung tumors (Figure 2B). The group no-C3G-HB mice, injected with NNK, and fed the control diet, had an average of 14.1 ± 1.7 tumors/mouse. NNK-injected mice in group pre-, conti.-, and post-C3G-HB, that were given the C3G-HB supplement had 8.7 ± 1.4, 10.2 ± 1.2, and 9.1 ± 1.4 tumors/mouse, respectively, and a reduction of tumor multiplicity by 38.3%, 22.8%, and 35.4%, respectively, in comparison to the no-C3G-HB group (Figure 2C). The inhibition of lung tumor multiplicity by continuous dietary supplementation of C3G-HB (conti.-C3G-HB) was not statistically significant ($p > 0.05$) from no-C3G-HB group.

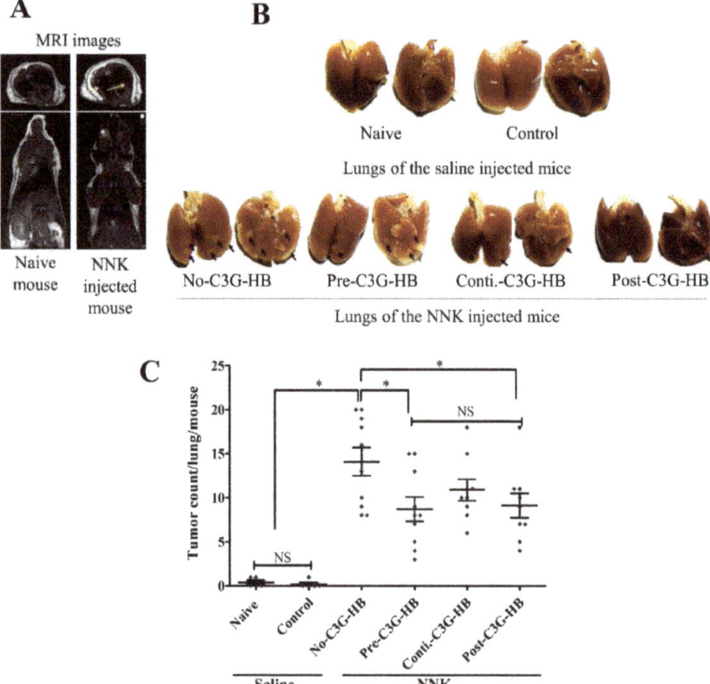

Figure 2. The effect of C3G-rich haskap berry supplement (C3G-HB) on NNK-induced lung tumorigenesis in A/JCr mice. Saline was injected as a sham for mice in naive and control groups. A single intraperitoneal injection of NNK (100 mg/kg body weight) was used to induce lung tumors in the rest of the mouse groups (pre-C3G-HB, conti.-C3G-HB, and post-C3G-HB). Naive mice were fed a regular mouse diet. Mice in control and NNK-injected groups were fed the C3G-HB supplemented diet, as presented in Figure 5. (**A**) The presence of lung tumors was confirmed by PET/MRI scan and a representative comparison between naive and no-C3G-HB groups ($n = 3$). (**B**) The number of peripheral tumors was counted in each lung under a dissecting microscope ($n = 5$ for naive and control groups, and $n = 10$ for NNK-injected groups). (**C**) The effect of C3G-HB dietary supplement on lung tumor multiplicity was analyzed by one-way ANOVA with Dunnett's test at $\alpha = 0.05$. No-C3G-HB, not given C3G-HB supplemented diet; Pre-C3G-HB, C3G-HB supplemented diet only before NNK injection; Conti.-C3G-HB, C3G-HB supplemented diet continuously before and after NNK injection; Post-C3G-HB, C3G-HB supplemented diet only after NNK injection. * Indicate statistical difference at $p \leq 0.05$ with mean ± SD. NS, Results do not significantly different.

The tumor incidence was not affected by the consumption of the C3G-HB dietary supplement. The tumor incidence of NNK-injected mice was 100% (10/10 and 9/9). Untreated mice (saline-injected) in naive group; 2 out of 5 mice (0.4 ± 0.2) and control; 1 out of 5 mice (0.2 ± 0.2) showed one "spontaneous" tumor on their lungs.

2.4. Lung Tumor Area

The lung tumor area was measured in three consecutive lung sections, representing three depths (top, middle, and bottom) of the lungs (Figure 3). The H and E-stained sections revealed the internal tumor area and tumor characteristics. Tumor lesions were less differentiated and composed of cells with higher nuclear crowding and cytological atypia. The H and E-stained sections indicated a significantly higher ($p < 0.0001$) tumor area in the no-C3G-HB group that received NNK and the control diet. The tumor area in each section was calculated using ImageJ software. The tumor burden in NNK-injected mice was 21.6 ± 4.1. Tumor area was significantly reduced in NNK-treated mice that received the C3G-HB-supplemented diet; 7.6 ± 2.8 (pre-C3G-HB), 7.1 ± 0.6 (conti.-C3G-HB), and 6.9 ± 0.6 (post-C3G-HB), and accordingly reduced by 64.7%, 67.3%, and 68.1%, respectively.

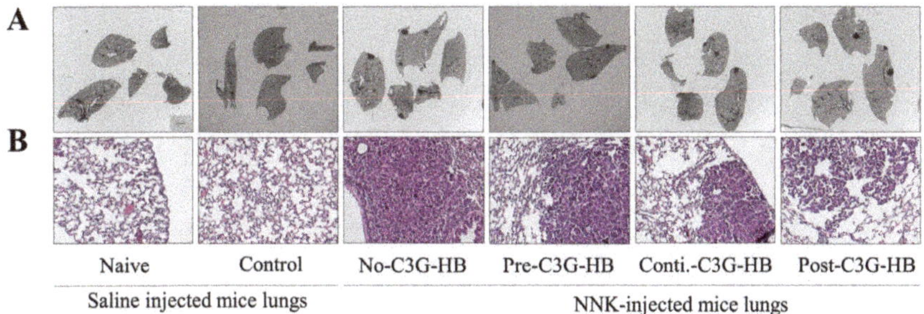

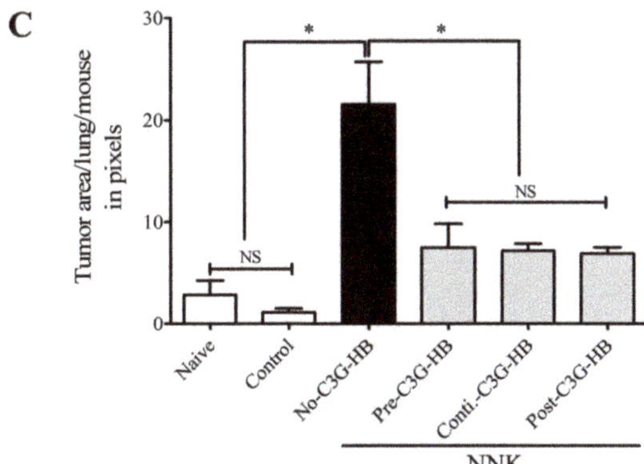

Figure 3. The effect of C3G-HB dietary supplementation on lung tumor area in A/JCr mice. Lung tumors were induced by a single intraperitoneal injection of NNK (100 mg/kg body weight). Saline was injected as a sham for the mice in naive and control groups. Naive mice were fed a regular mouse diet. Mice in control and NNK-injected groups were fed the C3G-HB supplemented diet, except for the mice in no-C3G-HB group (refer to Figure 5). Formalin-fixed lung sections were stained with H and E (3 sections/mouse). (**A**) The whole lung area was imaged, and the tumor area was measured by ImageJ software. (**B**) Representative H and E-stained sections, 200× magnification. (**C**) One-way ANOVA with Dunnett's test at $\alpha = 0.05$ was used for data analysis to compare the treatment effect. No-C3G-HB, not given C3G-HB supplemented diet; Pre-C3G-HB, C3G-HB supplemented diet only before NNK injection; Conti.-C3G-HB, C3G-HB supplemented diet continuously before and after NNK injection; Post-C3G-HB, C3G-HB supplemented diet only after NNK injection. * Indicate statistical difference at $p \le 0.05$ with mean $\pm$ SD. NS, Results do not significantly different.

2.5. Expression of PCNA and Ki-67

The expression of proteins involved in cell proliferation, PCNA and Ki-67, was determined in lung tissue (Figure 4). The cell proliferation markers, PCNA and Ki-67, were highly expressed ($p < 0.0001$) in the lungs of NNK-injected mice (no-C3G-HB, pre-C3G-HB, conti.-C3G-HB, and post-C3G-HB) relative to the saline-injected control mice (naive and control groups). The expression of PCNA was significantly higher compared to that of Ki-67. The level of PCNA and Ki-67 was significantly ($p < 0.0001$) reduced in the lungs of NNK-injected mice that were fed C3G-HB. As a percentage, the expression of PCNA was decreased by 41% to 64% (Figure 4A) and Ki-67 by 33% to 57% (Figure 4B), respectively. The results indicate a reduction of cell proliferation rate in lung tumors of mice fed with C3G-HB dietary supplement.

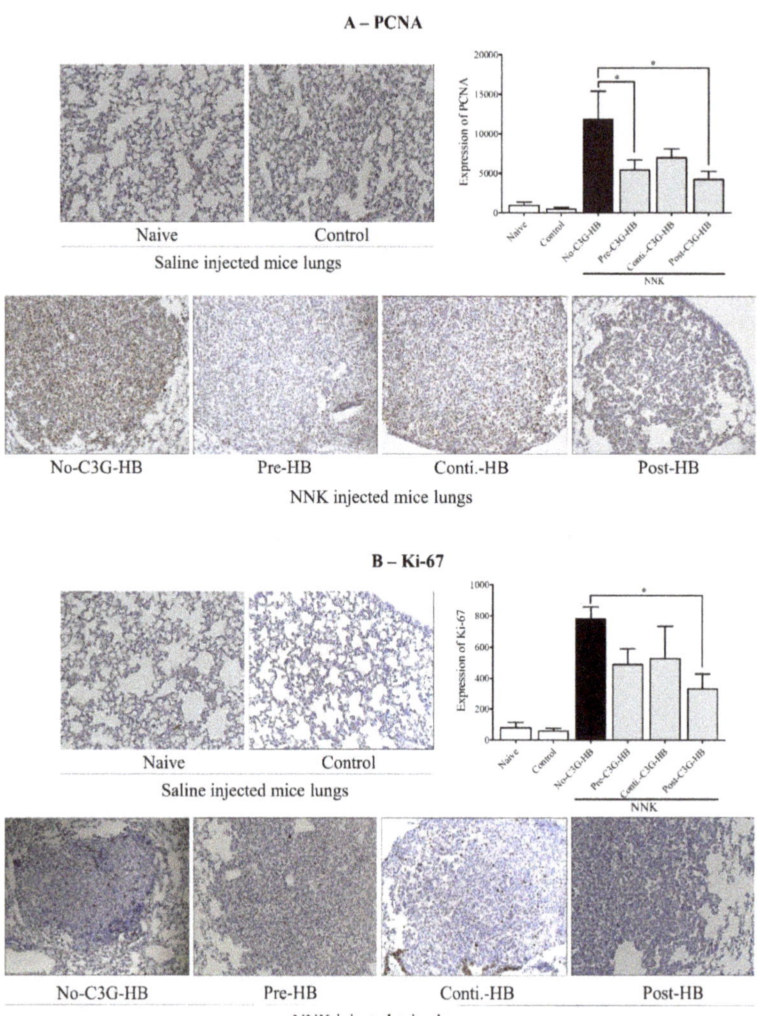

Figure 4. The expression of PCNA and Ki67 in lung tumors. Tumors were imaged using Zeiss Axioplan II and Axiocam HRC color camera at 200 × magnification. The number of (**A**) PCNA and (**B**) Ki-67 positive cells in each section (2 sections/mouse, $n = 5$) were counted using ImageJ software and presented as the expression in bar graphs. One-way ANOVA with Dunnett's test at $\alpha = 0.05$ and $\alpha = 0.1$ was used for data analysis. No-C3G-HB, not given C3G-HB supplemented diet; Pre-C3G-HB, C3G-HB supplemented diet only before NNK injection; Conti.-C3G-HB, C3G-HB supplemented diet continuously before and after NNK injection; Post-C3G-HB, C3G-HB supplemented diet only after NNK injection. * Indicate statistical difference at $p \leq 0.05$ with mean ± SD. NS, Results do not significantly different.

3. Discussion

We administered C3G-HB as a dietary supplement before, during, or after exposure to the pro-carcinogen NNK to evaluate the chemopreventive and chemotherapeutic effect of C3G-HB against NNK-induced lung tumorigenesis in A/JCr mice. The A/JCr mouse is recognized as an *in vivo* model for investigating carcinogen-induced lung tumorigenesis [22]. These mice develop spontaneous lung tumors over time [23]; hence, tumor observation in the naive (2/5 mice) and control (1/5 mice) groups is not surprising. To the best of our knowledge, this is the first study to investigate the chemopreventive effect of C3G-HB against NNK-induced lung tumorigenesis in vivo.

C3G is the most predominant anthocyanin in haskap berry extract. C3G represents about 90% of anthocyanins in haskap berry. Our primary goal is to develop a nutraceutical from haskap berry for use in preventing lung carcinogenesis. Developing pure C3G as a nutraceutical is not practical due to the cost of purification and loss of consumer perception as a natural health product. Hence, we did not test the effect of pure C3G; however, our previous *in vitro* study confirmed that pure C3G has a similar effect as of the extracts of the C3G-HB, and reduced the carcinogen-induced DNA damage and oxidative stress in BEAS-2B normal lung epithelial cells [21].

C3G-HB (6 mg of C3G in 0.2 g of lyophilized whole haskap berry power/mouse/day) significantly ($p < 0.05$) reduced the NNK-induced lung tumor multiplicity, the most sensitive indicator of potency [24], by 38% (pre-C3G-HB) and 35% (post-C3G-HB), respectively. However, administering C3G-HB continuously before and after NNK injection was less effective (only 22%) compared to the pre- and post-supplementation. In contrast, measurement of tumor area in histological samples revealed over 65% reduction upon C3G-HB ingestion. Similarly, Khan and colleagues reported a reduction of benzo(a)pyrene- and N-nitroso-tris-chloroethylurea-induced lung tumors by 53% and 74%, respectively, in A/JCr mice that were fed with 0.2% *w/v* C3G-rich pomegranate juice [10]. At tumor initiation stage, NNK is converted into electrophilic metabolites (via CYP450s enzymes) that covalently bind with DNA to form bulky DNA adducts leading to lung tumorigenesis [25,26]. Our previous findings confirmed that ethanolic and aqueous extracts of C3G-HB can attenuate the NNK-induced DNA double-strand breaks, suppress oxidative stress, and induce DNA damage repair proteins in normal lung epithelial BEAS-2B cells [21]. Oral administration of haskap berry attenuates oxidative stress in mice [19] and restores oxidative defense mechanisms by activating catalase, superoxide dismutase, glutathione peroxidase, and glutathione in mice that were exposed to ionizing radiation [27–29]. In addition, flavonoids have been reported to reversibly and irreversibly inhibit cytochrome 450 (CYP450) enzymes and interfere in xenobiotic metabolism [30]. The antioxidant activity of C3G-HB [21] may have attenuated the NNK-related electrophilic metabolites and hence inhibited the formation of DNA adducts that trigger lung tumorigenesis. Therefore, it is crucial to investigate the oxidative defense mechanism of the C3G-HB and its effect on CYP450 enzyme activity in the future.

PCNA is necessary for DNA synthesis (a processivity factor of DNA polymerases) and DNA repair (involved in nucleotide excision repair and base excision repair). PCNA is highly expressed during the active cell cycle; G_1 phase, peaks at S-phase and declines during G_2/M-phases. Ki-67 is expressed in G_1-, S-, and G_2-phases, but not in the G_0-phase of the cell cycle [31]. Similar to previous findings [3–5], NNK induced lung cell proliferation, which is indicated by the highly expressed PCNA (12.7-fold) and Ki-67 (10-fold) in lung tissues of the no-C3G-HB group comparison to the naive group. The C3G-HB dietary supplementation reduced cell proliferation markers, PCNA (40–60%) and Ki-67 (30–60%), respectively. C3G, as a pure compound (250 and 500 µM) and in fruit extracts (0.2% *w/v* C3G-rich pomegranate juice) inhibits cell proliferation through deactivating MAPK and PI3K/AKT signaling pathways, which are activated by NNK at cancer progression [10,13,32]. Thus, it is necessary to study the effect of C3G-HB in cell proliferation pathways to understand its tumor inhibitory mechanism.

Regardless of the feed intake, the C3G-HB supplement reduced the body weight of the mice by 3.5% comparison to the regular diet (control vs. naive). C3G enhances energy metabolism by upregulating brown adipose tissue mitochondrial function [33]. Ingestion of C3G-rich blood orange juice [34] and purple sweet potato [35] results in reduced lipogenesis, including triglycerides through activation of adenosine monophosphate-activated protein kinase (AMPK) signaling pathways. Therefore, enhanced metabolism and/or reduction of fat synthesis might account for the weight loss effect of C3G-HB in control mice over the naive group. AMPK is also identified as a metabolic tumor suppressor which reprograms cellular metabolism and prevent tumorigenesis [36]. Therefore, C3G-activated AMPK may have regulated energy levels that inhibit cell proliferation. These results suggest the beneficial health effect of C3G-HB against NNK-induced lung tumorigenesis *in vivo*.

In humans, C3G metabolism generates various metabolites such as C3G glucuronides, methylates of the C3G, i.e., peonidin-3-glucoside, and simple phenolic acids including protocatechuic acid, ploroglucinaldehyde, and hippuric acid [37–39]. Even though the metabolites of C3G are similar in humans and mice, the clearance rate of C3G in humans is slower than in mice. In humans, C3G metabolites are present in the blood plasma for ≤ 48 h. In contrast, in mice, a major fraction of C3G metabolites are excreted after 24 h [40,41]. Therefore, when considering C3G-HB as a nutraceutical for use in humans, the absorption, distribution, metabolism, and elimination of C3G should be evaluated using a properly designed human clinical study.

4. Materials and Methods

4.1. Materials

This study was performed at Dalhousie University's animal care facility, following the approval of the University Committee on Laboratory Animals (protocol 15–106). The carcinogen, 4-(methylnitrosamino)-1-(3-pyridyl)-1-butanone (NNK, MW. 207.23 g/mol, Cat No. M325750) was purchased from Toronto Research Chemicals Inc., Toronto, ON, Canada. Frozen haskap berry cv. Tundra was obtained from LaHave Natural Farm, Blockhouse, NS, Canada. Female A/JCr albino mice at 3–4 weeks age (n = 50) were purchased from Charles River Laboratories, Inc., Montreal, QC, Canada.

4.2. Preparation of C3G-HB and Analysis

Frozen haskap berries were lyophilized, ground to a fine powder, and stored at −80 °C. A representative sample was analyzed to determine the nutrient composition (Harlow Institute, Department of Agriculture, Truro, NS, Canada). C3G was quantified by high-performance liquid chromatography and mass spectrophotometry (HPLC/MS/MS, Waters Limited, Mississauga, ON, Canada) after extraction (1 mg/mL) using methanol containing 1% acetic acid and filtered through a 0.22 μm nylon filter [20].

4.3. Preparation of Dietary Supplement

Ingestion of 1.5 g polyphenols/day for a healthy adult of 70 kg body weight is considered to be a health-promoting dose [42,43]. A health-promoting C3G-HB dose, equivalent to the animal dose, was calculated as follows [44]:

$$\text{Human equivalent dose (mg/kg)} = \text{Animal dose (mg/kg)} \times \frac{\text{Animal Km Factor}*}{\text{Human Km Factor}} \quad (1)$$

Km factor = body weight (kg)/body surface area (m^2). The Km factors of mouse and adult human are 3 and 37, respectively [44].

Accordingly, the experimental diet/mouse/day consisted of 0.2 g C3G-HB (equivalent to 6 mg of C3G/mouse/day) and 5% Splenda®mixed into regular mouse chow (Prolab®RMH 3000 from LabDiet, St. Louis, MO, USA) and formed into a 2 g (dry weight) pellet. The control diet consisted of regular mouse chow containing 5% Splenda®. C3G-HB powder was mixed thoroughly for 20 min to obtain a homogeneous preparation for use in making pellets. Pellets were prepared every two days and stored in sealed containers in the dark at 4 °C.

4.4. Experimental Plan and Procedure

The experiment was designed to evaluate the early and late intervention of C3G-HB against NNK-induced lung tumorigenesis (Figure 5). Mice were housed individually in filter-topped plastic cages and maintained under 12-h light-dark cycles. After one week of adaptation, mice were randomly divided into six groups, n = 5 for saline-injected naive and control groups and n = 10 for NNK injected no-C3G-HB, pre-C3G-HB, conti.-C3G-HB and post-C3G-HB groups). Prolab® RMH 3000 diet and distilled water were provided ad libitum. The C3G-HB or control pellets were given daily as a dietary supplement. Briefly, mice in naive and no-C3G-HB groups were given control pellets, while the diet of control, pre-C3G-HB, conti.-C3G-HB and post-C3G-HB groups was supplemented with C3G-HB (Figure 5). Mice in pre-C3G-HB were fed the C3G-HB supplemented diet for three weeks, prior to NNK injection and then switching to the control diet after NNK injection. Post-C3G-HB group was given the control diet until the NNK injection and then switched to C3G-HB supplemented diet until the end of the experiment. Three weeks after the start of dietary supplementation, a single dose of NNK (100 mg/kg body weight in 0.2 mL saline) was injected into the peritoneal cavity of mice to induce lung tumors. Mice in naive and control groups, received an equivalent volume of saline. Once a week, body weight was measured until the end of the experiment.

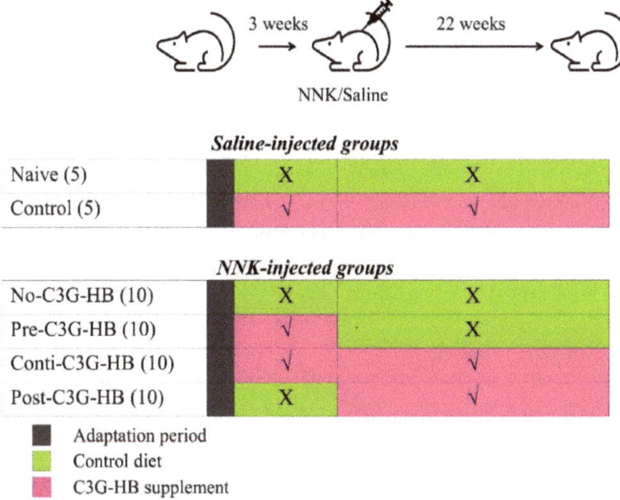

Figure 5. Experimental protocol for investigating the chemopreventive ability of C3G-HB in NNK-induced lung tumorigenesis in A/JCr mice. NNK (100 mg/kg body weight/mouse) or saline was injected three weeks after the adaptation period; n = 5 in saline-injected groups, naive and control, and n = 10 in NNK-injected groups (no-C3G-HB, pre-C3G-HB, conti.-C3G-HB and post-C3G-HB). No-C3G-HB, not given C3G-HB supplemented diet; Pre-C3G-HB, C3G-HB supplemented diet only before NNK injection; Conti.-C3G-HB, C3G-HB supplemented diet continuously before and after NNK injection; Post-C3G-HB, C3G-HB supplemented diet only after NNK injection.

4.5. Lung Tumor Assays

4.5.1. Positron Emission Tomography-Magnetic Resonance Imaging (PET-MRI)

Before euthanizing mice, lung tumorigenesis was confirmed by PET-MRI at the Biomedical Translational Imaging Center (BIOTIC) at the IWK Children's Hospital, Halifax, NS, Canada. Briefly, three mice from naive and no-C3G-HB groups were randomly selected and fasted for six hours and then injected with ^{18}F-fluorodeoxyglucose 100 μCi via the tail vein. After 30 min, the mice were anesthetized, and a PET-MRI scan was performed. Breathing pattern, heart rate, and body temperature were monitored throughout the scanning period.

Twenty-two weeks after NNK treatment, mice were anesthetized with isoflurane. Blood samples were collected by cardiac puncture, and a higher dose of isoflurane was used to sacrifice mice. Dissected lungs were perfused and washed in phosphate-buffered saline (PBS) before being fixed in 10% (*v/v*) acetate-buffered formalin. Peripheral lung tumors were enumerated using a dissecting microscope. Lungs were embedded in paraffin, and paraffin-embedded tissues were stored for histopathological examination.

4.5.2. Tumor Histology and Tumor Area

Paraffin-embedded lungs were cut into 5 μm thick sections (50 tissue sections/lung) using a microtome (Leica Rm 2255, Leica Biosystems, Concord, ON, Canada). Three sections representing three areas of the lungs were selected at predetermined depths and stained with hematoxylin and eosin (H and E). The H and E-stained lung sections were imaged under bright field microscopy (AxioPlan 11MOT AxioCam HRc, Carl Zeiss Canada Ltd., Toronto, ON, Canada), and lung tumor area was quantified using ImageJ software [45].

4.5.3. Immunohistochemistry

The expression of proliferating cell nuclear antigen (PCNA) and Ki-67 was evaluated by immunohistochemistry (IHC). Briefly, paraffin sections were deparaffinized in xylene, rehydrated through an ethanol solution gradient and washed carefully in running tap water. Antigen retrieval was carried out by heating sections in 0.01 M citrate buffer (pH 6.0) for 30 min in the decloaking chamber. Endogenous peroxidase activity was quenched by incubating the section in 3% H_2O_2 in Tris-buffered saline (TBS) for 10 min. Non-specific binding sites were blocked by incubation with rodent block (M) from Biocare Medical, Pacheco, CA, USA. The sections were incubated overnight at room temperature in a humid chamber with monoclonal antibodies against PCNA (1:6000 dilution) or Ki-67 (1:50 dilution). After several washes with TBS, the slides were incubated with anti-rabbit horseradish peroxidase-conjugated secondary antibody (EnVision, Dako North America Inc.Carpinteria, CA, USA) for 30 min, then washed three times with TBS and incubated with chromogen 3-diaminobenzidine (DAB Chromogen kit, Biocare Medical, Pacheo, Ca, USA) for 3 min. The slides were carefully rinsed under running tap water and counterstained with hematoxylin. The slides were observed under bright field microscopy (AxioPlan 11MOT AxioCam HRc, Carl Zeiss Canada Ltd., North York, ON, Canada) at 200× magnification.

4.6. Statistical Analysis

The observed differences in the tumor multiplicity, tumor area, PCNA and Ki-67 expression were tested for statistical significance using one-way Analysis of Variance (ANOVA). Tukey's pairwise comparison and Dunnett's test with a 95% confidence interval was used for comparisons among multiple groups. Minitab statistical software was used for data analysis.

5. Conclusions

In summary, we have demonstrated that dietary supplementation of C3G-HB can inhibit the NNK-induced lung tumorigenesis in A/JCr mice. C3G-HB may be a promising dietary supplement to suppress lung cancer development among high-risk populations such as smokers, possibly via effects on critical cellular signaling pathways that regulate cell proliferation. Future studies of the effects of C3G-HB on phase I and phase II metabolic enzymes and cell signaling pathways will elucidate the mode of action of C3G-HB against lung carcinogenesis.

Author Contributions: M.A. performed all the experiments, analyzed the data, and drafted the manuscript. H.P.V.R., the principal investigator, designed the study. All the authors made intellectual contributions to the manuscript, have read, edited, and approved the final manuscript.

Funding: This study was supported by the Discovery Grant of Natural Sciences and Engineering Research Council (NSERC) of Canada (HPVR; Grant number RGPIN2016-05369) and the Cancer Research Training Program of the Beatrice Hunter Cancer Research Institute (BHCRI) supported by the Saunders-Matthey cancer prevention foundation (MA; Grant number CRTP2018). APC was sponsored by MDPI.

Acknowledgments: The authors would like to recognize the intellectual contribution of Dr. Michael Johnston, former Director of the BHCRI, at the early stage of this project designing.

Conflicts of Interest: The authors declare that there is no conflict of interest.

References

1. Bray, F.; Ferlay, J.; Soerjomataram, I.; Siegel, R.L.; Torre, L.A.; Jemal, A. Global cancer statistics 2018: GLOBOCAN estimates of incidence and mortality worldwide for 36 cancers in 185 countries. *CA Cancer J. Clin.* **2018**, *68*, 394–424. [CrossRef]

2. Hecht, S.S. It Is Time to Regulate Carcinogenic Tobacco-Specific Nitrosamines in Cigarette Tobacco. *Cancer Prev. Res.* **2014**, *7*, 639–647. [CrossRef]

3. Hecht, S.S. Tobacco Smoke Carcinogens and Lung Cancer. *J. Natl. Cancer Inst.* **1999**, *91*, 1194–1210. [CrossRef]

4. Ronai, Z.A.; Gradia, S.; Peterson, L.A.; Hecht, S.S. SHORT COMMUNICATION: G to A transitions and G to T transversions in codon 12 of the Ki-ras oncogene isolated from mouse lung tumors induced by 4-(methylnitrosamino)-1-(3-pyridyl)-1-butanone (NNK) and relati DNA methylating and pyridyloxobutylating agents. *Carcinogenesis* **1993**, *14*, 2419–2422. [CrossRef]

5. Yamakawa, K.; Yokohira, M.; Nakano, Y.; Kishi, S.; Kanie, S.; Imaida, K. Activation of MEK1/2-ERK1/2 signaling during NNK-induced lung carcinogenesis in female A/J mice. *Cancer Med.* **2016**, *5*, 903–913. [CrossRef]

6. Taylor, K.M.; Wheeler, R.; Singh, N.; Vosloo, A.; Ray, D.W.; Sommer, P. The tobacco carcinogen NNK drives accumulation of DNMT1 at the GR promoter thereby reducing GR expression in untransformed lung fibroblasts. *Sci. Rep.* **2018**, *8*, 1–8. [CrossRef]

7. Huang, R.-Y.; Li, M.-Y.; Hsin, M.K.Y.; Underwood, M.J.; Ma, L.T.; Mok, T.S.-K.; Warner, T.D.; Chen, G.G. 4-Methylnitrosamino-1-3-pyridyl-1-butanone (NNK) promotes lung cancer cell survival by stimulating thromboxane A2 and its receptor. *Oncogene* **2010**, *30*, 106–116. [CrossRef]

8. Zhou, Y.; Zheng, J.; Li, Y.; Xu, D.-P.; Li, S.; Chen, Y.-M.; Li, H.-B. Natural Polyphenols for Prevention and Treatment of Cancer. *Nutrients* **2016**, *8*, 515. [CrossRef]

9. Amararathna, M.; Johnston, M.R.; Rupasinghe, H.P.V. Plant Polyphenols as Chemopreventive Agents for Lung Cancer. *Int. J. Mol. Sci.* **2016**, *17*, 1352. [CrossRef]

10. Khan, N.; Afaq, F.; Kweon, M.-H.; Kim, K.; Mukhtar, H. Oral Consumption of Pomegranate Fruit Extract Inhibits Growth and Progression of Primary Lung Tumors in Mice. *Cancer Res.* **2007**, *67*, 3475–3482. [CrossRef]

11. Charepalli, V.; Reddivari, L.; Radhakrishnnan, S.; Vadde, R.; Agarwal, R.; Vanamala, J.K.P. Anthocyanin-containing purple-fleshed potatoes suppress colon tumorigenesis via elimination of colon cancer stem cells. *J. Nutr. Biochem.* **2015**, *26*, 1641–1649. [CrossRef] [PubMed]

12. Mazewski, C.; Liang, K.; De Mejia, E.G. Inhibitory potential of anthocyanin-rich purple and red corn extracts on human colorectal cancer cell proliferation in vitro. *J. Funct. Foods* **2017**, *34*, 254–265. [CrossRef]

13. He, Y.; Hu, Y.; Jiang, X.; Chen, T.; Ma, Y.; Wu, S.; Sun, J.; Jiao, R.; Li, X.; Deng, L.; et al. Cyanidin-3-O-glucoside inhibits the UVB-induced ROS/COX-2 pathway in HaCaT cells. *J. Photochem. Photobiol. B Biol.* **2017**, *177*, 24–31. [CrossRef] [PubMed]

14. Rupasinghe, H.P.V.; Yu, L.J.; Bhullar, K.S.; Bors, B. Short Communication: Haskap (*Lonicera caerulea*): A new berry crop with high antioxidant capacity. *Can. J. Plant Sci.* **2012**, *92*, 1311–1317. [CrossRef]

15. Khattab, R.; Brooks, M.S.-L.; Ghanem, A. Phenolic Analyses of Haskap Berries (*Lonicera caerulea* L.): Spectrophotometry Versus High Performance Liquid Chromatography. *Int. J. Food Prop.* **2015**, *19*, 1708–1725. [CrossRef]

16. Rupasinghe, H.P.V.; Boehm, M.M.A.; Sekhon-Loodu, S.; Parmar, I.; Bors, B.; Jamieson, A.R. Anti-Inflammatory Activity of Haskap Cultivars is Polyphenols-Dependent. *Biomolecules* **2015**, *5*, 1079–1098. [CrossRef] [PubMed]

17. Wang, Y.; Li, B.; Ma, Y.; Wang, X.; Zhang, X.; Zhang, Q.; Meng, X. *Lonicera caerulea* berry extract attenuates lipopolysaccharide induced inflammation in BRL-3A cells: Oxidative stress, energy metabolism, hepatic function. *J. Funct. Foods* **2016**, *24*, 1–10. [CrossRef]

18. Wu, S.; He, X.; Wu, X.; Qin, S.; He, J.; Zhang, S.; Hou, D.-X. Inhibitory effects of blue honeysuckle (*Lonicera caerulea* L) on adjuvant-induced arthritis in rats: Crosstalk of anti-inflammatory and antioxidant effects. *J. Funct. Foods* **2015**, *17*, 514–523. [CrossRef]

19. Liu, M.; Tan, J.; He, Z.; He, X.; Hou, D.-X.; He, J.; Wu, S. Inhibitory effect of blue honeysuckle extract on high-fat-diet-induced fatty liver in mice. *Anim. Nutr.* **2018**, *4*, 288–293. [CrossRef]

20. De Silva, A.K.H.; Rupasinghe, H.P.V.; Kithma, A. Polyphenols composition and anti-diabetic properties in vitro of haskap (*Lonicera caerulea* L.) berries in relation to cultivar and harvesting date. *J. Food Compos. Anal.* **2020**, *88*, 103402. [CrossRef]

21. Amararathna, M.; Hoskin, D.; Rupasinghe, H.P.V. Anthocyanin-rich haskap (*Lonicera caerulea* L.) berry extracts reduce nitrosamine-induced DNA damage in human normal lung epithelial cells. *Food Chem. Toxicol.* **2020**, *141*, 11140. [CrossRef] [PubMed]

22. Ge, G.-Z.; Xu, T.-R.; Chen, C.-S. Tobacco carcinogen NNK-induced lung cancer animal models and associated carcinogenic mechanisms. *Acta Biochim. Biophys. Sin.* **2015**, *47*, 477–487. [CrossRef] [PubMed]

23. Zeidler-Erdely, P.C.; Kashon, M.L.; Battelli, L.A.; Young, S.-H.; Erdely, A.; Roberts, J.R.; Reynolds, S.H.; Antonini, J.M. Pulmonary inflammation and tumor induction in lung tumor susceptible A/J and resistant C57BL/6J mice exposed to welding fume. *Part. Fibre Toxicol.* **2008**, *5*, 1–16. [CrossRef] [PubMed]

24. Hecht, S.S.; Isaacs, S.; Trushin, N. Lung tumor induction in A/J mice by the tobacco smoke carcinogens 4-(methylnitrosamino)-l-(3-pyridyl)-l-butanone and benzo[a]pyrene: A potentially useful model for evaluation of chemopreventive agents. *Carcinogenesis* **1994**, *15*, 2721–2725. [CrossRef]

25. Maser, E.; Richter, E.; Friebertshäuser, J. The Identification of 11beta-hydroxysteroid Dehydrogenase as Carbonyl Reductase of the Tobacco-Specific Nitrosamine 4-(methylnitrosamino)-1-(3-pyridyl)-1-butanone. *Eur. J. Biochem.* **1996**, *238*, 484–489. [CrossRef]

26. Hecht, S.S.; Trushin, N.; Reid-Quinn, C.A.; Burak, E.S.; Jones, A.B.; Southers, J.L.; Gombar, C.T.; Carmella, S.G.; Anderson, L.M.; Rice, J.M. Metabolism of the tobacco-specific nitrosamine 4-(methylnitrosamino)-1-(3-pyridyl)-1-butanone in the patas monkey: Pharmacokinetics and characterization of glucuronide metabolites. *Carcinogenesis* **1993**, *14*, 229–236. [CrossRef]

27. Svobodová, A.R.; Galandáková, A.; Palikova, I.; Dolezal, D.; Kylarova, D.; Ulrichová, J.; Vostálová, J. Effects of oral administration of *Lonicera caerulea* berries on UVB-induced damage in SKH-1 mice. A pilot study. *Photochem. Photobiol. Sci.* **2013**, *12*, 1830. [CrossRef]

28. Vostálová, J.; Galandáková, A.; Palikova, I.; Ulrichová, J.; Dolezal, D.; Lichnovska, R.; Vrbkova, J.; Svobodová, A.R. *Lonicera caerulea* fruits reduce UVA-induced damage in hairless mice. *J. Photochem. Photobiol. B Biol.* **2013**, *128*, 1–11. [CrossRef]

29. Zhao, H.; Wang, Z.-Y.; Ma, F.; Yang, X.; Cheng, C.; Yao, L. Protective Effect of Anthocyanin from *Lonicera Caerulea* var. Edulis on Radiation-Induced Damage in Mice. *Int. J. Mol. Sci.* **2012**, *13*, 11773–11782. [CrossRef]

30. Bojić, M.; Kondža, M.; Rimac, H.; Benković, G.; Males, Z. The Effect of Flavonoid Aglycones on the CYP1A2, CYP2A6, CYP2C8 and CYP2D6 Enzymes Activity. *Molecules* **2019**, *24*, 3174. [CrossRef]

31. Juríková, M.; Danihel, Ľ.; Polák, Š.; Varga, I. Ki67, PCNA, and MCM proteins: Markers of proliferation in the diagnosis of breast cancer. *Acta Histochem.* **2016**, *118*, 544–552. [CrossRef] [PubMed]

32. Pratheeshkumar, P.; Son, Y.-O.; Wang, X.; Divya, S.P.; Joseph, B.; Hitron, J.A.; Wang, L.; Kim, D.; Yin, Y.; Roy, R.V.; et al. Cyanidin-3-glucoside inhibits UVB-induced oxidative damage and inflammation by regulating MAP kinase and NF-κB signaling pathways in SKH-1 hairless mice skin. *Toxicol. Appl. Pharmacol.* **2014**, *280*, 127–137. [CrossRef] [PubMed]

33. You, Y.; Yuan, X.; Liu, X.; Liang, C.; Meng, M.; Huang, Y.; Han, X.; Guo, J.; Ren, C.; Zhang, Q.; et al. Cyanidin-3-glucoside increases whole body energy metabolism by upregulating brown adipose tissue mitochondrial function. *Mol. Nutr. Food Res.* **2017**, *61*, 1–13.

34. Titta, L.; Trinei, M.; Stendardo, M.; Berniakovich, I.; Petroni, K.; Tonelli, C.; Riso, P.; Porrini, M.; Minucci, S.; Pelicci, P.G.; et al. Blood orange juice inhibits fat accumulation in mice. *Int. J. Obes.* **2009**, *34*, 578–588. [CrossRef] [PubMed]

35. Hwang, Y.P.; Choi, J.H.; Han, E.H.; Kim, H.G.; Wee, J.-H.; Jung, K.O.; Kwon, K.-I.; Jeong, T.C.; Chung, Y.C.; Jeong, H.G.; et al. Purple sweet potato anthocyanins attenuate hepatic lipid accumulation through activating adenosine monophosphate–activated protein kinase in human HepG2 cells and obese mice. *Nutr. Res.* **2011**, *31*, 896–906. [CrossRef]

36. Li, W.; Saud, S.M.; Young, M.R.; Chen, G.; Hua, B.-J. Targeting AMPK for cancer prevention and treatment. *Oncotarget* **2015**, *6*, 7365–7378. [CrossRef] [PubMed]

37. De Ferrars, R.M.; Czank, C.; Zhang, Q.; Botting, N.P.; Kroon, P.; Cassidy, A.; Kay, C.D. The pharmacokinetics of anthocyanins and their metabolites in humans. *Br. J. Pharmacol.* **2014**, *171*, 3268–3282. [CrossRef]

38. Marczylo, T.H.; Cooke, D.; Brown, K.; Steward, W.P.; Gescher, A.J. Pharmacokinetics and metabolism of the putative cancer chemopreventive agent cyanidin-3-glucoside in mice. *Cancer Chemother. Pharmacol.* **2009**, *64*, 1261–1268. [CrossRef]

39. Fornasaro, S.; Ziberna, L.; Gasperotti, M.; Tramer, F.; Vrhovšek, U.; Mattivi, F.; Passamonti, S. Determination of cyanidin 3-glucoside in rat brain, liver and kidneys by UPLC/MS-MS and its application to a short-term pharmacokinetic study. *Sci. Rep.* **2016**, *6*, 22815. [CrossRef]

40. Felgines, C.; Krisa, S.; Mauray, A.; Besson, C.; Lamaison, J.-L.; Scalbert, A.; Mérillon, J.-M.; Texier, O. Radiolabelled cyanidin 3-O-glucoside is poorly absorbed in the mouse. *Br. J. Nutr.* **2010**, *103*, 1738–1745. [CrossRef]

41. Czank, C.; Cassidy, A.; Zhang, Q.; Morrison, D.J.; Preston, T.; Kroon, P.; Botting, N.P.; Kay, C.D. Human metabolism and elimination of the anthocyanin, cyanidin-3-glucoside: A 13C-tracer study. *Am. J. Clin. Nutr.* **2013**, *97*, 995–1003. [CrossRef] [PubMed]

42. Scalbert, A.; Williamson, G. Dietary intake and bioavailability of polyphenols. *J. Nutr.* **2000**, *130*, 2073S–2085S. [CrossRef] [PubMed]

43. Taguchi, C.; Fukushima, Y.; Kishimoto, Y.; Suzuki-Sugihara, N.; Saita, E.; Takahashi, Y.; Kondo, K. Estimated Dietary Polyphenol Intake and Major Food and Beverage Sources among Elderly Japanese. *Nutrients* **2015**, *7*, 10269–10281. [CrossRef] [PubMed]

44. Nair, A.B.; Jacob, S. A simple practice guide for dose conversion between animals and human. *J. Basic Clin. Pharm.* **2016**, *7*, 27–31. [CrossRef] [PubMed]

45. Schindelin, J.; Arganda-Carreras, I.; Frise, E.; Kaynig, V.; Lonair, M.; Pietzsch, T.; Preibisch, S.; Rueden, C.; Saalfeld, S.; Schmid, B.; et al. Fiji: An open-source platform for biological-image analysis. *Nat. Methods* **2012**, *9*, 676–682. [CrossRef] [PubMed]

Sample Availability: C3G-HB is available from the authors for collaborative research.

 *molecules* 

Review

Do Aspirin and Flavonoids Prevent Cancer through a Common Mechanism Involving Hydroxybenzoic Acids?—The Metabolite Hypothesis

Ranjini Sankaranarayanan [1], D. Ramesh Kumar [2], Janki Patel [1] and G. Jayarama Bhat [1,*]

[1] Department of Pharmaceutical Sciences and Translational Cancer Research Center, South Dakota State University, College of Pharmacy and Allied Health Professions, Brookings, SD 57007, USA; ranjini.sankaranarayanan@sdstate.edu (R.S.); janki.patel@jacks.sdstate.edu (J.P.)

[2] Department of Entomology, University of Kentucky, Lexington, KY 40506, USA; rameshinsilico@gmail.com

* Correspondence: jayarama.gunaje@sdstate.edu; Tel.: +1-605-688-6894

Academic Editor: H.P. Vasantha Rupasinghe
Received: 30 March 2020; Accepted: 9 May 2020; Published: 10 May 2020

Abstract: Despite decades of research to elucidate the cancer preventive mechanisms of aspirin and flavonoids, a consensus has not been reached on their specific modes of action. This inability to accurately pinpoint the mechanism involved is due to the failure to differentiate the primary targets from its associated downstream responses. This review is written in the context of the recent findings on the potential pathways involved in the prevention of colorectal cancers (CRC) by aspirin and flavonoids. Recent reports have demonstrated that the aspirin metabolites 2,3-dihydroxybenzoic acid (2,3-DHBA), 2,5-dihydroxybenzoic acid (2,5-DHBA) and the flavonoid metabolites 2,4,6-trihydroxybenzoic acid (2,4,6-THBA), 3,4-dihydroxybenzoic acid (3,4-DHBA) and 3,4,5-trihydroxybenzoic acid (3,4,5-THBA) were effective in inhibiting cancer cell growth in vitro. Limited in vivo studies also provide evidence that some of these hydroxybenzoic acids (HBAs) inhibit tumor growth in animal models. This raises the possibility that a common pathway involving HBAs may be responsible for the observed cancer preventive actions of aspirin and flavonoids. Since substantial amounts of aspirin and flavonoids are left unabsorbed in the intestinal lumen upon oral consumption, they may be subjected to degradation by the host and bacterial enzymes, generating simpler phenolic acids contributing to the prevention of CRC. Interestingly, these HBAs are also abundantly present in fruits and vegetables. Therefore, we suggest that the HBAs produced through microbial degradation of aspirin and flavonoids or those consumed through the diet may be common mediators of CRC prevention.

Keywords: aspirin; flavonoids; cancer prevention; hydroxybenzoic acids; cell cycle; CDKs; colorectal cancer

1. Introduction

Cancer is a global disease, and more than 1 million cases of colorectal cancers (CRC) are diagnosed worldwide each year [1]. Due to the increasing prevalence of CRC in the recent years, there is an urgent need to develop effective strategies for its prevention. Efforts to discover chemo-preventive drugs have met with limited success although conventional drugs like aspirin have been shown to prevent CRC. In addition, an increasing body of evidence suggests that consumption of fruits/vegetables rich in phytochemicals can prevent the occurrences of CRC [2–4]. Interestingly, while aspirin is a widely used synthetic "drug", it is primarily a compound derived from the naturally occurring salicylic acid that is abundantly present in plant sources [5]. Flavonoids are another class of phytochemicals found in plants, fruits and vegetables that are also linked to a decrease in the occurrence of cancers [6–8]. Following the intake of aspirin or flavonoids, they are subjected to metabolism, both in the gut and liver, and this process produces several metabolites, some of which are hydroxybenzoic acids (HBAs) [9–12]. The chemistry and pathways of HBA generation have been well characterized; however, their role in cancer prevention has not been extensively studied. In the recent years, there has been an increased interest to understand their targets and roles in cancer prevention. This review aims to highlight the potential role of HBAs, generated through aspirin and flavonoid metabolism, in CRC prevention.

In this review we provide a brief overview first on aspirin's ability to prevent cancer, followed by a discussion on the known roles of flavonoids in cancer prevention. We then explain the pathways of aspirin and flavonoid degradation leading to the production of HBAs and the other sources of these compounds commonly found in the diet. In addition, we have also highlighted the in vitro and in vivo studies performed using HBAs, currently available in literature, demonstrating its chemo-preventive/therapeutic potential against numerous cancers. Finally, we propose the "metabolite hypothesis", to explain the cancer preventive effects of aspirin and flavonoids through the generation of HBAs.

2. Aspirin and Cancer Prevention

Aspirin has become one of the largest selling pharmaceutical compounds in the world since its first clinical introduction by Bayer in 1899 [5]. In addition to its well-known analgesic, anti-pyretic and anti-inflammatory actions, it has many beneficial health effects including the reduced risk for cardiovascular disease and CRC upon regular consumption [13–15]. Aspirins efficacy to reduce CRC is reported to be between 20%–40%, and the evidence for this effect comes from multiple epidemiological and clinical studies which showed that its intake for 5 or more years reduces the risk associated with colorectal adenomas and carcinomas [16–19]. This is also supported by animal studies where aspirin has been shown to decrease chemically induced carcinogenesis in colorectal tissues [20,21]. These observations and evidences prompted the United States Preventive Services Task Force (USPSTF), to recommend "initiating low dose aspirin use for the primary prevention of cardiovascular disease and colorectal cancer in adults aged 50–59 years" in 2016 [22]. Additionally, in view of these compelling evidences, numerous clinical trials have been launched to address its efficacy against CRC [18,23,24]. Though aspirin has been recommended for the primary prevention of CRC by the USPSTF, its role in secondary and tertiary prevention has not been clearly established [25,26].

The intriguing aspect of aspirins ability to prevent CRC is that low doses (75–300 mg/day) are as effective as higher doses (≥500 mg/day) [27]. Interestingly, aspirin is also more effective against CRC when compared to cancers of the other tissues [28,29]. Numerous theories have been proposed to explain the potential pathways of cancer prevention by aspirin; however, a consensus has not been reached. The most widely discussed among them is the "platelet hypothesis" that implicates the inhibition of cyclooxygenase-1 (COX-1) enzymes in platelets as the contributing factor to cancer prevention [14]. As COX-2 overexpression is an important step in colon tumorigenesis [30,31] and as aspirin is more specific to COX-1 (IC_{50} 1.67 µM) than to COX-2 (IC_{50} 278 µM), the direct inhibition of COX-1 by low-dose aspirin is insufficient to explain its observed anti-cancer effects [18,32]. The platelet hypothesis hence proposes that aspirin's chemopreventive effects may be attributed to the sequential inhibition of COX-1 and COX-2. The inhibition of COX-1 in platelets translates to prevention of both, platelet activation and release of cytokines/growth factors/lipid mediators at the site of gastrointestinal (GI) lesions, that eventually results in the inhibition of COX-2 expression in adjacent nucleated cells [14]. Though attractive, this hypothesis requires the orchestration of multiple events to exert the proposed preventive effects and is yet to be conclusively proven. Apart from the platelet hypothesis, other theories have been proposed including inhibition of mTOR signaling leading to the activation of AMP-kinase, inhibition of Wnt signaling, inhibition of NF-κB, inhibition of polyamine synthesis and modulation of EGFR expression, among others, and these have been reviewed extensively elsewhere [18,33–36]. A schematic of the pathways affected by aspirin is shown in Figure 1.

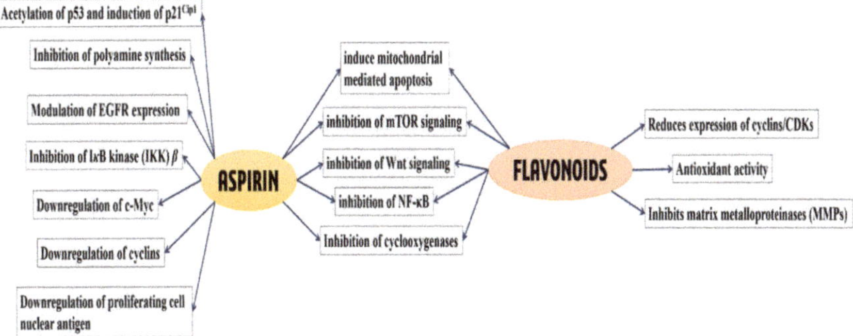

Figure 1. Classical pathways and cellular targets known to be affected by aspirin and flavonoids, leading to the prevention of various cancers. Aspirin and flavonoids affect numerous molecular pathways, some of which overlap. Pathways affected by aspirin alone are indicated in the left, shared pathways are shown in the middle, while pathways affected by flavonoids are shown in the right [6,13,14,33–42].

3. Flavonoids and Cancer Prevention

Epidemiological studies, short-term randomized controlled trials and preclinical studies in CRC patients have provided strong evidence in support of the cancer-preventive properties of flavonoids [2,6,7,43]. Flavonoids are subdivided into 6 categories based on their chemical structure—flavonols, flavan-3-ols, flavanones, flavones, anthocyanins and isoflavones [44]. These compounds are extremely sensitive to their environment and undergo rapid degradation in the presence of increased temperature and fluctuating pH. Additionally, they also degrade to simpler phenolic acids due to the actions of the gut microbes [44–47]. Most flavonoids are predicted to prevent the occurrence of cancer through various mechanisms that include their antioxidant properties, enzyme-receptor inhibition, regulation of apoptosis and the modification of signal transduction pathways which are described in several reviews. Interestingly, similar to aspirin, these compounds are also known to downregulate Akt/mTOR pathway, induce mitochondrial mediated apoptosis, inhibit NF-κB pathway, attenuate Wnt signaling, activate AMPK and suppress abnormal epithelial cell proliferation [6,37,38]. Despite these findings, it still remains unclear if the primary mediators of cancer prevention are the parent flavonoids or their degraded products. A schematic of the pathways affected by flavonoids is shown in Figure 1.

4. HBAs Are Generated through Aspirin and Flavonoid Metabolism

The reported half-life of aspirin is about 20 min [48] and its metabolism either through cytochrome P450 (CYP450) catalyzed reactions [9] or direct conjugation of salicylic acid by phase 2 enzymes [10] in the liver has been well documented. Once absorbed, intact aspirin is partially hydrolyzed to salicylic acid (half-life is 4–6 h) by esterases in the blood and liver [49]. Salicylic acid can then be directly excreted (1%–31%) or can be metabolized in a number of different ways for elimination through kidney or bile. It undergoes conjugation with glycine to form salicyluric acid which accounts for 20%–65% of the metabolites generated, whereas ether and ester glucuronides of salicylic acid constitute 1%–42% of metabolites following conjugation with glucuronic acid [50]. Additionally, CYP450 enzymes in the liver can also metabolize salicylic acid to 2,5-dihydroxybenzoic acid (2,5-DHBA; gentisic acid) and 2,3-dihydroxybenzoic acid (2,3-DHBA; pyrocatechuic acid) that accounts for 1%–8% of the dose. 2,5-DHBA can further undergo conjugation with glycine to form gentisuric acid [9]. Aspirin has also been reported to be metabolized in the gut by the resident microflora. In this regard, Kim et al. demonstrated the importance of human fecal microbiota to degrade aspirin to salicylic acid and hydroxylated salicylic acids [12]. They showed that when rats were administered aspirin along with ampicillin, the bioavailability of aspirin increased when compared to rats administered with aspirin alone. Supporting this observation, a very recent study by Zhang et al., in 2019, also showed that administration of aspirin to rats following amoxicillin treatment decreased aspirin metabolism in the intestine, as compared to rats treated with aspirin alone [51]. The authors of both studies have suggested an important role for the intestinal microflora in the biotransformation of aspirin before its absorption into circulation. It is also important to emphasize that intestinal epithelial cells also express CYP450 enzymes [52], although their capability to generate these HBAs is not well studied. A schematic of the HBAs generated from aspirin is shown in Figure 2A.

The degradation of flavonoids, on the other hand, can occur through changes in physical parameters (like pH) [47], through microbial action in the gut before absorption [11,44], and through host metabolism in the liver [53]. It is reported that the absorption of flavonoids is 1%–15% in the intestine, and that it is extensively metabolized in the liver through conjugation reactions for subsequent elimination in the body [46,54,55]. It is also reported that these conjugated intermediates may be returned to the intestine through the bile, where it further undergoes deconjugation, subjecting them to further degradation through microbial metabolism [11,56]. The basic backbone of flavonoids is highly conserved and comprises of a benzene A-ring bound to a heterocyclic C-ring, which in turn is attached to a second benzene B-ring (Figure 2B). Depending upon the class of flavonoids, these rings are appended with different functional groups that confers their characteristic properties [57]. The functional groups, their number and position on this backbone will also determine their stability and the metabolite(s) they generate [58]. Flavonoids are generally stable under acidic conditions but undergo rapid degradation to simpler phenolic compounds under alkaline conditions (like in the intestine). Additionally, multiple studies have documented the ability of gut microbes to degrade flavonoids into simpler phenolic acids, many of which are HBAs. The most commonly observed HBAs include, 3,4-dihydroxybenzoic acid (3,4-DHBA; protocatechuic acid), 3,4,5-trihydroxybenzoic acid (3,4,5-THBA; gallic acid), 4-hydroxybenzoic acid (4-HBA), 2,6-dihydroxybenzoic acid (2,6-DHBA), 2,4-dihydroxybenzoic acid (2,4-DHBA) and 2,4,6-trihydroxybenzoic acid (2,4,6-THBA; phloroglucinol carboxylic acid) [11,44,46,47,55,59,60]. A schematic of the HBAs generated from flavonoids is shown in Figure 2B.

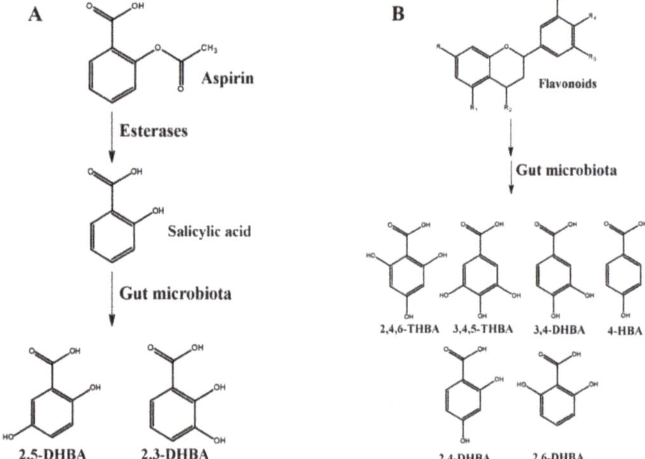

Figure 2. Metabolism of aspirin and flavonoids to generate hydroxybenzoic acids. (**A**) Aspirin metabolism generates 2,3-dihydroxybenzoic acid (2,3-DHBA) and 2,5-DHBA through CYP450 reactions in the liver [61]. DHBAs have also been shown to be generated through microbial metabolism of aspirin/salicylic acid [12]. (**B**) Flavonoid metabolism generates metabolites 2,4,6-trihydroxybenzoic acid (2,4,6-THBA), 3,4-DHBA, 3,4,5-THBA, 4-HBA, 2,4-DHBA, 2,6-DHBA through microbial degradation in the intestine [11,45,56,58]. R-R5 represent various functional groups (example -hydroxy, -ketone, -hydrogen, -methoxy, etc.) that are appended/attached to the flavonoid backbone to generate different groups of flavonoids.

5. HBAs of Aspirin and Flavonoid Origin Exhibit Anti-Proliferative Effects in Cancer Cells

Studies carried out in our laboratory have demonstrated that aspirin metabolites 2,3-DHBA and 2,5-DHBA are capable of inhibiting Cyclin Dependent Kinase (CDK) enzyme activity and cancer cell growth, suggesting their potential role in CRC prevention [62,63]. Our study also demonstrated that 2,5-DHBA was effective in inhibiting cell proliferation in HCT-116 and HT-29 cells [63]. It is important to note that HCT-116 cells do not express COX-2 and HT-29 cells have inactive COX-2 [64], indicating that a COX-independent mechanism is at play. Supporting our observations, an in vivo study by Altinoz et al. also demonstrated enhanced survival in Ehrlich breast ascites carcinoma bearing mice upon oral administration of 2,5-DHBA [65]. Interestingly, other direct targets including FGF-receptors have also been identified for 2,5-DHBA [66].

Increasing evidences are now beginning to support the hypothesis that the degraded products are more likely responsible for the cancer preventive actions of flavonoids than the parent molecules [47,55]. A study by Peiffer et al. showed that administration of 3,4-DHBA to rats effectively inhibited NMBA-induced esophageal cancer [67]; in another study, it was demonstrated that 3,4,5-THBA inhibited prostate tumor growth and progression in TRAMP mice [68]. Several other reports have also documented that 3,4,5-THBA was effective in inducing apoptosis in a variety of cancer cells [59,69]. Experiments carried out in our laboratory have shown that 2,4,6-THBA inhibits cancer cell growth in cells expressing a functional monocarboxylic acid transporter (MCT) SLC5A8 [57].

Although in these studies effective inhibition of cancer cell growth by these HBAs required micromolar concentrations, it could be argued that the high levels of phenolic acid content observed in the gut arising from flavonoid rich food or other dietary sources is achievable [56,70], and may be sufficient to reach pharmacologically relevant concentrations to exert the observed inhibitory effect. Similarly, salicylic acid generated from the hydrolysis of aspirin may also reach micromolar concentrations in the gut as ~50% of orally administered aspirin is left unabsorbed in the GI lumen [71,72]. Hence, upon consumption of an 81 mg aspirin tablet, its concentration in the gut will be in the range of 0.3 mM to 1.4 mM under fed (~750 mL GI volume) and fasting (~160 mL GI volume) conditions, respectively [73]. It is important to note that not all HBAs are effective, as 4-HBA, 2,4-DHBA and 2,6-DHBA failed to inhibit cancer cell growth, suggesting that HBAs are selective in their modes of action [57,63]. It is also crucial to highlight the role of transporters in the uptake of these HBAs. In this regard, while the uptake of 2,4,6-THBA has been demonstrated to occur through the MCT SLC5A8 [57], transporter requirement for other HBAs has not been determined. In addition, while our studies have identified CDKs as potential direct targets for 2,4,6-THBA, 2,3-DHBA and 2,5-DHBA, the targets for 3,4-DHBA and 3,4,5-THBA are yet to be identified (Table 1).

Table 1. Table showing the source of the hydroxybenzoic acid (HBA) metabolites, their potential to inhibit Cyclin Dependent Kinase (CDKs) and their ability to retard cancer cell growth.

Compound	Aspirin Metabolite	Flavonoid Metabolite	CDK Inhibition	Inhibition of Cancer Cell Growth	Reference
2,3-DHBA	+	-	+	+	[63]
2,5-DHBA	+	-	+	+	[63]
2,4,6-THBA	-	+	+	+ (in the presence of a functional SLC5A8)	[57]
3,4,5-THBA	-	+	-	+	[57,69]
3,4-DHBA	-	+	-	+	[57,67]
2,4-DHBA	-	+	-	-	[63]
2,6-DHBA	-	+	-	-	[63]
4-HBA	-	+	-	-	[57]

6. Other Dietary Sources of HBAs

Published reports indicate that fruits and vegetables are rich sources of HBAs [74]. These HBAs produced as secondary metabolites in plants, majorly as byproducts of the shikimate pathway, have been implicated in plant defense against invading pathogens and also act as signaling molecules and antioxidants [3,74,75]. Salicylic acid, an HBA and the precursor for 2,3-DHBA and 2,5-DHBA, is widely found in many foods [76–79] and it has been argued that consumption of the spices rich in salicylic acid may account for low cancer incidence in rural India [77]. 2,3-DHBA and 2,5-DHBA are also reported to be present in dietary sources which may provide a direct link between the consumption of fruits and vegetable and reduced cancer risk. 2,3-DHBA is present in medicinal herbs such as Madagascar rosy periwinkle, *Boreava orientalis*, fermented soy products, in a number of fruits such as batoko plum, avocados and cranberries [75]. 2,5-DHBA is also found in abundance in plants and vegetables such as grapes, citrus fruits, *Hibiscus rosa-sinensis*, sesame, avocados, batoko plum, kiwi fruits, apple, bitter melon and black berries [75,80,81]. The flavonoid metabolite 3,4-DHBA is widely distributed in buckwheat, mustard, kiwi fruits, blackberries, strawberries, chokeberries and mangoes. Additionally, it is also present in chicory, olives, dates, grapes, cauliflowers and lentils. 3,4,5-THBA is reported to be present abundantly in tea, grapes, berries and chestnut [75]. All of these HBAs that have been demonstrated to be effective against cancer cell growth is also interestingly reported to be present in red wine along with 2,4,6-THBA [82]. These reports suggest that the presence of these HBAs of plant origin may have a positive effect on gut health that includes prevention of CRC.

7. The Metabolite Hypothesis—A Common Mechanism for Cancer Prevention

The unstable nature of flavonoids, the rapid hydrolysis of aspirin in the gut and the reports on the cancer-preventive potential of their degraded products through inhibition of cancer cell growth collectively suggest that the HBAs generated from flavonoids and aspirin may be key contributors to their cancer prevention properties. Although the parent compounds may directly contribute to the observed chemopreventive effects through other pathways already reported in literature (Figure 1), we suggest that the contribution of HBAs should also be taken into account. Unabsorbed flavonoids and aspirin/salicylic acid may act as substrates for the gut microbial enzymes to convert them into simpler, pharmacologically active HBAs, like 2,3-DHBA, 2,5-DHBA from aspirin and 3,4-DHBA, 3,4,5-THBA and 2,4,6-THBA from flavonoids. The generation of such HBAs in the gut compel us to propose a central role for these molecules in cancer prevention by aspirin, flavonoids and the diet. As these HBAs are metabolites generated through biotransformation in the body and are also secondary metabolites in plants with the capacity to inhibit cancer cell growth, we would like to refer to this hypothesis as "metabolite hypothesis". A model depicting the convergence of the pathways generating HBAs from aspirin, flavonoids and diet through microbial/host enzymes is shown in Figure 3. The chemopreventive ability of aspirin and flavonoids have been established through many epidemiological studies and as the exact mechanisms of chemoprevention for these compounds have not been clearly established, we are suggesting through this review, that HBAs may be contributing to their chemopreventive actions. Though the data that was published previously on the ability of HBAs to prevent cancer cell growth were performed with cancerous cell lines, and therefore are more reflective of therapy than prevention, we believe that these observations can be extrapolated to cancer prevention. However, the extent to which HBAs are generated from these compounds and their overall contribution to the cancer prevention potential of the parent compounds require further investigation. Cancer prevention by HBAs is relatively underexplored, and further investigations are required to identify potential targets (intracellular vs. extracellular), their uptake mechanisms by cells, signaling pathways and target gene expression. Such studies would have tremendous implications in developing new strategies for cancer prevention.

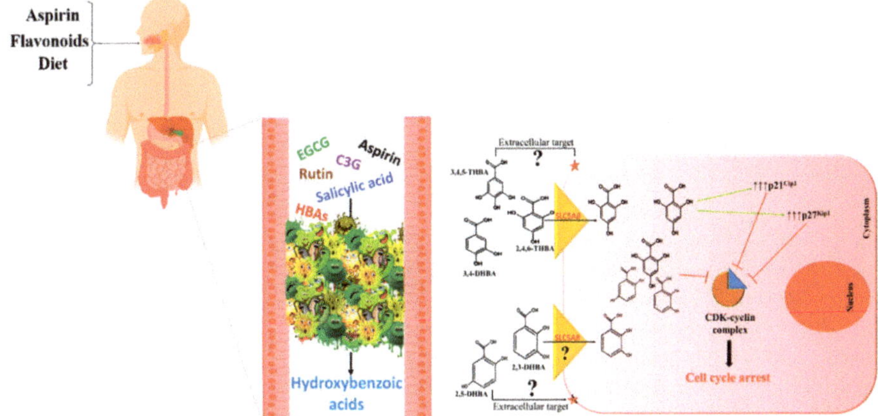

Figure 3. Metabolite hypothesis: Model depicting convergence of the pathways generating HBAs from parent compounds through host/microbial enzymes. We propose that actions of HBAs, generated through biotransformation of aspirin and flavonoids, retard rate of cell proliferation. This would provide an opportunity for (i) immune surveillance, leading to the destruction of cancer cells, or (ii) DNA repair in cells containing damaged DNA, providing genetic stability, both of which are important steps in the prevention of cancer. While 2,4,6-THBA is likely to retard cell proliferation through CDK inhibition and upregulation of p21^{Cip1} and p27^{Kip1}, the exact mechanisms of cell growth inhibition by other HBAs is still not understood [57,63]. EGCG—Epigallocatechin gallate, C3G—Cyanidin-3-glucoside.

8. Conclusions

It has been established over the course of time that increased consumption of fruits and vegetables has health benefits, making Hippocrates' dictum "Let food be your medicine and medicine your food" all the more significant. Identification of polyphenols and phenolic acids commonly found as a component of the diet has now led to the generation of nutraceutical and/or pharmaceutical compounds that have revolutionized the health sector. Considering the increasing incidence of CRC worldwide, it is now more important than ever to come up with viable preventive strategies against CRC. We believe that HBAs merit further studies to understand their role in cancer prevention as the proposed mechanism (metabolite hypothesis) involving HBAs is a simple and tenable explanation for CRC prevention by both aspirin and flavonoids. Since intestinal epithelial cells are the first to get exposed to HBAs following their consumption or metabolism (from aspirin/flavonoids), in our view CRC prevention is likely to be a local effect. The colorectal tissues may thus have an anatomical advantage as they are the first to get exposed to these HBAs, making them preferred targets, while those HBAs absorbed into circulation may affect cancer development in other tissues as well. We believe that effective cancer prevention requires the partnership between the parent compounds and the right microbial species responsible for HBA generation. We also propose that the inter-individual variability previously observed in many epidemiological studies [8,17,27] could be attributed to the lack of an appropriate microbial ecology necessary for their degradation. Therefore, a strategy involving supplementation of the diet with appropriate probiotics and aspirin/flavonoids, or directly consuming HBAs may prove useful in CRC prevention.

Author Contributions: R.S. and G.J.B. wrote the initial draft of the manuscript, D.R.K. and J.P. were involved in discussion, editing and revising of the manuscript. All authors have read and agreed to the published version of the manuscript.

Funding: This study was supported by funds from the Office of Research and the Department of Pharmaceutical Sciences, South Dakota State University.

Acknowledgments: Authors acknowledge Chaitanya Valiveti for useful discussions.

Conflicts of Interest: The authors declare no conflict of interest. The funders had no role in the design of the study; in the collection, analyses, or interpretation of data; in the writing of the manuscript, or in the decision to publish the results.

Abbreviations

CRC	Colorectal cancers
COX	Cyclooxygenase
CDK	Cyclin dependent kinase
CYP450	Cytochrome P450
2,3-DHBA	2,3-dihydroxybenzoic acid
2,5-DHBA	2,5-dihydroxybenzoic acid
3,4-DHBA	3,4-dihydroxybenzoic acid
HBA	Hydroxybenzoic acid
IC50	Inhibitory Concentration -50%
MCT	Monocarboxylate transporter
3,4,5-THBA	3,4,5-trihydroxybenzoic acid
2,4,6-THBA	2,4,6-trihydroxybenzoic acid
FGF	Fibroblast growth factor
GI	Gastrointestinal
USPSTF	United States Preventive Services Task Force

References

1. Terzic, J.; Grivennikov, S.; Karin, E.; Karin, M. Inflammation and colon cancer. *Gastroenterology* **2010**, *138*, 2101–2114.e2105. [CrossRef] [PubMed]
2. Bondonno, N.P.; Dalgaard, F.; Kyro, C.; Murray, K.; Bondonno, C.P.; Lewis, J.R.; Croft, K.D.; Gislason, G.; Scalbert, A.; Cassidy, A.; et al. Flavonoid intake is associated with lower mortality in the Danish Diet Cancer and Health Cohort. *Nat. Commun.* **2019**, *10*, 3651. [CrossRef] [PubMed]
3. Russell, W.; Duthie, G. Session 3: Influences of food constituents on gut health: Plant secondary metabolites and gut health: The case for phenolic acids. *Proc. Nutr. Soc.* **2011**, *70*, 389–396. [CrossRef] [PubMed]
4. Rosa, L.; Nja, S.; Soares, N.; Monterio, M.; Teodoro, A. Anticancer Properties of Phenolic Acids in Colon Cancer—A Review. *J. Nutr. Food Sci.* **2016**, *6*. [CrossRef]
5. Vane, J.R.; Botting, R.M. The mechanism of action of aspirin. *Thromb. Res.* **2003**, *110*, 255–258. [CrossRef]
6. Li, Y.; Zhang, T.; Chen, G.Y. Flavonoids and Colorectal Cancer Prevention. *Antioxidants* **2018**, *7*, 187. [CrossRef]
7. Wang, L.S.; Stoner, G.D. Anthocyanins and their role in cancer prevention. *Cancer Lett.* **2008**, *269*, 281–290. [CrossRef]
8. Zamora-Ros, R.; Guino, E.; Alonso, M.H.; Vidal, C.; Barenys, M.; Soriano, A.; Moreno, V. Dietary flavonoids, lignans and colorectal cancer prognosis. *Sci. Rep.* **2015**, *5*, 14148. [CrossRef]
9. Bojic, M.; Sedgeman, C.A.; Nagy, L.D.; Guengerich, F.P. Aromatic hydroxylation of salicylic acid and aspirin by human cytochromes P450. *Eur. J. Pharm. Sci.* **2015**, *73*, 49–56. [CrossRef]
10. Kuehl, G.E.; Bigler, J.; Potter, J.D.; Lampe, J.W. Glucuronidation of the aspirin metabolite salicylic acid by expressed UDP-glucuronosyltransferases and human liver microsomes. *Drug Metab. Dispos.* **2006**, *34*, 199–202. [CrossRef]
11. Stevens, J.F.; Maier, C.S. The Chemistry of Gut Microbial Metabolism of Polyphenols. *Phytochem. Rev.* **2016**, *15*, 425–444. [CrossRef] [PubMed]
12. Kim, I.S.; Yoo, D.H.; Jung, I.H.; Lim, S.; Jeong, J.J.; Kim, K.A.; Bae, O.N.; Yoo, H.H.; Kim, D.H. Reduced metabolic activity of gut microbiota by antibiotics can potentiate the antithrombotic effect of aspirin. *Biochem. Pharm.* **2016**, *122*, 72–79. [CrossRef] [PubMed]
13. Dovizio, M.; Bruno, A.; Tacconelli, S.; Patrignani, P. Mode of action of aspirin as a chemopreventive agent. *Recent Results Cancer Res.* **2013**, *191*, 39–65. [CrossRef] [PubMed]
14. Patrignani, P.; Patrono, C. Aspirin and Cancer. *J. Am. Coll. Cardiol.* **2016**, *68*, 967–976. [CrossRef]

15. Chan, A.T.; Arber, N.; Burn, J.; Chia, W.K.; Elwood, P.; Hull, M.A.; Logan, R.F.; Rothwell, P.M.; Schror, K.; Baron, J.A. Aspirin in the chemoprevention of colorectal neoplasia: An overview. *Cancer Prev. Res.* **2012**, *5*, 164–178. [CrossRef]

16. Cuzick, J.; Thorat, M.A.; Bosetti, C.; Brown, P.H.; Burn, J.; Cook, N.R.; Ford, L.G.; Jacobs, E.J.; Jankowski, J.A.; La Vecchia, C.; et al. Estimates of benefits and harms of prophylactic use of aspirin in the general population. *Ann. Oncol.* **2015**, *26*, 47–57. [CrossRef]

17. Rothwell, P.M.; Wilson, M.; Price, J.F.; Belch, J.F.; Meade, T.W.; Mehta, Z. Effect of daily aspirin on risk of cancer metastasis: A study of incident cancers during randomised controlled trials. *Lancet* **2012**, *379*, 1591–1601. [CrossRef]

18. Dovizio, M.; Tacconelli, S.; Sostres, C.; Ricciotti, E.; Patrignani, P. Mechanistic and pharmacological issues of aspirin as an anticancer agent. *Pharmaceuticals* **2012**, *5*, 1346–1371. [CrossRef]

19. Bosetti, C.; Santucci, C.; Gallus, S.; Martinetti, M.; La Vecchia, C. Aspirin and the risk of colorectal and other digestive tract cancers: An updated meta-analysis through 2019. *Ann. Oncol.* **2020**, *31*, 558–568. [CrossRef]

20. Shpitz, B.; Bomstein, Y.; Kariv, N.; Shalev, M.; Buklan, G.; Bernheim, J. Chemopreventive effect of aspirin on growth of aberrant crypt foci in rats. *Int. J. Colorectal Dis.* **1998**, *13*, 169–172. [CrossRef]

21. Tian, Y.; Ye, Y.; Gao, W.; Chen, H.; Song, T.; Wang, D.; Mao, X.; Ren, C. Aspirin promotes apoptosis in a murine model of colorectal cancer by mechanisms involving downregulation of IL-6-STAT3 signaling pathway. *Int. J. Colorectal Dis.* **2011**, *26*, 13–22. [CrossRef] [PubMed]

22. Bibbins-Domingo, K.; U.S. Preventive Services Task Force. Aspirin Use for the Primary Prevention of Cardiovascular Disease and Colorectal Cancer: U.S. Preventive Services Task Force Recommendation StatementAspirin Use for the Primary Prevention of CVD and CRC. *Ann. Intern. Med.* **2016**, *164*, 836–845. [CrossRef] [PubMed]

23. Drew, D.A.; Chin, S.M.; Gilpin, K.K.; Parziale, M.; Pond, E.; Schuck, M.M.; Stewart, K.; Flagg, M.; Rawlings, C.A.; Backman, V.; et al. ASPirin Intervention for the REDuction of colorectal cancer risk (ASPIRED): A study protocol for a randomized controlled trial. *Trials* **2017**, *18*, 50. [CrossRef] [PubMed]

24. Rothwell, P.M.; Price, J.F.; Fowkes, F.G.; Zanchetti, A.; Roncaglioni, M.C.; Tognoni, G.; Lee, R.; Belch, J.F.; Wilson, M.; Mehta, Z.; et al. Short-term effects of daily aspirin on cancer incidence, mortality, and non-vascular death: Analysis of the time course of risks and benefits in 51 randomised controlled trials. *Lancet* **2012**, *379*, 1602–1612. [CrossRef]

25. Brenner, H.; Chen, C. The colorectal cancer epidemic: Challenges and opportunities for primary, secondary and tertiary prevention. *Br. J. Cancer* **2018**, *119*, 785–792. [CrossRef]

26. Spratt, J.S. The primary and secondary prevention of cancer. *J. Surg. Oncol.* **1981**, *18*, 219–230. [CrossRef]

27. Rothwell, P.M.; Wilson, M.; Elwin, C.E.; Norrving, B.; Algra, A.; Warlow, C.P.; Meade, T.W. Long-term effect of aspirin on colorectal cancer incidence and mortality: 20-year follow-up of five randomised trials. *Lancet* **2010**, *376*, 1741–1750. [CrossRef]

28. Bosetti, C.; Rosato, V.; Gallus, S.; Cuzick, J.; La Vecchia, C. Aspirin and cancer risk: A quantitative review to 2011. *Ann. Oncol.* **2012**, *23*, 1403–1415. [CrossRef]

29. Rothwell, P.M.; Fowkes, F.G.; Belch, J.F.; Ogawa, H.; Warlow, C.P.; Meade, T.W. Effect of daily aspirin on long-term risk of death due to cancer: Analysis of individual patient data from randomised trials. *Lancet* **2011**, *377*, 31–41. [CrossRef]

30. Eberhart, C.E.; Coffey, R.J.; Radhika, A.; Giardiello, F.M.; Ferrenbach, S.; DuBois, R.N. Up-regulation of cyclooxygenase 2 gene expression in human colorectal adenomas and adenocarcinomas. *Gastroenterology* **1994**, *107*, 1183–1188. [CrossRef]

31. Marnett, L.J.; DuBois, R.N. COX-2: A target for colon cancer prevention. *Annu. Rev. Pharm. Toxicol.* **2002**, *42*, 55–80. [CrossRef] [PubMed]

32. Vane, J.R.; Bakhle, Y.S.; Botting, R.M. Cyclooxygenases 1 and 2. *Annu. Rev. Pharm. Toxicol.* **1998**, *38*, 97–120. [CrossRef] [PubMed]

33. Alfonso, L.; Ai, G.; Spitale, R.C.; Bhat, G.J. Molecular targets of aspirin and cancer prevention. *Br. J. Cancer* **2014**, *111*, 61–67. [CrossRef] [PubMed]

34. Li, H.; Zhu, F.; Boardman, L.A.; Wang, L.; Oi, N.; Liu, K.; Li, X.; Fu, Y.; Limburg, P.J.; Bode, A.M.; et al. Aspirin Prevents Colorectal Cancer by Normalizing EGFR Expression. *EBioMedicine* **2015**, *2*, 447–455. [CrossRef] [PubMed]

35. Martinez, M.E.; O'Brien, T.G.; Fultz, K.E.; Babbar, N.; Yerushalmi, H.; Qu, N.; Guo, Y.; Boorman, D.; Einspahr, J.; Alberts, D.S.; et al. Pronounced reduction in adenoma recurrence associated with aspirin use and a polymorphism in the ornithine decarboxylase gene. *Proc. Natl. Acad. Sci. USA* **2003**, *100*, 7859–7864. [CrossRef]

36. Zell, J.A. Clinical trials update: Tertiary prevention of colorectal cancer. *J. Carcinog* **2011**, *10*, 8. [CrossRef]

37. Afshari, K.; Haddadi, N.S.; Haj-Mirzaian, A.; Farzaei, M.H.; Rohani, M.M.; Akramian, F.; Naseri, R.; Sureda, A.; Ghanaatian, N.; Abdolghaffari, A.H. Natural flavonoids for the prevention of colon cancer: A comprehensive review of preclinical and clinical studies. *J. Cell Physiol.* **2019**, *234*, 21519–21546. [CrossRef]

38. Kikuchi, H.; Yuan, B.; Hu, X.; Okazaki, M. Chemopreventive and anticancer activity of flavonoids and its possibility for clinical use by combining with conventional chemotherapeutic agents. *Am. J. Cancer Res.* **2019**, *9*, 1517–1535.

39. Ribeiro, D.; Freitas, M.; Tome, S.M.; Silva, A.M.; Laufer, S.; Lima, J.L.; Fernandes, E. Flavonoids inhibit COX-1 and COX-2 enzymes and cytokine/chemokine production in human whole blood. *Inflammation* **2015**, *38*, 858–870. [CrossRef]

40. Koosha, S.; Alshawsh, M.A.; Looi, C.Y.; Seyedan, A.; Mohamed, Z. An Association Map on the Effect of Flavonoids on the Signaling Pathways in Colorectal Cancer. *Int. J. Med. Sci.* **2016**, *13*, 374–385. [CrossRef]

41. Hoensch, H.; Richling, E.; Kruis, W.; Kirch, W. [Colorectal cancer prevention by flavonoids]. *Med. Klin.* **2010**, *105*, 554–559. [CrossRef] [PubMed]

42. Murakami, A.; Ashida, H.; Terao, J. Multitargeted cancer prevention by quercetin. *Cancer Lett.* **2008**, *269*, 315–325. [CrossRef] [PubMed]

43. Khan, N.; Mukhtar, H. Tea polyphenols for health promotion. *Life Sci.* **2007**, *81*, 519–533. [CrossRef] [PubMed]

44. Braune, A.; Blaut, M. Bacterial species involved in the conversion of dietary flavonoids in the human gut. *Gut Microbes* **2016**, *7*, 216–234. [CrossRef] [PubMed]

45. Bermudezsoto, M.; Tomasbarberan, F.; Garciaconesa, M. Stability of polyphenols in chokeberry (Aronia melanocarpa) subjected to in vitro gastric and pancreatic digestion. *Food Chem.* **2007**, *102*, 865–874. [CrossRef]

46. Ozdal, T.; Sela, D.A.; Xiao, J.; Boyacioglu, D.; Chen, F.; Capanoglu, E. The Reciprocal Interactions between Polyphenols and Gut Microbiota and Effects on Bioaccessibility. *Nutrients* **2016**, *8*, 78. [CrossRef]

47. Seeram, N.P.; Bourquin, L.D.; Nair, M.G. Degradation Products of Cyanidin Glycosides from Tart Cherries and Their Bioactivities. *J. Agric. Food Chem.* **2001**, *49*, 4924–4929. [CrossRef]

48. Costello, P.B.; Green, F.A. Aspirin survival in human blood modulated by the concentration of erythrocytes. *Arthritis Rheum.* **1982**, *25*, 550–555. [CrossRef]

49. Williams, F.M.; Mutch, E.M.; Nicholson, E.; Wynne, H.; Wright, P.; Lambert, D.; Rawlins, M.D. Human liver and plasma aspirin esterase. *J. Pharm. Pharm.* **1989**, *41*, 407–409. [CrossRef]

50. Hutt, A.J.; Caldwell, J.; Smith, R.L. The metabolism of aspirin in man: A population study. *Xenobiotica* **1986**, *16*, 239–249. [CrossRef]

51. Zhang, J.; Sun, Y.; Wang, R.; Zhang, J. Gut Microbiota-Mediated Drug-Drug Interaction between Amoxicillin and Aspirin. *Sci. Rep.* **2019**, *9*, 16194. [CrossRef] [PubMed]

52. Xie, F.; Ding, X.; Zhang, Q.Y. An update on the role of intestinal cytochrome P450 enzymes in drug disposition. *Acta Pharm. Sin. B* **2016**, *6*, 374–383. [CrossRef]

53. van Duynhoven, J.; Vaughan, E.E.; Jacobs, D.M.; Kemperman, R.A.; van Velzen, E.J.; Gross, G.; Roger, L.C.; Possemiers, S.; Smilde, A.K.; Dore, J.; et al. Metabolic fate of polyphenols in the human superorganism. *Proc. Natl. Acad. Sci. USA* **2011**, *108* (Suppl. 1), 4531–4538. [CrossRef]

54. Ottaviani, J.I.; Borges, G.; Momma, T.Y.; Spencer, J.P.; Keen, C.L.; Crozier, A.; Schroeter, H. The metabolome of [2-(14)C](-)-epicatechin in humans: Implications for the assessment of efficacy, safety, and mechanisms of action of polyphenolic bioactives. *Sci. Rep.* **2016**, *6*, 29034. [CrossRef] [PubMed]

55. Hanske, L.; Engst, W.; Loh, G.; Sczesny, S.; Blaut, M.; Braune, A. Contribution of gut bacteria to the metabolism of cyanidin 3-glucoside in human microbiota-associated rats. *Br. J. Nutr.* **2013**, *109*, 1433–1441. [CrossRef] [PubMed]

56. Scalbert, A.; Williamson, G. Dietary intake and bioavailability of polyphenols. *J. Nutr.* **2000**, *130*, 2073S–2085S. [CrossRef] [PubMed]

57. Sankaranarayanan, R.; Valiveti, C.K.; Kumar, D.R.; Van Slambrouck, S.; Kesharwani, S.S.; Seefeldt, T.; Scaria, J.; Tummala, H.; Bhat, G.J. The Flavonoid Metabolite 2,4,6-Trihydroxybenzoic Acid Is a CDK Inhibitor and an Anti-Proliferative Agent: A Potential Role in Cancer Prevention. *Cancers* **2019**, *11*, 427. [CrossRef]

58. Simons, A.L.; Renouf, M.; Hendrich, S.; Murphy, P.A. Human Gut Microbial Degradation of Flavonoids: Structure–Function Relationships. *J. Agric. Food Chem.* **2005**, *53*, 4258–4263. [CrossRef]

59. Forester, S.C.; Choy, Y.Y.; Waterhouse, A.L.; Oteiza, P.I. The anthocyanin metabolites gallic acid, 3-O-methylgallic acid, and 2,4,6-trihydroxybenzaldehyde decrease human colon cancer cell viability by regulating pro-oncogenic signals. *Mol. Carcinog* **2014**, *53*, 432–439. [CrossRef]

60. Gao, K.; Xu, A.; Krul, C.; Venema, K.; Liu, Y.; Niu, Y.; Lu, J.; Bensoussan, L.; Seeram, N.P.; Heber, D.; et al. Of the Major Phenolic Acids Formed during Human Microbial Fermentation of Tea, Citrus, and Soy Flavonoid Supplements, Only 3,4-Dihydroxyphenylacetic Acid Has Antiproliferative Activity. *J. Nutr.* **2006**, *136*, 52–57. [CrossRef]

61. Grootveld, M.; Halliwell, B. 2,3-Dihydroxybenzoic acid is a product of human aspirin metabolism. *Biochem. Pharm.* **1988**, *37*, 271–280. [CrossRef]

62. Dachineni, R.; Kumar, D.R.; Callegari, E.; Kesharwani, S.S.; Sankaranarayanan, R.; Seefeldt, T.; Tummala, H.; Bhat, G.J. Salicylic acid metabolites and derivatives inhibit CDK activity: Novel insights into aspirin's chemopreventive effects against colorectal cancer. *Int. J. Oncol.* **2017**, *51*, 1661–1673. [CrossRef] [PubMed]

63. Sankaranarayanan, R.; Valiveti, C.K.; Dachineni, R.; Kumar, D.R.; Lick, T.; Bhat, G.J. Aspirin metabolites 2,3DHBA and 2,5DHBA inhibit cancer cell growth: Implications in colorectal cancer prevention. *Mol. Med. Rep.* **2019**. [CrossRef] [PubMed]

64. Hsi, L.C.; Baek, S.J.; Eling, T.E. Lack of cyclooxygenase-2 activity in HT-29 human colorectal carcinoma cells. *Exp. Cell Res.* **2000**, *256*, 563–570. [CrossRef]

65. Altinoz, M.A.; Elmaci, I.; Cengiz, S.; Emekli-Alturfan, E.; Ozpinar, A. From epidemiology to treatment: Aspirin's prevention of brain and breast-cancer and cardioprotection may associate with its metabolite gentisic acid. *Chem. Biol. Interact.* **2018**, *291*, 29–39. [CrossRef]

66. Fernandez, I.S.; Cuevas, P.; Angulo, J.; Lopez-Navajas, P.; Canales-Mayordomo, A.; Gonzalez-Corrochano, R.; Lozano, R.M.; Valverde, S.; Jimenez-Barbero, J.; Romero, A.; et al. Gentisic acid, a compound associated with plant defense and a metabolite of aspirin, heads a new class of in vivo fibroblast growth factor inhibitors. *J. Biol. Chem.* **2010**, *285*, 11714–11729. [CrossRef]

67. Peiffer, D.S.; Zimmerman, N.P.; Wang, L.S.; Ransom, B.W.; Carmella, S.G.; Kuo, C.T.; Siddiqui, J.; Chen, J.H.; Oshima, K.; Huang, Y.W.; et al. Chemoprevention of esophageal cancer with black raspberries, their component anthocyanins, and a major anthocyanin metabolite, protocatechuic acid. *Cancer Prev. Res.* **2014**, *7*, 574–584. [CrossRef]

68. Raina, K.; Rajamanickam, S.; Deep, G.; Singh, M.; Agarwal, R.; Agarwal, C. Chemopreventive effects of oral gallic acid feeding on tumor growth and progression in TRAMP mice. *Mol. Cancer* **2008**, *7*, 1258–1267. [CrossRef]

69. Verma, S.; Singh, A.; Mishra, A. Gallic acid: Molecular rival of cancer. *Environ. Toxicol. Pharm.* **2013**, *35*, 473–485. [CrossRef]

70. Jenner, A.M.; Rafter, J.; Halliwell, B. Human fecal water content of phenolics: The extent of colonic exposure to aromatic compounds. *Free Radic. Biol. Med.* **2005**, *38*, 763–772. [CrossRef]

71. Pedersen, A.K.; FitzGerald, G.A. Dose-related kinetics of aspirin. Presystemic acetylation of platelet cyclooxygenase. *N. Engl. J. Med.* **1984**, *311*, 1206–1211. [CrossRef] [PubMed]

72. Rowland, M.; Riegelman, S.; Harris, P.A.; Sholkoff, S.D. Absorption kinetics of aspirin in man following oral administration of an aqueous solution. *J. Pharm. Sci.* **1972**, *61*, 379–385. [CrossRef] [PubMed]

73. Schiller, C.; Frohlich, C.P.; Giessmann, T.; Siegmund, W.; Monnikes, H.; Hosten, N.; Weitschies, W. Intestinal fluid volumes and transit of dosage forms as assessed by magnetic resonance imaging. *Aliment. Pharm.* **2005**, *22*, 971–979. [CrossRef] [PubMed]

74. Tomás-Barberán, F.A.; Clifford, M.N. Dietary hydroxybenzoic acid derivatives—Nature, occurrence and dietary burden. *J. Sci. Food Agric.* **2000**, *80*, 1024–1032. [CrossRef]

75. Juurlink, B.H.; Azouz, H.J.; Aldalati, A.M.; AlTinawi, B.M.; Ganguly, P. Hydroxybenzoic acid isomers and the cardiovascular system. *Nutr. J.* **2014**, *13*, 63. [CrossRef]

76. Paterson, J.R.; Blacklock, C.; Campbell, G.; Wiles, D.; Lawrence, J.R. The identification of salicylates as normal constituents of serum: A link between diet and health? *J. Clin. Pathol.* **1998**, *51*, 502–505. [CrossRef]

77. Paterson, J.R.; Srivastava, R.; Baxter, G.J.; Graham, A.B.; Lawrence, J.R. Salicylic acid content of spices and its implications. *J. Agric. Food Chem.* **2006**, *54*, 2891–2896. [CrossRef]

78. Venema, D.P.; Hollman, P.C.H.; Janssen, K.P.L.T.M.; Katan, M.B. Determination of Acetylsalicylic Acid and Salicylic Acid in Foods, Using HPLC with Fluorescence Detection. *J. Agric. Food Chem.* **1996**, *44*, 1762–1767. [CrossRef]

79. Paterson, J.R.; Lawrence, J.R. Salicylic acid: A link between aspirin, diet and the prevention of colorectal cancer. *QJM* **2001**, *94*, 445–448. [CrossRef]

80. Abedi, F.; Razavi, B.M.; Hosseinzadeh, H. A review on gentisic acid as a plant derived phenolic acid and metabolite of aspirin: Comprehensive pharmacology, toxicology, and some pharmaceutical aspects. *Phytother. Res.* **2020**, *34*, 729–741. [CrossRef]

81. Altinoz, M.A.; Elmaci, İ.; Ozpinar, A. Gentisic Acid, a Quinonoid Aspirin Metabolite in Cancer Prevention and Treatment. New Horizons in Management of Brain Tumors and Systemic Cancers. *J. Cancer Res. Oncobiol.* **2018**, *1*. [CrossRef]

82. Barroso, C.G.; Torrijos, R.C.; Pérez-Bustamante, J.A. HPLC separation of benzoic and hydroxycinnamic acids in wines. *Chromatographia* **1983**, *17*, 249–252. [CrossRef]

Article

Flavonoid and Non-Flavonoid Compounds of Autumn Royal and Egnatia Grape Skin Extracts Affect Membrane PUFA's Profile and Cell Morphology in Human Colon Cancer Cell Lines

Valeria Tutino [1], Isabella Gigante [1], Rosa Anna Milella [2], Valentina De Nunzio [1], Riccardo Flamini [3], Mirko De Rosso [3], Maria Principia Scavo [4], Nicoletta Depalo [5], Elisabetta Fanizza [5,6], Maria Gabriella Caruso [7] and Maria Notarnicola [1,*]

[1] Laboratory of Nutritional Biochemistry, National Institute of Gastroenterology "S. de Bellis" Research Hospital, 70013 Castellana Grotte (BA), Italy; valeria.tutino@irccsdebellis.it (V.T.); isabella.gigante87@gmail.com (I.G.); valentinadx@hotmail.it (V.D.N.)

[2] Research Centre for Viticulture and Enology, Council for Agricultural Research and Economics, 70010 Turi (BA), Italy; rosaanna.milella@crea.gov.it

[3] Research Centre for Viticulture and Enology, Council for Agricultural Research and Economics, 31015 Conegliano (TV), Italy; riccardo.flamini@crea.gov.it (R.F.); mirko.derosso@crea.gov.it (M.D.R.)

[4] Personalized Medicine Laboratory, National Institute of Gastroenterology "S. de Bellis" Research Hospital, 70013 Castellana Grotte (BA), Italy; maria.scavo@irccsdebellis.it

[5] Institute for Chemical-Physical Processes (IPCF)-CNR SS Bari, 70125 Bari (BA), Italy; n.depalo@ba.ipcf.cnr.it (N.D.); elisabetta.fanizza@uniba.it (E.F.)

[6] Dipartimento di Chimica, Università degli Studi di Bari Aldo Moro, 70126 Bari (BA), Italy

[7] Ambulatory of Clinical Nutrition, National Institute of Gastroenterology "S. de Bellis" Research Hospital, 70013 Castellana Grotte (BA), Italy; gabriella.caruso@irccsdebellis.it

* Correspondence: maria.notarnicola@irccsdebellis.it; Tel.: +39-080-4994342

Academic Editor: H.P. Vasantha Rupasinghe
Received: 29 June 2020; Accepted: 21 July 2020; Published: 23 July 2020

Abstract: Grapes contain many flavonoid and non-flavonoid compounds with anticancer effects. In this work we fully characterized the polyphenolic profile of two grape skin extracts (GSEs), Autumn Royal and Egnatia, and assessed their effects on Polyunsaturated Fatty Acid (PUFA) membrane levels of Caco2 and SW480 human colon cancer cell lines. Gene expression of 15-lipoxygenase-1 (15-LOX-1), and peroxisome proliferator-activated receptor gamma (PPAR-γ), as well as cell morphology, were evaluated. The polyphenolic composition was analyzed by Ultra-High-Performance Liquid Chromatography/Quadrupole-Time of Flight mass spectrometry (UHPLC/QTOF) analysis. PUFA levels were evaluated by gas chromatography, and gene expression levels of 15-LOX-1 and PPAR-γ were analyzed by real-time Polymerase Chain Reaction (PCR). Morphological cell changes caused by GSEs were identified by field emission scanning electron microscope (FE-SEM) and photomicrograph examination. We detected a different profile of flavonoid and non-flavonoid compounds in Autumn Royal and Egnatia GSEs. Cultured cells showed an increase of total PUFA levels mainly after treatment with Autumn Royal grape, and were richer in flavonoids when compared with the Egnatia variety. Both GSEs were able to affect 15-LOX-1 and PPAR-γ gene expression and cell morphology. Our results highlighted a new antitumor mechanism of GSEs that involves membrane PUFAs and their downstream pathways.

Keywords: flavonoids; non-flavonoids; membrane PUFAs profile; cell morphology; human colon cancer cells

1. Introduction

Grape (*Vitis vinifera* L.) is a fruit rich in polyphenols, bioactive compounds able to prevent the occurrence of cancer, reduce tumorigenesis, and influence important cancer-related pathways [1–3]. The anticancer effects of polyphenols are closely related to their chemical structure and concentration, as well as to the type of cancer [1,2]. Polyphenolic compounds are mainly divided into two groups: flavonoids, based on the common C6-C3-C6 skeleton which consists of two phenyl rings (A and B) linked by a heterocyclic ring (C), and non-flavonoids such as stilbenes (C6-C2-C6) and phenolic acids (C6-C1) [3,4]. The most abundant classes of flavonoids present in the grape skin and seeds include anthocyanins, flavonols, flavan-3-ols, and proanthocyanidins [5]. Flavonoids exist either as glycosides with attached sugars or as aglycones with no attached sugars, and differ in the degree of hydroxylation and substitution. These functional hydroxyl groups mediate the antioxidant effects of flavonoids and their ability to interact with biological membranes [2,6]. Stilbenes, such as resveratrol, are phytoalexins synthesized by plants in response to mechanical injury, UV irradiation, and fungal attacks [5]. Benzoic acid and cinnamic acid represent the most common phenolic acids present in the grape skin [7].

The phenolic composition of the grapes mainly depends on genotype but can be affected by environmental factors and agronomic practices [5]. Several in vivo and in vitro studies have demonstrated the antimetastatic effects of some polyphenolic compounds [8,9]. Mantena S. K. et al. evaluated the chemoprotective efficacy of grape seed proanthocyanidins in both metastatic breast cancer cells and in Balb/c mice, a mouse model of breast cancer obtained after subcutaneous implantation of the highly invasive and metastatic 4T1 mouse breast cell line [10]. There are several mechanisms of action through which grape polyphenols are able to inhibit the invasion and progression of metastasis [11–13]. These natural compounds can act on the structural components of the cytoskeleton, on cellular adhesions, and on the composition of the membrane fatty acids [11,14].

Polyphenols are able to influence cell membrane fluidity and cell motility by the stearoyl-CoA desaturase-1 (SCD1) enzyme activity, given by the oleic acid/stearic acid ratio [15,16]. The increase in the oleic acid content in cell membranes, and consequently the up-regulation of SCD1 enzyme, are known to stimulate the process of invasion and metastasis in human cancer cells [17,18]. Rearrangements in the lipidomic profile of the membrane are an important feature that distinguish cancer cells, since phospholipids are directly involved in the morphological changes occurring in tumorigenesis and tumor progression [16,19,20]. Moreover, it is now known that an imbalance in the ratio of omega-6/omega-3 (n-6/n-3) Polyunsaturated Fatty Acids (PUFAs), in favor of n-6, is associated with the development of chronic inflammatory diseases, including colon cancer [21,22]. Tumorigenesis of the colon is strongly influenced by the oxidative metabolism of PUFAs regulated by different enzymes, as 15-lipoxygenase-1 (15-LOX-1) [23,24]. 15-LOX-1 enzyme is known to exert antioxidant and antimetastatic action by activating peroxisome proliferator-activated receptor gamma (PPAR-γ) [25,26]. The preferred substrate for 15-LOX-1 is the essential fatty acid, namely linoleic acid (LA). This n-6 fatty acid with its metabolite, the 13-HODE (13-S-hydroxyoctadecadienoic acid), are down-expressed in human colon cancer [27].

Based on these assumptions, the main aims of the present study are: (1) to fully characterize the polyphenolic content of two table grape skin extracts (GSEs), Autumn Royal, a seedless black grape with healthy properties, and Egnatia, a new red seedless genotype obtained by breeding programs carried out by our research group; (2) to evaluate the effects of two GSEs on the membrane PUFA profile in two human colon cancer cell lines at different grade of differentiation, Caco2 and SW480; (3) to evaluate in the same treated cells the gene expression of 15-LOX-1 enzyme and its downstream factor PPAR-γ, as well as the possible changes in cell morphology.

2. Results and Discussion

2.1. Total Polyphenolic Content in Autumn Royal and Egnatia GSEs

The total content of polyphenols was determined in Autumn Royal and Egnatia GSEs by a spectrophotometric assay. The results obtained confirmed that Autumn Royal is richer in total polyphenols, expressed as milligrams of gallic acid equivalents per gram of dry weight of skin (GAE/g dw), than Egnatia (53.10 ± 1.99 mg GAE/g dw versus 37.45 ± 0.73 mg GAE/g dw, *p*-value < 0.05, respectively) and that the anthocyanins were the most represented polyphenols in both varieties [28]. Polyphenolic profiles were determined by Ultra-High-Performance Liquid Chromatography/Quadrupole-Time of Flight mass spectrometry (UHPLC/QTOF) analysis. Anthocyanins were identified by positive ionization (Figure 1), the other polyphenols in negative ionization mode (Figure 2). Main MS/MS fragments and mean signal intensity of the metabolites identified in the samples are reported in the supplementary materials (Table S1). Positive and negative extract ion chromatograms (EIC), showing the signals of compounds identified, are reported in the supplementary materials (Figures S1A and S1B, respectively). More than 100 flavonoid and non-flavonoid compounds were identified in the both GSEs.

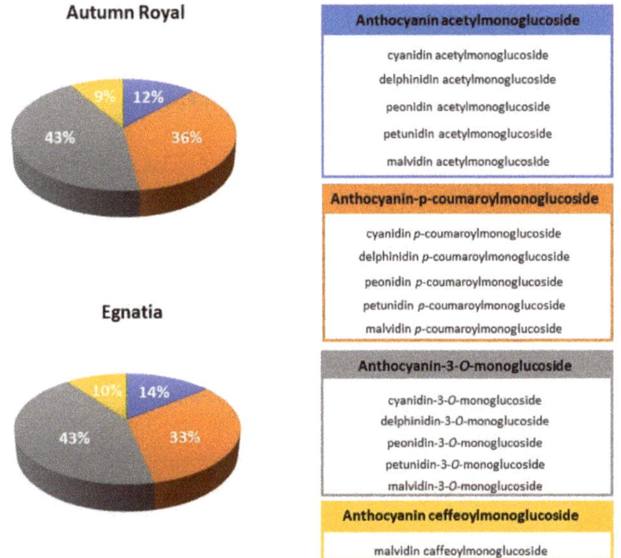

Figure 1. Anthocyanin composition of Autumn Royal and Egnatia table grape skin extracts expressed as the relative percentages (%) of $M^{+\bullet}$ signals intensity in the positive-ion Ultra-High-Performance Liquid Chromatography/Quadrupole-Time of Flight mass spectrometry (UHPLC/QTOF) chromatogram.

The anthocyanin profiles of Autumn Royal and Egnatia GSEs were very similar and characterized by a higher percentage of anthocyanin-3-*O*-monoglucoside compounds (43% in both samples) (Figure 1). These molecules have different antitumor effects depending on the B-ring substituents [29]. Anthocyanins, whose structure contains an *o*-dihydroxy (catechol) B-ring, such as cyanidin-3-*O*-monoglucoside and delphinidin-3-*O*-monoglucoside, are characterized by tumorigenesis inhibition activity [30].

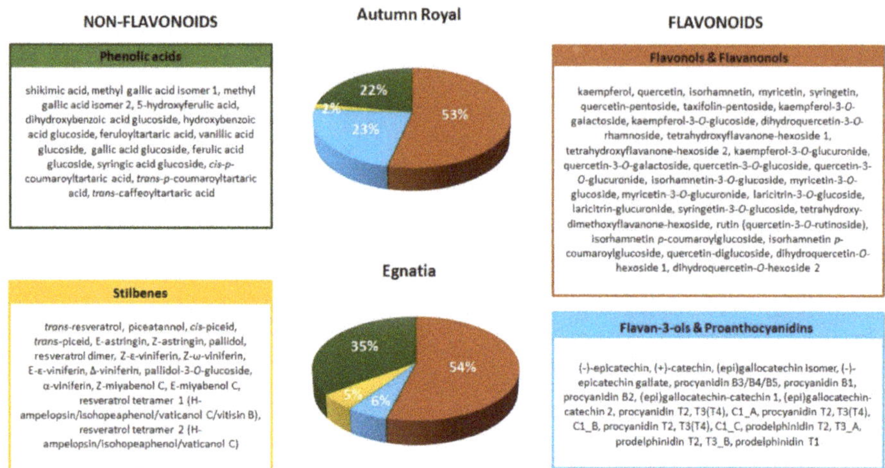

Figure 2. Flavonoid and non-flavonoid composition of Autumn Royal and Egnatia table grape skin extracts expressed as the relative percentages (%) of the total $[M - H]^-$ signal intensity in the negative-ion Ultra-High-Performance Liquid Chromatography/Quadrupole-Time of Flight mass spectrometry (UHPLC/QTOF) chromatogram.

Signals of phenolic acids were relevant in both varieties (22% and 35% for Autumn Royal and Egnatia, respectively), while total stilbene signals intensity was 2% in Autumn Royal and 5% in Egnatia extracts (Figure 2). Both samples showed flavonols and flavanonols as main signals, with total signal intensity 53% and 54% in Autumn Royal and Egnatia, respectively. In particular, high intensity of the B-ring trisubstituted flavonols, such as myricetin and syringetin glycosides, was found. Instead, signals of flavan-3-ols and proanthocyanidins were higher in Autumn Royal (23%) compared to Egnatia extract (6%). Several studies showed that flavan-3-ol oligomers (proanthocyanidins) are potent antioxidants and free radical scavengers also characterized by anticancer properties [31,32]. In our previous study, we found a higher antioxidant activity in Autumn Royal compared to Egnatia, and this result could also be due to the greater content of flavan-3-ols and proanthocyanidins found in Autumn Royal GSEs [28].

Within each class of flavonoids, a structural variation exists in their basic 15-carbon skeleton, leading to different physicochemical properties. An improvement in the antitumor biological activities of flavonoids is due to the 3′- and 4′-hydroxyl groups (*o*-diphenol groups), known as catechol groups, present on the B-ring [33]. Compared to the Egnatia variety, Autumn Royal presented a higher catechol percentage, calculated by summing the intensities of the signals of those metabolites whose structure includes one or more *o*-diphenol groups (100% versus 57.6%, respectively). Table 1 shows the *o*-diphenol compounds identified in both GSEs.

The number and position of hydroxyl groups also influence the interactions of flavonoids with the cell membrane lipid bilayer. The hydrophilic flavonoids, which contain more hydroxyl groups, interact with the polar head groups by hydrogen bonds, inducing membrane rigidification. On the contrary, the hydrophobic flavonoids pass through cell membranes, causing a modification of their permeability and fluidity [6,34,35]. Therefore, the contribution of each phenolic compound can be different, and their synergistic/antagonist interactions might influence their biological effects.

Table 1. Catechol derivatives identified in Autumn Royal and Egnatia grape skin extracts (GSEs).

	o-Diphenol Compounds
Phenolic acids	shikimic acid, methyl gallic acid, 5-hydroxy-ferulic acid, gallic acid glucoside, *trans*-caffeoyltartaric acid
Flavonols and Flavanonols	quercetin glycosides, myricetin glycosides, laricitrin glycosides, taxifolin-pentoside, dihydroquercetin-3-*O*-rhamnoside
Flavan-3-ols and Proanthocyanidins	(+)-catechin/(−)-epicatechin, (epi)gallocatechins, (−)-epicatechin gallate, procyanidins/prodelphinidins
Anthocyanins	cyanidin glycosides, delphinidin glycosides, petunidin glycosides

2.2. Membrane PUFA Profile After GSE Treatment in Human Colon Cancer Cell Lines

To investigate the effects of Autumn Royal and Egnatia GSEs on the n-3 and n-6 PUFAs membrane composition, Caco2 and SW480 human colon cancer cell lines were treated with increasing concentrations of GSEs (20, 50 and 80 µg/mL), and the lipidomic profile was analyzed after 48 h of treatment (Table 2a,b). The choice to use these GSEs concentrations and this experimental time (48 h), was dictated by our previous study, in which we demonstrated that the greatest antiproliferative effect of GSEs in Caco2 and SW480 cell lines was observed in these experimental conditions [16].

Compared to the untreated control group (CTR), in Caco2 cells, the treatment with Autumn Royal and Egnatia GSEs caused an increase of both essential fatty acids (EFAs), linoleic acid (LA) and α-linolenic acid (ALA), already starting from the concentration of 20 µg/mL (Table 2a), and these increases did not correspond to a modification in the n-3 and n-6 fatty acids downstream pathways (Table 2a). Moreover, the increase in total PUFA levels, already at the lowest concentration (20 µg/mL) of both GSEs, was essentially due to the contribution of LA and ALA (Table 2a).

In the SW480 cell line, the treatment with Autumn Royal GSE increased LA levels at 20 µg/mL of concentration, whereas for Egnatia GSE, a higher concentration (50 µg/mL) was necessary (Table 2b). Compared to CTR, only a decrease in arachidonic acid (AA) levels was observed after exposure of both GSEs at 50 µg/mL (Table 2b). As regards the n-3 PUFAs pathway, ALA levels were induced after exposure to the two grape polyphenols, in particular after treatment with 80 µg/mL of Autumn Royal GSE and 20 µg/mL of Egnatia GSE (Table 2b). The increase in total PUFAs obtained after Autumn Royal treatment, at the concentration of 20 µg/mL, was due exclusively to the contribution of LA (Table 2b). The reduction of AA levels found at the concentration of 50 µg/mL of Autumn Royal was more pronounced than the increase of LA obtained at the same concentration of extract, thus balancing the levels of total PUFAs at 50 and 80 µg/mL (Table 2b).

Moreover, no alteration in n-6/n-3 ratio levels was found in all two cell lines and with both treatments (data not shown).

We used two human colon adenocarcinoma cell lines with different degrees of differentiation, Caco2 and SW480, in order to investigate the variations of the membrane PUFA levels induced by the quality of polyphenols contained in the two table grape varieties used. Cancer cells are in continuous proliferation and need large quantities of fatty acids and phospholipids to generate new cellular membranes [19,36]. Previously, we demonstrated the ability of Autumn Royal and Egnatia GSEs to influence membrane fluidity in Caco2 and SW480, through the inhibition of the enzyme SCD1 [16]. This enzyme, that converts saturated fatty acids (SFAs) into monounsaturated fatty acids (MUFAs), was reduced mainly in the Caco2 cell line after GSEs treatment. In this study, we demonstrated that Autumn Royal and Egnatia GSEs were also able to influence the membrane levels of total PUFAs. Different basal levels of PUFAs were found in untreated Caco2 and SW80 cells, probably due to cell type, developmental and growth stage of cells [37]. In SW480 cells with a lower degree of differentiation, the high AA levels found contribute to increasing the total PUFAs levels present in these cells, preparing the cellular pathways towards more inflammatory outcomes.

Table 2. (a) Mean percentage of n-6 and n-3 Polyunsaturated Fatty Acids (PUFAs) in the Caco2 membrane cell line treated with increasing concentrations of Autumn Royal and Egnatia GSEs (20, 50, and 80 µg/mL) after 48 h of treatment; (b) mean percentage of n-6 and n-3 Polyunsaturated Fatty Acids (PUFAs) in the SW480 membrane cell line treated with increasing concentrations of Autumn Royal and Egnatia GSEs (20, 50, and 80 µg/mL) after 48 h of treatment. All data are expressed as the mean ± Standard Deviation (SD) of three consecutive experiments. p-value was determined by ANOVA with Dunnett's post-test. * $p < 0.05$ versus the untreated control group (CTR).

a	Caco2 PUFAs (%)		Autumn Royal				Egnatia	
		CTR	20 µg/mL	50 µg/mL	80 µg/mL	20 µg/mL	50 µg/mL	80 µg/mL
n-6 PUFAs	linoleic acid (LA)	2.20 ± 0.16	2.89 ± 0.37 *	3.20 ± 0.33 *	4.60 ± 0.39 *	3.58 ± 0.18 *	4.14 ± 0.32 *	6.40 ± 0.39 *
	γ-linolenic acid (GLA)	0.14 ± 0.04	0.22 ± 0.07	0.12 ± 0.05	0.21 ± 0.03	0.18 ± 0.11	0.13 ± 0.04	0.15 ± 0.05
	dihomo-γ-linoleic acid (DGLA)	0.38 ± 0.10	0.44 ± 0.04	0.30 ± 0.05	0.29 ± 0.10	0.37 ± 0.08	0.34 ± 0.10	0.30 ± 0.14
	arachidonic acid (AA)	3.87 ± 0.51	4.27 ± 0.26	4.26 ± 0.42	4.25 ± 0.33	4.60 ± 0.64	3.52 ± 0.63	3.65 ± 0.61
n-3 PUFAs	α-linolenic acid (ALA)	0.19 ± 0.04	0.57 ± 0.09 *	0.81 ± 0.07 *	1.34 ± 0.21 *	0.64 ± 0.06 *	0.93 ± 0.07 *	1.74 ± 0.03 *
	eicosapentaenoic acid (EPA)	0.93 ± 0.06	1.03 ± 0.14	0.78 ± 0.08	0.89 ± 0.11	0.92 ± 0.13	0.90 ± 0.08	0.94 ± 0.11
	docosaepentaenoic acid (DPA)	1.32 ± 0.10	1.52 ± 0.06	1.26 ± 0.10	1.29 ± 0.09	1.58 ± 0.12	1.53 ± 0.13	1.52 ± 0.09
	docosaehenanoic acid (DHA)	3.21 ± 0.50	3.64 ± 0.37	3.06 ± 0.21	3.26 ± 0.41	3.54 ± 0.95	3.20 ± 0.67	3.63 ± 0.77
	Total PUFAs	12.89 ± 0.24	15.74 ± 0.51 *	14.24 ± 0.71*	16.34 ± 0.44 *	15.73 ± 0.50 *	15.29 ± 0.70 *	18.88 ± 0.42 *

b	SW480 PUFAs (%)		Autumn Royal				Egnatia	
		CTR	20 µg/mL	50 µg/mL	80 µg/mL	20 µg/mL	50 µg/mL	80 µg/mL
n-6 PUFAs	linoleic acid (LA)	3.00 ± 0.36	3.91 ± 0.37 *	4.20 ± 0.41 *	4.36 ± 0.35 *	3.82 ± 0.26	4.69 ± 0.51 *	5.64 ± 0.63 *
	γ-linolenic acid (GLA)	0.20 ± 0.02	0.18 ± 0.01	0.21 ± 0.03	0.23 ± 0.02	0.16 ± 0.03	0.21 ± 0.03	0.22 ± 0.06
	dihomo-γ-linoleic acid (DGLA)	0.06 ± 0.02	0.04 ± 0.02	0.04 ± 0.02	0.06 ± 0.01	0.03 ± 0.01	0.11 ± 0.04	0.06 ± 0.06
	arachidonic acid (AA)	6.28 ± 0.36	5.92 ± 0.45	4.68 ± 0.32 *	4.41 ± 0.26 *	5.69 ± 0.37	4.73 ± 0.70 *	4.37 ± 0.38 *
n-3 PUFAs	α-linolenic acid (ALA)	0.15 ± 0.06	0.12 ± 0.07	0.15 ± 0.02	0.52 ± 0.06 *	0.32 ± 0.05 *	0.46 ± 0.09 *	0.75 ± 0.10 *
	eicosapentaenoic acid (EPA)	0.63 ± 0.14	0.51 ± 0.22	0.65 ± 0.20	0.73 ± 0.07	0.54 ± 0.06	0.77 ± 0.02	0.79 ± 0.02
	docosaepentaenoic acid (DPA)	2.21 ± 0.41	2.89 ± 0.11	2.53 ± 0.11	2.55 ± 0.11	2.82 ± 0.17	2.64 ± 0.35	2.57 ± 0.60
	docosaehenanoic acid (DHA)	4.60 ± 0.47	5.31 ± 1.03	5.04 ± 0.09	4.92 ± 0.78	5.05 ± 0.80	4.89 ± 0.53	4.60 ± 0.39
	Total PUFAs	18.58 ± 0.62	19.87 ± 0.14 *	18.00 ± 0.15	18.52 ± 0.49	19.51 ± 0.59	18.49 ± 0.49	19.30 ± 0.29

Both GSEs induced the levels of EFAs, LA and ALA, in all two cell lines studied. EFAs are important structural components of cell membranes that the animal cells must exclusively obtain from their environment [37]. Therefore, the increase of LA and ALA detected in treated Caco2 and SW80 cells was certainly attributable to the fatty acids contained in the grape skins.

Several studies have shown that LA, γ-linolenic acid (GLA), and dihomo-γ-linoleic acid (DGLA) have anticancer effects, unlike AA, which has been associated with the inflammation and with onset of the tumor [38–41]. Treatment with increasing concentrations of both extracts increased LA levels in SW480 cells. In addition, in these cell lines, both extracts led to a reduction in AA compared to untreated cells, exerting an anti-inflammatory effect inside cells. As regards the n-3 fatty acids pathway, no variation was observed. Hanikoglu A. et al. found differences in the reorganization of fatty acids in cell membranes of two different breast cancer cell lines, MCF-7 and MDA-MB231, after treatment with somatostatin, curcumin, and quercetin, alone or in combination [14]. Cancer cells, according to the degree of differentiation, behave differently to treatment with drugs and/or natural compounds, and this feature could explain the different response of SW480 to treatment with GSEs, with respect to Caco2 cells.

2.3. Effects of GSE Treatments on the Gene Expression of 15-LOX-1 and PPAR-γ in Human Colon Cancer Cell Lines

To better investigate the effects of Autumn Royal and Egnatia GSEs on membrane PUFA levels and their antitumoral and antimetastatic action, the gene expression of 15-LOX-1 and PPAR-γ, markers involved in the onset of colorectal cancer (CRC), was studied. Figure 3 shows the effects of increasing Autumn Royal and Egnatia GSE concentrations (20, 50, and 80 μg/mL) on the mRNA levels of 15-LOX-1 (Figure 3a) and PPAR-γ (Figure 3b) in Caco2 and SW480 cell lines after 48 h of treatment. Compared to CTR, Autumn Royal GSE induced a significant up-regulation of 15-LOX-1 gene expression in both cell lines studied, starting from the lowest concentration (20 μg/mL) (Figure 3a). Regarding treatment with Egnatia GSE, in Caco2 cells a higher concentration (80 μg/mL) was needed to observe a statistically significant increase in the gene expression levels of 15-LOX-1 compared to CTR (Figure 3a). For SW480 cells, the increase in the gene expression of 15-LOX-1 was already visible at 50 μg/mL of Egnatia GSE (Figure 3a). 15-LOX-1, through its product 13-S-HODE, activates PPAR-γ by inhibiting colorectal tumorigenesis. Therefore, possible changes in PPAR-γ gene expression after exposure to increasing concentrations of GSEs were investigated. Compared to CTR, Autumn Royal GSE treatment exerted an up-regulation of PPAR-γ mRNA levels, starting from 20 μg/mL in both Caco2 and SW480 cells (Figure 3b), whereas a higher concentration of Egnatia (80 μg/mL) was need to obtain the same significant increase in PPAR-γ gene expression (Figure 3b).

Carcinogenesis is known to be also caused by changes in PUFA levels of the cell membrane, including colon cancer formation [19]. 15-LOX-1 is able to oxygenate both n-3 and n-6 PUFAs. The main substrate of 15-LOX-1 is represented by LA, leading to the formation of 13(S)-HODE that activates PPAR-γ, an antimetastatic and anti-inflammatory factor in CRC [26,27].

The increase in expression of 15-LOX-1 found in the cell lines studied after GSE treatment confirms the antitumor effect exerted by these extracts, and the data obtained highlight a new mechanism of action through which GSEs inhibit colon tumorigenesis. Previously, we demonstrated the ability of GSEs to block cell migration and motility by inhibiting SCD1 and some components of the cytoskeleton [16]. The data obtained in this work show that the antiproliferative effect of GSEs also occurs through the induction of the expression of 15-LOX-1 that can be used for therapeutic purposes in CRC.

Cimen I. et al. have shown that 15-LOX-1 indirectly inhibits NF-kB through 13(S)-HODE-mediated PPAR-γ activation in HCT-116 and HT29 CRC cell lines, thus blocking cell proliferation [26]. Moreover, again in HCT-116 and HT29 cell lines, the expression of 15-LOX-1 reduced the ability of cells to adhere to fibronectin, thus inhibiting cell motility [42].

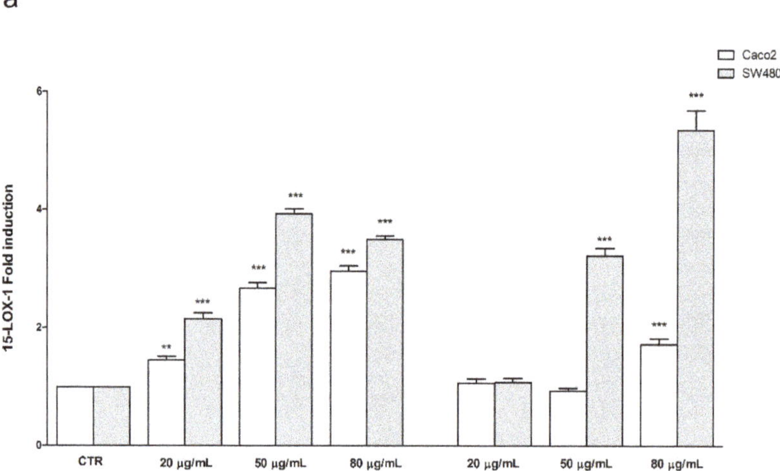

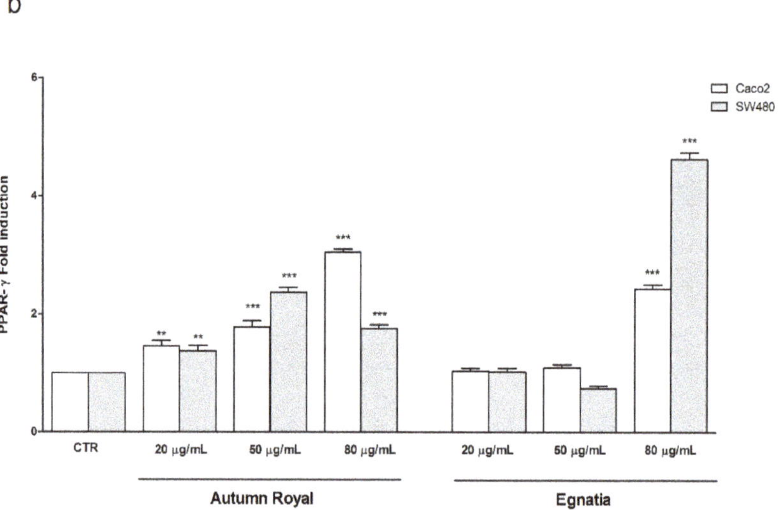

Figure 3. (a) 15-LOX-1 gene expression levels detected in Caco2 and SW480 cells treated with increasing concentrations (20, 50, 80 µg/mL) of Autumn Royal and Egnatia GSEs for 48 h of incubation; (b) PPAR-γ gene expression levels detected in Caco2 and SW480 cells treated with increasing concentrations (20, 50, 80 µg/mL) of Autumn Royal and Egnatia GSEs for 48 h of incubation. All data are expressed as mean ± Standard Deviation (SD) of three consecutive experiments. p-value was determined by ANOVA with Dunnett's posttest. ** $p < 0.03$ and *** $p < 0.01$ versus untreated control group (CTR).

Previous in vitro studies have shown that grape extracts can act differently on proliferation and apoptotic pathways [43,44]. These different biological effects of GSEs could depend both on the type of cancer cell and on the different polyphenolic content of grape extracts [43,45,46]. In fact, it is known that there are cell lines more sensitive to treatment with polyphenols than others in relation to cellular differentiation degree. Moreover, certain flavonoids and non-flavonoids contained in grape extracts can act together synergistically to give particular antiproliferative effects on cancer cells [46,47]. These considerations suggest that the quality of the polyphenolic content in a grape cultivar is an important factor that must be considered.

2.4. Autumn Royal and Egnatia GSEs Induce Cell Morphological Changes

Figures 4 and 5 show the cell morphology of Caco2 and SW480 cell lines, respectively, treated with increasing concentrations of Autumn Royal and Egnatia GSEs (10, 20, 50, and 80 µg/mL) for 24 (T1) and 48 h (T2), analyzed by field emission scanning electron microscope (FE-SEM). To highlight the cellular morphological differences before and after each time GSEs exposure, untreated cells were used as control (CTR) at T0, T1, and T2 (Figures 4 and 5).

Caco2 CTR cells appeared firmly adherent and covered with abundant microvilli, with a visible cytoplasm and at the center a notable nucleus region without shrinkage (Figure 4). Both at T1 and T2, treatment with Autumn Royal GSE induced visible characteristic morphological changes, such as shrinkage of membrane cells, starting from low concentrations of extract (10 µg/mL), showing a typical state of cell suffering (Figure 4a). In comparison to CTR, the increase of the Autumn Royal concentration induced evident and characteristic changes in the cells, as the cytoplasmic contraction and cell membrane collapse. Moreover, at the highest concentrations of Autumn Royal GSE, it was no longer possible to distinguish cell structures (Figure 4a). As regards the treatment with Egnatia GSE, at T1, the maximum concentration (80 µg/mL) was necessary to observe clear signs of apoptosis, while at T2, already at 50 µg/mL, membrane blebbing and cell shrinkage were noted (Figure 4b).

The FE-SEM micrographs reported in Figure 5 show the untreated controls (CTR) of the SW480 cell line that appeared flat and adherent to the substrate with an evident central nucleus. The morphological changes induced by Autumn Royal GSE on SW480 cells were visible at T1 starting from the highest treatment concentrations (50 and 80 µg/mL), given that at the lowest concentrations the cells were morphologically similar to the untreated cells (Figure 5a). At T2, the proapoptotic effect of the polyphenols contained in Autumn Royal GSE was appreciated, starting from the lowest concentrations (10 µg/mL) (Figure 5a). At the concentration of 80 µg/mL of GSE, the cellular structures did not appear very detailed and appreciable, when compared to the CTR; in addition, the apparent break of the surface of the cell membrane caused cell death (Figure 5a). Both at T1 and T2, SW480 cells treated with low concentrations of Egnatia GSE were visibly adherent with a round and abundant cytoplasm (Figure 5b). Furthermore, these cells were connected with neighboring cells and extended in all directions. The higher concentrations of GSE (50 and 80 µg/mL) induced characteristic changes, as well as the reduction of cell cytoplasm and a decrease of surface microvilli, which led to cellular apoptosis (Figure 5b).

To our knowledge, this is the first study describing cell morphological changes induced by polyphenols using FE-SEM micrographs. A previous study of Wang S. et al. demonstrated the ability of Trollius chinensis flavonoids to induce apoptosis in human breast cancer MCF-7 cells using SEM analysis [48]. In these cells treated with high concentrations of flavonoids, the microvilli on the cellular surface completely disappeared and cell membranes collapsed. Other morphological changes, in particular cell shrinkage and membrane blebbing, have been found in HCT-15 colon cancer cells after treatment with diet-derived gallic acid [49]. However, our study confirms previous data obtained about antiproliferative and proapoptotic effects in human colon cancer cell lines treated with both GSEs [28] and, for the first time, demonstrates that the beneficial effects of GSE polyphenols are also due to their ability to induce morphological changes in cancer cells, preventing their growth and proliferation.

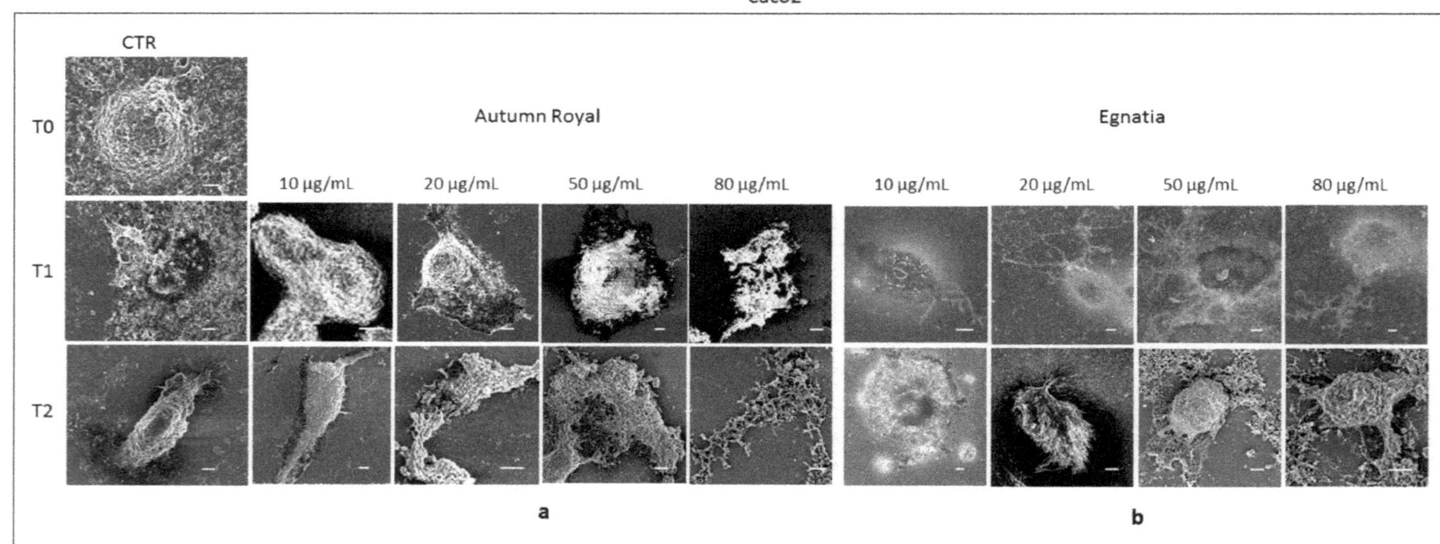

Figure 4. (**a**) Representative field emission scanning electron microscope (FE-SEM) micrographs (scale bar 5 µm, acquisition voltage 3 kV) of Caco2 cell line treated with increasing concentrations (10, 20, 50, 80 µg/mL) of Autumn Royal GSE after 24 (T1) and 48 (T2) h of incubation; (**b**) Representative FE-SEM micrographs (scale bar 5 µm, acquisition voltage 3 kV) of Caco2 cell line treated with increasing concentrations (10, 20, 50, 80 µg/mL) of Egnatia GSE after 24 (T1) and 48 (T2) h of incubation. The FE-SEM micrographs were selected as representative of a series of images collected on each sample. Untreated cells were used as control (CTR) at T0, T1, and T2. Each experiment was performed in triplicate.

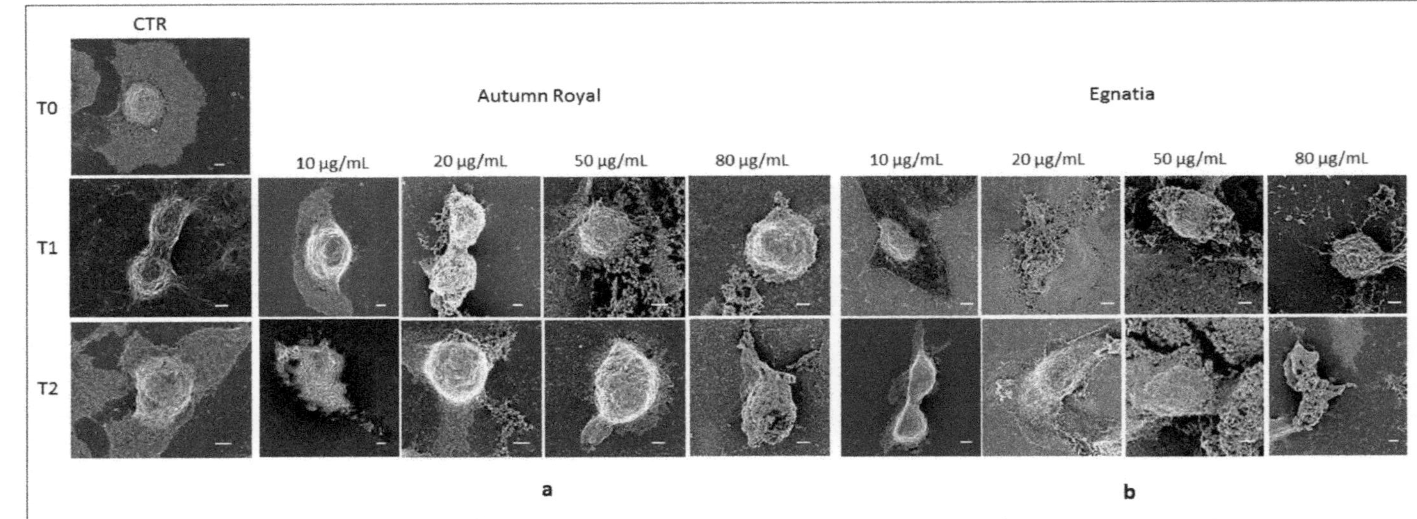

Figure 5. (**a**) Representative field emission scanning electron microscope (FE-SEM) micrographs (scale bar 5 µm, acquisition voltage 3 kV) of SW480 cell line treated with increasing concentrations (10, 20, 50, 80 µg/mL) of Autumn Royal GSE after 24 (T1) and 48 (T2) h of incubation; (**b**) Representative FE-SEM micrographs (scale bar 5 µm, acquisition voltage 3 kV) of SW480 cell line treated with increasing concentrations (10, 20, 50, 80 µg/mL) of Egnatia GSE after 24 (T1) and 48 (T2) h of incubation. The FE-SEM micrographs were selected as representative of a series of images collected on each sample. Untreated cells were used as control (CTR) at T0, T1, and T2. Each experiment was performed in triplicate.

3. Materials and Methods

3.1. Preparation of the GSEs

Table grapes cultivar Autumn Royal, a seedless black grape variety, and Egnatia, a new red seedless genotype, were planted and grown in Apulia region at the Research Center for Viticulture and Enology of the Council for Agricultural Research and Economy (CREA-VE, Turi, BA, Italy). Grape samples were harvested at maturity in summer 2019 and berries were randomly collected and frozen at −20 °C. Approximately 100 frozen berries were manually peeled. To prepare the extracts, 250 mg of dry skin powder were mixed with 5 mL extraction solution of ethanol:water:hydrogen chloride 37% (70:30:1 *v/v/v*). After 24 h of complete darkness, the mixture was centrifuged, and the supernatant recovered, concentrated in a SpeedVac concentrator (Savant®SPD131DDA, Thermo Fisher Scientific, Waltham, MA, USA) for 90 min at 25 °C and 1.5 atmospheres of pressure and analyzed.

3.2. Total Polyphenolic Content

Total phenolic content was determined by Folin–Ciocalteu micro scale protocol with slight modification, as previously described [50]. Briefly, 1 mL of water, 0.02 mL of extract sample, 0.2 mL of the Folin reagent, and 0.8 mL of 10% sodium carbonate solution were mixed and brought to 3 mL. The absorbance was measured at 765 nm after 90 min. Results were expressed as mg GAE/g dw using calibration curves with standard gallic acid.

3.3. UHPLC/QTOF Mass Spectrometry

The extracts were three-fold diluted with H_2O/CH_3CN 95:5 (*v/v*) and analyzed using an Ultra-High Performance Liquid Chromatography (UHPLC) Agilent 1290 Infinity coupled to Agilent 1290 Infinity Autosampler (G4226A) and Agilent 6540 accurate-mass Quadrupole-Time of Flight (Q-TOF) Mass Spectrometer (nominal resolution 40.000) with Dual Agilent Jet Stream Ionization source (Agilent Technologies, Santa Clara, CA, USA). For each sample, two analyses in both positive and negative ionization mode by recording data in full scan acquisition mode were performed. Chromatographic separation was performed by using a Zorbax reverse-phase column (RRHD SB-C18 3×150 mm, 1.8 µm) (Agilent Technologies) and mobile phase composed by (A) 0.1% *v/v* aqueous formic acid and (B) 0.1% *v/v* formic acid in acetonitrile. Gradient elution program: 5% B isocratic for 8 min, from 5% to 45% B in 10 min, from 45% to 65% B in 5 min, from 65% to 90% in 4 min, 90% B isocratic for 10 min. Flow rate: 0.4 mL/min; sample injection: 10 µL; column temperature: 35 °C. After each sample a blank composed by the two mobile phases 1:1 *v/v* was analyzed to check the absence of false positives.

QTOF conditions used: sheath gas nitrogen 10 L/min at 400 °C; dehydration gas 8 L/min at 350 °C; nebulizer pressure 60 psi; nozzle voltage 0 kV (negative mode) and 1 kV (positive mode); capillary voltage ± 3.5 kV in positive and negative ion modes. Signals in the *m/z* 100–1700 range at acquisition rate 2 spectra/s were recorded. Mass calibrations were performed with standard mix G1969-85000 (Supelco Inc., Bellefonte, WA, USA), residual error for the expected masses between ± 0.2 ppm. Negative ionization lock masses: TFA anion at *m/z* 112.9856 and HP-0921 at *m/z* 966.0007 (ion [M + HCOO]⁻); positive ionization lock masses: purine at *m/z* 121.0509 and HP-0921 at *m/z* 922.0098. MS/MS fragmentation of the parent ions selected in the *m/z* 100–1700 range by using collision energies between 20 and 60 eV. Acquisition rate: 2 spectra/s.

Data acquisition software Agilent MassHunter version B.04.00 (B4033.2). Data analysis performed by using Agilent MassHunter Qualitative Analysis software B.05.00 (5.0.519.0). The overall identification score of compounds was calculated by the weighted average of the isotopic pattern signals (exact masses, relative abundances, *m/z* distances: $W_{mass} = 100$, $W_{abundance} = 60$, $W_{spacing} = 50$). Targeted identification of metabolites was performed by using the homemade database *GrapeMetabolomics* [51].

3.4. Cell Culture and Treatment

Human colon adenocarcinoma derived Caco2 cell line (ATCC: HTB 37) (well-differentiated) (G1–2) (from adenocarcinoma) and SW480 cell line (ATCC: CCL 228) (poorly-differentiated) (G3–4) (from adenocarcinoma grades III–IV) were purchased from the American Type Culture Collection (ATCC) Cell Bank (Manassas, VA, USA). Cells were cultured in Roswell Park Memorial Institute (RPMI) 1640 medium, for Caco2 cells, and Dulbecco's Modified Eagle Medium (DMEM) for SW480 cells. All cell culture medium and reagents were purchased from Gibco, Life Technologies Limited, Paisley, UK. Culture medium was supplemented with 10% fetal bovine serum (FBS), 2 mM glutamine, 100 U/mL penicillin, and 100 µg/mL streptomycin and incubated at 37 °C in a humidified atmosphere containing 5% CO_2 in air. Autumn Royal and Egnatia GSEs were added to the medium at increasing concentrations dissolved in 100 µL of solvent, composed by ethanol:water:hydrogen chloride 37% (70:30:1 *v/v/v*) (Sigma Aldrich, Milan, Italy), whereas the cells of control group received the same amount of the solvent. The cells were then incubated at 37 °C in a humidified 5% CO_2 incubator for 24 and 48 h.

3.5. Lipids Extraction and PUFAs Analysis

Cell membrane fatty acids were extracted after 48 h of Autumn Royal or Egnatia GSEs treatment at 20, 50, and 80 µg/mL of concentrations. Untreated cells were used as control. Lipids from cell lysate were extracted using the Folch extraction method with some modifications [52]. PUFA analysis was assessed by a gas chromatograph (ThermoFisher Scientific, Focus GC, Milan, Italy) using ChromQuest 4.1 software (Thermo Fisher Scientific, Focus GC, Milan, Italy), as previously described [16].

3.6. RNA Extraction and Quantitative Real-Time PCR

After 48 h of treatment with Autumn Royal or Egnatia GSEs (20, 50, and 80 µg/mL), total RNA was extracted from Caco2 and SW480 cells using the Qiagen RNeasy Mini Kit (Qiagen, Hilden, Germany), according to the manufacturer's instructions. The control sample was represented by untreated cells. Samples were retro-transcribed and analyzed using real-time Polymerase Chain Reaction (PCR) for the evaluation of 15-LOX-1 and PPAR-γ expressions on a CFX96 Touch Real-Time PCR Detection System (Bio-Rad Laboratories, Hercules, CA, USA) according to the manufacturer's instructions. Table 3 shows gene-specific primer sets used (Bio-Rad Laboratories); β-actin gene was chosen as the reference gene. The ΔΔCt method was used for relative quantification by CFX Manager software 2.1 (Bio-Rad Laboratories).

Table 3. Primers for quantitative real-time PCR.

Target Genes	Gene Symbol	Gene Aliases	ID Assay
Arachidonate 15-lipoxygenase	ALOX15	15-LOX-1, 15LOX-1	qHsaCED0045954
Peroxisome proliferator-activated receptor gamma	PPARG	CIMT1, GLM1, NR1C3, PPARG1, PPARG2, PPARgamma	qHsaCID0011718
Actin, beta	ACTB	PS1TP5BP1	qHsaCED0036269

3.7. FE-SEM Investigation

Caco2 and SW480 cells were seeded on 1 cm^2 silicon chips (Ted Pella Inc., Redding, CA, USA) at a density of 10^4 cells per well in eight-well chamber slides and, at a sub-confluent density, untreated control cells (CTR) were fixed at T0. To observe cell morphological changes, cell lines were exposed to increasing concentrations of Autumn Royal and Egnatia GSEs (10, 20, 50, and 80 µg/mL) for 24 (T1) and 48 h (T2). Cell lines were treated as previously described [53]. Briefly, each experimental time had a CTR group to highlight the cellular morphological differences before and after each time GSE exposure. As regards fixation procedures, the cells were treated with 3% glutaraldehyde in PBS for 1 h at 4 °C and then incubated for 1 h at room temperature in 1% osmium tetroxide (OsO$_4$). Samples were washed several times with 0.005 M of sodium cacodylate pH 7.2, and the dehydration was completed by passing the samples in increasing concentrations of acetone, from 20% to 100%, for 10 min each step. After completing the drying, samples were coated with a thin Au film using a sputter coater (208HR High Resolution Sputter Coater, Ted Pella Inc.). The fixed cells were observed by using a Zeiss Sigma FE-SEM (Carl Zeiss, Oberkochen, Germany), equipped with an in-lens secondary electron detector. The samples deposited on silicon chips were fixed to stainless-steel sample holders by using double-sided carbon tape. A uniform Au metal coating of few nanometers was deposited on the samples placed on silicon chips by using a turbomolecular pumped SC7620 Mini Sputter/Glow Discharge System of Quorum Technologies (Quorum Technologies Ltd, Lewes, UK). The overall FE-SEM measurements on the samples were acquired at constant Extra-High Tension (EHT) value of 3 kV and at working distance (WD) ranging from 1.8 to 3 mm. The FE-SEM micrographs presented for each experiment were selected as representative of a series of images collected on each sample. Each experiment was performed in triplicate.

3.8. Statistical Analysis

Data on total polyphenolic content of GSEs were analyzed by paired Student *t*-test. The significance of the differences between the control and each treated experimental group was evaluated with one-way analysis of variance (ANOVA) and Dunnett's posttest. Differences were considered as statistically significant with a *p*-value < 0.05. Data, expressed as mean ± Standard Deviation (SD), were analyzed using STATA statistical software, version 15.1 (StataCorp, 4905 Lakeway Drive, College Station, TX 77845, USA).

4. Conclusions

Our in vitro data wanted to emphasize the effects of table grape polyphenolic compounds on some molecular mechanisms involved in tumorigenesis. Moreover, the lipidomic approach followed in this study provided valuable information for understanding the protective effect of two GSEs studied on human cell metabolism. Given their ability to influence cell morphology, the flavonoid and non-flavonoid compounds present in table grapes could become a promising dietary source for cancer prevention and treatment.

Supplementary Materials: The following are available online at http://www.mdpi.com/1420-3049/25/15/3352/s1, Figure S1A: Positive extract ion chromatogram (EIC) of M$^+$ signals of the anthocyanins identified, Figure S1B: Negative extract ion chromatogram (EIC) of [M − H]$^-$ signals of the phenolic compounds identified, Table S1: Main MS/MS fragments and mean signal intensity of the metabolites identified in the samples.

Author Contributions: Conceptualization, V.T. and M.N.; methodology, V.T., I.G., R.A.M., V.D.N., R.F., M.D.R., M.P.S., N.D., and E.F.; data curation, V.T., I.G., R.A.M., R.F., and V.D.N.; writing—original draft preparation, V.T., I.G., R.A.M., R.F., and M.P.S.; writing—review and editing, V.T., M.G.C., and M.N.; supervision, M.N. All authors have read and agreed to the published version of the manuscript.

Funding: This research was funded by RC 2020-2022, Linea 2, Prg. 15 (D.D.G. n. 2 – 10.01.2020) and "PO Puglia FESR 2007-2013, Asse I, Linea 1.2, (codice n. 47)".

Conflicts of Interest: The authors declare no conflict of interest.

References

1. Magrone, T.; Magrone, M.; Russo, M.A.; Jirillo, E. Magrone Recent Advances on the Anti-Inflammatory and Antioxidant Properties of Red Grape Polyphenols: In Vitro and In Vivo Studies. *Antioxidants* **2019**, *9*, 35. [CrossRef]

2. Mojzer, E.B.; Hrnčič, M.K.; Škerget, M.; Knez, Ž.; Bren, U. Polyphenols: Extraction Methods, Antioxidative Action, Bioavailability and Anticarcinogenic Effects. *Molecules* **2016**, *21*, 901. [CrossRef] [PubMed]

3. Durazzo, A.; Lucarini, M.; Souto, E.B.; Cicala, C.; Caiazzo, E.; Izzo, A.A.; Novellino, E.; Santini, A. Polyphenols: A concise overview on the chemistry, occurrence, and human health. *Phytother. Res.* **2019**, *33*, 2221–2243. [CrossRef] [PubMed]

4. Pereira, D.M.; Valentão, P.; Pereira, J.A.; Andrade, P.B. Phenolics: From Chemistry to Biology. *Molecules* **2009**, *14*, 2202–2211. [CrossRef]

5. Xia, E.-Q.; Deng, G.-F.; Guo, Y.-J.; Li, H.-B. Biological Activities of Polyphenols from Grapes. *Int. J. Mol. Sci.* **2010**, *11*, 622–646. [CrossRef] [PubMed]

6. Hendrich, A.B. Flavonoid-membrane interactions: Possible consequences for biological effects of some polyphenolic compounds1. *Acta Pharmacol. Sin.* **2006**, *27*, 27–40. [CrossRef] [PubMed]

7. Yang, J.; Xiao, Y.-Y. Grape Phytochemicals and Associated Health Benefits. *Crit. Rev. Food Sci. Nutr.* **2013**, *53*, 1202–1225. [CrossRef]

8. Lee, H.S.; Ha, A.W.; Kim, W.K. Effect of resveratrol on the metastasis of 4T1 mouse breast cancer cells in vitro and in vivo. *Nutr. Res. Pr.* **2012**, *6*, 294–300. [CrossRef]

9. Wang, L.; Ling, Y.; Chen, Y.; Li, C.-L.; Feng, F.; You, Q.-D.; Lu, N.; Guo, Q. Flavonoid baicalein suppresses adhesion, migration and invasion of MDA-MB-231 human breast cancer cells. *Cancer Lett.* **2010**, *297*, 42–48. [CrossRef]

10. Mantena, S.K.; Baliga, M.S.; Katiyar, S.K. Grape seed proanthocyanidins induce apoptosis and inhibit metastasis of highly metastatic breast carcinoma cells. *Carcinogenesis* **2006**, *27*, 1682–1691. [CrossRef]

11. Ramos, S. Cancer chemoprevention and chemotherapy: Dietary polyphenols and signalling pathways. *Mol. Nutr. Food Res.* **2008**, *52*, 507–526. [CrossRef]

12. Valenzuela, M.; Bastias, L.; Montenegro, I.; Werner, E.; Madrid, A.; Godoy, P.; Párraga, M.; Villena, J. Autumn Royal and Ribier Grape Juice Extracts Reduced Viability and Metastatic Potential of Colon Cancer Cells. *Evid.-Based Complement. Altern. Med.* **2018**, *2018*, 1–7. [CrossRef] [PubMed]

13. Avtanski, D.; Poretsky, L. Phyto-polyphenols as potential inhibitors of breast cancer metastasis. *Mol. Med.* **2018**, *24*, 29. [CrossRef] [PubMed]

14. Hanikoglu, A.; Kucuksayan, E.; Hanikoglu, F.; Ozben, T.; Menounou, G.; Sansone, A.; Chatgilialoglu, C.; Di Bella, G.; Ferreri, C. Effects of somatostatin, curcumin, and quercetin on the fatty acid profile of breast cancer cell membranes. *Can. J. Physiol. Pharmacol.* **2020**, *98*, 131–138. [CrossRef]

15. Ortinau, L.C.; Pickering, R.T.; Nickelson, K.J.; Stromsdorfer, K.L.; Naik, C.Y.; Haynes, R.A.; Bauman, D.E.; Rector, R.S.; Fritsche, K.L.; Ii, J.P. Sterculic Oil, a Natural SCD1 Inhibitor, Improves Glucose Tolerance in Obese ob/ob Mice. *ISRN Endocrinol.* **2012**, *2012*, 947323. [CrossRef] [PubMed]

16. Tutino, V.; Gigante, I.; Scavo, M.P.; Refolo, M.G.; De Nunzio, V.; Milella, R.A.; Caruso, M.G.; Notarnicola, M. Stearoyl-CoA Desaturase-1 Enzyme Inhibition by Grape Skin Extracts Affects Membrane Fluidity in Human Colon Cancer Cell Lines. *Nutrients* **2020**, *12*, 693. [CrossRef] [PubMed]

17. Scaglia, N.; Igal, R.A. Stearoyl-CoA Desaturase Is Involved in the Control of Proliferation, Anchorage-independent Growth, and Survival in Human Transformed Cells. *J. Boil. Chem.* **2005**, *280*, 25339–25349. [CrossRef] [PubMed]

18. Ran, H.; Zhu, Y.; Deng, R.; Zhang, Q.; Liu, X.; Feng, M.; Zhong, J.; Lin, S.; Tong, X.; Su, Q. Stearoyl-CoA desaturase-1 promotes colorectal cancer metastasis in response to glucose by suppressing PTEN. *J. Exp. Clin. Cancer Res.* **2018**, *37*, 54. [CrossRef]

19. Maria, N.; Maria, G.C.; Valeria, T.; Valentina, D.N.; Isabella, G.; Giampiero, D.L.; Nicola, V.; Ornella, R.; Rosa, R.; Elisa, S.; et al. Nutrition and lipidomic profile in colorectal cancers. *Acta Bio Med. Atenei Parm.* **2018**, *89*, 87–96.

20. Long, J.; Zhang, C.-J.; Zhu, N.; Du, K.; Yin, Y.-F.; Tan, X.; Liao, D.-F.; Qin, L. Lipid metabolism and carcinogenesis, cancer development. *Am. J. Cancer Res.* **2018**, *8*, 778–791.

21. Garcia, L.C.; Achón-Tuñón, M.; González-González, M.P. The influence of the polyunsaturated fatty acids in the prevention and promotion of cancer. *Nutr. Hosp.* **2015**, *32*, 41–49.

22. Tutino, V.; De Nunzio, V.; Caruso, M.G.; Veronese, N.; Lorusso, D.; Di Masi, M.; Benedetto, M.L.; Notarnicola, M. Elevated AA/EPA Ratio Represents an Inflammatory Biomarker in Tumor Tissue of Metastatic Colorectal Cancer Patients. *Int. J. Mol. Sci.* **2019**, *20*, 2050. [CrossRef] [PubMed]

23. Tian, R.; Zuo, X.; Jaoude, J.; Mao, F.; Colby, J.; Shureiqi, I. ALOX15 as a suppressor of inflammation and cancer: Lost in the link. *Prostaglandins Other Lipid Mediat.* **2017**, *132*, 77–83. [CrossRef] [PubMed]

24. Serhan, C.N. Pro-resolving lipid mediators are leads for resolution physiology. *Nature* **2014**, *510*, 92–101. [CrossRef] [PubMed]

25. Bhattacharya, S.; Mathew, G.; Jayne, D.; Pelengaris, S.; Khan, M. 15-Lipoxygenase-1 in Colorectal Cancer: A Review. *Tumor Boil.* **2009**, *30*, 185–199. [CrossRef]

26. Çimen, I.; Astarci, E.; Banerjee, S. 15-lipoxygenase-1 exerts its tumor suppressive role by inhibiting nuclear factor-kappa B via activation of PPAR gamma. *J. Cell. Biochem.* **2011**, *112*, 2490–2501. [CrossRef]

27. Mao, F.; Wang, M.; Wang, J.; Xu, W. The role of 15-LOX-1 in colitis and colitis-associated colorectal cancer. *Inflamm. Res.* **2015**, *64*, 661–669. [CrossRef]

28. Gigante, I.; Milella, R.A.; Tutino, V.; DeBiase, G.; Notarangelo, L.; Giannandrea, M.A.; De Nunzio, V.; Orlando, A.; D'Alessandro, R.; Caruso, M.; et al. Autumn Royal and Egnatia Grape Extracts differently modulate Cell Proliferation in Human Colorectal Cancer Cells. *Endocr. Metab. Immune Disord. Drug Targets* **2020**, *20*, 1–19. [CrossRef]

29. Lin, B.-W.; Gong, C.-C.; Song, H.-F.; Cui, Y.-Y. Effects of anthocyanins on the prevention and treatment of cancer. *Br. J. Pharmacol.* **2016**, *174*, 1226–1243. [CrossRef]

30. Hou, D.-X.; Kai, K.; Li, J.-J.; Lin, S.; Terahara, N.; Wakamatsu, M.; Fujii, M.; Young, M.R.; Colburn, N. Anthocyanidins inhibit activator protein 1 activity and cell transformation: Structure-activity relationship and molecular mechanisms. *Carcinogenesis* **2004**, *25*, 29–36. [CrossRef]

31. Unusan, N. Proanthocyanidins in grape seeds: An updated review of their health benefits and potential uses in the food industry. *J. Funct. Foods* **2020**, *67*, 103861. [CrossRef]

32. Nandakumar, V.; Singh, T.; Katiyar, S.K. Multi-targeted prevention and therapy of cancer by proanthocyanidins. *Cancer Lett.* **2008**, *269*, 378–387. [CrossRef] [PubMed]

33. Wang, T.-Y.; Li, Q.; Bi, K.-S. Bioactive flavonoids in medicinal plants: Structure, activity and biological fate. *Asian J. Pharm. Sci.* **2018**, *13*, 12–23. [CrossRef] [PubMed]

34. Oteiza, P.I.; Erlejman, A.G.; Verstraeten, S.V.; Keen, C.L.; Fraga, C.G. Flavonoid-membrane Interactions: A Protective Role of Flavonoids at the Membrane Surface? *Clin. Dev. Immunol.* **2005**, *12*, 19–25. [CrossRef]

35. Mandić, L.; Sadžak, A.; Strasser, V.; Baranović, G.; Jurašin, D.D.; Sikirić, M.D.; Šegota, S. Enhanced Protection of Biological Membranes during Lipid Peroxidation: Study of the Interactions between Flavonoid Loaded Mesoporous Silica Nanoparticles and Model Cell Membranes. *Int. J. Mol. Sci.* **2019**, *20*, 2709. [CrossRef]

36. Pakiet, A.; Kobiela, J.; Stepnowski, P.; Sledzinski, T.; Mika, A. Changes in lipids composition and metabolism in colorectal cancer: A review. *Lipids Health Dis.* **2019**, *18*, 29. [CrossRef]

37. Else, P.L. The highly unnatural fatty acid profile of cells in culture. *Prog. Lipid Res.* **2020**, *77*, 101017. [CrossRef] [PubMed]

38. Fujiwara, F.; Todo, S.; Imashuku, S. Antitumor effect of gamma-linolenic acid on cultured human neuroblastoma cells. *Prostaglandins Leukot. Med.* **1986**, *23*, 311–320. [CrossRef]

39. Das, U.N. Essential fatty acids and their metabolites and cancer. *Nutrition* **1999**, *15*, 239–240.

40. Ramchurren, N.; Karmali, R. Effects of gamma-linolenic and dihomo-gamma-linolenic acids on 7,12-dimethylbenz(α)anthracene-induced mammary tumors in rats. *Prostaglandins Leukot. Essent. Fat. Acids* **1995**, *53*, 95–101. [CrossRef]

41. Xu, Y.; Qian, S.Y. Anti-cancer activities of omega-6 polyunsaturated fatty acids. *Biomed. J.* **2014**, *37*, 112–119. [CrossRef] [PubMed]

42. Çimen, I.; Tunçay, S.; Banerjee, S. 15-Lipoxygenase-1 expression suppresses the invasive properties of colorectal carcinoma cell lines HCT-116 and HT-29. *Cancer Sci.* **2009**, *100*, 2283–2291. [CrossRef] [PubMed]

43. Derry, M.; Raina, K.; Agarwal, R.; Agarwal, C. Differential effects of grape seed extract against human colorectal cancer cell lines: The intricate role of death receptors and mitochondria. *Cancer Lett.* **2012**, *334*, 69–78. [CrossRef] [PubMed]

44. DiNicola, S.; Cucina, A.; Pasqualato, A.; Proietti, S.; D'Anselmi, F.; Pasqua, G.; Santamaria, A.R.; Coluccia, P.; Lagana, A.; Antonacci, D.; et al. Apoptosis-inducing factor and caspase-dependent apoptotic pathways triggered by different grape seed extracts on human colon cancer cell line Caco-2. *Br. J. Nutr.* **2010**, *104*, 824–832. [CrossRef] [PubMed]

45. Kaur, M.; Mandair, R.; Agarwal, R.; Agarwal, C. Grape Seed Extract Induces Cell Cycle Arrest and Apoptosis in Human Colon Carcinoma Cells. *Nutr. Cancer* **2008**, *60*, 2–11. [CrossRef] [PubMed]

46. DiNicola, S.; Cucina, A.; Pasqualato, A.; D'Anselmi, F.; Proietti, S.; Lisi, E.; Pasqua, G.; Antonacci, D.; Bizzarri, M. Antiproliferative and Apoptotic Effects Triggered by Grape Seed Extract (GSE) versus Epigallocatechin and Procyanidins on Colon Cancer Cell Lines. *Int. J. Mol. Sci.* **2012**, *13*, 651–664. [CrossRef] [PubMed]

47. Gao, Y.; Tollefsbol, T.O. Combinational Proanthocyanidins and Resveratrol Synergistically Inhibit Human Breast Cancer Cells and Impact Epigenetic–Mediating Machinery. *Int. J. Mol. Sci.* **2018**, *19*, 2204. [CrossRef]

48. Wang, S.; Tian, Q.; An, F. Growth inhibition and apoptotic effects of total flavonoids from Trollius chinensis on human breast cancer MCF-7 cells. *Oncol. Lett.* **2016**, *12*, 1705–1710. [CrossRef]

49. Subramanian, A.P.; Jaganathan, S.K.; Mandal, M.; Supriyanto, E.; Muhamad, I.I. Gallic acid induced apoptotic events in HCT-15 colon cancer cells. *World J. Gastroenterol.* **2016**, *22*, 3952–3961. [CrossRef]

50. Milella, R.A.; Basile, T.; Alba, V.; Gasparro, M.; Giannandrea, M.A.; DeBiase, G.; Genghi, R.; Antonacci, D. Optimized ultrasonic-assisted extraction of phenolic antioxidants from grape (*Vitis vinifera* L.) skin using response surface methodology. *J. Food Sci. Technol.* **2019**, *56*, 4417–4428. [CrossRef]

51. Flamini, R.; De Rosso, M.; De Marchi, F.; Vedova, A.D.; Panighel, A.; Gardiman, M.; Maoz, I.; Bavaresco, L. An innovative approach to grape metabolomics: Stilbene profiling by suspect screening analysis. *Metabolomics* **2013**, *9*, 1243–1253. [CrossRef]

52. Folch, J.; Lees, M.; Stanley, G.H.S. A simple method for the isolation and purification of total lipides from animal tissues. *J. Boil. Chem.* **1957**, *226*, 497–509.

53. DePalo, N.; Fanizza, E.; Vischio, F.; Denora, N.; Laquintana, V.; Cutrignelli, A.; Striccoli, M.; Giannelli, G.; Agostiano, A.; Curri, M.L.; et al. Imaging modification of colon carcinoma cells exposed to lipid based nanovectors for drug delivery: A scanning electron microscopy investigation. *RSC Adv.* **2019**, *9*, 21810–21825. [CrossRef]

Sample Availability: Samples of the compounds are not available from the authors.

 *molecules*

Article

Inhibitory Effects of Tangeretin, a Citrus Peel-Derived Flavonoid, on Breast Cancer Stem Cell Formation through Suppression of Stat3 Signaling

Yu-Chan Ko [1,†], Hack Sun Choi [2,†], Ren Liu [1], Ji-Hyang Kim [1], Su-Lim Kim [1], Bong-Sik Yun [3] and Dong-Sun Lee [1,2,3,4,5,*]

[1] Interdisciplinary Graduate Program in Advanced Convergence Technology and Science, Jeju National University, Jeju 63243, Korea; koyuchan94@gmail.com (Y.-C.K.); liuren0308@gmail.com (R.L.); seogwi12@naver.com (J.-H.K.); ksl1101@naver.com (S.-L.K.)

[2] Subtropical/Tropical Organism Gene Bank, Jeju National University, Jeju 63243, Korea; choix074@jejunu.ac.kr

[3] Division of Biotechnology, College of Environmental and Bioresource Sciences, Jeonbuk National University, Gobong-ro 79, Iksan 54596, Korea; bsyun@jbnu.ac.kr

[4] Practical Translational Research Center, Jeju National University, Jeju 63243, Korea

[5] Faculty of Biotechnology, College of Applied Life Sciences, Jeju National University, SARI, Jeju 63243, Korea

* Correspondence: dongsunlee@jejunu.ac.kr

† Yu-Chan Ko and Hack Sun Choi contributed equally to this work.

Academic Editor: H.P. Vasantha Rupasinghe

Received: 20 May 2020; Accepted: 2 June 2020; Published: 3 June 2020

Abstract: Breast cancer stem cells (BCSCs) are responsible for tumor chemoresistance and recurrence. Targeting CSCs using natural compounds is a novel approach for cancer therapy. A CSC-inhibiting compound was purified from citrus extracts using silica gel, gel filtration and high-pressure liquid chromatography. The purified compound was identified as tangeretin by using nuclear magnetic resonance (NMR). Tangeretin inhibited cell proliferation, CSC formation and tumor growth, and modestly induced apoptosis in CSCs. The frequency of a subpopulation with a CSC phenotype (CD44$^+$/CD24$^-$) was reduced by tangeretin. Tangeretin reduced the total level and phosphorylated nuclear level of signal transducer and activator of transcription 3 (Stat3). Our results in this study show that tangeretin inhibits the Stat3 signaling pathway and induces CSC death, indicating that tangeretin may be a potential natural compound that targets breast cancer cells and CSCs.

Keywords: tangeretin; cancer stem cells; Stat3; citrus; CD44$^+$/CD24$^-$; phytochemicals

1. Introduction

Breast cancer is the most frequently diagnosed cancer and the leading cause of cancer-related death among females [1]. Triple-negative breast cancer (TNBC) accounts for approximately 10–15% of all diagnosed breast cancers [2], and is defined as ER-, PR- and HER2-negative breast cancer. Women with TNBC have a high risk of recurrence within three years of diagnosis, and the mortality rate is increased for five years after diagnosis [3]. Cancer stem cells (CSCs) are relatively resistant to chemotherapy, radiotherapy and hormone therapy [4]. CSCs have functional roles in self-renewal and differentiation [5]. CSCs are responsible for the processes of cancer initiation, metastasis and cancer relapse [6]. Therefore, breast CSCs may contribute to drug resistance and relapse [7]. The CD44$^+$/CD24$^-$ being the most common cell-surface phenotype of breast CSCs can facilitate invasion, migration and proliferation [8].

Signal transducer and activator of transcription 3 (Stat3) plays a role in the inflammatory response and is a member of the seven-member Stat protein family (Stat1, 2, 3, 4, 5a, 5b, and 6) that is activated by growth factors [9]. Stat3 undergoes alternative splicing into Stat3α (92 kDa) and the isoform Stat3β

(83 kDa). Stat3 plays a role in human oncogenesis. The activation of Stat3 signaling is associated with proliferation, antiapoptotic effects and cellular transformation [10,11]. Additionally, Stat3 plays a crucial role in differentiation under normal physiological conditions [6]. Inflammatory cytokines play key roles in regulating the interaction between CSCs and the IL6 inflammatory feedback loop, leading to the expansion of the CSCs population [12]. IL-8 interacts with the CXCR1 receptor on BCSCs (breast cancer stem cells), which promotes their cellular activities such as self-renewal and invasion. IL-8 signaling is a key pathway for regulating BCSCs [13].

Citrus species are natural products containing phytochemicals, which are promising for development into cancer therapies [14,15]. Citrus flavonoids, including nobiletin, hesperidin, tangeretin and naringin, have many biological activities, including a strong antioxidant and radical scavenging activity [16]. Tangeretin, also known as 4,5,6,7,8-pentamethoxyflavone, is a major compound in citrus peels. It has been shown to possess a variety of pharmacological activities, including anti-oxidative, anti-inflammatory and anticancer properties [17]. Inflammatory cytokine IL-6 plays an important role in mediating the interaction between CSCs and the microenvironment, which can influence tumor growth by regulating CSC subpopulations. However, no studies have shown the mechanisms underlying the targeted effects of tangeretin on CSC formation and Stat3 signaling in BCSCs. In our study, we showed that tangeretin had antiproliferative effects on breast cancer cells and reduced BCSC proliferation or prevalence through a decrease in Sox2 expression by inhibiting Stat3 signaling.

2. Results

2.1. CSC Inhibitor Derived from Citrus

A CSC-inhibiting compound derived from citrus was purified by bioassay-guided isolation, as shown in Figure 1A and Supplementary Figure S10. The compound was isolated using organic solvent extraction, silica gel, gel filtration, TLC and preparatory HPLC. The isolated sample was determined to be a single compound using HPLC (Figure 1B). We assayed mammosphere formation using a purified sample (Figure 1C). The structural name of the purified compound was identified as tangeretin (Figure 2).

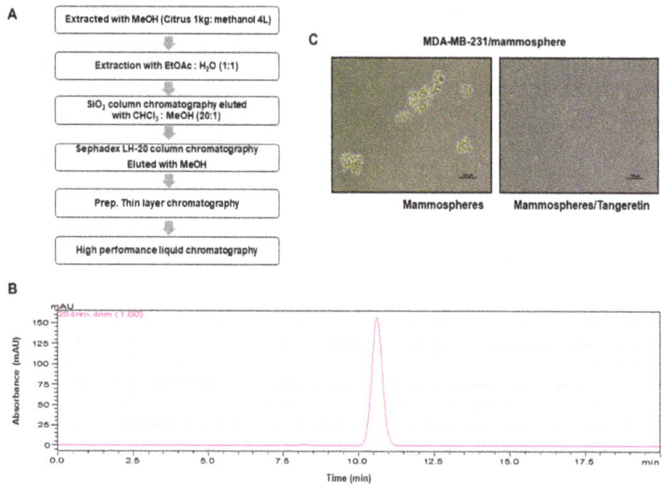

Figure 1. Purification protocol for a CSC inhibitor derived from citrus peels and a mammosphere formation assay using a purified sample. (**A**) Flowchart for the isolation of the mammosphere inhibitor. (**B**) HPLC chromatogram of the inhibitor purified from citrus. (**C**) Assay for mammosphere formation in the presence of the HPLC-purified sample. Cancer cells were treated with the HPLC-purified sample. Images show representative mammospheres, and were imaged by microscopy (scale bar: 100 µm).

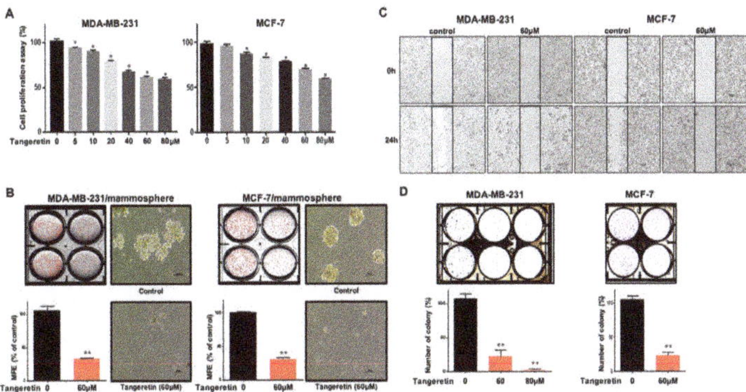

Figure 2. Chemical structure of the compound purified from citrus. Chemical structure of tangeretin.

2.2. Tangeretin Suppresses the Proliferation of MDA-MB-231 and MCF-7 Cells and the Formation of Mammospheres

Breast cancer cells were incubated with various concentrations of tangeretin for 24 h. The antiproliferative function of tangeretin was assayed. Tangeretin inhibited cell proliferation, as shown in Figure 3A. To test whether tangeretin can suppress mammosphere formation, we treated cancer cells with tangeretin. Tangeretin inhibited mammosphere formation, as shown in Figure 3B. Tangeretin suppressed the migration and colony formation of cancer cell lines (Figure 3C,D). Our data indicated that tangeretin inhibited cancer hallmarks (proliferation, migration and colony formation) and mammosphere formation.

Figure 3. The effects of tangeretin on cell proliferation and mammosphere formation. (**A**) MDA-MB-231 and MCF-7 cells were treated with tangeretin for 24 h in a medium supplemented with 10% FBS. The cytotoxicity of tangeretin was measured with an EZ-Cytox kit. (**B**) Tangeretin inhibits the formation of mammospheres. To establish mammospheres, 1×10^4 MDA-MB-231 cells and 5×10^4 MCF-7 cells were seeded in ultralow-attachment 6-well plates with CSC culture medium. The mammospheres were incubated with increasing concentrations of tangeretin. Photos show representative mammospheres, and were captured by microscopy (scale bar: 100 μm). Mammosphere formation efficiency (MFE) was examined. (**C**) The effect of tangeretin on the migration of breast cancer cells (MDA-MB-231 and MCF-7 cells) was evaluated. Migration with or without tangeretin was captured at 0 and 24 h (scale bar: 100 μm). (**D**) Tangeretin inhibits colony formation by cancer cells. Breast cancer cells (MDA-MB-231 and MCF-7 cells) were incubated and treated with tangeretin. Representative data were collected. The data from triplicate experiments are represented as the mean ± SD; * $p < 0.05$, ** $p < 0.01$.

2.3. Tangeretin Modestly Induces Apoptosis in Mammospheres and Inhibits Mammosphere Proliferation

Late apoptosis in mammospheres was modestly induced by 80 μM of tangeretin (Figure 4A). Tangeretin reduced the transcript level of stem cell marker genes (Oct3/4, Sox2, and Nanog gene) (Figure 4B). To test whether tangeretin suppresses mammosphere growth, we treated mammospheres with tangeretin and counted the number of cancer cells derived from mammospheres. Tangeretin treatment inhibited mammosphere growth (Figure 4C). Our data showed that tangeretin, which disregulates the Stat3/Sox2 signaling pathway, was essential for inhibiting the proliferation of BCSCs.

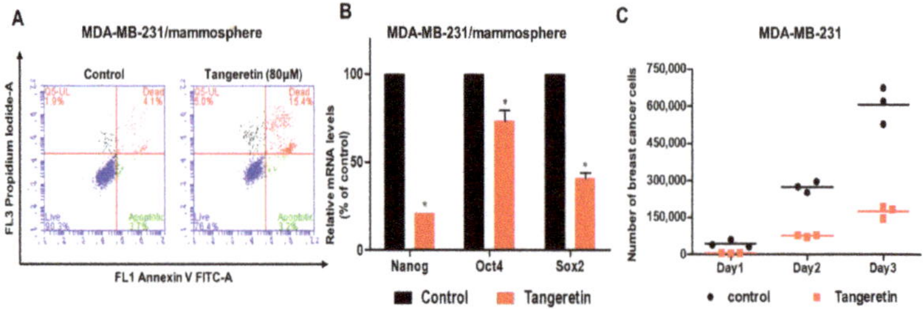

Figure 4. The effects of tangeretin on apoptosis and mammosphere growth. (**A**) Tangeretin modestly induced apoptosis in mammospheres treated with tangeretin. We induced mammosphere formation and treated mammospheres with tangeretin. After treatment, apoptotic cells were examined using Annexin V/PI staining. (**B**) Transcription levels of CSC markers, including the Nanog, Sox2 and Oct4 genes, determined in tangeretin- and DMSO-treated mammospheres using CSC marker-specific primers and real-time PCR. β-actin acts as an internal control. The data shown represents the mean ± SD of three independent experiments. * $p < 0.05$ vs. DMSO-treated control. (**C**) Tangeretin inhibited mammosphere growth. Mammospheres with or without tangeretin were dissociated into single cells, and the single cells were plated in 6 cm dishes in equal numbers. The cells were examined one, two and three days later.

2.4. Tangeretin Decreases Tumor Growth In Vivo

As tangeretin has antiproliferative effects, we examined whether tangeretin suppresses tumor formation in a nude mouse model. There was no significant body weight difference between control and tangeretin-treated mice (Figure 5A). At each time point, the tumor volume (Figure 5B) and weight (Figure 5C) of the tangeretin-treated nude mice were smaller than those of the untreated nude mice.

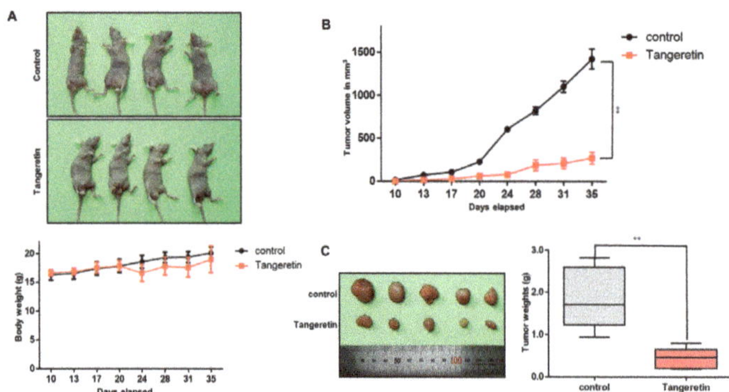

Figure 5. The effect of tangeretin on tumor growth in a xenograft model. (**A**) MDA-MB-231 cells (2×10^6 cells/mouse) were inoculated into the mammary fat pad of female nude mice and treated with tangeretin (2.5 mg/kg) or DMSO (n = 6). Nude mice were intraperitoneally injected with/without Tangeretin (2.5 mg/kg) once a week for a total of four time injections. The body weights of the tangeretin-treated group were comparable to those of the control group. (**B**) Tumor volume was calculated as (width^2 × length)/2 at the indicated time points. (**C**) The tumor weights of the control and tangeretin-treated mice were assayed after sacrifice at day 35. The data are presented as the mean ± SD of three independent experiments. ** $p < 0.05$ versus the DMSO-treated control group.

2.5. Tangeretin Treatment Modestly Reduces the CD44⁺/CD24⁻ Population Size

MDA-MB-231 and MCF-7 cells were treated with tangeretin, and we analyzed the BCSC marker-expressing $CD44^+/CD24^-$ subpopulation. Tangeretin modestly reduced the $CD44^+/CD24^-$ cell population size from 84.1% to 56.8% in MDA-MB-231 cells and from 2.2% to 0.9% in MCF-7 cells (Figure 6). This result shows that tangeretin reduces the frequency of a BCSC trait.

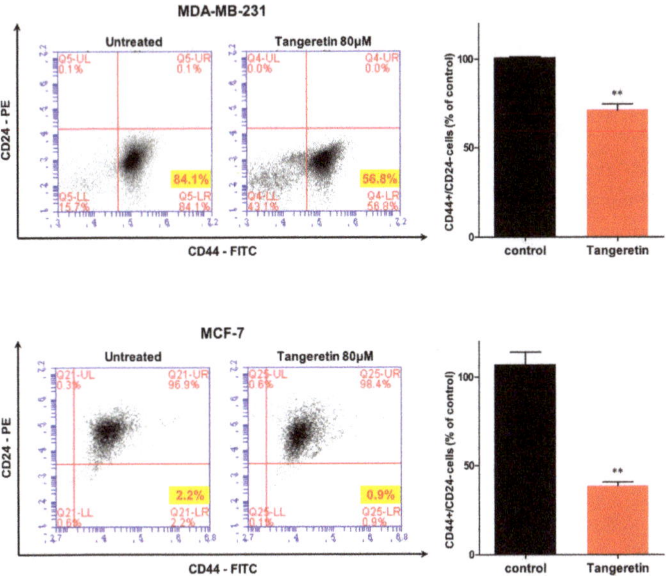

Figure 6. The effect of tangeretin on the $CD44^{high}/CD24^{low}$ cell proportion. The $CD44^{high}/CD24^{low}$ subpopulation within an MDA-MB-231 and MCF-7 cell population treated with tangeretin (80 μM) or DMSO for 24 h was analyzed by flow cytometry. For flow cytometry analysis, 2×10^4 cells were acquired. The gating was based on the binding of an antibody without tangeretin (red cross). The data from triplicate experiments are represented as the mean ± SD; ** $p < 0.01$.

2.6. Tangeretin Inhibits the Stat3 Signaling Pathway and Reduces the Sox2 Level in Mammospheres

To investigate the cellular mechanism by which tangeretin inhibits mammosphere formation, we assessed the expression levels of Stat3 and pStat3 in mammospheres. Our results showed that tangeretin decreased the total protein levels of Stat3 and pStat3 in BCSCs (Figure 7A). The protein level of phospho-Stat3 was significantly reduced in the cytosol and nucleus of mammosphere cells. The Stat3 protein level was also decreased, as shown in Figure 7B. Additionally, we investigated Stat3 probe DNA binding to tangeretin-treated nuclear extracts by EMSA. We examined Stat3 probe DNA binding to mammosphere nuclear proteins using a Stat probe. Tangeretin reduced Stat3 DNA binding (Figure 7C, # 3). The specificity of Stat3 binding was determined using a self-competitor (100×) (Figure 7C, # 4) or a mutated Stat oligo (100×) (Figure 7C, # 5). The band indicated by arrows is Stat3 and the specific DNA complex. To examine the effect on Stat3 on mammosphere formation, we performed mammosphere formation using siRNA of Stat3. Our data showed that Stat3 reduction decreased mammosphere formation (Figure 7D). To analyze the cellular function of tangeretin, after tangeretin treatment, we checked the Sox2 level because Stat3 dimer activated Sox2 gene [18]. Our data showed that tangeretin reduced the level of transcripts and protein of Sox2 (Figure 7E). Sox2 plays a role in CSC progression [19]. Our data showed that the Stat3/Sox2 signal is important for mammosphere formation. Our data showed that tangeretin, which disregulates the Stat3/Sox2 signaling pathway, was essential for inhibiting the proliferation of BCSCs (Figure 8).

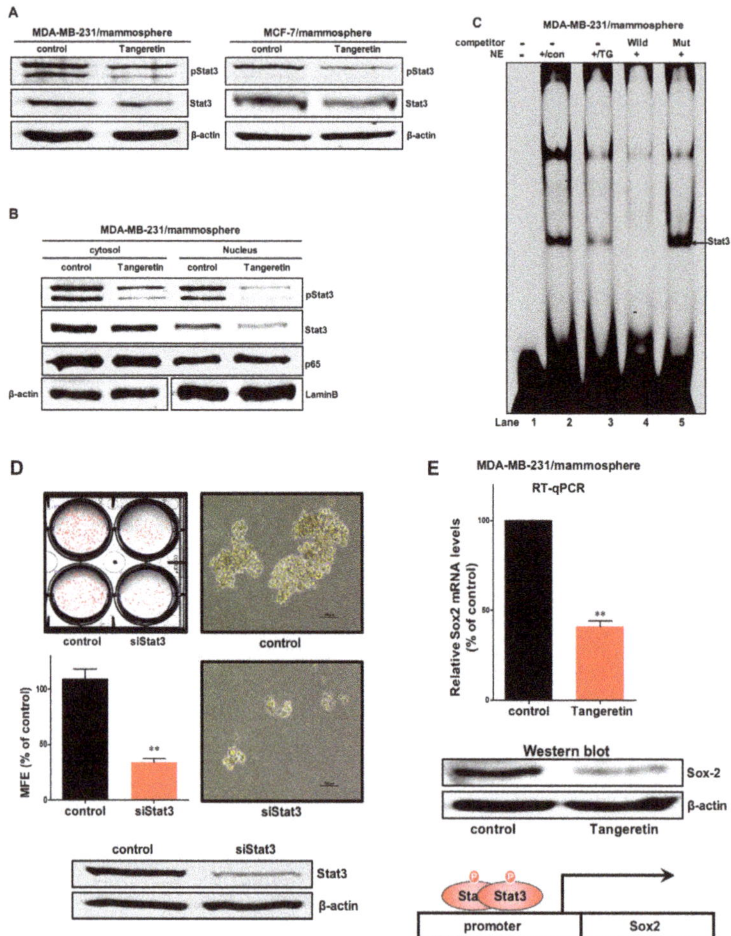

Figure 7. Tangeretin regulates Stat3 signaling and Sox2 regulation. (**A**) The total levels of Stat3 and pStat3 were measured in mammospheres composed of MDA-MB-231 cells and MCF-7 cells after treatment with tangeretin (0 or 60 μM) for 48 h using western blotting. Total lysates were subjected to immunoblot analysis with anti-Stat3 and anti-pStat3 antibodies. β-actin was used as an internal control. (**B**) The levels of Stat3 and pStat3 in the cytosolic and nuclear protein fractions were measured in mammospheres composed of MDA-MB-231 cells after treatment with tangeretin for 48 h using western blotting. Nuclear and cytosolic proteins were run on SDS-PAGE gels, followed by immunoblotting with anti-Stat3, anti-pStat3, anti-β-actin and anti-Lamin B antibodies. (**C**) EMSA was used to analyze mammosphere nuclear proteins after treatment with tangeretin. Nuclear extracts were reacted with a Stat3 probe and subjected to 6% native PAGE. Lane 1: Stat3 probe; lane 2: nuclear extracts with the Stat3 probe; lane 3: tangeretin-treated nuclear proteins with the Stat3 probe; lane 4: nuclear proteins incubated with a self-competitor oligo (100×); and lane 5: nuclear extracts incubated with a mutated-stat3 probe (100×). The arrow indicates the DNA/Stat3 complex in the mammosphere nuclear lysates. (**D**) Effect of Stat3 on mammosphere formation using siRNA of Stat3. Stat3-downregulated MDA-MB-231 cells were cultured for seven days using mammosphere media. Images were obtained by microscrope at 100× magnification. (**E**) The transcriptional level of the Sox2 gene in MDA-MB-231 was determined in tangeretin-treated mammospheres. A Sox2-specific primer was used for real-time RT-qPCR. Western blot analysis of mammosphere under tangeretin. β-actin served as an internal control. The data are presented as the mean ± SD of three independent experiments. ** $p < 0.01$ versus the DMSO-treated control group.

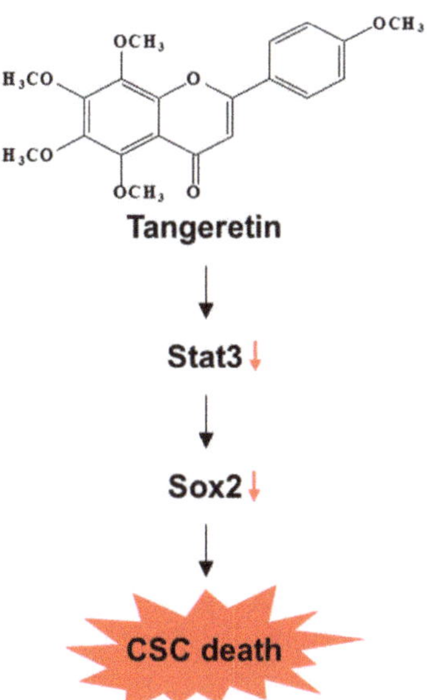

Figure 8. The proposed model for CSC death induced by tangeretin is shown.

3. Discussion

It has been postulated that high consumption of fruits can prevent more than 20% of all cancer cases [20]. This preventive effect is predominantly mediated by phytochemicals interacting with specific target proteins that play important roles in cancer [21–23]. Citrus, one of the most important food sources of phytochemicals with health benefits, has many biological properties and controls key pathways involved in pathologies such as cancer [24–26]. The combination CD44$^+$/CD24$^-$ has emerged as the most important marker for BCSC isolation, and the CD24 population of MDA-MB-231 is low [27].

First, we purified a CSC inhibitor from citrus. Assay-based fractionation and several chromatographic methods isolated one compound from a citrus powder, tangeretin. Tangeretin is the major flavonoid of citrus. It also has antioxidant, anti-inflammatory and anticancer properties [17]. Tangeretin modestly induces apoptosis in bladder cancer cells through mitochondrial dysfunction [28]. Tangeretin and nobiletin induces G1 cell cycle arrest but not apoptosis in breast and colon cancers [29]. Nobiletin inhibits CD36-dependent tumor angiogenesis, migration, invasion and sphere formation through the CD36/Stat3/Nf-Kb signaling axis [30,31]. Quercetin suppresses breast CSCs through its inhibition of the PI3K/Akt/mTOR signaling pathway [32]. Despite numerous studies, there are no studies on tangeretin-induced antiproliferative and anti-CSC effects. Our results showed that tangeretin suppresses the proliferation of BCSCs. Tangeretin inhibited mammosphere formation in breast cancer cells (Figure 3) and modestly induced apoptosis in CSCs (Figure 4).

Our data showed the reduction of CD44$^+$/CD24$^-$ subpopulation, mammosphere formation, colony formation and tumor formation. It is known that tangeretin did not inflict damaging effects sufficient to result in a reduced capacity to survive and proliferate. However, inhibition of the growth of breast cancers without inducing cancer cell death may be advantageous in treating breast tumors. Breast cancer cells resumed growth similar to untreated control within a day of tangeretin removal [29]. The Stat3 protein is essential for the maintenance of BCSCs [4]. The acetylated derivative of tangeretin (5-AcTMF) had anticancer effects on human glioblastoma multiforme cells through blockade of Stat3 signaling [33]. Extracellular IL-8 protein is a factor for BCSCs formation [13]. We investigated Stat3 signaling under tangeretin treatment. Tangeretin suppressed the total protein levels of Stat3 and pStat3.

The nuclear protein levels of Stat3 and pStat3 were also decreased by tangeretin. We assessed Sox2 transcript levels in BCSCs under tangeretin treatment and confirmed that the Sox2 mRNA level was decreased in the treated samples. In addition, the protein level of Sox2 was decreased in treated cells compared with control cells (Figure 7). Finally, tangeretin had an inhibitory effect on tumor growth in a breast cancer xenograft model. The tangeretin-treated group had a smaller tumor size and lower tumor weight than the control group.

In our study, tangeretin inhibited BCSC formation and targeted BCSCs by inhibiting the Stat3/Sox2 signaling pathway. Tangeretin is a possible therapeutic agent for breast cancer and BCSCs.

4. Materials and Methods

4.1. Reagents

Silica gel 60 and TLC plates were purchased from MERK (Darmstadt, Germany) and Sephadex LH-20 was obtained from Pharmacia (Uppsala, Sweden). High-pressure liquid chromatography was performed with a Shimadzu application system (Shimadzu, Kyoto, Japan). Cell viability was measured using the EZ-Cytox Cell Viability Assay Kit (DoGenBio, Seoul, Korea). Tangeretin was obtained from ChemFaces Co. (Hubei, China).

4.2. Plant Materials

Citrus peel was collected from Jeju Island, South Korea. The citrus peel was freeze-dried, and the dried citrus was ground. The samples (No. 2017_030) were deposited in the Department of Biomaterials, Jeju National University, JeJu-Si, South Korea.

4.3. Extraction and Isolation of an Inhibitor

Citrus powder was extracted with 100% methanol. The bioassay-based isolation protocol is summarized in Figure 1A. The methanol extracts were mixed with water, and the methanol was evaporated. The water-suspended samples were extracted with equal volumes of ethyl acetate. The ethyl acetate-concentrated part was loaded onto a silica gel column (3 × 35 cm) and fractionated with solvent (chloroform-methanol, 20:1) (Figure S1). The three parts were divided and assayed by evaluating mammosphere formation. The #2 part potentially inhibited mammosphere formation. The #2 part was loaded onto a Sephadex LH-20 open column (2.5 × 30 cm) and eluted in three fractions (Figure S2). The three parts were obtained and assayed by evaluating mammosphere formation. Part # 3 showed inhibition of mammosphere formation. Part #3 was isolated using preparatory TLC (glass plate; 20 × 20 cm) and developed in a TLC glass chamber. Individual bands were separated from the silica gel plate. Each fraction was assayed by evaluating mammosphere formation (Figure S3). The #1 fraction was loaded onto a Shimadzu HPLC instrument (Shimadzu, Tokyo, Japan). HPLC used an ODS 10 × 250 mm C18 column (flow rate; 3 mL/min). The mobile phase was water and acetonitrile. For elution, the acetonitrile proportion was initially set at 20%, increased to 80% at 20 min and finally increased to 100% at 40 min (Figure S4).

4.4. Structural Analysis of the Purified Sample

The chemical structures of the isolated compounds were determined by mass and nuclear magnetic resonance (NMR) measurements. The molecular weight was established as 372 by ESI-mass spectrometry, which showed a quasimolecular ion peak at m/z 373.3 [M + H]$^+$ in the positive mode (Figure S9). The ^{1}H NMR spectrum in CDCl$_3$ exhibited signals due to four aromatic methine protons at δ 7.86 (2H) and 7.01 (2H), which are attributable to a 1,4-disubstituted benzene, one aromatic singlet methine at δ 6.59, and five methoxy groups at δ 4.09, 4.01, 3.94, 3.93 and 3.87. In the ^{13}C NMR spectrum, there were 20 carbon peaks, including a carbonyl carbon at δ 177.3; nine sp^2 quaternary carbons at δ 162.3, 161.2, 151.4, 148.4, 147.7, 144.0, 138.0, 123.8 and 114.8; five sp^2 methine carbons at δ 127.7 (×2), 114.5 (×2) and 106.7; and five methoxy carbons at δ 62.2, 62.0, 61.8, 61.6 and 55.5 (Figure S5).

All proton-bearing carbons were assigned by the HMQC spectrum, and the ^{1}H-^{1}H COSY spectrum established a partial structure, 1,4-disubstituted benzene (Figures S6, S7 and S9). The chemical structure was determined to be from the HMBC spectrum, which exhibited long-range correlations from the methine proton at δ 6.59 to the carbons at δ 177.3, 161.2, 123.8 and 114.8, and from the methine protons at δ 7.86 to the carbon at δ 161.2, implying that this compound was a polymethoxylated flavone (Figures S8 and S9). Finally, long-range correlations from the five methoxy proton peaks to the oxygenated sp^2 quaternary carbons established the structure of this compound as that of tangeretin (Figure 2).

4.5. Culture of Breast Cancer Cells and Mammospheres

Two breast cancer cell lines, MCF-7 and MDA-MB-231, were purchased from the American Type Culture Collection (Rockville, MD, USA) and maintained in DMEM supplemented with 10% fetal bovine serum (FBS; HyClone Fisher Scientific, CA, USA) and 1% penicillin/streptomycin (Gibco, Thermo Fisher Scientific, CA, USA). Cancer cells (5×10^4 or 1×10^4) were incubated in an ultralow-attachment 6-well plate with MammoCult^{TM} culture medium (STEMCELL Technologies, Vancouver, BC, Canada). All cells were cultured in a humidified 5% CO_2 incubator at 37 °C for 7 days. The formation of mammospheres was assessed by the NICE program [34]. Mammosphere formation was examined by examining mammosphere formation efficiency (MFE) (%) [35].

4.6. Cell Viability Assay

MDA-MB-231 cells (2×10^4 cells/well) were seeded in a 96-well plate for 24 h and treated with various concentrations (0, 5, 10, 20, 40, 60 and 80 μM) of tangeretin for 24 h in a medium with 10% FBS (proliferation assay). Then, we followed the manufacturer's protocol for the EZ-Cytox Kit (DoGenBio, Seoul, Korea). The OD at a wavelength of 460 nm was measured using a GloMax® Explorer microplate reader (Promega, Madison, WI, USA).

4.7. Colony Formation Assay

MDA-MB-231 and MCF-7 cells (2×10^3 cells/well) were seeded in a 6-well plate, treated with different concentrations of tangeretin in DMEM and maintained for 7 days at 37 °C in a 5% CO_2 incubator. The grown colonies were washed with 1× PBS three times, fixed for 10 min using 3.7% formaldehyde, treated for 20 min and stained for 20 min with 0.05% crystal violet.

4.8. Wound-Healing Assay

MDA-MB-231 cells were seeded in a 6-well plate at 2×10^6 cells/well. A scratch was made by using a microtip after the cells had grown into a monolayer. After the cells were washed two times with 1× PBS, the cancer cells were cultured with tangeretin in fresh DMEM for 24 h. Photomicrographs of the wounded areas were acquired using a light microscope [36].

4.9. Flow Cytometry Analysis

After incubating with tangeretin for 24 h, cancer cells were dissociated using 1× trypsin/EDTA. We used a previously described method [36]. In total, 1×10^6 cells were cultured with anti-CD44-FITC and anti-CD24-PE antibodies (BD, San Jose, CA, USA) on ice for 30 min. The cancer cells were centrifuged and washed two times with 1× FACS buffer and analyzed on an Accuri C6 (BD, San Jose, CA, USA).

4.10. Gene Expression Analysis

Total RNA was extracted from cancer cells and purified, and real-time RT-qPCR was performed using a real-time One-step RT-qPCR kit (Enzynomics, Daejeon, Korea). We used a previously described method [37]. The specific primers are listed in Table S1.

4.11. Western Blot Analysis

Protein samples were extracted from mammospheres and cancer cells. After electrophoresis on a 12% SDS-PAGE gel, proteins were transferred to polyvinylidene fluoride (PVDF) membranes (Millipore, Burlington, MA, USA). The membranes were blocked in Odyssey blocking buffer at room temperature for 1 h and then incubated overnight with primary antibodies. The antibodies were anti-phospho-Stat3, anti-lamin B (Cell Signaling, Danvers, MA, USA), anti-Stat3, anti-Sox2 and anti-β-actin (Santa Cruz Biotechnology, Dallas, TA, USA). After membranes were washed three times using Tris-buffered saline/Tween 20, all membranes were incubated with IRDye 680RD- and IRDye 800W-labeled secondary antibodies for 1 h at room temperature, and the signal images were determined with an Odyssey CLx (LI-COR, Lincoln, NE, USA).

4.12. Electrophoretic Mobility Shift Assays (EMSAs)

Nuclear proteins were prepared as described previously [38]. EMSAs were performed with an Odyssey Infrared EMSA kit (LI-COR, Lincoln, NE, USA) according to the manufacturer's instructions. IRDye 700-labeled forward and reverse strands of the Stat3 oligonucleotide were annealed. The IRDye 700-labeled Stat3 oligonucleotide was incubated with nuclear extracts in a final volume of 20 µL at room temperature for 20 min. The samples were electrophoresed on a 6% polyacrylamide nondenaturing gel, and EMSA data were visualized on an ODYSSEY CLx system (LI-COR, Lincoln, NE, USA).

4.13. Xenograft Transplantation

Twelve female nude mice were injected with two million MDA-MB-231 cells with or without an additional tangeretin (2.5 mg/kg) injection. Tumor volume was estimated for 35 days using the formula (width^2 × length)/2. The mouse experiments were performed as described previously [39]. Animal care and animal experiments were performed in accordance with protocols approved by the Institutional Animal Care and Use Committee (JNU-IACUC; Approval Number 2017-003) of Jeju National University. Female nude mice (4 weeks old) were purchased from OrientBio (Daejeon, South Korea) and cultured in mouse facilities for 1 week.

4.14. SiRNA of Stat3

To examine the inhibitory function of Stat3 on mammosphere formation, breast cancer cells were transfected with siRNAs targeting human Stat3 (Bioneer, Daejeon, South Korea). The Stat3 siRNAs (NM_1145658) and a scrambled siRNA were purchased from Bioneer (Daejeon Cor., South Korea). For siRNA transfection, cancer cells were cultured and transfected using Lipofectamine 2000 (Invitrogen, Carlsbad, CA, USA) according to the manufacturer's instructions. The protein levels of Stat3 were determined via immunoblot analysis.

4.15. Statistical Analysis

All data were analyzed with GraphPad Prism 7.0 software (GraphPad Prism, Inc., San Diego, CA, USA). All data from three independent experiments are reported as the mean ± standard deviation (SD). Data were analyzed using one-way ANOVA. A p-value of less than 0.05 was considered significant.

5. Conclusions

A CSC-inhibiting compound from citrus extracts was purified using silica gel, gel filtration, TLC, and HPLC. The compound was identified as tangeretin. Tangeretin inhibited cell proliferation, CSC formation and tumor growth, and modestly induced apoptosis in CSCs. The size of the CSC subpopulation (CD44$^+$/CD24$^-$) was reduced by tangeretin. Tangeretin reduced the total level and phosphorylated nuclear level of Stat3 Tangeretin decreased the transcript levels of Sox2 and Sox2 protein in mammospheres. Our results in this study show that tangeretin inhibits the Stat3/Sox2 signaling pathway and induces CSC death, indicating that tangeretin may be a potential natural compound targeting breast cancer.

Supplementary Materials: The following are available online. Specific primer sequences for real-time RT-PCR are described in Table S1. Isolation and structure analysis of CSC inhibitor are described at Figures S1–S9.

Author Contributions: H.S.C., and Y.-C.K. designed this study, participated in all the experiments, and H.S.C., and Y.-C.K. wrote the manuscript. J.-H.K., S.-L.K., R.L., and B.-S.Y., helped design and perform the experiments. D.-S.L. supervised the study. All authors have read and agree to the published version of manuscript.

Funding: This research was supported by the Basic Science Research Program through the National Research Foundation of Korea (NRF) funded by the Ministry of Education (NRF-2016R1A6A1A03012862, NRF-2020R1A2C1006316) and NRF-2018R1D1A1B07045261).

Conflicts of Interest: There are no conflicts to declare.

References

1. Torre, L.A.; Bray, F.; Siegel, R.L.; Ferlay, J.; Lortet-Tieulent, J.; Jemal, A. Global cancer statistics, 2012. *CA Cancer J. Clin.* **2015**, *65*, 87–108. [CrossRef] [PubMed]

2. Mendez, O.; Perez, J.; Soberino, J.; Racca, F.; Cortes, J.; Villanueva, J. Clinical Implications of Extracellular HMGA1 in Breast Cancer. *Int. J. Mol. Sci.* **2019**, *20*, 5950. [CrossRef] [PubMed]

3. Boyle, P. Triple-negative breast cancer: Epidemiological considerations and recommendations. *Ann. Oncol.* **2012**, *23*, 7–12. [CrossRef]

4. Kim, J.H.; Choi, H.S.; Kim, S.L.; Lee, D.S. The PAK1-Stat3 Signaling Pathway Activates IL-6 Gene Transcription and Human Breast Cancer Stem Cell Formation. *Cancers* **2019**, *11*, 1527. [CrossRef] [PubMed]

5. Ailles, L.E.; Weissman, I.L. Cancer stem cells in solid tumors. *Curr. Opin. Biotechnol.* **2007**, *18*, 460–466. [CrossRef] [PubMed]

6. Bromberg, J.; Darnell, J.E., Jr. The role of STATs in transcriptional control and their impact on cellular function. *Oncogene* **2000**, *19*, 2468–2473. [CrossRef]

7. Liu, S.; Wicha, M.S. Targeting breast cancer stem cells. *J. Clin. Oncol.* **2010**, *28*, 4006–4012. [CrossRef]

8. Kusoglu, A.; Biray Avci, C. Cancer stem cells: A brief review of the current status. *Gene* **2019**, *681*, 80–85. [CrossRef]

9. Johnston, P.A.; Grandis, J.R. STAT3 signaling: Anticancer strategies and challenges. *Mol. Interv.* **2011**, *11*, 18–26. [CrossRef]

10. Burke, W.M.; Jin, X.; Lin, H.J.; Huang, M.; Liu, R.; Reynolds, R.K.; Lin, J. Inhibition of constitutively active Stat3 suppresses growth of human ovarian and breast cancer cells. *Oncogene* **2001**, *20*, 7925–7934. [CrossRef]

11. Kunigal, S.; Lakka, S.S.; Sodadasu, P.K.; Estes, N.; Rao, J.S. Stat3-siRNA induces Fas-mediated apoptosis in vitro and in vivo in breast cancer. *Int. J. Oncol.* **2009**, *34*, 1209–1220. [PubMed]

12. Korkaya, H.; Liu, S.; Wicha, M.S. Regulation of cancer stem cells by cytokine networks: Attacking cancer's inflammatory roots. *Clin. Cancer Res.* **2011**, *17*, 6125–6129. [CrossRef] [PubMed]

13. Singh, J.K.; Simoes, B.M.; Howell, S.J.; Farnie, G.; Clarke, R.B. Recent advances reveal IL-8 signaling as a potential key to targeting breast cancer stem cells. *Breast Cancer Res.* **2013**, *15*, 210. [CrossRef]

14. Meiyanto, E.; Hermawan, A.; Anindyajati, A. Natural products for cancer-targeted therapy: Citrus flavonoids as potent chemopreventive agents. *Asian Pac. J. Cancer Prev.* **2012**, *13*, 427–436. [CrossRef] [PubMed]

15. Giacosa, A.; Barale, R.; Bavaresco, L.; Gatenby, P.; Gerbi, V.; Janssens, J.; Johnston, B.; Kas, K.; La Vecchia, C.; Mainguet, P.; et al. Cancer prevention in Europe: The Mediterranean diet as a protective choice. *Eur. J. Cancer Prev.* **2013**, *22*, 90–95. [CrossRef]

16. Gattuso, G.; Barreca, D.; Gargiulli, C.; Leuzzi, U.; Caristi, C. Flavonoid composition of Citrus juices. *Molecules* **2007**, *12*, 1641–1673. [CrossRef]

17. Zheng, J.; Shao, Y.; Jiang, Y.; Chen, F.; Liu, S.; Yu, N.; Zhang, D.; Liu, X.; Zou, L. Tangeretin inhibits hepatocellular carcinoma proliferation and migration by promoting autophagy-related BECLIN1. *Cancer Manag. Res.* **2019**, *11*, 5231–5242. [CrossRef]

18. Zhu, X.; Zhou, W. The Emerging Regulation of VEGFR-2 in Triple-Negative Breast Cancer. *Front. Endocrinol. (Lausanne)* **2015**, *6*, 159. [CrossRef]

19. Domenici, G.; Aurrekoetxea-Rodriguez, I.; Simoes, B.M.; Rabano, M.; Lee, S.Y.; Millan, J.S.; Comaills, V.; Oliemuller, E.; Lopez-Ruiz, J.A.; Zabalza, I.; et al. A Sox2-Sox9 signalling axis maintains human breast luminal progenitor and breast cancer stem cells. *Oncogene* **2019**, *38*, 3151–3169. [CrossRef]

20. Amin, A.R.; Kucuk, O.; Khuri, F.R.; Shin, D.M. Perspectives for cancer prevention with natural compounds. *J. Clin. Oncol.* **2009**, *27*, 2712–2725. [CrossRef]

21. Batra, P.; Sharma, A.K. Anti-cancer potential of flavonoids: Recent trends and future perspectives. *3 Biotech.* **2013**, *3*, 439–459. [CrossRef] [PubMed]

22. Magne Nde, C.B.; Zingue, S.; Winter, E.; Creczynski-Pasa, T.B.; Michel, T.; Fernandez, X.; Njamen, D.; Clyne, C. Flavonoids, Breast Cancer Chemopreventive and/or Chemotherapeutic Agents. *Curr. Med. Chem.* **2015**, *22*, 3434–3446.

23. Ravishankar, D.; Rajora, A.K.; Greco, F.; Osborn, H.M. Flavonoids as prospective compounds for anti-cancer therapy. *Int. J. Biochem. Cell Biol.* **2013**, *45*, 2821–2831. [CrossRef] [PubMed]

24. Cirmi, S.; Maugeri, A.; Ferlazzo, N.; Gangemi, S.; Calapai, G.; Schumacher, U.; Navarra, M. Anticancer Potential of Citrus Juices and Their Extracts: A Systematic Review of Both Preclinical and Clinical Studies. *Front. Pharmacol.* **2017**, *8*, 420. [CrossRef] [PubMed]

25. So, F.V.; Guthrie, N.; Chambers, A.F.; Moussa, M.; Carroll, K.K. Inhibition of human breast cancer cell proliferation and delay of mammary tumorigenesis by flavonoids and citrus juices. *Nutr. Cancer* **1996**, *26*, 167–181. [CrossRef] [PubMed]

26. Zou, Z.; Xi, W.; Hu, Y.; Nie, C.; Zhou, Z. Antioxidant activity of Citrus fruits. *Food Chem.* **2016**, *196*, 885–896. [CrossRef] [PubMed]

27. Wu, K.; Jiao, X.; Li, Z.; Katiyar, S.; Casimiro, M.C.; Yang, W.; Zhang, Q.; Willmarth, N.E.; Chepelev, I.; Crosariol, M.; et al. Cell fate determination factor Dachshund reprograms breast cancer stem cell function. *J. Biol. Chem.* **2011**, *286*, 2132–2142. [CrossRef]

28. Lin, J.J.; Huang, C.C.; Su, Y.L.; Luo, H.L.; Lee, N.L.; Sung, M.T.; Wu, Y.J. Proteomics Analysis of Tangeretin-Induced Apoptosis through Mitochondrial Dysfunction in Bladder Cancer Cells. *Int. J. Mol. Sci.* **2019**, *20*, 1017. [CrossRef]

29. Morley, K.L.; Ferguson, P.J.; Koropatnick, J. Tangeretin and nobiletin induce G1 cell cycle arrest but not apoptosis in human breast and colon cancer cells. *Cancer Lett.* **2007**, *251*, 168–178. [CrossRef]

30. Goh, J.X.H.; Tan, L.T.; Goh, J.K.; Chan, K.G.; Pusparajah, P.; Lee, L.H.; Goh, B.H. Nobiletin and Derivatives: Functional Compounds from Citrus Fruit Peel for Colon Cancer Chemoprevention. *Cancers* **2019**, *11*, 867. [CrossRef]

31. Sp, N.; Kang, D.Y.; Kim, D.H.; Park, J.H.; Lee, H.G.; Kim, H.J.; Darvin, P.; Park, Y.M.; Yang, Y.M. Nobiletin Inhibits CD36-Dependent Tumor Angiogenesis, Migration, Invasion, and Sphere Formation Through the Cd36/Stat3/Nf-Kappab Signaling Axis. *Nutrients* **2018**, *10*, 772. [CrossRef]

32. Li, X.; Zhou, N.; Wang, J.; Liu, Z.; Wang, X.; Zhang, Q.; Liu, Q.; Gao, L.; Wang, R. Quercetin suppresses breast cancer stem cells (CD44(+)/CD24(-)) by inhibiting the PI3K/Akt/mTOR-signaling pathway. *Life Sci.* **2018**, *196*, 56–62. [CrossRef] [PubMed]

33. Cheng, Y.P.; Li, S.; Chuang, W.L.; Li, C.H.; Chen, G.J.; Chang, C.C.; Or, C.R.; Lin, P.Y.; Chang, C.C. Blockade of STAT3 Signaling Contributes to Anticancer Effect of 5-Acetyloxy-6,7,8,4'-Tetra-Methoxyflavone, a Tangeretin Derivative, on Human Glioblastoma Multiforme Cells. *Int. J. Mol. Sci.* **2019**, *20*, 3366. [CrossRef] [PubMed]

34. Clarke, M.L.; Burton, R.L.; Hill, A.N.; Litorja, M.; Nahm, M.H.; Hwang, J. Low-cost, high-throughput, automated counting of bacterial colonies. *Cytometry A* **2010**, *77*, 790–797. [CrossRef] [PubMed]

35. Choi, H.S.; Kim, D.A.; Chung, H.; Park, I.H.; Kim, B.H.; Oh, E.S.; Kang, D.H. Screening of breast cancer stem cell inhibitors using a protein kinase inhibitor library. *Cancer Cell Int.* **2017**, *17*, 25. [CrossRef] [PubMed]

36. Choi, H.S.; Kim, S.L.; Kim, J.H.; Deng, H.Y.; Yun, B.S.; Lee, D.S. Triterpene Acid (3-O-p-Coumaroyltormentic Acid) Isolated from Aronia Extracts Inhibits Breast Cancer Stem Cell Formation through Downregulation of c-Myc Protein. *Int. J. Mol. Sci.* **2018**, *19*, 2528. [CrossRef]

37. Choi, H.S.; Kim, J.H.; Kim, S.L.; Deng, H.Y.; Lee, D.; Kim, C.S.; Yun, B.S.; Lee, D.S. Catechol derived from aronia juice through lactic acid bacteria fermentation inhibits breast cancer stem cell formation via modulation Stat3/IL-6 signaling pathway. *Mol. Carcinog.* **2018**, *57*, 1467–1479. [CrossRef]

38. Choi, H.S.; Hwang, C.K.; Kim, C.S.; Song, K.Y.; Law, P.Y.; Wei, L.N.; Loh, H.H. Transcriptional regulation of mouse mu opioid receptor gene: Sp3 isoforms (M1, M2) function as repressors in neuronal cells to regulate the mu opioid receptor gene. *Mol. Pharmacol.* **2005**, *67*, 1674–1683. [CrossRef]

39. Kim, S.L.; Choi, H.S.; Kim, J.H.; Jeong, D.K.; Kim, K.S.; Lee, D.S. Dihydrotanshinone-Induced NOX5 Activation Inhibits Breast Cancer Stem Cell through the ROS/Stat3 Signaling Pathway. *Oxid. Med. Cell Longev.* **2019**, *2019*, 9296439. [CrossRef]

Sample Availability: Samples of the compounds are available from the authors.

 *molecules* 

Article

Chalcone Methoxy Derivatives Exhibit Antiproliferative and Proapoptotic Activity on Canine Lymphoma and Leukemia Cells

Aleksandra Pawlak [1,*], Marta Henklewska [1], Beatriz Hernández Suárez [1], Mateusz Łużny [2], Ewa Kozłowska [2], Bożena Obmińska-Mrukowicz [1] and Tomasz Janeczko [2]

[1] Department of Pharmacology and Toxicology, Wrocław University of Environmental and Life Sciences, C.K. Norwida 31, 50-375 Wrocław, Poland; marta.henklewska@upwr.edu.pl (M.H.); beatriz.hernandez-suarez@upwr.edu.pl (B.H.S.); b.mrukowicz@gmail.com (B.O.-M.)

[2] Department of Chemistry, Wrocław University of Environmental and Life Sciences, Norwida 25, 50-375 Wrocław, Poland; mateusz.luzny@upwr.edu.pl (M.Ł.); ewa.kozlowska1@upwr.edu.pl (E.K.); tomasz.janeczko@upwr.edu.pl (T.J.)

* Correspondence: aleksandra.pawlak@upwr.edu.pl

Academic Editor: H.P. Vasantha Rupasinghe

Received: 4 September 2020; Accepted: 21 September 2020; Published: 23 September 2020

Abstract: Chalcones are interesting candidates for anti-cancer drugs due to the ease of their synthesis and their extensive biological activity. The study presents antitumor activity of newly synthesized chalcone analogues with a methoxy group on a panel of canine lymphoma and leukemia cell lines. The antiproliferative effect of the 2′-hydroxychalcone and its methoxylated derivatives was evaluated in MTT assay after 48 h of treatment in different concentrations. The proapoptotic activity was studied by cytometric analysis of cells stained with Annexin V/FITC and propidium iodide and by measure caspases 3/7 and 8 activation. The DNA damage was evaluated by Western blot analysis of phosphorylated histone H2AX. The new compounds had selective antiproliferative activity against the studied cell lines, the most effective were the 2′-hydroxy-2″,5″-dimethoxychalcone and 2′-hydroxy-4′,6′-dimethoxychalcone. 2′-Hydroxychalcone and the two most active derivatives induced apoptosis and caspases participation, but some percentage of necrotic cells was also observed. Comparing phosphatidylserine externalization after treatment with the different compounds it was noted that the addition of two methoxy groups increased the proapoptotic potential. The most active compounds triggered DNA damage even in the cell lines resistant to chalcone-induced apoptosis. The results confirmed that the analogues could have anticancer potential in the treatment of canine lymphoma or leukemia.

Keywords: chalcones; apoptosis; DNA damage; anticancer activity; canine cancer cell lines

1. Introduction

The search for novel natural or synthetic compounds with potential antitumor activity is one of the major goals of contemporary oncology. The research involves both known compounds with proven biological activity as well as newly synthesized molecules that are structural analogues of biologically active substances already used in the treatment of cancer. Interest is aroused by compounds that may constitute the basis of chemotherapy in the future, as well as compounds that may support therapy or sensitize cancer cells to classic cytostatic drugs. Because the high toxicity of many anticancer substances (particularly bone marrow toxicity) significantly reduces the effectiveness of the treatment, to maximize the efficacy and reduce adverse effects, chemotherapy is usually based on a combination of drugs with various molecular mechanisms of action and different side effects.

With regard to this principle, research focusing on the antitumor activity of new natural compounds or their derivatives is highly justified. Such a potentially interesting group of compounds may be chalcones, an essential group of natural compounds classified as flavonoids [1]. Their typical structural feature is an open, α,β-unsaturated three-carbon fragment connecting two aromatic rings [2]. They are yellow [3] and commonly occur in the world of plants: in citrus fruits, vegetables and spices [4,5]. Chalcones (1-(2′-hydroxyphenyl)-3-phenylprop-2-en-1-ones) and their derivatives are often obtained from natural sources or as a result of chemical synthesis [6,7]. The most commonly used chemical synthesis method is the Claisen–Schmidt condensation of the corresponding aromatic aldehyde and 2′-hydroxyacetophenones in alkaline or acid catalysis [8–10].

The great interest in chalcones is influenced by the ease of synthesis of these compounds and the extensive biological activity characteristics they exhibit: in particular antiviral, anthelmintic, antibacterial, antiprotozoal, insecticidal, antiulcer, cytotoxic, and anticancer [11–13]. Due to the confirmed therapeutic effect of chalcones on human cancer, we decided to check their activity against canine cancer cells.

Cancer in dogs is a major challenge in modern veterinary medicine and research into the search for new cancer therapies is highly needed [14]. On the other hand, of the various animal species that develop cancer (e.g., cats, horses, rabbits, ferrets), dogs turned out to be the most appropriate model for comparative research due to their body size, life expectancy, cancer incidence and sharing of living conditions with humans. Moreover, dogs together with mice and rats, are popular laboratory animals necessary for different types of toxicological studies [15].

In conclusion, the presented study examined the antitumor activity of new chalcone analogues with the methoxy group on selected canine lymphoma and leukemia cell lines. After the hydroxyl group, the methoxy group is the second most commonly identified group in natural flavonoid compounds, and because of this, we decided to test a series of chalcones containing a different amount and places of methoxy group substitution. The cytotoxic, antiproliferative and proapoptotic effects of newly synthesized compounds were analyzed. The study showed that the obtained analogues could have anticancer potential in the treatment of canine lymphoma or leukemia.

2. Results

2.1. Chalcone Analogues with the Methoxy Group Have Selective Antiproliferative Activity on Canine Cancer Cell Lines

First, an assay was performed to determine whether the tested chalcones have antiproliferative activity against normal cells, and if the position of the substituents in the structure has an influence on the potency of such effects. The study showed that normal cells, represented by the NIH/3T3 and J774E.1 cell line, were less sensitive to the antiproliferative action of the tested chalcones than cancer cells. The IC50 value (concentration that inhibits the proliferation of 50% of cells) was determined only for two compounds in case of the NIH/3T3 cell line and four compounds in case of the J77.4.E cell line. These four compounds (**1, 2, 6** and **7**) proved to be the most active also in research with the use of cancer cell lines. The curves showing the inhibition of normal cell growth by individual compounds are shown in Figure 1A,B.

While the tested compounds did not show a significant inhibitory effect on the growth of normal cells, their potency in relation to cancer cell lines was visibly different. All the tested compounds had a particularly strong effect on the CLBL-1 cell line, representing the most common type of lymphoma in dogs—diffuse large B-cell lymphoma (DLBCL). Other B lymphocytes, from chronic lymphocytic leukemia (CLB70) were also more sensitive than the others. Natural killer (NK) cells obtained from the rare and refractory form of canine lymphoma were found to be the most resistant. Similarly, but slightly more strongly, the tested compounds acted on the GL-1 cell line, representing canine acute leukemia.

Detailed results in the form of a table containing IC50 values for all tested compounds on all tested cell lines are presented in Table 1. Growth inhibition curves of canine cancer cell lines are shown in Figure 1C–F. % start a new page

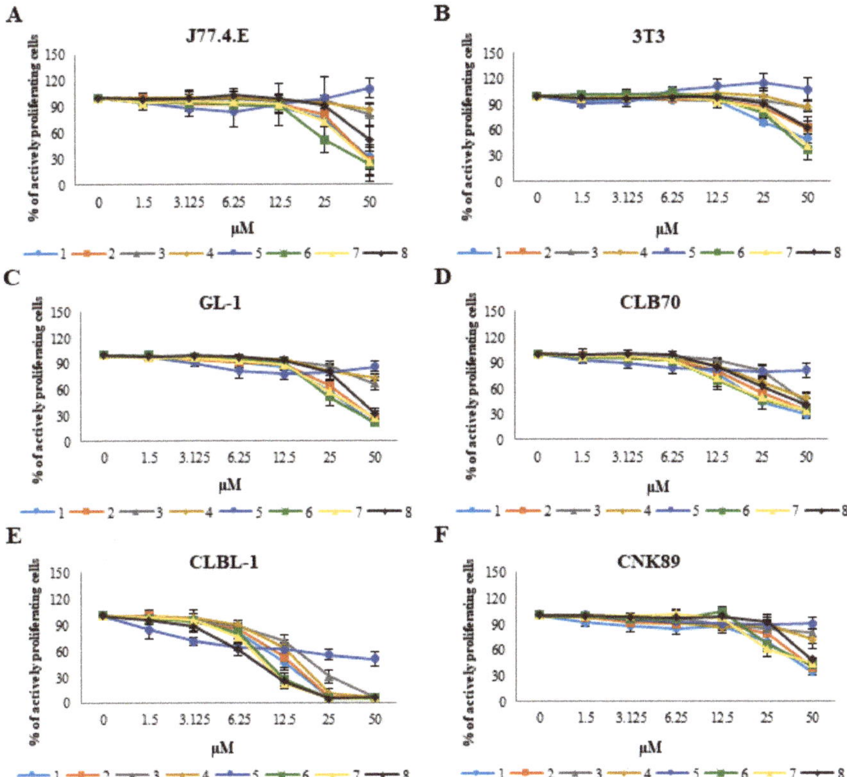

Figure 1. Concentration-dependent curves presenting the effects of the tested chalcones on viability of normal (**A**,**B**) and canine cancer cell lines (**C–F**) after 48 h of incubation with different concentrations of the individual chalcones (1.5, 3.125, 6.25, 12.5, 25 and 50 μM). The values are means from three independent experiments.

Table 1. IC_{50} values (μM concentration of the tested compounds that inhibits the proliferation of 50% of cells) for all tested compounds and all cell lines used in the study, obtained by the MTT test after 48 h treatment. The results are presented as mean $\pm$ standard deviation (SD) of three separate experiments, with three wells each. Statistical differences were analyzed using one-way ANOVA followed by Tukey's multiple comparison test. Values without common letters (a, b, c, d) in the superscript differ statistically ($p < 0.05$). The table shows differences in the strength of action of each compound on selected cell lines. N.A.—not achieved.

Compound/Cell Line	GL-1	CLBL-1	CNK-89	CLB-70	3T3	J774
2′-hydroxy-2″-methoxychalcone (1)	30.33 ± 5.51 [ac]	11.88 ± 1.58 [b]	37.30 ± 6.57 [a]	24.09 ± 6.23 [c]	N.A.	37.83 ± 3.40 [a]
2′-hydroxy-3″-methoxychalcone (2)	33.89 ± 7.00 [ac]	13.34 ± 3.04 [b]	42.61 ± 3.60 [a]	30.66 ± 9.39 [c]	N.A.	36.50 ± 3.87 [ac]
2′-hydroxy-4″-methoxychalcone (3)	N.A.	18.78 ± 2.36 [a]	N.A.	43.81 ± 0.66 [b]	N.A.	N.A.
2′-hydroxy-3″,4″,5″-trimethoxychalcone (4)	N.A.	15.62 ± 1.52	N.A.	N.A.	N.A.	N.A.
2′-hydroxy-4′,6′,3″,4″,5″-pentamethoxychalcone (5)	N.A.	33.57 ± 9.96	N.A.	N.A.	N.A.	N.A.
2′-hydroxy-2″,5″-dimethoxychalcone (6)	27.52 ± 4.59 [a]	9.76 ± 0.99 [b]	40.83 ± 3.00 [c]	24.47 ± 5.26 [a]	40.40 ± 3.02 [c]	29.82 ± 10.04 [a]
2′-hydroxy-4′,6′-dimethoxychalcone (7)	31.18 ± 6.32 [ac]	9.18 ± 0.75 [b]	38.63 ± 8.05 [ad]	26.78 ± 9.03 [c]	46.11 ± 4.66 [d]	32.17 ± 8.69 [ac]
2′-hydroxychalcone (8)	40.58 ± 3.92 [a]	8.04 ± 1.08 [b]	48.51 ± 1.52 [c]	38.94 ± 6.02 [a]	N.A.	N.A.

2.2. Chalcone Analogues with the Methoxy Group, at Least Partially, Kill Cancer Cells through Apoptosis

The substances with the strongest activity in the MTT test were selected for further research on the nature of cell death induced by the studied compounds. To determine the type of cell death induced by the tested chalcones, in the first place, the assay for detection of early signs of apoptosis, which is the externalization of phosphatidylserine, was performed. The study used the two cell lines which were the most sensitive to the tested compounds: CLBL-1 and CLB70 and the compounds with the strongest action: 2′-hydroxy-2″,5″-dimethoxychalcone (compound **6**), 2′-hydroxy-4′,6′-dimethoxychalcone (compound **7**) and initial compound (compound **8**). It was shown that cell death takes place by apoptosis, as confirmed by the statistical analysis of the results. After 48 h of incubation with the tested chalcone analogues at relatively low concentrations of 5 and 10 µM (concentrations harmless to normal cells), both the CLBL-1 and CLB70 cells were found to have a positive reaction with Annexin V, which is evidence of phosphatidylserine externalization and thus the ongoing process of apoptosis. As a result of the actions of the tested compounds, in addition to the population of apoptotic cells, the presence of cells that have already lost the integrity of the cell membrane was found (such a population of cells can also be referred as necrotic). The presence of cells staining positively with propidium iodide, a compound permanently staining the DNA of dead cells, can be explained by the relatively long incubation time used in the study (some of the cells that died by apoptosis had already lost their integrity too). Another explanation assumes that apoptosis is only one of the processes that cells undergo under the influence of the tested chalcones. Detailed results and representative dot plots are shown in Figure 2.

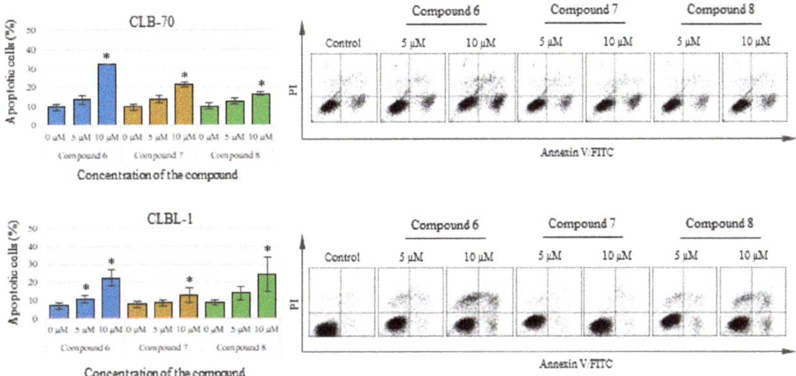

Figure 2. Percentage of annexin V positive (apoptotic) cells after 48 h of incubation with the culture medium alone (0) or 5 and 10 µM of the tested chalcones. On the right: representative dot-plots of annexin V/PI staining. The values are means of four independent experiments. * Considered significant in comparison to the control ($p < 0.05$).

To better explain the type of cell death that cancer cells undergo after treatment with the tested compounds, specific assays were carried out to determine the activity of caspases—the most important enzymes involved in the process of apoptosis. By flow cytometry, it was found that in CLBL-1 and CLB70 cells, after treatment with the tested chalcones, activation of the effector caspase 3 and initiator caspase 8 occur. The observed activation of caspase 3, the most important caspase in the caspase-dependent pathway of apoptosis was not massive and covered only a small percent of cells (about 15%). This result confirms the induction of apoptosis in the cells, but at the same time, indicates that apoptosis is not the only type of cell death of cells treated with the tested compounds. This result is consistent with the result of the Annexin V staining, in which similar percentage of necrotic cells was observed. 2′-Hydroxy-2″,5″-dimethoxychalcone (**6**) and initial compound activated caspase

3 the most strongly, whereas 2′-hydroxy-4′,6′-dimethoxychalcone (**7**) had a slightly weaker effect. More pronounced activation of caspase 3 occurred in the CLBL-1 cell line. Similar results were obtained for caspase 8, which was more strongly activated in the CLBL-1 cells, showing a statistically significantly higher level after treatment with all three compounds. Detailed results along with representative histograms are presented in Figure 3 (caspase 3 activation) and Figure 4 (caspase 8 activation).

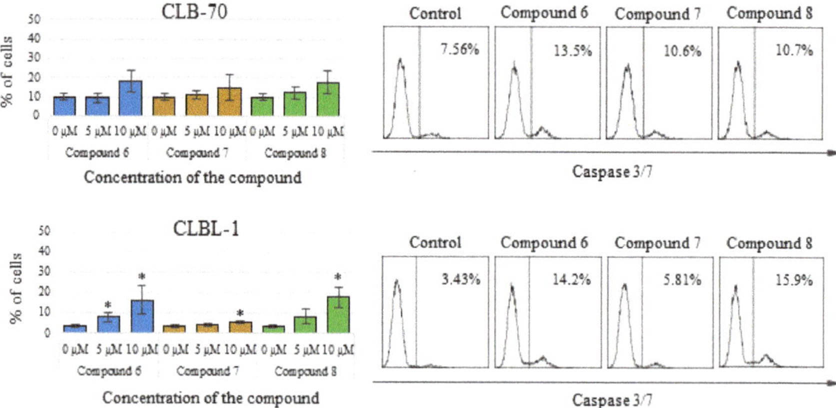

Figure 3. Percentage of cells with active caspase 3/7 after 48 h of incubation with the culture medium alone (0) or 5 and 10 µM of the tested chalcones. On the right: representative histograms for caspase 3/7 activation. The values are means of four independent experiments. * considered significant in comparison to the control ($p < 0.05$).

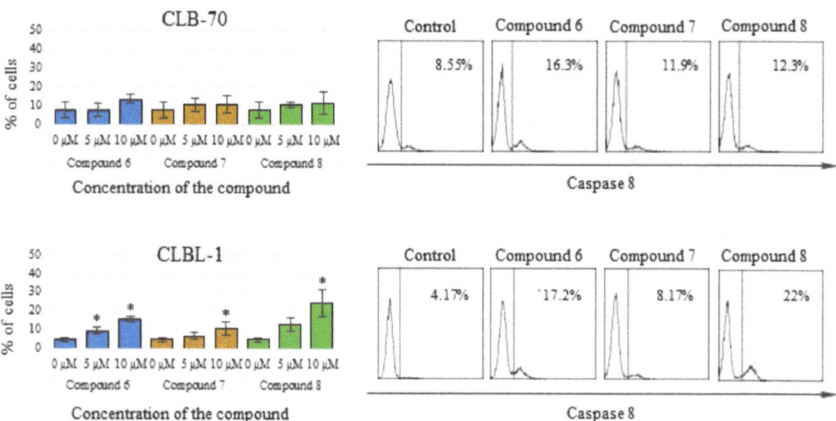

Figure 4. Percentage of cells with active caspase 8 after 48 h of incubation with the culture medium alone (0) or 5 and 10 µM of the tested chalcones. On the right: representative histograms for caspase 8 activation. The values are means of four independent experiments. * considered significant in comparison to the control ($p < 0.05$).

2.3. Tested Chalcone Analogues Trigger DNA Damage in Canine Lymphoma/Leukemia Cell Lines

Because the degree of apoptosis induction detected in Annexin V binding assay was lower than expected and did not correlate perfectly with the results of cytotoxicity tests, we decided to check the level of DNA damage as additional marker of potentially lethal events in the cells (leading to cell death and the initiation of the apoptosis process). The study showed that the incubation of

CLB70, CNK89 and even the resistant GL-1 cell lines with 2′-hydroxy-2″,5″-dimethoxychalcone (**6**), 2′-hydroxy-4′,6′-dimethoxychalcone (**7**) and initial 2′-hydroxychalcone (**8**) provoked the DNA damaged manifested by increased histone H2AX phosphorylation. Detailed results are presented in Figure 5.

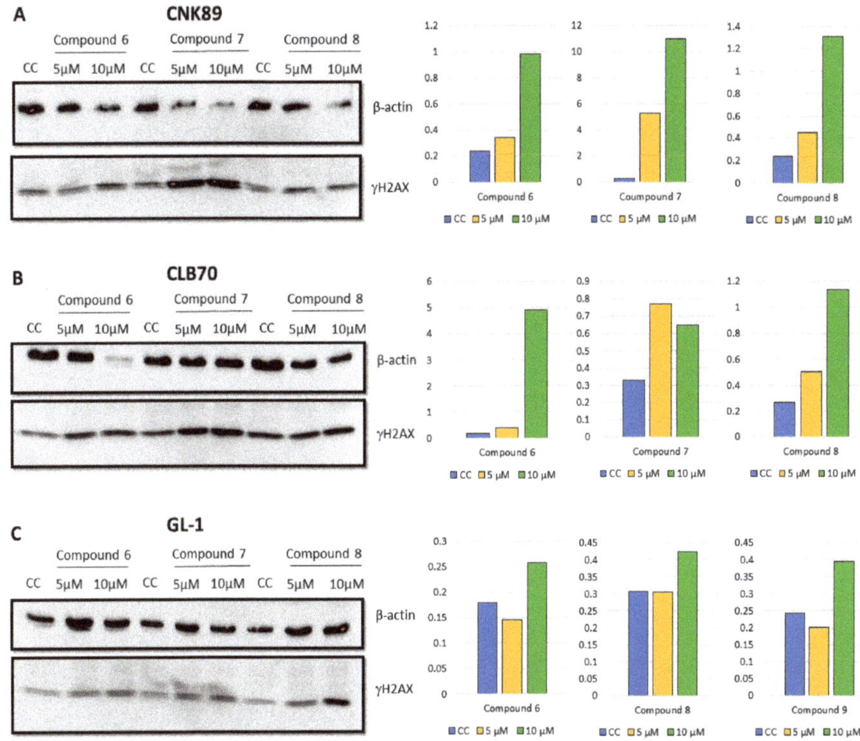

Figure 5. Western blot analysis for phosphorylated H2AX of CNK89, CLB70 and GL-1 cell lines after 48 h of incubation with different concentrations (5 and 10 µM) of compound **6–8** (**A–C**).

3. Discussion

The aim of the presented study was to obtain chalcone analogues with better anticancer properties compared to the parent compound. Because a significant problem in the use of various flavonoids is their poor availability, it was decided to focus on improving properties such as the ability to enter the cell. Therefore, the modification that was made was to create analogues containing a methoxy group instead of a hydroxyl group. Such a change can have a dual effect. Often, the conversion of a hydroxyl to a methoxy group results in a decrease in the biological activity of the flavonoid compound in vitro [16,17]. On the other hand, it was proven that methoxylated flavonoids have the ability to easily penetrate the cells [18]. For example, it was demonstrated that after transferring lung cell lines to the medium containing dimethoxyflavones, within five minutes, the cells accumulated 30–50 times more test compounds than were present in the surrounding buffer [19]. The intracellular transport of 5,7-dimethoxyflavone was approximately 10-fold higher than that of chrysin (5,7-dihydroxyflavone). Moreover, chrysin was rapidly metabolized by the human liver (S9 fraction), with no parent compound remaining after a 20-min incubation. In contrast, 5,7-dimethoxyflavone was metabolically stable over the whole 60-min time-course studied [20,21]. Better absorption and distribution, as well as the prolonged metabolism of methoxylated flavonoids, makes them even more promising compounds. Furthermore, their oral bioavailability exceeds the bioavailability of hydroxylated flavonoids, which makes them easy to administer [22]. Based on the knowledge of this phenomenon, we decided to create derivatives with a methoxy group and see how such modification affects the antitumor activity in vitro.

In our research, we first decided to check whether the obtained analogues show selective antiproliferative activity against cancer cells. As shown in Figure 2, it turned out that in fact, canine leukemia/lymphoma cells are more sensitive to the action of the obtained chalcones than the normal cell lines used in the study. The potency of the antiproliferative activity of the tested compounds was higher in relation to the cancer cell lines, but the same compounds that most strongly inhibited the proliferation of cancer cells were also the most toxic to normal cells. These compounds were: 2′-hydroxy-2″-methoxychalcone (**1**), 2′-hydroxy-3″-methoxychalcone (**2**), 2′-hydroxy-2″,5″-dimethoxychalcone (**6**) and 2′-hydroxy-4′,6′-dimethoxychalcone (**7**), the most active of which were the last two. Analyzing the differences in the chemical structure of individual compounds, at this stage of research, it can be concluded that the presence of two methoxy groups, regardless of their position, is responsible for the strength of the cytotoxic activity on normal cell lines. Regarding cancer cell lines, the differences in the potency of all four compounds mentioned above are too small to find conclusions. It seems, however, that the strongest antitumor activity is shown by compounds containing one methoxy group, but only in the position 2 or 3 of the B ring. Derivatives containing a methoxy group in 4 position or containing 3 or more such groups – are characterized by weaker antiproliferative activity. Looking at the table with IC50 values, one more interesting fact can be seen—the lowest IC50 value was observed in relation to the CLBL-1 cell line, for the parent compound—2′-hydroxychalcone (**8**). This effect was observed only for the CLBL-1 cell line and this value does not differ statistically from the values for 2′-hydroxy-2″,5″-dimethoxychalcone (**6**) and 2′-hydroxy-4′,6′-dimethoxychalcone (**7**). The explanation for this observation may be the fact that the CLBL-1 cell line was characterized by very strong sensitivity to all the tested compounds, therefore it is impossible in this case to capture the influence of small differences in the structure of the compounds on their activity against this cell line. However, the observed high sensitivity of the CLBL-1 cell line to the flavonoids may be a valuable therapeutic indication. This cell line was obtained from a dog with the most common type of lymphoma, and if it reflects the sensitivity of the cells of this type of disease then it may be possible to consider more thorough studies on the potential use of this group of compounds in therapy in dogs.

The proapoptotic effect of flavonoids or specifically chalcones is still a frequently discussed subject of scientific research. Numerous scientific reports describe the different strengths and mechanisms of proapoptotic action of compounds from this group [23–26]. In our research, we focused not on studying the mechanism of the induced apoptosis, but on comparing the strength with which various methoxy derivatives of chalcones induce suicidal cell death. Comparing proapoptotic activity 2′-hydroxy-2″,5″-dimethoxychalcone (**6**), 2′-hydroxy-4′,6′-dimethoxychalcone (**7**) and initial 2′-hydroxychalcone (**8**) it was noted that the presence of methoxy groups in positions 2 and 5 in the B ring, affects the potency of the proapoptotic action of the obtained derivatives. 2′-Hydroxy-2″,5″-dimethoxychalcone (**6**) showed a clearly stronger proapoptotic activity than the parent compound both in the annexin V and in caspase 3/7 and 8 activity tests. In the case of the CLBL-1 cell line, this difference was less visible, and in relation to caspase 8 the parent compound more strongly caused its cleavage. The presence of two methoxy groups, but at positions 4′ and 6′ (compound **7**) also increased the proapoptotic activity compared to the parent compound, as clearly seen in studies using the CLB70 cell line. However, this difference could not be confirmed by test using the CLBL-1 cell line. Another important observation was that the potency of the cytotoxic activity of the obtained derivatives with the methoxy group observed in the MTT test significantly exceeded that observed in the apoptosis tests. Because the MTT assay assesses the antiproliferative activity of the tested compounds, it can be concluded that the obtained chalcone derivatives strongly inhibited the proliferation of canine lymphoma/leukemia cells and the induction of apoptosis itself was slightly weaker. Such action of various chalcones has already been described. In addition to apoptosis, the compounds belonging to the chalcones have also been found to cause cell cycle arrest, which may be both the cause and the effect of the proapoptotic action of this group of compounds [27–30]. Because the mechanism of antitumor action of chalcones, apart from inhibiting cell proliferation and induction of apoptosis, also includes regulation of cell survival, invasiveness or angiogenesis [31,32], we decided to additionally check how the introduction of the methoxy group impact DNA damage. One of the mechanisms identified for this action is related to the chalcones' ability to selectively target of the ubiquitin-proteasome system (UPS) [31,32]. Published evidence suggests that the presence of an α,β-unsaturated carbonyl group is the key molecular determinant conferring UPS- and DUBs (deubiquitinating enzymes)-inhibitory activity of different chalcones [33]. Because DUBs are crucial in regulating a variety of cellular pathways, including cell growth and proliferation, apoptosis, protein quality control, DNA repair and transcription [34], the ability to inhibit them is an important feature of a potential anti-cancer compound. Inhibiting DUBs activity impairs 20S proteasome proteolytic activities and the cellular deubiquitinating enzymes, leading to increased accumulation of ubiquitinated proteins and consequently causes endoplasmic reticulum stress and cell death [35,36]. The effect may be also DNA damage. In the present study it has been clearly demonstrated that both the parent compound and the two strongest derivatives cause DNA damage observed as an increase in histone H2AX phosphorylation. Despite the use in this assay of cell lines (CNK89 and GL-1) slightly less sensitive to the action of chalcones it cannot be unequivocally stated that the obtained derivatives acted more strongly than the parent compound. Such observations can be explained by the lack of influence of methoxy groups on the formation of DNA damage (the presence of an α,β-unsaturated carbonyl group is rather responsible for such action) and the fact that the observed DNA damage was to some extent an effect of the ongoing apoptosis. However, the strength and extent of the DNA damage found, even in the cancer cell lines more resistant to chalcones, indicates that at least in part, the DNA damage was the cause of the death of the canine lymphoma/leukemia cells.

4. Materials and Methods

4.1. Compounds

The chalcone analogues used in the study are numbered from 1 to 8. All were obtained via a Claisen–Schmidt condensation of an appropriate 2-hydroxyacetophenone and a substituted benzaldehyde dissolved in methanol in an alkaline environment at high temperature according to the procedure described previously [37–39]. The structures of the tested chalcones are shown in Figure 6.

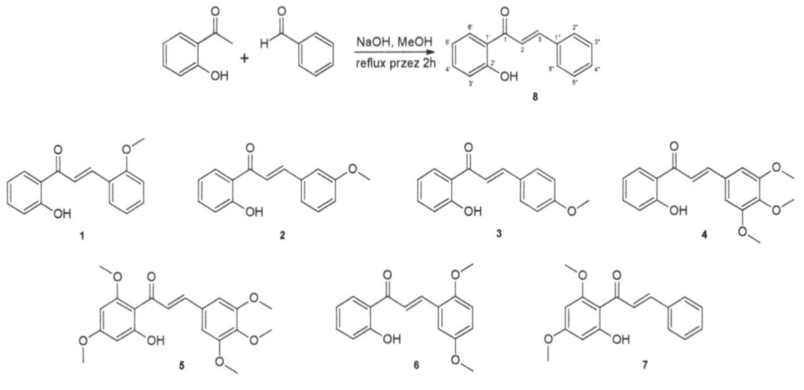

Figure 6. General scheme for the synthesis of chalcones and structures of obtained methoxy derivatives. The numbering of carbon atoms has been placed on compound **8**.

The resulting compounds were characterized by the following NMR spectral data (Supplementary Materials Figures S1–S29):

2′-Hydroxy-2″-methoxychalcone (**1**). ^{1}H NMR (600 MHz) (CDCl$_3$) δ (ppm): 3.94 (s 3H, -OCH_3), 6.94 (td, 1H, *J* = 7.8, 0.6 Hz, H-5′), 6.96 (d, 1H, *J* = 8.8 Hz, H-3″), 7.01 (t, 1H, *J* = 7.5 Hz, H-5″), 7.04 (dd, 1H, *J* = 8.3, 0.6 Hz, H-3′), 7.41 (ddd, 1H, *J* = 8.6, 7.9, 1.5 Hz, H-4″), 7.49 (t, 1H, *J* =8.8, 7.8, 1.4 Hz, H-4′); 7.65 (dd, 1H, *J* = 7.6, 1.3 Hz, H-6″), 7.78 (d, 1H, *J* = 15.6 Hz, H-2), 7.93 (dd, 1H, *J* = 8.0, 1.3 Hz, H-6′), 8.23 (d, 1H, *J* = 15.6 Hz, H-3); 12.95 (s, 1H, -O*H*). ^{13}C NMR (151 MHz, CDCl$_3$) δ = 55.73 (-OC*H*$_3$), 111.45 (C-3″), 118.67 (C-3′), 118.88 (C-5′), 120.33 (C-1′), 120.94 (C-2), 120.94 (C-5″), 123.76 (C-1″), 129.77 (C-6′), 129.84 (C-6″), 132.34 (C-4″), 136.28 (C-4′), 141.27 (C-3), 159.17 (C-2″), 163.71 (C-2′), 194.44 (C-1).

2′-Hydroxy-3″-methoxychalcone (**2**). ^{1}H NMR (600 MHz) (CDCl$_3$) δ (ppm): 3.87 (s, 3H, -OCH_3), 6.95 (ddd, 1H, *J* = 8.1, 7.1, 1.1 Hz, H-5′), 6.99 (dd, 1H, *J* = 8.2, 2.4 Hz, H-4″), 7.04 (dd, 1H, *J* = 8.4, 0.9 Hz, H-3′), 7.17 (dd, 1H, *J* = 2.2, 1.1 Hz, H-2″), 7.27 (d, 1H, *J* = 7.7 Hz, H-6″), 7.36 (t, 1H, *J* = 7.9 Hz, H-5″), 7.51 (ddd, 1H, *J* = 8.5, 7.2, 1.3 Hz, 1H, H-4′), 7.64 (d, 1H, *J* = 15.5 Hz, H-2), 7.89 (d, 1H, *J* = 15.5 Hz, H-3), 7.92 (dd, 1H, *J* = 8.1, 1.4 Hz, H-6′), 12.80 (s, 1H, -O*H*). ^{13}C NMR (151 MHz, CDCl$_3$) δ = 55.55 (-OC*H*$_3$), 113.87 (C-2″), 116.76 (C-4″), 118.80 (C-3′), 119.01 (C-5′), 120.16 (C-1′), 120.58 (C-2), 121.43 (C-6″), 129.81 (C-6′), 130.19 (C-5″), 136.12 (C-1″), 136.58 (C-4′) 145.53 (C-3), 160.14 (C-3″), 163.75 (C-2′), 193.86 (C-1).

2′-Hydroxy-4″-methoxychalcone (**3**). ^{1}H NMR (600 MHz) (CDCl$_3$) δ (ppm): 3.87 (s, 3H, -OCH_3), 6.94 (ddd, 1H, *J* = 8.1, 7.1, 1.0 Hz, H-5′), 6.94–6.97 (m, 2H, H-3″, H-5″), 7.02 (dd, 1H, *J* = 8.3, 0.8 Hz, H-3′), 7.49 (ddd, 1H, *J* = 8.5, 7.1, 1.5 Hz, H-4′), 7.54 (d, 1H, *J* = 15.4 Hz, H-2), 7.63 (m, 2H, H-2″, H-6″), 7.90 (d, 1H, *J* = 15.0 Hz, H-3), 7.93 (dd, 1H, *J* = 8.0, 1.3 Hz, H-6′), 12.95 (s, 1H, -O*H*). ^{13}C NMR (151 MHz, CDCl$_3$) δ = 55.59 (-OC*H*$_3$), 114.66 (C-3″,-5″), 117.74 (C-2), 118.72 (C-3′), 118.89 (C-5′), 120.26 (C-1′), 127.49 (C-1″), 129.67 (C-6′), 130.69 (C-2″, C-6″), 136.28 (C-4′), 145.50 (C-3), 162.17 (C-4″), 163.69 (C-2′), 193.82 (C-1).

2′-Hydroxy-3″,4″,5″-trimethoxychalcone (**4**). ^{1}H NMR (600 MHz) (CDCl$_3$) δ (ppm): 3.91 (s, 3H, C-4″-OCH_3), 3.94 (s, 6H, C-3″-OCH_3 and C-5″-OCH_3), 6.89 (s, 2H, H-2″ and H-6″), 6.96 (ddd, 1H, *J* = 8.0, 7.2, 1.0 Hz, H-5′), 7.04 (dd, 1H, *J* = 8.4, 1.0 Hz, H-3′), 7.51 (ddd, 1H, *J* = 8.3, 7.1, 1.4 Hz, H-4′), 7.54 (d, 1H, *J* = 15.4 Hz, H-2), 7.85 (d, 1H, *J* = 15.5 Hz, H-3), 7.93 (dd, 1H, *J* = 8.1, 1.5 Hz, H-6′), 12.84 (s, 1H, -O*H*). ^{13}C NMR (151 MHz, CDCl$_3$) δ = 56.42 (C-2″-OCH_3 and C-6″-OCH_3), 61.19 (C-4″-OCH_3), 106.08 (C-2″, C-6″), 118.81 (C-3′), 118.95 (C-5′), 119.41 (C-2), 120.16 (C-1′), 129.74 (C-6′), 130.20 (C-1″), 136.52 (C-4′), 140.97 (C-4″), 145.79 (C-3), 153.68 (C-3″, C-5″), 163.73 (C-2′), 193.67 (C-1).

2′-Hydroxy-4′,6′,3″,4″,5″-pentamethoxychalcone (**5**). ^{1}H NMR (600 MHz) (CDCl$_3$) δ (ppm): 3.84 (s, 3H, C-4′-OCH_3), 3.90 (s, 3H, C-4″-OCH_3), 3.91 (s, 3H, C-6′-OCH_3), 3.91 (s, 6H, C-2″-OCH_3 and C-6″-OCH_3), 5.96 (d, 1H, *J* = 2.3 Hz, H-5′), 6.11 (d, 1H, *J* = 2.3 Hz, H-3′), 6.84 (s, 2H, H-2″, H-6″), 7.70 (d, 1H, *J* = 15.5 Hz, H-3), 7.80 (d, 1H, *J* = 15.5 Hz, H-2), 14.31 (s, 1H, -O*H*). ^{13}C NMR (151 MHz, CDCl$_3$) δ = 55.75 (C-4′-OCH_3), 55.94 (C-6′-OCH_3), 56.27 (C-2″-OCH_3 and C-6″-OCH_3), 61.15 (C-4″-OCH_3), 91.46 (C-5′), 93.97 (C-3′), 105.69 (C-2″, C-6″), 106.45 (C-1′), 127.06 (C-2), 131.28 (C-1″), 140.21 (C-4″), 142.54 (C-3), 153.54 (C-3″, C-5″), 162.53 (C-6′), 166.34 (C-4′), 169.56 (C-2′), 192.49 (C-1).

2′-Hydroxy-2″,5″-dimethoxychalcone (**6**). ^{1}H NMR (600 MHz) (CDCl$_3$) δ (ppm): 3.83 (s, 3H, C-5″-OCH_3), 3.90 (s, 3H, C-2″-OCH_3), 6.90 (d, 1H, *J* = 9.0 Hz, H-3″), 6.94 (ddd, 1H, *J* = 9.0, 7.2, 0.9 Hz, H-5′), 6.97 (dd, 1H, *J* = 9.0, 3.1 Hz, H-4″), 7.02 (dd, 1H, *J* = 8.4, 1.0 Hz, H-3′), 7.17 (d, 1H, *J* = 3.1 Hz, H-6″), 7.49 (ddd, 1H, *J* = 8.3, 7.1, 1.5 Hz, H-4′), 7.75 (d, 1H, *J* = 15.6 Hz, H-2), 7.92 (dd, 1H, *J* = 8.1, 1.6 Hz, H-6′), 8.19 (d, 1H, *J* = 15.6 Hz, H-3), 12.92 (s, 1H, -O*H*). ^{13}C NMR (151 MHz, CDCl$_3$) δ = 56.02 (C-5″-OCH_3), 56.27 (C-2″-OCH_3), 112.63 (C-3″), 114.32 (C-6″), 117.80 (C-4″), 118.70 (C-3′), 118.91 (C-5′), 120.31 (C-1′), 121.17 (C-2), 124.31 (C-1″), 129.85 (C-6′), 136.34 (C-4′), 141.05 (C-3), 153.65 (C-5″), 153.73 (C-2), 163.71 (C-2′), 194.35 (C-1).

2′-Hydroxy-4′,6′-dimethoxychalcone (**7**). ^{1}H NMR (600 MHz) (CDCl$_3$) δ (ppm): 3.83 (s, 3H, C-4′-OCH_3), 3.92 (s, 3H, C-6′-OCH_3), 5.97 (d, 1H, *J* = 2.4 Hz, H-5′), 6.11 (d, 1H, *J* = 2.4 Hz, H-3′), 7.36–7.43 (m, 3H,

H-3″, H-4″, H-5″), 7.58–7.62 (m, 2H, H-2″, H-6″), 7.79 (d, 1H, *J* = 15.6 Hz, H-3), 7.91 (d, 1H, *J* = 15.6 Hz, H-2), 14.29 (s, 1H, -*OH*). ^{13}C NMR (151 MHz, CDCl$_3$) δ = 55.72 (C-4″-OCH$_3$), 55.99 (C-6′-OCH$_3$), 91.41 (C-5′), 93.93 (C-3′), 106.47 (C-1′), 127.66 (C-2), 128.48 (C-2″ and C-6″), 129.00 (C-3″ and C-5″), 130.18 (C-4″), 135.70 (C-1″), 142.45 (C-3), 162.64 (C-6′), 166.37 (C-4′), 168.53 (C-2′), 192.77 (C-1).

2′-Hydroxychalcone (**8**). ^{1}H NMR (600 MHz) (CDCl$_3$) δ (ppm): 3.94 (s 3H, -OCH$_3$), 6.94 (td, 1H, *J* = 7.8, 0.6 Hz, H-5′), 6.96 (d, 1H, *J* = 8.8 Hz, H-3″), 7.01 (t, 1H, *J* = 7.5 Hz, H-5″), 7.04 (dd, 1H, *J* = 8.3, 0.6 Hz, H-3′), 7.41 (ddd, 1H, *J* = 8.6, 7.9, 1.5 Hz, H-4″), 7.49 (t, 1H, *J* =8.8, 7.8, 1.4 Hz, H-4′); 7.65 (dd, 1H, *J* = 7.6, 1.3 Hz, H-6″), 7.78 (d, 1H, *J* = 15.6 Hz, H-2), 7.93 (dd, 1H, *J* = 8.0, 1.3 Hz, H-6′), 8.23 (d, 1H, *J* = 15.6 Hz, H-3); 12.95 (s, 1H, -*OH*). ^{13}C NMR (151 MHz, CDCl$_3$) δ = 55.73 (-OCH$_3$), 111.45 (C-3″), 118.67 (C-3′), 118.88 (C-5′), 120.33 (C-1′), 120.94 (C-2), 120.94 (C-5″), 123.76 (C-1″), 129.77 (C-6′), 129.84 (C-6″), 132.34 (C-4″), 136.28 (C-4′), 141.27 (C-3), 159.17 (C-2″), 163.71 (C-2′), 194.44 (C-1).

4.2. Cell Lines and Cell Culture

The study involved the following normal, non-cancerous NIH/3T3 (fibroblasts) and J774E.1 (macrophages) murine cell lines and a panel of canine cancer cell lines: CLBL-1 (B-cell lymphoma), GL-1 (B/T-cell leukemia), CLB70 (B-cell chronic lymphocytic leukemia) and CNK89 (NK-cell lymphoma). The NIH/3T3 and J774E.1 cell lines were bought from the American Type Culture Collection (ATCC, Rockville, MD, USA). CLBL-1 was obtained from Barbara C. Ruetgen from the Institute of Immunology, Department of Pathobiology, University of Veterinary Medicine, Vienna, Austria [40], GL-1 was obtained from Yasuhito Fujino and Hajime Tsujimoto from the University of Tokyo, Department of Veterinary Internal Medicine [41,42] while CLB70 [43] and CNK89 (Grudzień et al., in press) were established in our laboratory.

The cell lines were maintained in RPMI 1640 (J774E.1, NIH/3T3, CLBL-1 and GL-1) (Institute of Immunology and Experimental Therapy, Polish Academy of Sciences, Wrocław, Poland) or Advanced RPMI (Gibco, Grand Island, NY, USA) (CLB70 and CNK89) culture medium supplemented with 2 mM L-glutamine (Sigma Aldrich, Steinheim, Germany), 100 U/mL penicillin, 100 μg/mL streptomycin (Sigma Aldrich, Steinheim, Germany), and 10–20% heat-inactivated fetal bovine serum—FBS (Gibco, Grand Island, NY, USA).

4.3. Cell Proliferation Assay

The cell proliferation was determined using the MTT test (Sigma Aldrich, Steinheim, Germany). In brief, 1×10^4 (NIH/3T3 and J77.4.E) or 1×10^5 (canine cancer cell lines) cells per well were seeded in a 96-well-plate (Thermo Fisher Scientific, Denmark), and the tested chalcones were added in the increasing concentrations (1.5, 3.125, 6.25, 12.5, 25 and 50 μM). After incubation for 48 h, 20 μL of MTT solution (5 mg/mL) were added to each well. After the contents dissolved, the optical density of wells was measured with a microplate reader (Spark, Tecan) at a reference wavelength of 570 nm. The values were means from three independent experiments (three wells each).

4.4. Western Blotting

For Western blot analysis the cells were seeded in a total of 5×10^6 cells per 25 cm^2 cell culture flasks and two selected concentration of the compounds were added. After 48 h incubation the cells were harvested, rinsed with cold PBS, suspended in a lysis buffer (50 mM Tris–HCl pH 7.5, 100 mM NaCl, 1% NP-40 and protease inhibitors set) and incubated for 20 min on ice. The suspensions were centrifuged at 10,000 rpm at 4 ∘C for 12 min. Then, a sodium dodecyl sulfate (SDS) sample buffer was added to clear supernatants, and the samples were boiled at 95 ∘C for 5 min and subjected to SDS-PAGE on 10–15% gel. For the tests, the resolved proteins were transferred to a PVDF membrane (Millipore, Billerica, MA, USA), using Semidry Transfer Cell (Bio-Rad, Hercules, CA, USA). After the transfer, the membrane was blocked overnight with 1% casein in TBS at 4 ∘C, and then incubated with primary antibody (dilution 1:2000) (Santa Cruz Biotechnology, USA) at room temperature for 1 h, followed by secondary horseradish peroxidase-labelled antibody (Dako, Denmark). The bound antibodies were visualized using ChemiDoc Touch Instruments (BioRad, Hercules, CA, USA). The anti-γH2A.X (ab26350) antibody was from Abcam (Cambridge, UK) while the anti-β actin (C-4) antibody was from Santa Cruz Biotechnology (Santa Cruz, CA, USA).

4.5. Apoptosis Assays by Flow Cytometry

After seeding at the same density as for cytotoxicity tests in 96-well plates (TPP, Trasadingen, Switzerland) the cells were incubated for 48 h with two selected concentrations of the tested chalcones (5 and 10 μM). For the phosphatidylserine externalization and membrane integrity test, cells were collected, suspended in a binding buffer and stained with Annexin V-FITC and PI (final PI concentration 1 μg/mL). To evaluate caspases 3/7 and 8 the cells were harvested, washed twice with PBS and stained according to the manufacturer's instructions. Briefly, for active caspase 3/7 detection CellEventCaspase-3/7 Green Detection Reagent was added to the samples. For the detection of active caspase 8, FITC-IETD-fmk was added to the samples and the cells were incubated for 0.5 h. Then, the cells were washed twice and re-suspended in wash buffer. Flow cytometric analysis was immediately performed using a flow cytometer (FACS Calibur; Becton Dickinson, Biosciences, San Jose, CA, USA). CellQuest 3.lf. Software (Becton Dickinson, San Jose, CA, USA) was used for data analysis.

4.6. Statistical Analysis

All data are shown as means with standard deviations (SD). Statistical differences were analyzed using one-way ANOVA followed by Tukey's multiple comparison test. Statistical analysis was performed with STATISTICA version 13.3 software (TIBCO Software Inc., Palo Alto, CA, USA). The results were considered significant at $p < 0.05$.

5. Conclusions

Modification of the chemical structure of 2'-hydroxychalcone leads to the formation of a derivatives with various antitumor activity in vitro. Such modification consists of creation the analogues containing a methoxy group instead of a hydroxyl group caused both a reduction and an increase in the antitumor strength of the action depending on the number of groups added and their positions. However, the basic mechanism of action did not change, and all the derivatives obtained exerted antiproliferative and proapoptotic activity and, have the capacity to induce DNA damage, but to varying degrees. The cytotoxic effect of the parent compound and derived derivatives was stronger in relation to cancer cell lines than to normal ones. Further research is needed into the possibility of using chalcones as an adjuvant treatment of canine lymphoma or leukemia.

Supplementary Materials: The following are available online. Figure S1: ^{1}H NMR spectral of 2'-hydroxy-2''-methoxychalcone (**1**) (CDCl3, 600 MHz), Figure S2: Part of the ^{1}H NMR spectral 2'-hydroxy-2''-methoxychalcone (**1**) (CDCl3, 600 MHz), Figure S3: ^{13}C NMR spectral of 2'-hydroxy-2''-methoxychalcone (**1**) (CDCl3, 151 MHz), Figure S4: COSY spectral of 2'-hydroxy-2''-methoxychalcone (**1**) (CDCl3, 151 MHz), Figure S5: HSQC spectral of 2'-hydroxy-2''-methoxychalcone (**1**) (CDCl3, 151 MHz), Figure S6: ^{1}H NMR spectral of 2'-hydroxy-3''-methoxychalcone (**2**) (CDCl3, 600 MHz), Figure S7: Part of the ^{1}H NMR spectral 2'-hydroxy-3''-methoxychalcone (**2**) (CDCl3, 600 MHz), Figure S8: ^{13}C NMR spectral of 2'-hydroxy-3''-methoxychalcone (**2**) (CDCl3, 151 MHz), Figure S9: HSQC spectral of 2'-hydroxy-3''-methoxychalcone (**2**) (CDCl3, 151 MHz), Figure S10: ^{1}H NMR spectral of 2'-hydroxy-4''-methoxychalcone (**3**) (CDCl3, 600 MHz), Figure S11: Part of the ^{1}H NMR spectral 2'-hydroxy-4''-methoxychalcone (**3**) (CDCl3, 600 MHz), Figure S12: ^{13}C NMR spectral of 2'-hydroxy-4''-methoxychalcone (**3**) (CDCl3, 151 MHz), Figure S13: COSY spectral of 2'-hydroxy-4''-methoxychalcone (**3**) (CDCl3, 151 MHz), Figure S14: HSQC spectral of 2'-hydroxy-4''-methoxychalcone (**3**) (CDCl3, 151 MHz), Figure S15: ^{1}H NMR spectral of 2'-hydroxy-3'',4'',5''-trimethoxychalcone (**4**) (CDCl3, 600 MHz), Figure S16: Part of the ^{1}H NMR spectral 2'-hydroxy-3'',4'',5''-trimethoxychalcone (**4**) (CDCl3, 600 MHz), Figure S17: ^{13}C NMR spectral of 2'-hydroxy-3'',4'',5''-trimethoxychalcone (**4**) (CDCl3, 151 MHz), Figure S18: COSY spectral of 2'-hydroxy-3'',4'',5''-trimethoxychalcone (**4**) (CDCl3, 151 MHz), Figure S19: HSQC spectral of 2'-hydroxy-3'',4'',5''-trimethoxychalcone (**4**) (CDCl3, 151 MHz), Figure S20: ^{1}H NMR spectral of 2'-hydroxy-4',6',3'',4'',5''-pentamethoxychalcone (**5**) (CDCl3, 600 MHz),

Figure S21: Part of the ^{1}H NMR spectral 2′-hydroxy-4′,6′,3″,4″,5″-pentamethoxychalcone (**5**) (CDCl3, 600 MHz), Figure S22: ^{13}C NMR spectral of 2′-hydroxy-4′,6′,3″,4″,5″-pentamethoxychalcone (**5**) (CDCl3, 151 MHz), Figure S23: HSQC spectral of 2′-hydroxy-4′,6′,3″,4″,5″-pentamethoxychalcone (**5**) (CDCl3, 151 MHz), Figure S24 HMBC spectral of 2′-hydroxy-4′,6′,3″,4″,5″-pentamethoxychalcone (**5**) (CDCl3, 151 MHz), Figure S25: ^{1}H NMR spectral of 2′-hydroxy-2″,5″-dimethoxychalcone (**6**) (CDCl3, 600 MHz), Figure S26: Part of the ^{1}H NMR spectral 2′-hydroxy-2″,5″-dimethoxychalcone (**6**) (CDCl3, 600 MHz), Figure S27: ^{13}C NMR spectral of 2′-hydroxy-2″,5″-dimethoxychalcone (**6**) (CDCl3, 151 MHz), Figure S28: COSY spectral of 2′-hydroxy-2″,5″-dimethoxychalcone (**6**) (CDCl3, 151 MHz), Figure S29: HSQC spectral of 2′-hydroxy-2″,5″-dimethoxychalcone (**6**) (CDCl3, 151 MHz).

Author Contributions: Conceptualization, methodology software validation, formal analysis, data curation, A.P.; investigation, A.P., M.H., B.H.S. and M.Ł.; resources, M.Ł. and E.K.; writing-original draft preparation, A.P. and E.K.; writing—review and editing, T.J.; visualization, M.H. and B.H.S.; supervision, B.O.-M. and T.J.; project administration, A.P.; funding acquisition, A.P. All authors have read and agreed to the published version of the manuscript.

Funding: This publication was financed under the Leading Research Groups support project from the subsidy increased for the period 2020–2025 in the amount of 2% of the subsidy referred to in Art. 387 (3) of the Law of 20 July 2018 on Higher Education and Science, obtained in 2019.

Acknowledgments: We would like to thank B.C. Ruetgen (Institute of Immunology, Department of Pathobiology, University of Veterinary Medicine Vienna) for providing CLBL-1 cell line and Y. Fujino and H. Tsujimoto (University of Tokyo, Department of Veterinary Internal Medicine) for providing GL-1 cell line.

Conflicts of Interest: The authors disclose nonfinancial or personal relationships with other people or organizations that could influence (bias) their work.

References

1. Kozłowska, J.; Potaniec, B.; Żarowska, B.; Anioł, M. Microbial transformations of 4′-methylchalcones as an efficient method of obtaining novel alcohol and dihydrochalcone derivatives with antimicrobial activity. *RSC Adv.* **2018**, *8*, 30379–30386. [CrossRef]

2. Mahapatra, D.K.; Asati, V.; Bharti, S.K. Chalcones and their therapeutic targets for the management of diabetes: Structural and pharmacological perspectives. *Eur. J. Med. Chem.* **2015**, *92*, 839–865. [CrossRef] [PubMed]

3. Kim, S.; Jones, R.; Yoo, K.; Pike, L. Gold color in onions (Allium cepa): A natural mutation of the chalcone isomerase gene resulting in a premature stop codon. *Mol. Genet. Genom.* **2004**, *272*, 411–419. [CrossRef] [PubMed]

4. Orlikova, B.; Tasdemir, D.; Golais, F.; Dicato, M.; Diederich, M. Dietary chalcones with chemopreventive and chemotherapeutic potential. *Genes Nutr.* **2011**, *6*, 125–147. [CrossRef]

5. di Carlo, G.; Mascolo, N.; Izzo, A.A.; Capasso, F. Flavonoids: Old and new aspects of a class of natural therapeutic drugs. *Life Sci.* **1999**, *65*, 337–353. [CrossRef]

6. Hseu, Y.C.; Huang, Y.C.; Thiyagarajan, V.; Mathew, D.C.; Lin, K.Y.; Chen, S.C.; Liu, J.Y.; Hsu, L.S.; Li, M.L.; Yang, H.L. Anticancer activities of chalcone flavokawain B from Alpinia pricei Hayata in human lung adenocarcinoma (A549) cells via induction of reactive oxygen species-mediated apoptotic and autophagic cell death. *J. Cell. Physiol.* **2019**, *234*, 17514–17526. [CrossRef]

7. Zhang, C.; Yao, X.; Ren, H.; Wang, K.; Chang, J. Isolation and characterization of three chalcone synthase genes in pecan (Carya illinoinensis). *Biomolecules* **2019**, *9*, 236. [CrossRef] [PubMed]

8. Łużny, M.; Krzywda, M.; Kozłowska, E.; Kostrzewa-Susłow, E.; Janeczko, T. Effective Hydrogenation of 3-(2″-furyl)-and 3-(2″-thienyl)-1-(2′-hydroxyphenyl)-prop-2-en-1-one in Selected Yeast Cultures. *Molecules* **2019**, *24*, 3185. [CrossRef]

9. Gładkowski, W.; Siepka, M.; Janeczko, T.; Kostrzewa-Susłow, E.; Popłoński, J.; Mazur, M.; Żarowska, B.; Łaba, W.; Maciejewska, G.; Wawrzeńczyk, C. Synthesis and Antimicrobial Activity of Methoxy-Substituted γ-Oxa-ε-lactones Derived from Flavanones. *Molecules* **2019**, *24*, 4151. [CrossRef]

10. Gaonkar, S.L.; Vignesh, U. Synthesis and pharmacological properties of chalcones: A review. *Res. Chem. Intermed.* **2017**, *43*, 6043–6077. [CrossRef]

11. Zhuang, C.; Zhang, W.; Sheng, C.; Zhang, W.; Xing, C.; Miao, Z. Chalcone: A privileged structure in medicinal chemistry. *Chem. Rev.* **2017**, *117*, 7762–7810. [CrossRef] [PubMed]

12. Rammohan, A.; Reddy, J.S.; Sravya, G.; Rao, C.N.; Zyryanov, G.V. Chalcone synthesis, properties and medicinal applications: A review. *Environ. Chem. Lett.* **2020**, *18*, 433–458. [CrossRef]

13. Dan, W.; Dai, J. Recent developments of chalcones as potential antibacterial agents in medicinal chemistry. *Eur. J. Med. Chem.* **2020**, *187*, 111980. [CrossRef] [PubMed]

14. Marconato, L.; Gelain, M.E.; Comazzi, S. The dog as a possible animal model for human non-Hodgkin lymphoma: A review. *Hematol. Oncol.* **2013**, *31*, 1–9. [CrossRef] [PubMed]

15. Clark, D.L.; Andrews, P.A.; Smith, D.D.; DeGeorge, J.J.; Justice, R.L.; Beitz, J.G. Predictive value of preclinical toxicology studies for platinum anticancer drugs. *Clin. Cancer Res. Off. J. Am. Assoc. Cancer Res.* **1999**, *5*, 1161–1167.

16. Wang, T.-y.; Li, Q.; Bi, K.-s. Bioactive flavonoids in medicinal plants: Structure, activity and biological fate. *Asian J. Pharm. Sci.* **2018**, *13*, 12–23. [CrossRef]

17. Mahfoudi, R.; Djeridane, A.; Benarous, K.; Gaydou, E.M.; Yousfi, M. Structure-activity relationships and molecular docking of thirteen synthesized flavonoids as horseradish peroxidase inhibitors. *Bioorg. Chem.* **2017**, *74*, 201–211. [CrossRef]

18. Dymarska, M.; Janeczko, T.; Kostrzewa-Susłow, E. Glycosylation of 3-hydroxyflavone, 3-methoxyflavone, quercetin and baicalein in fungal cultures of the genus Isaria. *Molecules* **2018**, *23*, 2477. [CrossRef]

19. Tsuji, P.A.; Walle, T. Inhibition of benzo [a] pyrene-activating enzymes and DNA binding in human bronchial epithelial BEAS-2B cells by methoxylated flavonoids. *Carcinogenesis* **2006**, *27*, 1579–1585. [CrossRef]

20. Wen, X.; Walle, T. Methylation protects dietary flavonoids from rapid hepatic metabolism. *Xenobiotica* **2006**, *36*, 387–397. [CrossRef]

21. Wen, X.; Walle, T. Methylated flavonoids have greatly improved intestinal absorption and metabolic stability. *Drug Metab. Dispos.* **2006**, *34*, 1786–1792. [CrossRef]

22. Walle, T. Methoxylated flavones, a superior cancer chemopreventive flavonoid subclass? *Semin. Cancer Boil.* **2007**, *17*, 354–362. [CrossRef] [PubMed]

23. Chowdhury, S.A.; Kishino, K.; Satoh, R.; Hashimoto, K.; Kikuchi, H.; Nishikawa, H.; Shirataki, Y.; Sakagami, H. Tumor-specificity and apoptosis-inducing activity of stilbenes and flavonoids. *Anticancer Res.* **2005**, *25*, 2055–2063. [PubMed]

24. Quintin, J.; Desrivot, J.; Thoret, S.; le Menez, P.; Cresteil, T.; Lewin, G. Synthesis and biological evaluation of a series of tangeretin-derived chalcones. *Bioorg. Med. Chem. Lett.* **2009**, *19*, 167–169. [CrossRef] [PubMed]

25. Tiamas, S.G.; Audet, F.; Samra, A.A.; Bignon, J.; Litaudon, M.; Fourneau, C.; Ariffin, A.; Awang, K.; Desrat, S.; Roussi, F. Asymmetric Total Synthesis and Biological Evaluation of Proapoptotic Natural Myrcene-Derived Cyclohexenyl Chalcones. *Eur. J. Org. Chem.* **2018**, *2018*, 5830–5835. [CrossRef]

26. Sharma, R.; Kumar, R.; Kodwani, R.; Kapoor, S.; Khare, A.; Bansal, R.; Khurana, S.; Singh, S.; Thomas, J.; Roy, B. A review on mechanisms of anti tumor activity of chalcones. *Anti-Cancer Agents Med. Chem.* **2016**, *16*, 200–211. [CrossRef]

27. Drutovic, D.; Chripkova, M.; Pilatova, M.; Kruzliak, P.; Perjesi, P.; Sarissky, M.; Lupi, M.; Damia, G.; Broggini, M.; Mojzis, J. Benzylidenetetralones, cyclic chalcone analogues, induce cell cycle arrest and apoptosis in HCT116 colorectal cancer cells. *Tumor Biol.* **2014**, *35*, 9967–9975. [CrossRef]

28. Pilatova, M.; Varinska, L.; Perjesi, P.; Sarissky, M.; Mirossay, L.; Solar, P.; Ostro, A.; Mojzis, J. In vitro antiproliferative and antiangiogenic effects of synthetic chalcone analogues. *Toxicol. In Vitro* **2010**, *24*, 1347–1355. [CrossRef]

29. Boumendjel, A.; Ronot, X.; Boutonnat, J. Chalcones derivatives acting as cell cycle blockers: Potential anti cancer drugs? *Curr. Drug Targets* **2009**, *10*, 363–371.

30. Bailon-Moscoso, N.; Cevallos-Solorzano, G.; Romero-Benavides, J.C.; Orellana, M.I.R. Natural Compounds as Modulators of Cell Cycle Arrest: Application for Anticancer Chemotherapies. *Curr. Genom.* **2017**, *18*, 106–131. [CrossRef]

31. Yang, H.; Landis-Piwowar, K.R.; Chen, D.; Milacic, V.; Dou, Q.P. Natural Compounds with Proteasome Inhibitory Activity for Cancer Prevention and Treatment. *Curr. Protein Pept. Sci.* **2008**, *9*, 227–239. [CrossRef]

32. Gupta, S.C.; Kim, J.H.; Prasad, S.; Aggarwal, B.B. Regulation of survival, proliferation, invasion, angiogenesis, and metastasis of tumor cells through modulation of inflammatory pathways by nutraceuticals. *Cancer Metastasis Rev.* **2010**, *29*, 405–434. [CrossRef] [PubMed]

33. Issaenko, O.A.; Amerik, A.Y. Chalcone-based small-molecule inhibitors attenuate malignant phenotype via targeting deubiquitinating enzymes. *Cell Cycle* **2012**, *11*, 1804–1817. [CrossRef] [PubMed]

34. Sowa, M.E.; Bennett, E.J.; Gygi, S.P.; Harper, J.W. Defining the Human Deubiquitinating Enzyme Interaction Landscape. *Cell* **2009**, *138*, 389–403. [CrossRef]

35. Mullally, J.; Fitzpatrick, F. Pharmacophore model for novel inhibitors of ubiquitin isopeptidases that induce p53-independent cell death. *Mol. Pharmacol.* **2002**, *62*, 351–358. [CrossRef] [PubMed]

36. Milacic, V.; Banerjee, S.; Landis-Piwowar, K.R.; Sarkar, F.H.; Majumdar, A.P.N.; Dou, Q.P. Curcumin Inhibits the Proteasome Activity in Human Colon Cancer Cells In vitro and In vivo. *Cancer Res.* **2008**, *68*, 7283–7292. [CrossRef] [PubMed]

37. Janeczko, T.; Gładkowski, W.; Kostrzewa-Susłow, E. Microbial transformations of chalcones to produce food sweetener derivatives. *J. Mol. Catal. B Enzym.* **2013**, *98*, 55–61. [CrossRef]

38. Janeczko, T.; Dymarska, M.; Siepka, M.; Gniłka, R.; Leśniak, A.; Popłoński, J.; Kostrzewa-Susłow, E. Enantioselective reduction of flavanone and oxidation of cis-and trans-flavan-4-ol by selected yeast cultures. *J. Mol. Catal. B: Enzym.* **2014**, *109*, 47–52. [CrossRef]

39. Janeczko, T.; Popłoński, J.; Kozłowska, E.; Dymarska, M.; Huszcza, E. EKostrzewa-Susłow, Application of α-and β-naphthoflavones as monooxygenase inhibitors of Absidia coerulea KCh 93, Syncephalastrum racemosum KCh 105 and Chaetomium sp. KCh 6651 in transformation of 17α-methyltestosterone. *Bioorg. Chem.* **2018**, *78*, 178–184. [CrossRef]

40. Rutgen, B.C.; Hammer, S.E.; Gerner, W.; Christian, M.; de Arespacochaga, A.G.; Willmann, M.; Kleiter, M.; Schwendenwein, I.; Saalmuller, A. Establishment and characterization of a novel canine B-cell line derived from a spontaneously occurring diffuse large cell lymphoma. *Leuk. Res.* **2010**, *34*, 932–938. [CrossRef]

41. Nakaichi, M.; Taura, Y.; Kanki, M.; Mamba, K.; Momoi, Y.; Tsujimoto, H.; Nakama, S. Establishment and characterization of a new canine B-cell leukemia cell line. *J. Vet. Med Sci.* **1996**, *58*, 469–471. [CrossRef] [PubMed]

42. Momoi, Y.; Okai, Y.; Watari, T.; Goitsuka, R.; Tsujimoto, H.; Hasegawa, A. Establishment and characterization of a canine T-lymphoblastoid cell line derived from malignant lymphoma. *Vet. Immunol. Immunopathol.* **1997**, *59*, 11–20. [CrossRef]

43. Pawlak, A.; Rapak, A.; Zbyryt, I. BObminska-Mrukowicz, The effect of common antineoplastic agents on induction of apoptosis in canine lymphoma and leukemia cell lines. *Vivo* **2014**, *28*, 843–850.

Sample Availability: Samples of the compounds **1–8** are available from the authors.

Article

Flavonoids from *Acer okamotoanum* Inhibit Adipocyte Differentiation and Promote Lipolysis in the 3T3-L1 Cells

Ji Hyun Kim [1], Sanghyun Lee [2] and Eun Ju Cho [1,*]

[1]   Department of Food Science and Nutrition & Kimchi Research Institute, Pusan National University, Busan 46241, Korea; kjjjjhh11@naver.com
[2]   Department of Plant Science and Technology, Chung-Ang University, Anseong 17546, Korea; slee@cau.ac.kr
*   Correspondence: ejcho@pusan.ac.kr; Tel.: +82-51-510-2837

Academic Editor: H.P. Vasantha Rupasinghe
Received: 31 March 2020; Accepted: 20 April 2020; Published: 21 April 2020

Abstract: Flavonoids, quercitrin, isoquercitrin (IQ), and afzelin, were isolated from ethyl acetate fraction of *Acer okamotoanum*. We investigated anti-obesity effects and mechanisms of three flavonoids from *A. okamotoanum* in the differentiated 3T3-L1 cells. The differentiated 3T3-L1 cells increased triglyceride (TG) contents, compared with non-differentiated normal group. However, treatments of three flavonoids from *A. okamotoanum* decreased TG contents without cytotoxicity. In addition, they showed significant down-regulation of several adipogenic transcription factors, such as γ-cytidine-cytidine-adenosine-adenosine-thymidine/enhancer binding protein -α, -β, and peroxisome proliferator-activated receptor-γ, compared with non-treated control group. Furthermore, treatment of the flavonoids inhibited expressions of lipogenesis-related proteins including fatty acid synthase, adipocyte protein 2, and glucose transporter 4. Moreover, IQ-treated group showed significant up-regulation of lipolysis-related proteins such as adipose triglyceride lipase and hormone-sensitive lipase. In addition, flavonoids significantly activated 5′-adenosine monophosphate-activated protein kinase (AMPK) compared to control group. In particular, IQ showed higher inhibition of TG accumulation by regulation of pathways related with both adipogenesis and lipolysis, than other flavonoids. The present results indicated that three flavonoids of *A. okamotoanum* showed anti-obesity activity by regulation of adipocyte differentiation, lipolysis, and AMPK signaling, suggesting as an anti-obesity functional agents.

Keywords: *Acer okamotoanum*; afzelin; isoquercitrin; obesity; quercitrin

---

1. Introduction

Obesity has been associated with various degenerative diseases, such as cardiovascular disease, type 2 diabetes mellitus, atherosclerosis, and Alzheimer's disease [1,2]. The environmental factors such as consumption of food and lifestyle play the most important role among other causes of obesity, in particular, excessive energy intake leads to increase of the adipose tissue and accumulation of triglyceride (TG) [3]. Adipose tissue acts energy metabolism by secreting adipokines, but its excessive expansion is closely related to the development of obesity [4]. The differentiation of preadipocytes (adipogenesis) results in intracellular lipid accumulation in adipose tissue of patients with obesity [5].

Adipogenesis is the process to become adipocytes mature from preadipocytes, and it is regulated by adipogenic key transcription factors including family of CCAAT/enhancer-binding proteins (C/EBPs) and peroxisome proliferator-activated receptors (PPARs) [6]. During adipogenesis process, adipogenic key transcription factors enhance the lipogenic enzymes such as fatty acid synthase (FAS) and adipocytes protein 2 (aP2) [7,8]. Lipolysis is the process that hydrolyzes TG and diglyceride by activation of hydrolyzing enzymes including hormone-sensitive lipase (HSL) and adipose triglyceride lipase (ATGL), resulting in the production of fatty acids and glycerol [9]. In addition, lipolysis is decreased by over-expression of adipogenic key transcription factors in the adipogenesis [9]. Moreover, phosphorylation of 5′-adenosine monophosphate-activated protein kinase (AMPK) is important for progression of obesity [10]. Therefore, many researchers are focused on treatment of obesity by regulation of AMPK pathway [10,11]. Recently, anti-obesity agents are popular and attracted much attention, but they have several side effects including diarrhea, headache, and gastrointestinal discomfort [12]. Therefore, finding of natural products without side effects for anti-obesity agents by inhibition of adipogenesis and promotion of lipolysis is one of strategies to prevent and treat obesity.

Acer okamotoanum is widely distributed in Ulleung-do, Republic of Korea, as one variety of Korean endemic species [13]. *A. okamotoanum* has been reported to exert some biological activities such as anti-oxidant, anti-hypertension and immune improvement effect [14–16]. *A. okamotoanum* contains biological compounds such as cleomiscosin A and C, gallic acid, and β-amyrin [17]. We previously isolated flavonoids such as quercitrin (QU; quercetin-3-rhamoside), isoquercitrin (IQ; quercetin-3-glucoside), and afzelin (AF; kaempferol-3-rhamoside) from ethyl acetate (EtOAc) fraction of *A. okamotoanum* [18]. Therefore, in the present study, we investigated anti-obesity effects of three flavonoids from *A. okamotoanum* including QU, IO, and AF in the differentiated 3T3-L1 cells. In addition, molecular mechanisms related to anti-obesity effects of flavonoids from *A. okamotoanum* on adipogenesis and lipolysis was also observed.

2. Results

2.1. Effects of Flavonoids from A. okamotoanum on Differentiation of Preadipocytes and Lipid Accumulation

We investigated cytotoxicity of flavonoids from *A. okamotoanum* at the doses (1–10 µg/mL) in the 3T3-L1 adipocytes. As shown in Figure 1, treatment of three flavonoids at concentrations up to 10 µg/mL had no significant cytotoxicity on 3T3-L1 adipocytes, compared with non-treated normal group. Therefore, we used flavonoids at the concentration up to 10 µg/mL in this study.

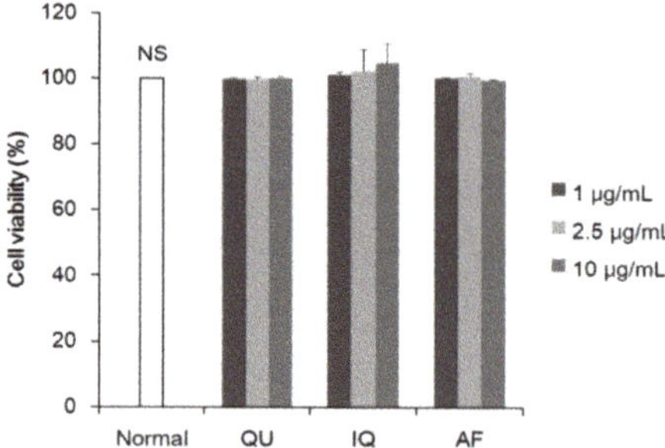

Figure 1. Effects of flavonoids from *Acer okamotoanum* on the cell viability in 3T3-L1 adipocytes. The 3T3-L1 adipocytes were pretreated with various concentrations (1–10 µg/mL) of flavonoids from *A. okamotoanum* for 72 h. Data are expressed as the mean ± standard deviation. NS: Non-significance; QU: Quercitrin; IQ: Isoquercitrin; AF: Afzelin.

To evaluate the effects of flavonoids from *A. okamotoanum* on differentiation of preadipocytes and lipid accumulation, we conducted Oil Red O staining, and then visualized cell morphology by light microscopy (Figure 2). The control group showed cell differentiation and lipid droplets induced by treatment of 3-isobutyl-1-methylxanthine, dexamethasone, and insulin (MDI), compared with normal group. However, treatment of three flavonoids such as QU, IQ, and AF at 10 µg/mL inhibited differentiation of preadipocytes and lipid droplets production, compared with control group. In particular, IQ inhibited more effectively differentiation and lipid droplets among other flavonoids.

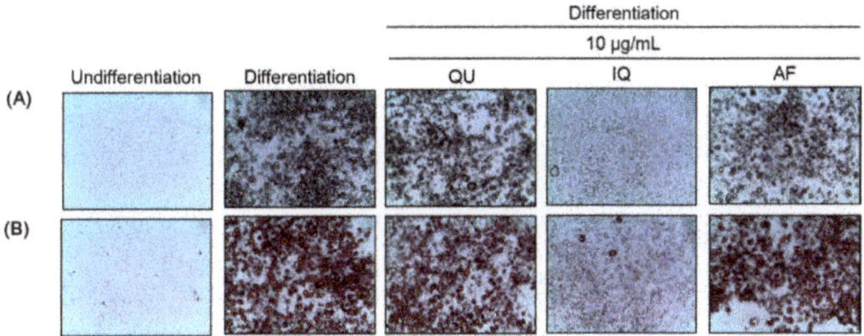

Figure 2. Effects of flavonoids from *Acer okamotoanum* on cell differentiation in differentiated 3T3-L1 cells. Adipocyte differentiation was induced by treatment with 3-isobutyl-1-methylxanthine, dexamethasone, and insulin (MDI) media in the absence or presence of flavonoids from *A. okamotoanum* during 2 days. The MDI media was then replaced with insulin media, and it was changed four times for every 2 days. The cells were confirmed by light microscopy (magnification, ×100) (**A**). Cells were fixed and stained with Oil Red O staining to visualize the lipid droplets by light microscopy (magnification, ×100) (**B**). Normal group indicates non-differentiated cells, whereas control group indicates the differentiated cells by treatment of MDI media. QU: Quercitrin; IQ: Isoquercitrin; AF: Afzelin.

We also measured intracellular TG contents by Oil Red O quantification (Figure 3). Non-differentiated normal group showed 7.64% TG contents, while control group showed 100.00% of TG contents. On the other hand, TG content of the flavonoids-treated groups such as QU, IQ, and AF at 10 µg/mL is decreased to 83.30%, 18.88%, and 90.17%, respectively. In particular, IQ-treated group inhibited accumulation of TG the most effectively.

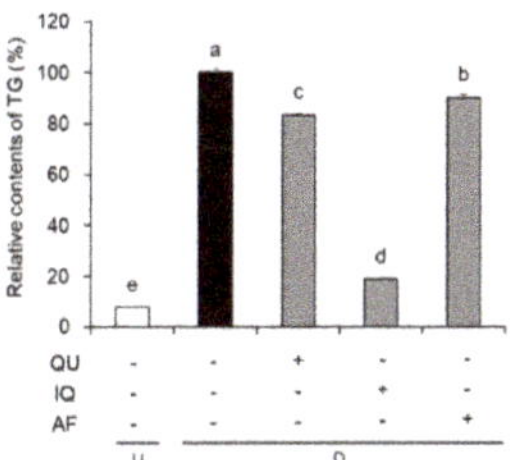

Figure 3. Effects of flavonoids from *Acer okamotoanum* (10 µg/mL) on intracellular triglyceride (TG) accumulation in differentiated 3T3-L1 cells. Adipocyte differentiation was induced by treatment with MDI media in the absence or presence of flavonoids from *A. okamotoanum* during 2 days. The MDI media was then replaced with insulin media, and it was changed four times for every 2 days. Data are expressed as the mean ± standard deviation. [a-e] Means with different letters indicate significant differences ($P < 0.05$) by Duncan's multiple range test. Normal group indicates non-differentiated cells, whereas control group indicates the differentiated cells by treatment of MDI media. U: Undifferentiation; D: Differentiation; QU: Quercitrin; IQ: Isoquercitrin; AF: Afzelin.

2.2. Effects of Flavonoids from A. okamotoanum on Expressions of Adipogenic Key Transcription Factors

To confirm the effects of flavonoids such as QU, IQ, and AF on adipogenic key transcription factors, we measured the protein expressions of C/EBPs family such as C/EBPα, C/EBPβ, and PPARs family including PPARγ. As shown in Figure 4, MDI-stimulated control group cells significantly increased these adipogenic key transcription factors such as C/EBPα, C/EBPβ, and PPARγ. In the C/EBPs family expressions, IQ- and AF-treated group showed significant down-regulation of C/EBPα and C/EBPβ levels, compared with control group. In PPARs expression, three flavonoids-treated group significantly decreased PPARγ level compared with control group. Especially, the treatment of IQ inhibited protein expressions of both C/EBPs and PPARs family.

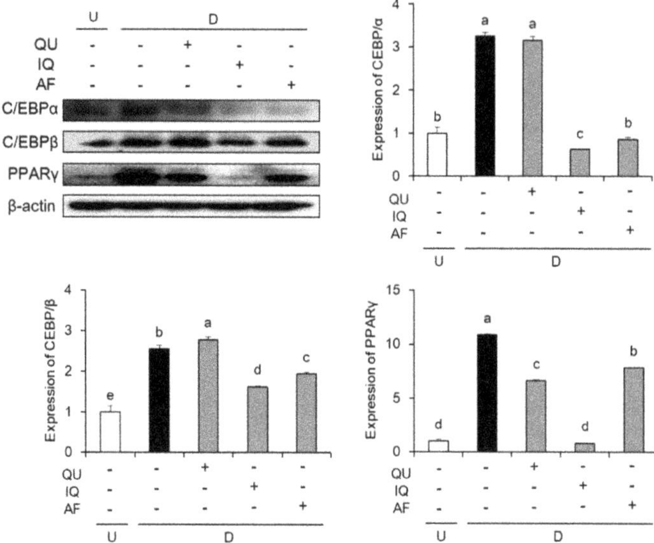

Figure 4. Effects of flavonoids from *Acer okamotoanum* (10 μg/mL) on adipogenic key transcription factors in differentiated 3T3-L1 cells. Adipocyte differentiation was induced by treatment with MDI media in the absence or presence of flavonoids from *A. okamotoanum* during 2 days. The MDI media was then replaced with insulin media, and it was changed four times for every 2 days. Data are expressed as the mean ± standard deviation. [a–e] Means with different letters indicate significant differences ($P < 0.05$) by Duncan's multiple range test. β-actin was used as a loading control. Relative expression levels were normalized to the β-actin levels and the normal group. Normal group indicates non-differentiated cells, whereas control group indicates the differentiated cells by treatment of MDI media. U: Undifferentiation; D: Differentiation; QU: Quercitrin; IQ: Isoquercitrin; AF: Afzelin.

2.3. Effects of Flavonoids from A. okamotoanum on Lipogenesis-Related Protein Expressions

To examine the effects of flavonoids such as QU, IQ, and AF on lipogenesis-related protein expressions, we carried out the measurement of protein expressions such as FAS, aP2, and glucose transporter 4 (GLUT4). In our results (Figure 5), the stimulation of MDI significantly increased the lipogenesis-related protein expressions such as FAS, aP2, and GLUT4. Treatment with 10 μg/mL of three flavonoids including QU, IQ, and AF resulted in significant down-regulation of protein expressions involved in lipogenesis. In addition, treatment with IQ showed higher decrease in protein expressions of FAS and aP2, whereas QU- or IQ-treated group showed down-regulation of GLUT4 protein expression.

2.4. Effects of Flavonoids from A. okamotoanum on Lipolysis-Associated Proteins

As shown in Figure 6, we examined the effects of flavonoids including QU, IQ, and AF on protein expressions associated with lipolysis such as ATGL and HSL. The protein expression of ATGL was significantly decreased by MDI stimulation, whereas ATGL expression was increased by treatment of QU or IQ in the differentiated 3T3-L1 cells. In addition, MDI stimulation significantly decreased phosphorylation of HSL, compared with MDI-non-stimulation. However, treatment with IQ significantly increased phosphorylation of HSL in the differentiated 3T3-L1 cells. In particular, IQ significantly up-regulated lipolysis- associated protein expressions, both ATGL and HSL.

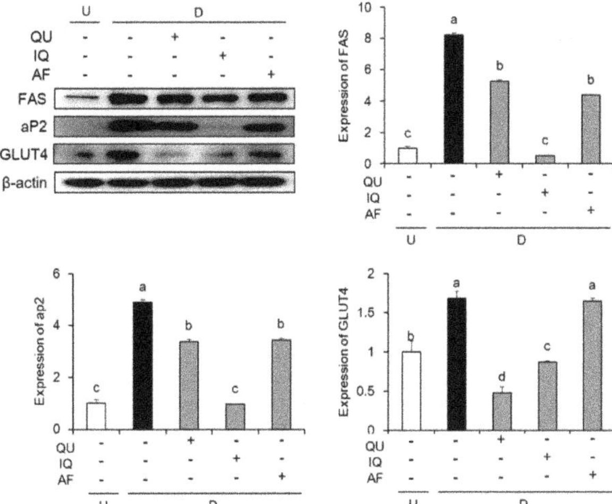

Figure 5. Effects of flavonoids from *Acer okamotoanum* (10 µg/mL) on lipogenesis-related protein expressions in differentiated 3T3-L1 cells. Adipocyte differentiation was induced by treatment with MDI media in the absence or presence of flavonoids from *A. okamotoanum* during 2 days. The MDI media was then replaced with insulin media, and it was changed four times for every 2 days. Data are expressed as the mean ± standard deviation. [a–d] Means with different letters indicate significant differences ($P < 0.05$) by Duncan's multiple range test. β-actin was used as a loading control. Relative expression levels were normalized to the β-actin levels and the normal group. Normal group indicates non-differentiated cells, whereas control group indicates the differentiated cells by treatment of MDI media. U: Undifferentiation; D: Differentiation; QU: Quercitrin; IQ: Isoquercitrin; AF: Afzelin.

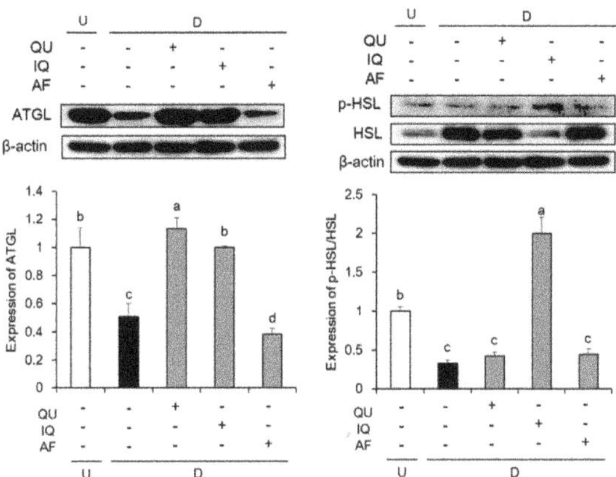

Figure 6. Effects of flavonoids from *Acer okamotoanum* (10 µg/mL) on lipolysis associated protein expressions in differentiated 3T3-L1 cells. Adipocyte differentiation was induced by treatment with MDI media in the absence or presence of flavonoids from *A. okamotoanum* during 2 days. The MDI media was then replaced with insulin media, and it was changed four times for every 2 days. Data are expressed as the mean ± standard deviation. [a–d] Means with different letters indicate significant differences ($P < 0.05$) by Duncan's multiple range test. β-actin was used as a loading control. Relative expression levels were normalized to the β-actin levels and the normal group. Normal group indicates non-differentiated cells, whereas control group indicates the differentiated cells by treatment of MDI media. U: Undifferentiation; D: Differentiation; QU: Quercitrin; IQ: Isoquercitrin; AF: Afzelin.

2.5. Effects of Flavonoids from A. okamotoanum on Activation of AMPK Signaling

We further investigated the effect of flavonoids on AMPK signaling pathway. The expressions of AMPK and ACC phosphorylation were determined by western blotting (Figure 7). The levels of AMPK phosphorylation was significantly lower in differentiated 3T3-L1 cells. On the other hand, treatment with flavonoids including QU and IQ showed significant high levels of AMPK phosphorylation compared with control group. In addition, three flavonoids-treated groups also significantly elevated the levels of ACC phosphorylation, indicating that three flavonoids up-regulated downstream target of AMPK pathway.

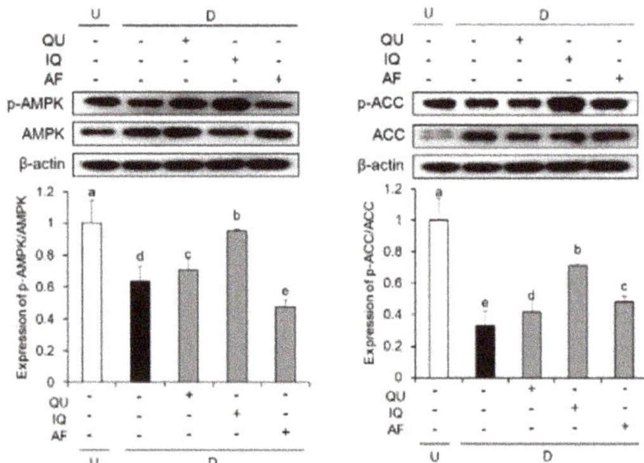

Figure 7. Effects of flavonoids from *Acer okamotoanum* (10 µg/mL) on AMPK signaling in differentiated 3T3-L1 cells. Adipocyte differentiation was induced by treatment with MDI media in the absence or presence of flavonoids from *A. okamotoanum* during 2 days. The MDI media was then replaced with insulin media, and it was changed four times for every 2 days. Data are expressed as the mean ± standard deviation. [a–e] Means with different letters indicate significant differences ($P < 0.05$) by Duncan's multiple range test. β-actin was used as a loading control. β-actin was used as a loading control. Relative expression levels were normalized to the β-actin levels and the normal group. Normal group indicates non-differentiated cells, whereas control group indicates the differentiated cells by treatment of MDI media. U: Undifferentiation; D: Differentiation; QU: Quercitrin; IQ: Isoquercitrin; AF: Afzelin.

3. Discussion

Obesity is caused by an excessive accumulation of lipid in adipose tissue. Adipocytes function for lipid homeostasis and energy balance by storing TG or releasing free fatty acids [19]. However, abnormal increase in number and size of differentiated adipocytes from preadipocytes has resulted in obesity by increasing of adipose tissue mass [20]. To prevent obesity, balance of adipogenesis and lipolysis in adipocytes plays important roles. Inhibition of preadipocyte differentiation as well as inductions of lipolysis has been focused on therapeutic strategies for treating obesity [21]. In particular, differentiation of 3T3-L1 preadipocytes induced by MDI media showed the characteristics of mature adipocyte in the growth, metabolism, and lipid accumulation [22]. Therefore, 3T3-L1 preadipocyte is widely used as an obesity-related study.

Recently, various dietary supplements are focused on the treatment of obesity. In particular, safe and effective natural products have been used worldwide as preventive agents for obesity and its related metabolic diseases [23]. *A. okamotoanum* is an endemic natural product and contains various bioactive compounds such as flavonoids [13]. We previously isolated three flavonoid glycosides such as QU, IQ, and AF from EtOAc fraction of *A. okamotoanum* [18]. QU and IQ are quercetin containing rhamnoside and glucoside, respectively, and AF is kaempferol with rhamnoside. However, anti-obesity effect and mechanisms of these flavonoids from *A. okamotoanum* including QU, IQ, and AF have not been investigated. In this study, we investigated the effects of flavonoids from *A. okamotoanum* on lipid accumulation, adipogenesis, lipolysis, and AMPK signaling under differentiated 3T3-L1 cells. Differentiated 3T3-L1 cells induced by MDI media showed significant increase of lipid accumulation, compared with non-differentiated 3T3-L1 cells. However, treatment of three flavonoids significantly decreased contents of TG. It indicated the promising role as an anti-obesity agent by inhibition of lipid accumulation.

To determine the concentration of flavonoids, the cytotoxicity assay was carried out. Three flavonoids used in the present study had no significant cytotoxicity up to concentrations of 10 μg/mL (Figure 1). We previously investigated that three flavonoids had no cytotoxicity up to concentrations of 50 μg/mL in 3T3-L1 cells (data not shown). In addition, previous studies examined anti-adipogenic effects of IQ at concentrations of 5–100 μM in 3T3-L1 cells, indicating no cytotoxicity under higher concentrations used in this study [24,25]. Therefore, in this study, inhibition of triglyceride accumulation by flavonoids was probably related to regulation of adipogenesis from 3T3-L1 preadipocytes to differentiated 3T3-L1 cells without cytotoxicity. In addition, we designed experimental schedule in measurement of cytotoxicity with reference to previous other studies [26,27].

C/EBPs and PPARs family are essential factors for adipogenesis [6]. Adipogenesis accelerates differentiation of preadipocytes into adipocytes, resulting in intracellular lipid accumulation [5,6]. During early stage of adipogenesis, cAMP agonist (IBMX) and glucocorticoids (dexamethasone) in the MDI directly induce expression of C/EBPβ [28,29]. C/EBPβ induced by MDI activates the expression of two major adipogenic transcription factors such as C/EBPα and PPARγ [30]. C/EBPα is intimately related to PPARγ activity, and also deficiency of C/EBPα in the adipocytes inhibits adipogenesis [28,31]. PPARγ is a nuclear receptor related with differentiation of adipose tissue, and it regulates the transcription of target genes associated with lipid homeostasis [32]. In addition, it is widely known that differentiated 3T3-L1 cells by treatment with MDI activate adipogenic key transcription factors including C/EBPα, C/EBPβ, and PPARγ [29]. In our results, we confirmed that differentiated 3T3-L1 cells showed up-regulation of adipogenic key transcription factors including C/EBPα, C/EBPβ, and PPARγ, whereas treatment with flavonoids showed down-regulation of these adipogenic transcription factors. In particular, IQ-treated group in differentiated 3T3-L1 cells reduced more effectively the expressions of three adipogenic transcription factors including C/EBPα, C/EBPβ, and PPARγ than other flavonoids. It indicated that IQ could inhibit the lipid accumulation by down-regulations of the adipogenic transcription factors.

Adipogenic key transcription factors such as C/EBPs and PPARs family activate the adipocyte specific proteins such as FAS, aP2, and GLUT4. C/EBPα and PPARγ activation is known to regulate the lipid metabolism-related genes such as FAS and aP2 [33]. FAS is one of the lipogenesis associated enzymes to facilitate the synthesis and cytoplasmic storage of massive amounts of TG during differentiation process [7,8]. C/EBPs activate expression of GLUT4, that is required during insulin-dependent glucose uptake [34,35]. Expression of GLUT4 is increased by activation of adipogenic key transcription factor such as C/EBPα in adipocyte differentiation [36]. Previous study reported that decrease of C/EBPα inhibited expression of GLUT, which would decrease glucose transport in adipocytes [36,37]. In our study, treatment of IQ decreased the expression of C/EBPα and GLUT4. Therefore, IQ inhibited the accumulation of TGs by decreasing C/EBPα-activated GLUT4 expression. GLUT4 is a member of glucose transporter in adipose tissue, and involved in the insulin-stimulated glucose uptake [38,39]. On the basis of these evidences, the inhibitory effect of IQ on TG accumulation is related to GLUT4 by regulation of C/EBPα expressions. In addition, differentiated 3T3-L1 cells treated with IQ effectively reduced expressions of aP2 and FAS by down-regulation of adipogenic key transcription factors among other flavonoids.

Lipolysis in the adipose tissue is regulated by ATGL and HSL that promote lipolysis by acting on the lipid droplet [9]. Expression of ATGL initiates lipolysis by cleaving the fatty acid from triacylglycerol, and then HSL catalyzes the hydrolysis of diacylglycerol [40]. Translocation of HSL is regulated by increase of cAMP and activation of protein kinase A, especially regulation of HSL is a major rate-limiting step in adipocytes during lipolysis [40,41]. Therefore, activation of ATGL and HSL are the strategy for treatment of obesity by activation of lipolysis. In this study, QU and IQ significantly up-regulated expression of ATGL. Furthermore, differentiated 3T3-L1 cells treated with IQ significantly elevated ATGL as well as phosphorylation of HSL. These results indicated that QU and IQ promote lipolysis by regulation of ATGL and HSL.

AMPK is a common regulator of lipid metabolism in adipocytes [10,42]. AMPK involves in various lipid metabolisms such as adipogenesis, lipolysis, glucose uptake, fatty acid β-oxidation, and adipokine secretion in the adipose tissue [42]. In the adipogenesis, activation of AMPK can inhibit adipocyte differentiation via reductions of transcription key factors such as PPARγ and C/EBPα [43,44]. It has been reported that activation of AMPK directly down-regulated adipogenic key transcription factors and inhibited adipocyte differentiation [43,44]. In the lipolysis process, activation of AMPK promotes lipolysis by phosphorylation of HSL and ATGL, resulting in fatty acid degradation [10,45]. In addition, AMPK enhances inactivation of ACC by phosphorylation and decrease in the free fatty acid production [11]. ACC is known as reduction of fatty acid synthesis and elongation in adipocytes [46,47]. Therefore, activation of AMPK leads to attenuation of adipogenesis and increase of lipolysis, thereby regulation of AMPK signaling pathway plays important roles for prevention and treatment of obesity [48]. In our study, IQ-treated group significantly increased phosphorylation of both AMPK and ACC among other flavonoids from *A. okamotoanum* in the differentiated 3T3-L1 cells. Therefore, we suggest that IQ effectively suppresses the adiopgenesis and promotes lipolysis via regulation of AMPK pathway, resulting in inhibition of lipid accumulation.

Dietary flavonoids from natural products have anti-obesity effects via regulation of various molecular pathways [49]. Flavonoids are the most abundant polyphenols in natural products, and they can be classified into six groups including flavones, flavonols, flavanones, flavanonols, flavanols, and antocyanin [50]. Dietary flavonoids from natural products exert anti-obesity effects via regulation of various molecular pathways in 3T3-L1 cells [49]. Flavones such as apigenin, balcalein, and lueteolin inhibited lipid accumulation by downregulation of adipogenesis, activation of AMPK, reduction of Akt-C/EBPa-GLUT4 signaling-medicated glucose uptake, and inflammatory responses [35,51,52]. Flavanones including hesperitin and naringin exert antiobesity effects via activating PPAR and up-regulating adiponectin in 3T3-L1 cells [53]. Anthocyanidins isolated from black soybean such as cyanidine, peonidin, and its glucoside reduced preadipocyte differentiation. Flavan-3-ols such as epigalloocatechin gallate in the green tea inhibited adipogenesis via regulation of Wnt/β-catenin pathway in 3T3-L1 cells [49].

In our study, flavonoids from *A. okamotoanum* including QU, IQ, and AF are classified as flavonols. Flavonols including quercetin, kaempferol, and rutin are flavonoids with a ketone group in onions, tomatoes, apples, and berries [50]. Flavonols such as quercetin, kaempferol, and its glycosides inhibited adipogenesis or promoted AMPK signaling in adipocytes [54,55]. In comparison of aglycone and its glycosides, quercetin glucoside showed strong activity in reduction of lipid accumulation and inhibited adipogenic factors such as C/EBP-β, -α, and aP2 than its aglycone, quercetin [54]. Wnt/β-catenin signaling is mandatory in adipogenesis, and it inhibited preadipocyte differentiation by regulation of β-catenin [25]. Lee et al. reported that IQ suppressed the adipogenesis in 3T3-L1 cells via the inhibition of Wnt/β-catenin signaling, regulation of lipid metabolism-related factors such as PPARγ, C/EBPα, SREBP-1, adiponectin, resistin, visfatin, and improvement of insulin resistance [54,56]. In addition, our results indicated the inhibitory effect of lipid accumulation of flavonoids from *A. okamotoanum* including QU, IQ, and AF, in particular IQ, by regulation of adipogenesis, lipolysis, and AMPK signaling. IQ has hydroxyl group at R1 position and glycosylated glucoside at R2 position. We suggest that presence of hydroxyl group at R1 position and glycosylation of glucoside at R2 position are higher anti-obesity effects compared with flavonols non-linked hydroxyl group at R1 and glucoside at R2 such as QU and AF. Previous study demonstrated that extract of *A. okamotoanum* leaf inhibited the anti-adipogenesis via down-regulation of PPARγ and C/EBPα [14]. However, anti-obesity effects and molecular mechanisms of flavonoids isolated from *A. okamotoanum* have not been demonstrated. This study indicated that the flavonoids of *A. okamotoanum*, in particular IQ, exerted anti-obesity effects by regulation of adipogenesis and lipolysis.

4. Materials and Methods

4.1. Sample Preparation

The isolation and identification of three flavonoids such as QU, IQ, and AF were based on our previous study [18]. In brief, *A. okamotoanum* was obtained from Ulleung-do, Gyeongsangbuk-do, Korea. A voucher specimen has been deposited at the Department of Plant Science and Technology, Chung-Ang University, Anseong, Korea (Voucher No. LEE 2014-04). Aerial parts of *A. okamotoanum* (995.4 g) were extracted with methanol (MeOH) by filtering and evaporation *in vaccum*. EtOAc fraction of *A. okamotoanum* (35.0 g) was partitioned from MeOH extract, and then the fraction of *A. okamotoanum* was isolated by silica gel column chromatography. Three flavonoids were identified as QU, IQ, and AF by ^{1}H-NMR, ^{13}C-NMR, and MS data. In addition, quantitative analysis of three flavonoids such as QU (48.26 µg/g), IQ (5.84 µg/g), and AF (2.66 µg/g) was determined by HPLC/UV using 0.45 µM syringe filter. These three flavonoids including QU, IQ, and AF were isolated from *A. okamotoanum* of EtOAc fraction [18]. Figure 8 and Table 1 expressed the structures of three flavonoids from *A. okamotoanum* such as QU, IQ, and AF.

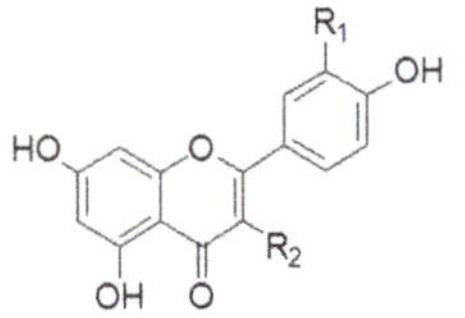

Figure 8. The structures of flavonoids from *Acer okamotoanum*.

Table 1. The Flavonoids from *Acer okamotoanum*.

Compound	R_1	R_2
QU	OH	*O*-Rhamnoside
IQ	OH	*O*-Glucoside
AF	H	*O*-Rhamnoside

4.2. Reagents

Fetal bovine serum (FBS), bovine calf serum (BCS), Dulbecco's modified eagle medium (DMEM), penicillin-streptomycin, and trypsin-EDTA solution were purchased from Welgene (Daegu, Korea). Dexamethasone, 3-isobutyl-1-methylxanthine (IBMX), and insulin were purchased from Sigma Aldrich (St. Louis, MO, USA). 3-(4,5-Dimethyl-2-thiazolyl)-2,5-diphenyl-2H-tetrazolium bromide (MTT) was acquired from Bio Basic (Toronto, ON, Canada), dimethyl sulfoxide (DMSO) was purchased from Bio Pure (Burlington, ON, Canada), and primary and secondary antibodies were purchased from Cell Signaling Technology (Beverly, MA, USA).

4.3. Cell Culture and Differentiation

The mouse 3T3-L1 preadipocytes were purchased from American Type Culture Collection (Maryland, OH, USA). The cells were grown in DMEM supplemented with 10% BCS and 1% penicillin-streptomycin, at 37 °C in 5% CO_2 atmosphere incubator (Thermo Electron Corporation, Milford, MA, USA). The differentiation was induced by 0.5 mM IBMX, 1 µM dexamethasone, and 5 µg/mL insulin in DMEM containing 10% FBS (MDI media) under the presence or absence of flavonoids including QU, IQ, and AF. After 2 days, MDI media was replaced with DMEM supplemented with 10% FBS and 5 µg/mL insulin (insulin media). After during 8 days, the cell culture media was replaced with insulin media 4 times every 2 day.

4.4. Cell Viability

The cell viability was evaluated using a MTT assay [57]. The 3T3-L1 preadipocytes were seeded at a density of 1 X 10^5 cells/mL in the 24 well plate, and incubated for 24 h. Flavonoids such as QU, IQ, and AF were treated in the wells at various concentrations (1, 2.5, 10 µg/mL), respectively, and then incubated at 37 °C for 72 h. Cell culture media was removed, and then MTT solution was added in the wells. The formazan crystals were dissolved in DMSO and the absorbance was measured at 540 nm using microplate reader (Thermo Fisher Scientific, Vantaa, Finland).

4.5. Oil Red O Staining

After 8 days of differentiation, differentiated 3T3-L1 cells were washed with phosphate buffered saline (PBS), fixed with 10% formalin for 10 min and washed with PBS and 60% isopropanol. The fixed cells were stained with 0.6% Oil Red O solution for 20 min in the dark, washed 4 times with PBS and 60% isopropanol. The lipid droplets within the differentiated 3T3-L1 cells were visualized and photographed by microscopy. For quantitative analysis under Oil Red O staining was carried out by eluting with 100% isopropanol and measuring absorbance at 500 nm [58].

4.6. Western Blot Analysis

The differentiated 3T3-L1 cells were harvested using a cell scraper, lysed with radioimmunoprecipitation assay buffer containing protease inhibitor cocktail at 4 °C for 1 h. The protein concentration was determined using Bio-Rad protein assay (Bio-Rad, Hercules, CA, USA). Equal amounts of extracted proteins were separated on 8–13% sodium dodecyl sulfate-polyacrylamide gel electrophoresis, transferred onto a polyvinylidene fluoride membrane. The each membrane was blocked 5% skim milk at room temperature for 1 h, and then incubated primary antibodies such as β-actin, C/EBPα, C/EBPβ, PPARγ, FAS, ap2, GLUT4, ATGL, phospho-HSL, HSL, phospho-AMPK, AMPK, phospho-acetyl-CoA carboxylase (ACC), and ACC at 4 °C for overnight. The next day, each membrane was washed with PBS-T, and then incubated with secondary antibodies at room temperature for 1 h. The protein bands were activated with enhanced chemiluminescence solution, visualized with Davinch-chemi^{TM} Chemiluminescence Imaging System (Core Bio, Seoul, Korea), and quantified with Image-J program (National Institutes of Health, Bethesda, MD, USA).

4.7. Statistical Analysis

Values are expressed as mean ± standard deviation (SD). SPSS software (IBM SPSS Inc., Chicago, IL, USA) was used to perform statistics analysis. The results were compared by one-way analysis of variance (ANOVA) followed by Duncan's multiple range analysis among the groups using SPSS program. Statistically significance was considered as $P < 0.05$.

5. Conclusions

In conclusion, this study demonstrated that three flavonoids from *A. okamotoanum* such as QU, IQ, and AF suppressed adipogenesis by inhibition of adipogenic key transcription factors and lipogenesis-related proteins, resulting in decrease of TG accumulation in differentiated 3T3-L1 cells. In addition, flavonoids from *A. okamotoanum* promoted lipolysis by down-regulation of lipolytic genes expressions. In particular, IQ effectively suppressed adipogenesis and induced lipolysis by regulation of AMPK signaling among other flavonoids of *A. okamotoanum*. The present study suggests that the flavonoids from *A. okamotoanum* can be the promising agents for the prevention and treatment of obesity.

Author Contributions: J.H.K. carried out experiments and wrote the paper. S.L. contributed sample preparation and analysis. E.J.C. supervised the work and revised the manuscript. All authors read and approved the manuscript in its current form.

Funding: This work was supported by the Basic Science Research Program through the National Research Foundation of Korea (NRF) funded by the Ministry of Education (2015R1D1A1A01058868), Republic of Korea. This research was supported by Global PH.D Fellowship Program through the National Research Foundation of Korea (NRF) funded by the Ministry of Education (2016_H1A2A1906940).

Conflicts of Interest: The authors declare no conflict of interest.

References

1. Alford, S.; Patel, D.; Perakakis, N.; Mantzoros, C.S. Obesity as a risk factor for Alzheimer's disease: Weighing the evidence. *Obes. Rev.* **2018**, *19*, 269–280. [CrossRef] [PubMed]
2. Yatsuya, H.; Li, Y.; Hilawe, E.H.; Ota, A.; Wang, C.; Chiang, C.; Zhang, Y.; Uemura, M.; Osako, A.; Ozaki, Y.; et al. Global trend in overweight and obesity and its association with cardiovascular disease incidence. *Circ. J.* **2014**, *78*, 2807–2818. [CrossRef] [PubMed]
3. Malik, V.S.; Willett, W.C.; Hu, F.B. Global obesity: Trends, risk factors and policy implications. *Nat. Rev. Endocrinol.* **2013**, *9*, 13–27. [CrossRef] [PubMed]
4. Luo, L.; Liu, M. Adipose tissue in control of metabolism. *J. Endocrinol.* **2016**, *231*, 77–99. [CrossRef]
5. Antonopoulos, A.S.; Tousoulis, D. The molecular mechanisms of obesity paradox. *Cardiovasc. Res.* **2017**, *113*, 1074–1086. [CrossRef]
6. de Sá, P.M.; Richard, A.J.; Hang, H.; Stephens, J.M. Transcriptional regulation of adipogenesis. *Compr. Physiol.* **2017**, *7*, 635–674.
7. Moseti, D.; Regassa, A.; Kim, W.K. Molecular regulation of adipogenesis and potential anti-adipogenic bioactive molecules. *Int. J. Mol. Sci.* **2016**, *17*, 124. [CrossRef]
8. Ali, A.T.; Hochfeld, W.E.; Myburgh, R.; Pepper, M.S. Adipocyte and adipogenesis. *Eur. J. Cell Biol.* **2013**, *92*, 229–236. [CrossRef]
9. Nielsen, T.S.; Jessen, N.; Jørgensen, J.O.; Møller, N.; Lund, S. Dissecting adipose tissue lipolysis: Molecular regulation and implications for metabolic disease. *J. Mol. Endocrinol.* **2014**, *52*, 199–222. [CrossRef]
10. Jeon, S.M. Regulation and function of AMPK in physiology and diseases. *Exp. Mol. Med.* **2016**, *48*, e245. [CrossRef]
11. Park, H.; Kaushik, V.K.; Constant, S.; Prentki, M.; Przybytkowski, E.; Ruderman, N.B.; Saha, A.K. Coordinate regulation of malonyl-CoA decarboxylase, sn-glycerol-3-phosphate acyltransferase, and acetyl-CoA carboxylase by AMP-activated protein kinase in rat tissues in response to exercise. *J. Biol. Chem.* **2002**, *277*, 32571–32577. [CrossRef]
12. Khera, R.; Murad, M.H.; Chandar, A.K.; Dulai, P.S.; Wang, Z.; Prokop, L.J.; Loomba, R.; Camilleri, M.; Singh, S. Association of pharmacological treatments for obesity with weight loss and adverse events: A systematic review and meta-analysis. *JAMA* **2016**, *315*, 2424–2434. [CrossRef] [PubMed]
13. Takayama, K.; Sun, B.Y.; Stuessy, T.F. Genetic consequences of anagenetic speciation in *Acer okamotoanum* (Sapindaceae) on Ullung Island, Korea. *Ann. Bot.* **2012**, *109*, 321–330. [CrossRef] [PubMed]
14. Kim, E.J.; Kang, M.J.; Seo, Y.B.; Nam, S.W.; Kim, G.D. *Acer okamotoanum* Nakai leaf extract inhibits adipogenesis via suppressing expression of PPARγ and C/EBPα in 3T3-L1 cells. *J. Microbiol. Biotechnol.* **2018**, *28*, 1645–1653. [CrossRef] [PubMed]
15. Yang, H.; Hwang, I.; Koo, T.H.; Ahn, H.J.; Kim, S.; Park, M.J.; Choi, W.S.; Kang, H.Y.; Choi, I.G.; Choi, K.C.; et al. Beneficial effects of *Acer okamotoanum* sap on L-NAME-induced hypertension-like symptoms in a rat model. *Mol. Med. Rep.* **2012**, *5*, 427–431.
16. An, B.S.; Kang, J.H.; Yang, H.; Yang, M.P.; Jeung, E.B. Effects of *Acer okamotoanum* sap on the function of polymorphonuclear neutrophilic leukocytes in vitro and in vivo. *Mol. Med. Rep.* **2013**, *7*, 654–658. [CrossRef]
17. Jin, W.; Thuong, P.T.; Su, N.D.; Min, B.S.; Son, K.H.; Chang, H.W.; Kim, H.P.; Kang, S.S.; Sok, D.E.; Bae, K. Antioxidant activity of cleomiscosins A and C isolated from *Acer okamotoanum*. *Arch. Pharm. Res.* **2007**, *30*, 275–281. [CrossRef]
18. Lee, J.; Lee, D.G.; Rodriguez, J.P.; Park, J.Y.; Cho, E.J.; Jacinto, S.D.; Lee, S. Determination of flavonoids in Acer okamotoanum and their aldose reductase inhibitory activities. *Hort. Environ. Biotechnol.* **2018**, *59*, 131–137. [CrossRef]
19. Rosen, E.D.; Spiegelman, B.M. Adipocytes as regulators of energy balance and glucose homeostasis. *Nature* **2006**, *444*, 847–853. [CrossRef]

20. Trujillo, M.E.; Scherer, P.E. Adipose tissue-derived factors: Impact on health and disease. *Endocr. Rev.* **2006**, *27*, 762–778. [CrossRef]

21. Billon, N.; Dani, C. Developmental origins of the adipocyte lineage: New insights from genetics and genomics studies. *Stem Cell Rev.* **2012**, *8*, 55–66. [CrossRef] [PubMed]

22. Poulos, S.P.; Dodson, M.V.; Hausman, G.J. Cell line models for differentiation: Preadipocytes and adipocytes. *Exp. Biol. Med. (Maywood)* **2010**, *235*, 1185–1193. [CrossRef] [PubMed]

23. Marimoutou, M.; Le Sage, F.; Smadja, J.; d'Hellencourt, C.L.; Gonthier, M.P.; Robert-Da Silva, C. Antioxidant polyphenol-rich extracts from the medicinal plants *Antirhea borbonica, Doratoxylon apetalum* and *Gouania mauritiana* protect 3T3-L1 preadipocytes against H_2O_2, TNFα and LPS inflammatory mediators by regulating the expression of superoxide dismutase and NF-κB genes. *J. Inflamm. (Lond.)* **2015**, *12*, 10.

24. Cai, H.D.; Su, S.L.; Guo, S.; Zhu, Y.; Qian, D.W.; Tao, W.W.; Duan, J.A. Effect of flavonoids from *Abelmoschus manihot* on proliferation, differentiation of 3T3-L1 preadipocyte and insulin resistance. *Zhongguo Zhong Yao Za Zhi* **2016**, *41*, 4635–4641. [PubMed]

25. Lee, S.H.; Kim, B.; Oh, M.J.; Yoon, J.; Kim, H.Y.; Lee, K.J.; Lee, J.D.; Choi, K.Y. *Persicaria hydropiper* (L.) spach and its flavonoid components, isoquercitrin and isorhamnetin, activate the Wnt/β-catenin pathway and inhibit adipocyte differentiation of 3T3-L1 cells. *Phytother. Res.* **2011**, *25*, 1629–1635. [CrossRef] [PubMed]

26. Hsu, H.F.; Tsou, T.C.; Chao, H.R.; Kuo, Y.T.; Tsai, F.Y.; Yeh, S.C. Effects of 2,3,7,8-tetrachlorodibenzo-p-dioxin on adipogenic differentiation and insulin-induced glucose uptake in 3T3-L1 cells. *J. Hazard Mater.* **2010**, *182*, 649–655. [CrossRef]

27. Han, M.H.; Jeong, J.S.; Jeong, J.W.; Choi, S.H.; Kim, S.O.; Hong, S.H.; Park, C.; Kim, B.W.; Choi, Y.H. Ethanol extracts of *Aster yomena* (Kitam.) Honda inhibit adipogenesis through the activation of the AMPK signaling pathway in 3T3-L1 preadipocytes. *Drug Discov. Ther.* **2017**, *11*, 281–287. [CrossRef]

28. Farmer, S.R. Transcriptional control of adipocyte formation. *Cell Metab.* **2006**, *4*, 263–273. [CrossRef]

29. Chen, J.; Bao, C.; Kim, J.T.; Cho, J.S.; Qiu, S.; Lee, H.J. Sulforaphene inhibition of adipogenesis via hedgehog signaling in 3T3-L1 adipocytes. *J. Agric. Food Chem.* **2018**, *66*, 11926–11934. [CrossRef]

30. Lefterova, M.I.; Zhang, Y.; Steger, D.J.; Schupp, M.; Schug, J.; Cristancho, A.; Feng, D.; Zhuo, D.; Stoeckert, C.J.; Liu, X.S.; et al. PPARgamma and C/EBP factors orchestrate adipocyte biology via adjacent binding on a genome-wide scale. *Genes Dev.* **2008**, *22*, 2941–2952. [CrossRef]

31. Wu, Z.; Rosen, E.D.; Brun, R.; Hauser, S.; Adelmant, G.; Troy, A.E.; McKeon, C.; Darlington, G.J.; Spiegelman, B.M. Cross-regulation of C/EBP alpha and PPAR gamma controls the transcriptional pathway of adipogenesis and insulin sensitivity. *Mol. Cell* **1999**, *3*, 151–158. [CrossRef]

32. Rosen, E.D.; Sarraf, P.; Troy, A.E.; Bradwin, G.; Moore, K.; Milstone, D.S.; Spiegelman, B.M.; Mortensen, R.M. PPAR gamma is required for the differentiation of adipose tissue in vivo and in vitro. *Mol. Cell* **1999**, *4*, 611–617. [CrossRef]

33. Rosen, E.D.; Walkey, C.J.; Puigserver, P.; Spiegelman, B.M. Transcriptional regulation of adipogenesis. *Genes Dev.* **2000**, *14*, 1293–1307. [PubMed]

34. Lee, J.O.; Lee, S.K.; Kim, J.H.; Kim, N.; You, G.Y.; Moon, J.W.; Kim, S.J.; Park, S.H.; Kim, H.S. Metformin regulates glucose transporter 4 (GLUT4) translocation through AMP-activated protein kinase (AMPK)-mediated Cbl/CAP signaling in 3T3-L1 preadipocyte cells. *J. Biol. Chem.* **2012**, *287*, 44121–44129. [CrossRef]

35. Nakao, Y.; Yoshihara, H.; Fujimori, K. Suppression of very early stage of adipogenesis by baicalein, a plant-derived flavonoid through reduced Akt-C/EBPα-GLUT4 signaling-mediated glucose uptake in 3T3-L1 adipocytes. *PLoS ONE* **2016**, *11*, e0163640. [CrossRef]

36. Kaestner, K.H.; Christy, R.J.; Lane, M.D. Mouse insulin-responsive glucose transporter gene: Characterization of the gene and trans-activation by the CCAAT/enhancer binding protein. *Proc. Natl. Acad. Sci. USA* **1990**, *87*, 251–255. [CrossRef]

37. Watanabe, M.; Hisatake, M.; Fujimori, K. Fisetin suppresses lipid accumulation in mouse adipocytic 3T3-L1 cells by repressing GLUT4-mediated glucose uptake through inhibition of mTOR-C/EBPα signaling. *J. Agric. Food Chem.* **2015**, *63*, 4979–4987. [CrossRef]

38. Abel, E.D.; Peroni, O.; Kim, J.K.; Kim, Y.B.; Boss, O.; Hadro, E.; Minnemann, T.; Shulman, G.I.; Kahn, B.B. Adipose-selective targeting of the GLUT4 gene impairs insulin action in muscle and liver. *Nature* **2001**, *409*, 729–733. [CrossRef]

39. Qiu, J.; Wang, Y.M.; Shi, C.M.; Yue, H.N.; Qin, Z.Y.; Zhu, G.Z.; Cao, X.G.; Ji, C.B.; Cui, Y.; Guo, X.R. NYGGF4 (PID1) effects on insulin resistance are reversed by metformin in 3T3-L1 adipocytes. *J. Bioenergy Biomembr.* **2012**, *44*, 665–671. [CrossRef]

40. Gaidhu, M.P.; Anthony, N.M.; Patel, P.; Hawke, T.J.; Ceddia, R.B. Dysregulation of lipolysis and lipid metabolism in visceral and subcutaneous adipocytes by high-fat diet: Role of ATGL, HSL, and AMPK. *Am. J. Physiol. Cell Physiol.* **2010**, *298*, 961–971. [CrossRef]

41. Arner, P. Human fat cell lipolysis: Biochemistry, regulation and clinical role. *Best Pract. Res. Clin. Endocrinol. Metab.* **2005**, *19*, 471–482. [CrossRef] [PubMed]

42. Daval, M.; Foufelle, F.; Ferré, P. Functions of AMP-activated protein kinase in adipose tissue. *J. Physiol.* **2006**, *574*, 55–62. [CrossRef] [PubMed]

43. Habinowski, S.A.; Witters, L.A. The effects of AICAR on adipocyte differentiation of 3T3-L1 cells. *Biochem. Biophys. Res. Commun.* **2001**, *286*, 852–856. [CrossRef] [PubMed]

44. Lee, H.; Kang, R.; Bae, S.; Yoon, Y. AICAR, an activator of AMPK, inhibits adipogenesis via the WNT/β-catenin pathway in 3T3-L1 adipocytes. *Int. J. Mol. Med.* **2011**, *28*, 65–71.

45. Xie, M.; Roy, R. AMP-activated kinase regulates lipid droplet localization and stability of adipose triglyceride lipase in *C. elegans* Dauer Larvae. *PLoS ONE* **2015**, *10*, e0130480. [CrossRef]

46. Russo, G.L.; Russo, M.; Ungaro, P. AMP-activated protein kinase: A target for old drugs against diabetes and cancer. *Biochem. Pharmacol.* **2013**, *86*, 339–350. [CrossRef]

47. Saha, A.K.; Ruderman, N.B. Malonyl-CoA and AMP-activated protein kinase: An expanding partnership. *Mol. Cell Biochem.* **2003**, *253*, 65–70. [CrossRef]

48. Luo, Z.; Zang, M.; Guo, W. AMPK as a metabolic tumor suppressor: Control of metabolism and cell growth. *Future Oncol.* **2010**, *6*, 457–470. [CrossRef]

49. Khalilpourfarshbafi, M.; Gholami, K.; Murugan, D.D.; Sattar, M.Z.A.; Abdullah, N.A. Differential effects of dietary flavonoids on adipogenesis. *Eur. J. Nutr.* **2019**, *58*, 5–25. [CrossRef]

50. Panche, A.N.; Diwan, A.D.; Chandra, S.R. Flavonoids: An overview. *J. Nutr. Sci.* **2016**, *5*, e47. [CrossRef]

51. Ono, M.; Fujimori, K. Antiadipogenic effect of dietary apigenin through activation of AMPK in 3T3-L1 cells. *J. Agric. Food Chem.* **2011**, *59*, 13346–13352. [CrossRef] [PubMed]

52. Park, H.S.; Kim, S.H.; Kim, Y.S.; Ryu, S.Y.; Hwang, J.T.; Yang, H.J.; Kim, G.H.; Kwon, D.Y.; Kim, M.S. Luteolin inhibits adipogenic differentiation by regulating PPARgamma activation. *Biofactors* **2009**, *35*, 373–379. [CrossRef] [PubMed]

53. Liu, L.; Shan, S.; Zhang, K.; Ning, Z.Q.; Lu, X.P.; Cheng, Y.Y. Naringenin and hesperetin, two flavonoids derived from Citrus aurantium up-regulate transcription of adiponectin. *Phytother. Res.* **2008**, *22*, 1400–1403. [CrossRef]

54. Lee, C.W.; Seo, J.Y.; Lee, J.; Choi, J.W.; Cho, S.; Bae, J.Y.; Sohng, J.K.; Kim, S.O.; Kim, J.; Park, Y.I. 3-O-Glucosylation of quercetin enhances inhibitory effects on the adipocyte differentiation and lipogenesis. *Biomed. Pharmacother.* **2017**, *95*, 589–598. [CrossRef] [PubMed]

55. Torres-Villarreal, D.; Camacho, A.; Castro, H.; Ortiz-Lopez, R.; de la Garza, A.L. Anti-obesity effects of kaempferol by inhibiting adipogenesis and increasing lipolysis in 3T3-L1 cells. *J. Physiol. Biochem.* **2019**, *75*, 83–88. [CrossRef]

56. Kennell, J.A.; MacDougald, O.A. Wnt signaling inhibits adipogenesis through beta-catenin-dependent and -independent mechanisms. *J. Biol. Chem.* **2005**, *280*, 24004–24010. [CrossRef]

57. Mosmann, T. Rapid colorimetric assay for cellular growth and survival: Application to proliferation and cytotoxicity assays. *J. Immunol. Meth.* **1983**, *65*, 55–63. [CrossRef]

58. Ramírez-Zacarías, J.L.; Castro-Muñozledo, F.; Kuri-Harcuch, W. Quantitation of adipose conversion and triglycerides by staining intracytoplasmic lipids with Oil red O. *Histochemistry* **1992**, *97*, 493–497. [CrossRef]

Sample Availability: Samples of the compounds are available from the authors.

 *molecules*

Article

Vasodilatory Effects and Mechanisms of Action of *Bacopa monnieri* Active Compounds on Rat Mesenteric Arteries

Natakorn Kamkaew [1,2], Tamkeen Urooj Paracha [3], Kornkanok Ingkaninan [4], Neti Waranuch [5] and Krongkarn Chootip [1,*]

[1] Department of Physiology, Faculty of Medical Science, Naresuan University, Phitsanulok 65000, Thailand; natakornk@gmail.com
[2] Division of Physiology, School of Medical Sciences, University of Phayao, Phayao 56000, Thailand
[3] Department of Pharmacy Practice, Faculty of Pharmaceutical Sciences, Naresuan University, Phitsanulok 65000, Thailand; ctamz@hotmail.com
[4] Department of Pharmaceutical Chemistry and Pharmacognosy, Faculty of Pharmaceutical Sciences and Center of Excellence for Innovation in Chemistry, Naresuan University, Phitsanulok 65000, Thailand; kornkanoki@nu.ac.th
[5] Cosmetics and Natural Products Research Center, Department of Pharmaceutical Technology and Center of Excellence for Innovation in Chemistry, Faculty of Pharmaceutical Sciences, Naresuan University, Phitsanulok 65000, Thailand; netiw@nu.ac.th
*   Correspondence: krongkarnc@nu.ac.th; Tel.: +66-55-964658

Academic Editor: H.P. Vasantha Rupasinghe
Received: 17 May 2019; Accepted: 11 June 2019; Published: 15 June 2019

Abstract: *B. monnieri* extract (BME) is an abundant source of bioactive compounds, including saponins and flavonoids known to produce vasodilation. However, it is unclear which components are the more effective vasodilators. The aim of this research was to investigate the vasorelaxant effects and mechanisms of action of saponins and flavonoids on rat isolated mesenteric arteries using the organ bath technique. The vasorelaxant mechanisms, including endothelial nitric oxide synthase (eNOS) pathway and calcium flux were examined. Saponins (bacoside A and bacopaside I), and flavonoids (luteolin and apigenin) at 0.1–100 μM caused vasorelaxation in a concentration-dependent manner. Luteolin and apigenin produced vasorelaxation in endothelial intact vessels with more efficacy (E_{max} 99.4 $\pm$ 0.7 and 95.3 $\pm$ 2.6%) and potency (EC_{50} 4.35 $\pm$ 1.31 and 8.93 $\pm$ 3.33 μM) than bacoside A and bacopaside I (E_{max} 83.6 $\pm$ 2.9 and 79.9 $\pm$ 8.2%; EC_{50} 10.8 $\pm$ 5.9 and 14.6 $\pm$ 5.4 μM). Pretreatment of endothelial intact rings, with L-NAME (100 μM); an eNOS inhibitor, or removal of the endothelium reduced the relaxant effects of all compounds. In K^+-depolarised vessels suspended in Ca^{2+}-free solution, these active compounds inhibited $CaCl_2$-induced contraction in endothelial denuded arterial rings. Moreover, the active compounds attenuated transient contractions induced by 10 μM phenylephrine in Ca^{2+}-free medium containing EGTA (1 mM). Thus, relaxant effects occurred in both endothelial intact and denuded vessels which signify actions through both endothelium and vascular smooth muscle cells. In conclusion, the flavonoids have about twice the potency of saponins as vasodilators. However, in the BME, there is ~20 $\times$ the amount of vaso-reactive saponins and thus are more effective.

Keywords: luteolin; apigenin; bacoside A; bacopaside I; vasorelaxation

---

1. Introduction

Bacopa monnieri (L.) Wettst. or Brahmi, is an Ayurvedic medicine traditionally used as a memory enhancer. Along with memory improvement, it is known to promote mental health, as a neurotonic and cardiotonic agent. *B. monnieri* extract (BME) clearly has a cognitive enhancing potential and neuroprotective effects [1–16]. It has been shown to be antioxidant in rat brain [17,18] and to possess several pharmacological actions such as anti-depressant [19–21], anti-dementia [9], anti-cholinesterase [8,9], anti-hyperglycaemic [22] and anti-hyperlipidaemia [23]. *B. monnieri* appears to be non-toxic using haematological and blood biochemical diagnostics [24–26]. BME demonstrated cardioprotection, improved coronary blood flow, and protection against myocardial ischemia reperfusion injury [27,28]. Our recent work showed that BME acted as a vasodilator by releasing nitric oxide (NO) from endothelium and inhibiting Ca^{2+} influx and Ca^{2+} release from the sarcoplasmic reticulum (SR). These mechanisms mediated an acute decrease in blood pressure [29]. Also, daily oral BME (40 mg/kg) in rats for 8 weeks showed a significant increase in cerebral blood flow [30], which implies cerebrovascular dilation.

BME contains an abundance of bioactive compounds. They include dammarane-type triterpenoid saponins, jujubogenin and pseudojujubogenin glycosides. These saponins are predominantly bacopaside I and bacoside A, a mixture of bacoside A_3, bacopaside II, jujubogenin isomer of bacopasaponin C, and bacopasaponin C [31–33]. Other than saponins, flavonoids, essentially luteolin and apigenin are also present in *B. monnieri* [10,34–36]. Bacoside A_3 and bacopaside II relax rat mesenteric arteries [29] but the mechanism(s) of their relaxation are presently unknown. The flavonoids found in *B. monnieri* also relax rat aortae [37–41] but these experiments used a variety of protocols and vascular preparations. Therefore, it is important to make a side-by-side comparison of these flavonoids with the *B. monnieri* saponins using a resistance vessel type. For this we choose the mesenteric artery which better exemplifies actions on regional blood flow and systemic blood pressure than the aorta. This work provides evidence to clarify the effective *B. monnieri* components for vasorelaxation which could be related to the improvement of blood flow or memory enhancement.

2. Results

2.1. Vasorelaxant Effects of the B. monnieri Active Compounds

Mesenteric arteries of rats were isolated and mounted in an organ bath via intraluminal wire hooks connected to a force transducer. The vessels were pre-contracted with 10 μM phenylephrine (PE), before adding *B. monnieri* compounds including flavonoids (luteolin and apigenin), bacopaside I, and the saponin mixture (bacoside A) at 0.1–100 μM. *B. monnieri* compounds caused vasorelaxation of endothelial intact arteries (+EC) in a concentration-dependent manner (Figure 1) with EC_{50} and E_{max} values shown in Table 1.

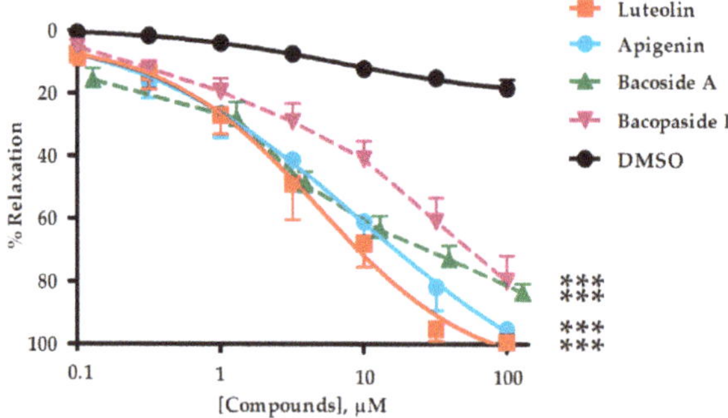

Figure 1. Relaxations induced by luteolin, apigenin, bacoside A, and bacopaside I (0.1–100 μM) and vehicle (DMSO) in endothelial intact mesenteric arteries precontracted with phenylephrine (10 μM). Values are mean ± SEM of 6–9 individual arterial rings. *** indicates $p < 0.001$ comparing relaxation for each compound with the control (DMSO) using two-way ANOVA (n = 6–9). Lines were fitted by non-linear regression.

Table 1. The EC_{50} and E_{max} of *B. monnieri* active compounds on relaxation of endothelial intact rat mesenteric arteries.

Active Compounds		EC_{50} (μM)	E_{max} (%)	n	*p*-Value Whole Graph Curves
Flavonoids	Luteolin	4.35 ± 1.31	99.4 ± 0.7	6	-
	Apigenin	8.93 ± 3.33	95.3 ± 2.6	9	NS
Saponins	Bacoside A	10.8 ± 5.9	83.6 ± 2.9 ††	7	< 0.05 †
	Bacopaside I	14.6 ± 5.4	79.9 ± 8.2 †	7	< 0.01 ††
Vehicle	DMSO	-	17.4 ± 3.1 ††	7	< 0.01 ††

Significantly different compared with luteolin † $p < 0.05$, †† $p < 0.01$ using unpaired Student's *t*-test (n = 6–9).

2.2. Mechanisms of Vasorelaxation by B. monnieri Compounds

All the *B. monnieri* compounds caused vasorelaxation in both endothelial intact (+EC) and endothelial denuded (-EC) mesenteric arterial rings. The relaxations were reduced by the removal of endothelium, implying that these compounds acted via an effect on endothelial vasodilators. However, the compounds still produced some vasorelaxations of the endothelial denuded arterial rings due to a direct action on vascular smooth muscle cells. For intact vessels, L-NAME (inhibitor of endothelial NO synthase; eNOS inhibitor), also reduced the vasorelaxations (Figure 2, Table 2). These reductions suggest that some or all the vasorelaxations were mediated through production and release of NO by endothelial cells.

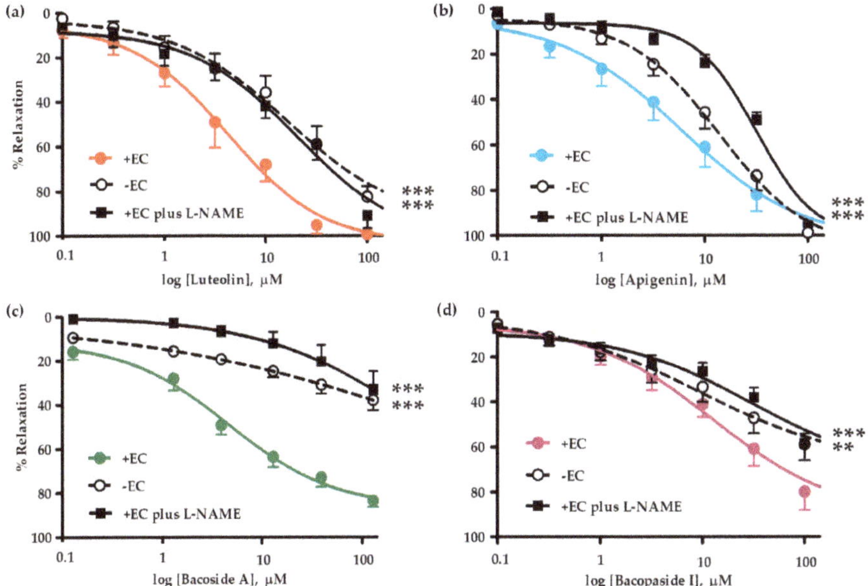

Figure 2. Cumulative concentration-response curves of (**a**) luteolin, (**b**) apigenin, (**c**) bacoside A and (**d**) bacopaside I in concentrations (0.1–100 µM) in endothelial intact (+EC), denuded (-EC) mesenteric arterial rings and endothelial intact vessels pre-incubated in L-NAME (100 µM). The graphs are expressed as %relaxation of vessel pre-contracted with 10 µM PE. Values are mean ± SEM of 6–9 individual arteries. ** $p < 0.01$, *** $p < 0.001$ each compound compared with intact vessels (+EC) using two-way ANOVA (n = 6–9).

Table 2. The EC_{50} and E_{max} of *B. monnieri* compounds on relaxations of endothelial intact (+EC), denuded (-EC) mesenteric arterial rings or endothelial intact arteries with L-NAME.

Active Compounds	EC_{50} (μM)	Emax (%)	n
Luteolin			
+EC	4.35 ± 1.31	99.35 ± 0.66	6
-EC	21.90 ± 5.86 †	82.42 ± 4.65 ††	6
+EC plus L-NAME	14.99 ± 3.56 †	90.85 ± 5.85	6
Apigenin			
+EC	8.93 ± 3.33	95.27 ± 2.61	9
-EC	12.80 ± 2.54	98.81 ± 1.19	8
+EC plus L-NAME	25.62 ± 3.38 ††	94.40 ± 2.10	7
Bacoside A			
+EC	10.81 ± 5.95	83.60 ± 2.86	7
-EC	14.50 ± 6.30	37.90 ± 4.72 ††	6
+EC plus L-NAME	33.81 ± 6.25 †	33.16 ± 8.41 ††	5
Bacopaside I			
+EC	14.63 ± 5.36	79.94 ± 8.17	7
-EC	17.29 ± 4.75	58.97 ± 7.05 †	7
+EC plus L-NAME	25.38 ± 4.33	58.45 ± 4.21 †	7

Comparison of EC_{50} or E_{max} of each component +EC vs. -EC or +EC plus L-NAME. † $p < 0.05$, †† $p < 0.01$ using unpaired Student's *t*-test.

2.3. *B. monnieri* Compounds and Ca^{2+} Influx

Voltage-operated Ca^{2+} channels (VOCCs) were activated by depolarising denuded vessels with 80 mM K^+ in Ca^{2+}-free Krebs' solution. Then vascular contraction elicited by $CaCl_2$ accumulatively added at increasing concentrations (0.01–10 mM). In the same vessel, the protocol was repeated by pre-incubation with 10 μM *B. monnieri* compounds for 15 min and these $CaCl_2$-induced contractions were inhibited and seen as a rightward shift of the plots and reduced E_{max} from control (Figure 3).

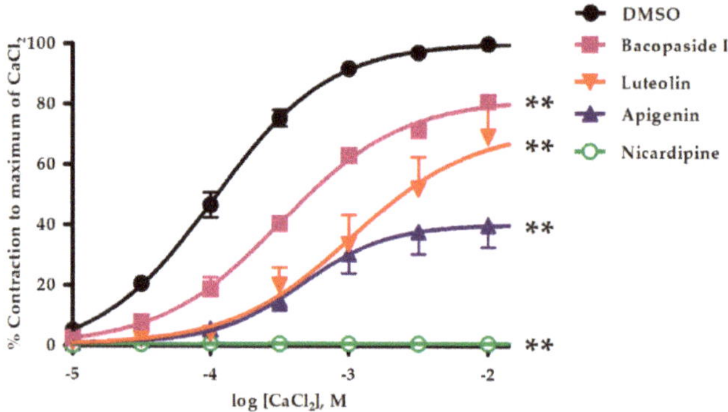

Figure 3. $CaCl_2$-induced contractions of denuded mesenteric arteries pre-incubated in high K^+, Ca^{2+}-free media in the conditions of pre-incubation with DMSO (negative control), 10 μM bacopaside I, 10 μM luteolin, 10 μM apigenin, and 1 μM nicardipine (positive control). *Y*-axis, % contraction compared to the contraction achieved with the highest Ca^{2+} concentration during the initial run without a *B. monnieri* compound in the same vessel. Values are mean ± SEM of 4–6 individual arteries. ** $p < 0.01$ each of the active compounds compared to DMSO using two-way ANOVA (n = 4–6).

The maximum contraction (E_{max}) of control, bacopaside I, luteolin and apigenin were 100 ± 1.3, 81.9 ± 1.7, 72.0 ± 6.7 and 40.2 ± 3.5%, respectively. Positive control, L-type Ca^{2+}-channel blocker, nicardipine (1 μM) completely abolished this $CaCl_2$-induced vasoconstriction (Figure 3).

2.4. B. monnieri Compounds and Intracellular Ca^{2+} Release

The release of intracellular Ca^{2+} from the sarcoplasmic reticulum is another important trigger of vascular contraction. Denuded arterial rings were pre-incubated in Ca^{2+}-free Krebs' solution for 10 min and then 10 μM PE added thereby eliciting a transient contraction. Then the protocol was repeated with the same arterial ring in the presence of the test compounds (control, apigenin, luteolin, bacoside A and bacopaside I) producing reduced contractions (98.8 ± 1.2, 50.1 ± 8.5, 54.3 ± 14.9, 85.8 ± 7.2 and 66.2 ± 2.9%, respectively) (Figure 4). Luteolin, apigenin and bacopaside I caused significant decrease in PE-induced contraction compared to the vehicle control ($p < 0.001$, <0.01 and <0.001, respectively).

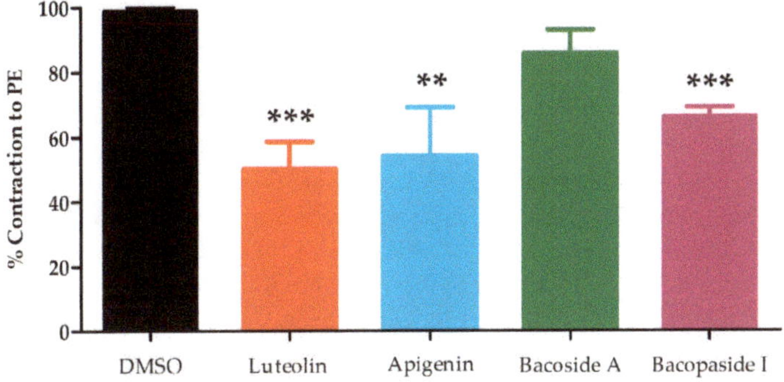

Figure 4. PE-induced contraction induced by Ca^{2+} release from sarcoplasmic reticulum of endothelial denuded mesenteric arteries in the presence of DMSO (control), 10 μM of luteolin, apigenin, bacoside A and bacopaside I. The data is % contraction to 10 μM PE induced contraction compared to contractions produced by the initial protocol without test compound. Values are mean ± SEM of 5–6 individual arteries. ** $p < 0.01$, *** $p < 0.001$ each of the active compounds compared with control using unpaired Student's *t*-test (n = 5–6).

3. Discussion

This is the first study comparing the vasodilatory mechanisms elicited by saponins (particularly bacoside A and bacopaside I) and the principal flavonoids (luteolin and apigenin) were the most potent (EC_{50} 4.4 and 8.9 μM) (Figure 1). However, these are present in BME at only about 1/20th the contents of the bacoside A saponins and bacopaside I (Figure S1 and Table S1) [42]. Thus in terms of the overall actions of the complete BME, the saponins would be expected to make a larger contribution to the vasorelaxation than the flavonoids.

However, higher potency of aglycone flavonoids compared to saponin glycosides may be due to sugar moieties interfering with the molecule interacting with the binding sites responsible for the vasorelaxation as suggested by previously, i.e., lipophilic groups in the ring skeleton of flavonoids increased their vasorelaxant activity [43]. This provides a basis for study of the molecular mechanisms of vasorelaxation of flavonoids.

We investigated the mechanisms of flavonoid- and saponin-induced relaxation by endothelial denudation in mesenteric arterial rings which impaired vasorelaxation (Figure 2). Role of NO was investigated using the eNOS inhibitor (L-NAME) with the test compounds. L-NAME increased EC_{50} and reduced E_{max} which imitated the effect of endothelial denudation, suggesting the relaxation was mainly medicated by NO. This accords with observations made by Jin et al. that a cyclooxygenase (COX) inhibitor did not affect the relaxation induced by apigenin [44], and consistent with our previous study of *B. monnieri* extract, where indomethacin had no effect on vasorelaxation [29]. There were some important concentration dependent differences between flavonoids and saponins. Firstly, denudation or blockade of eNOS reduced the effect of bacoside A more than bacopaside I, luteolin and apigenin.

Perhaps this was a reflection of bacoside A being a mixture of saponins. However, curiously the responses of luteolin and apigenin to denudation and L-NAME where the latter had a greater effect.

Vascular smooth muscle express plasma membrane L-type Ca^{2+} channels that allow depolarisation dependent Ca^{2+} entry to trigger contraction. All three compounds (luteolin, apigenin and bacopaside I) tested in denuded vessels depressed this mechanism of contraction that can also explain in part, the vasorelaxant effect. But here, apigenin appeared to be more effective than luteolin while it was less effective in relaxation studies suggesting some heterogeneity in the mechanism of flavonoid action.

Ca^{2+} release from intracellular stores also regulates contraction via inositol trisphosphate (IP_3) or ryanodine receptors (RyR) associated channels in the SR membranes. IP_3 associated channels are commonly activated by plasma membrane G-protein coupled receptors including α_1-receptors which are activated by PE. RyR channels are activated by Ca^{2+} itself. The three pure compounds also inhibited Ca^{2+} released from stores which can account for at least some vasorelaxation of vessels precontracted by PE. However, the bacoside A was without clear effect again suggesting some heterogeneity between the four test substances. Other Ca^{2+}-channels may also be involved, for example T-channels and TRP channels, especially TRPC4 which is activated by α_1-receptor activation.

K^+ channels also play a role in regulation of vascular tone, i.e., voltage-dependent K^+ (K_v) channels open upon depolarization of the plasma membrane in vascular smooth muscle cells, and thus inhibits Ca^{2+} influx through VOCCs, resulting in vasodilation [45]. Jiang et al. also reported that luteolin inhibited Ca^{2+} channels, inhibited release of stored Ca^{2+} while K^+ channels were activated, specifically via K_{ATP}, K_{Ca}, K_V and K_{IR} [40] therefore the effects of apigenin, bacoside A and bacopaside I involving K^+ channels deserve further investigation. Our findings support those of Si et al. that luteolin can directly act on vascular endothelial cells, by inducing eNOS phosphorylation at Ser1177, leading to NO production [41]. The flavonoids evoke relaxations and also protect endothelial dependent vasorelaxation against oxidative stress [44,46,47] and diabetes [48], however vasoprotective effects of saponins needs further comprehensive investigation.

4. Materials and Methods

4.1. General Information

Tissues were from male Wistar rats (200–300 g) which were obtained from Nomura Siam International Co. Ltd. (Bangkok, Thailand). Experiments were approved by the Naresuan University Animal Care and Use Committee (NUACUC), protocol number NU-AE 600710. The rats were housed under the environmental conditions at 22 ± 1 °C, 12-h light and dark cycle, fed with standard rodent diet and tap water in Naresuan University Center for Animal Research (NUCAR) according to the guidelines for care and use of laboratory animals (Institute of Laboratory Animal Research, eighth edition 2011. Rats were anesthetized by intraperitoneal injection of thiopental sodium (100 mg/kg BW) and killed. The mesenteric arteries were excised, cleaned of surrounding loose connective tissue and cut into rings of 3–5 mm width. In some experiments, endothelial cells were mechanically removed by gently rubbing the lumen with a stainless steel wire. The mesenteric rings were mounted on a pair of intraluminal wires in organ chambers containing physiological Krebs' solution (mM): NaCl, 122; KCl, 5; [N-(2-hydroxyethyl) piperazine N'-(2-ethanesulfonic acid)] HEPES, 10; KH_2PO_4, 0.5; NaH_2PO_4, 0.5; $MgCl_2$, 1; glucose, 11; and $CaCl_2$, 1.8 (pH 7.3), at 37 °C and aerated [29,49–51]. The vessel segments were allowed to equilibrate for 1-h at a resting tension of 1–1.3 g during which the solution was replaced every 15 min. Changes in isometric tension were measured using force transducer lever (CB Sciences Inc., Milford, MA, USA) connected to a MacLab A/D converter (Chart V7; A.D. Instruments, Castle Hill, NSW, Australia), stored and displayed on a personal computer. Following stabilization, the arterial rings were tested for viability by the application of 10 µM PE. Upon development of a steady contraction, the endothelium status was tested with 10 µM ACh. The vessel was considered endothelial intact when the ACh induced >70% relaxation. After establishing the status of the endothelium, the rings were then rinsed with Krebs' solution for 30 min and one of the following protocols was initiated.

Luteolin (lot 126M4061V) and apigenin (lot WE445301/1) were purchased from Sigma Aldrich (St. Louis, MO, USA). Bacoside A (lot 00002005-003) and bacopaside I (lot 00002002-T17H) were purchased from ChromaDex, Inc. (Irvine, CA, USA).

4.2. Vasorelaxant Effects of B. monnieri Active Compounds on Endothelial Intact Arteries

Following stabilization, endothelial intact rings of mesenteric arteries were pre-contracted with 10 μM PE. After the contraction had become constant, the *B. monnieri* active compounds (0.1–100 μM), including luteolin, apigenin, bacoside A or bacopaside I were added cumulatively.

4.3. Vasorelaxant Effects of B. monnieri Active Compounds on Endothelial Denuded Arteries

Successful endothelial denudation was confirmed by the absence of relaxation upon addition of 10 μM ACh. For investigation of the role of endothelium in 0.1–100 μM *B. monnieri* active compounds (luteolin, apigenin, bacoside A or bacopaside I) induced vasorelaxation, endothelial denuded arteries were used. The data of effect of active compounds were presented as %relaxation.

4.4. Study of Vasorelaxant Mechanisms of B. monnieri Active Compounds via eNOS Pathway

The role of the endothelial relaxing factor, NO, in *B. monnieri* active compounds (luteolin, apigenin, bacoside A or bacopaside I) induced vasorelaxation were evaluated in endothelial intact ring pre-treated with N^G-nitro-L-arginine methyl ester (L-NAME, 100 μM), an inhibitor of eNOS, for 30 min prior to 10 μM PE exposure.

4.5. Study of Vasorelaxant Mechanisms of B. monnieri Active Compounds on Extracellular Ca^{2+} Influx

Endothelial denuded mesenteric arteries were equilibrated in Ca^{2+}-free Krebs' solution (containing (mM): ethylene glycol-bis (ß-aminoethyl ether)-N,N,N,N tetra acetic acid (EGTA), 0.01; NaCl, 122; KCl, 5; HEPES, 10; KH_2PO_4, 0.5; NaH_2PO_4, 0.5; $MgCl_2$, 1 and glucose, 11 (pH 7.3)) for 30 min followed by replacing with Ca^{2+}-free Krebs' solution containing 80 mM K^+ for 10 min which depolarizes the vascular smooth muscle cells, thus opening VOCCs. Various concentrations of $CaCl_2$ were then added (0.01–10 mM) in a logarithmic progression. After obtaining the maximum response, the baths were washed out and replenished with Ca^{2+}-free Krebs' solution for 30 min. The Ca^{2+}-free 80 mM K^+ solution was then re-applied following pre-incubation for 10 min with either: 10 μM active compounds or 1 μM nicardipine (antagonist of VOCCs). Concentration-response curves to cumulative addition of $CaCl_2$ were then repeated and compared with maximum contraction evoked by previous control $CaCl_2$ challenges.

4.6. Study of Vasorelaxant Mechanisms of B. monnieri Active Compounds on Intracellular Ca^{2+} Release

To stimulate initial Ca^{2+} loading of the SR Ca^{2+} stores, endothelial denuded mesenteric arteries were exposed to 80 mM K^+ solution for 5 min, and then washed out with Ca^{2+}-free Krebs' solution containing 1 mM EGTA for 10 min. The arterial rings were then challenged with 10 μM PE (acting through phospholipase C/IP$_3$ signaling) which release Ca^{2+} from the SR thereby eliciting a transient contraction [29]. The same protocol was then repeated to ensure that similar transient contractions to PE could be obtained. Then, the arterial rings were challenged again with 80 mM K^+ solution for 5 min, and washed out with Ca^{2+}-free Krebs' solution containing 1 mM EGTA and 10 μM active compounds for 10 min. The arterial rings were again challenged with 10 μM PE. The PE-induced contractions were compared in the presence or absence of active compounds.

4.7. Statistical Analyses

Statistical analyses used GraphPad Prism version 5.00 for Windows, (GraphPad Software Inc., La Jolla, CA, USA). Data from each concentration-effect curve was analysed using non-repeated two-way ANOVA. Curve fitting in the figures was generated by the same software using non-linear regression.

EC_{50} and E_{max} were compared using unpaired Student's *t* test. Values are expressed as mean $\pm$ SEM. A *p*-value < 0.05 was considered significant. 'n' is the number of vascular rings used, each ring originating from a different animal.

5. Conclusions

This study demonstrated that *B. monnieri* active components, including both saponins and flavonoids, produced vasodilatory effects on rat isolated mesenteric arteries partially via endothelial dependent release of vasodilators and also by direct effects on vascular smooth muscle cells via blockade of Ca^{2+} influx and its release from SR. This study for the first time reports the comparative vasodilatory effects of saponins and flavonoids found in *B. monnieri* extract. However, *B. monnieri* extract, flavonoids i.e., luteolin and apigenin would be more potent vasodilators but saponins have a greater effect because of their greater contents. Accordingly, the clinical benefits on enhanced blood flow and cognitive function may arise from a combination of flavonoids and particularly the saponins.

Supplementary Materials: Supplementary materials are available online. Figure S1: Representative HPLC-UV chromatogram of mixed seven standards at 20 μg/mL for 1 and 2 and 100 μg/mL for 3–7 (A) and Brahmi extract (2 mg/mL) (B); luteolin (1), apigenin (2), bacoside A3 (3), bacopaside II (4), bacopaside X (5), bacopasaponin C (6) and bacopaside I (7), Table S1: Amount of each compound in 95% ethanolic extract of Brahmi analyzed by HPLC.

Author Contributions: Conceptualization, K.I., N.W. and K.C.; methodology and experimental design, N.K., T.U.P. and K.C.; software, N.K.; validation, N.K., T.U.P. and K.C.; formal analysis, N.K. and K.C.; investigation, N.K.; resources, K.I., N.W. and K.C.; data curation and interpretation, N.K. and K.C.; writing—original draft preparation, N.K.; manuscript writing—review and editing, T.U.P. and K.C.; visualization, K.I., N.W. and K.C.; supervision, K.I., N.W. and K.C.; project administration, K.C.; funding acquisition, K.C.

Funding: This research was funded by National Research Council of Thailand (NRCT), grant No. R2561B032.

Acknowledgments: We would like to acknowledge the Center of Excellence for Innovation in Chemistry (PERCH-CIC) and the International Research Network (IRN61W0005) on research facility support. We would like to thank C. Norman Scholfield for his constructive criticism of the manuscript.

Conflicts of Interest: The authors declared no conflict of interest.

References

1.   *Bacopa monniera*. Monograph. *Altern. Med. Rev. J. Clin. Ther.* **2004**, *9*, 79–85.
2.   Russo, A.; Borrelli, F. *Bacopa monniera*, a reputed nootropic plant: An overview. *Phytomedicine* **2005**, *12*, 305–317. [CrossRef] [PubMed]
3.   Kumar, V. Potential medicinal plants for CNS disorders: An overview. *Phytother. Res. PTR* **2006**, *20*, 1023–1035. [CrossRef]
4.   Rajan, K.E.; Preethi, J.; Singh, H.K. Molecular and functional characterization of *Bacopa monniera*: A retrospective review. *Evid.-Based Complement. Altern. Med. eCAM* **2015**, *2015*, 945217. [CrossRef] [PubMed]
5.   Aguiar, S.; Borowski, T. Neuropharmacological review of the nootropic herb *Bacopa monnieri*. *Rejuvenation Res.* **2013**, *16*, 313–326. [CrossRef] [PubMed]
6.   Singh, H.K. Brain enhancing ingredients from Ayurvedic medicine: Quintessential example of *Bacopa monniera*, a narrative review. *Nutrients* **2013**, *5*, 478–497. [CrossRef] [PubMed]
7.   Mathur, D.; Goyal, K.; Koul, V.; Anand, A. The molecular links of re-emerging therapy: A review of evidence of Brahmi (*Bacopa monniera*). *Front. Pharmacol.* **2016**, *7*, 44. [CrossRef]
8.   Das, A.; Shanker, G.; Nath, C.; Pal, R.; Singh, S.; Singh, H. A comparative study in rodents of standardized extracts of *Bacopa monniera* and *Ginkgo biloba*: Anticholinesterase and cognitive enhancing activities. *Pharmacol. Biochem. Behav.* **2002**, *73*, 893–900. [CrossRef]
9.   Dhanasekaran, M.; Tharakan, B.; Holcomb, L.A.; Hitt, A.R.; Young, K.A.; Manyam, B.V. Neuroprotective mechanisms of ayurvedic antidementia botanical *Bacopa monniera*. *Phytother. Res. PTR* **2007**, *21*, 965–969. [CrossRef]
10. Limpeanchob, N.; Jaipan, S.; Rattanakaruna, S.; Phrompittayarat, W.; Ingkaninan, K. Neuroprotective effect of *Bacopa monnieri* on beta-amyloid-induced cell death in primary cortical culture. *J. Ethnopharmacol.* **2008**, *120*, 112–117. [CrossRef]

11. Uabundit, N.; Wattanathorn, J.; Mucimapura, S.; Ingkaninan, K. Cognitive enhancement and neuroprotective effects of *Bacopa monnieri* in Alzheimer's disease model. *J. Ethnopharmacol.* **2010**, *127*, 26–31. [CrossRef] [PubMed]

12. Vollala, V.R.; Upadhya, S.; Nayak, S. Enhancement of basolateral amygdaloid neuronal dendritic arborization following *Bacopa monniera* extract treatment in adult rats. *Clinics (Sao Paulo Brazil)* **2011**, *66*, 663–671. [CrossRef] [PubMed]

13. Vollala, V.R.; Upadhya, S.; Nayak, S. Learning and memory-enhancing effect of *Bacopa monniera* in neonatal rats. *Bratisl. Lek. Listy* **2011**, *112*, 663–669. [PubMed]

14. Vollala, V.R.; Upadhya, S.; Nayak, S. Enhanced dendritic arborization of hippocampal CA3 neurons by Bacopa monniera extract treatment in adult rats. *Rom. J. Morphol. Embryol.* **2011**, *52*, 879–886. [PubMed]

15. Vollala, V.R.; Upadhya, S.; Nayak, S. Enhanced dendritic arborization of amygdala neurons during growth spurt periods in rats orally intubated with *Bacopa monniera* extract. *Anat. Sci. Int.* **2011**, *86*, 179–188. [CrossRef] [PubMed]

16. Kongkeaw, C.; Dilokthornsakul, P.; Thanarangsarit, P.; Limpeanchob, N.; Norman Scholfield, C. Meta-analysis of randomized controlled trials on cognitive effects of *Bacopa monnieri* extract. *J. Ethnopharmacol.* **2014**, *151*, 528–535. [CrossRef] [PubMed]

17. Anbarasi, K.; Vani, G.; Balakrishna, K.; Devi, C.S. Effect of bacoside A on brain antioxidant status in cigarette smoke exposed rats. *Life Sci.* **2006**, *78*, 1378–1384. [CrossRef] [PubMed]

18. Jyoti, A.; Sharma, D. Neuroprotective role of *Bacopa monniera* extract against aluminium-induced oxidative stress in the hippocampus of rat brain. *Neurotoxicology* **2006**, *27*, 451–457. [CrossRef]

19. Mannan, A.; Abir, A.B.; Rahman, R. Antidepressant-like effects of methanolic extract of *Bacopa monniera* in mice. *BMC Complement. Altern. Med.* **2015**, *15*, 337. [CrossRef]

20. Sairam, K.; Dorababu, M.; Goel, R.K.; Bhattacharya, S.K. Antidepressant activity of standardized extract of *Bacopa monniera* in experimental models of depression in rats. *Phytomedicine* **2002**, *9*, 207–211. [CrossRef]

21. Kadali, S.R.M.; Das, M.C.; Rao, A.S.; Sri, G.K. Antidepressant activity of brahmi in albino mice. *J. Clin. Diagn. Res. JCDR* **2014**, *8*, 35–37. [CrossRef] [PubMed]

22. Udhaya Lavinya, B.; Sabina, E.P. Anti-hyperglycaemic effect of Brahmi (*Bacopa monnieri* L.) in streptozotocininduced diabetic rats: A study involving antioxidant, biochemical and haematological parameters. *J. Chem. Pharm. Res.* **2015**, *7*, 531–534.

23. Kamesh, V.; Sumathi, T. Antihypercholesterolemic effect of *Bacopa monniera* Linn. on high cholesterol diet induced hypercholesterolemia in rats. *Asian Pac. J. Trop. Med.* **2012**, *5*, 949–955. [CrossRef]

24. Sireeratawong, S.; Jaijoy, K.; Khonsung, P.; Lertprasertsuk, N.; Ingkaninan, K. Acute and chronic toxicities of *Bacopa monnieri* extract in Sprague-Dawley rats. *BMC Complement. Altern. Med.* **2016**, *16*, 249. [CrossRef] [PubMed]

25. Joshua Allan, J.; Damodaran, A.; Deshmukh, N.S.; Goudar, K.S.; Amit, A. Safety evaluation of a standardized phytochemical composition extracted from *Bacopa monnieri* in Sprague–Dawley rats. *Food Chem. Toxicol.* **2007**, *45*, 1928–1937. [CrossRef] [PubMed]

26. Pravina, K.; Ravindra, K.R.; Goudar, K.S.; Vinod, D.R.; Joshua, A.J.; Wasim, P.; Venkateshwarlu, K.; Saxena, V.S.; Amit, A. Safety evaluation of BacoMind in healthy volunteers: A phase I study. *Phytomedicine* **2007**, *14*, 301–308. [CrossRef] [PubMed]

27. Srimachai, S.; Devaux, S.; Demougeot, C.; Kumphune, S.; Ullrich, N.D.; Niggli, E.; Ingkaninan, K.; Kamkaew, N.; Scholfield, C.N.; Tapechum, S.; et al. *Bacopa monnieri* extract increases rat coronary flow and protects against myocardial ischemia/reperfusion injury. *BMC Complement. Altern. Med.* **2017**, *17*, 117. [CrossRef]

28. Nandave, M.; Ojha, S.K.; Sujata, J.; Kumari, S.; Arya, D.S. Cardioprotective effect of *Bacopa monneira* against isoproterenol-induced myocardial necrosis in rats. *Int. J. Pharmacol.* **2007**, *3*, 385–392.

29. Kamkaew, N.; Scholfield, C.N.; Ingkaninan, K.; Maneesai, P.; Parkington, H.C.; Tare, M.; Chootip, K. *Bacopa monnieri* and its constituents is hypotensive in anaesthetized rats and vasodilator in various artery types. *J. Ethnopharmacol.* **2011**, *137*, 790–795. [CrossRef]

30. Kamkaew, N.; Norman Scholfield, C.; Ingkaninan, K.; Taepavarapruk, N.; Chootip, K. *Bacopa monnieri* increases cerebral blood flow in rat independent of blood pressure. *Phytother. Res. PTR* **2013**, *27*, 135–138. [CrossRef]

31. Phrompittayarat, W.; Putalun, W.; Tanaka, H.; Jetiyanon, K.; Wittaya-Areekul, S.; Ingkaninan, K. Determination of pseudojujubogenin glycosides from Brahmi based on immunoassay using a monoclonal antibody against bacopaside I. *Phytochem. Anal. PCA* **2007**, *18*, 411–418. [CrossRef] [PubMed]

32. Phrompittayarat, W.; Putalun, W.; Tanaka, H.; Wittaya-Areekul, S.; Jetiyanon, K.; Ingkaninan, K. An enzyme-linked immunosorbant assay using polyclonal antibodies against bacopaside I. *Anal. Chim. Acta* **2007**, *584*, 1–6. [CrossRef] [PubMed]

33. Nuengchamnong, N.; Sookying, S.; Ingkaninan, K. LC-ESI-QTOF-MS based screening and identification of isomeric jujubogenin and pseudojujubogenin aglycones in *Bacopa monnieri* extract. *J. Pharm. Biomed. Anal.* **2016**, *129*, 121–134. [CrossRef] [PubMed]

34. Honnegowda, S.; Bagul, M.S.; Padh, H.; Rajani, M. A rapid densitometric method for the quantification of luteolin in medicinal plants using HPTLC. *Chromatographia* **2004**, *60*, 131–134.

35. Deepak, M.; Sangli, G.K.; Arun, P.C.; Amit, A. Quantitative determination of the major saponin mixture bacoside A in *Bacopa monnieri* by HPLC. *Phytochem. Anal. PCA* **2005**, *16*, 24–29. [CrossRef] [PubMed]

36. Rajasekaran, A. Simultaneous estimation of luteolin and apigenin in methanol leaf extract of *Bacopa monnieri* Linn by HPLC. *Br. J. Pharm. Res.* **2014**, *4*, 1629–1637. [CrossRef]

37. Chan, E.C.; Pannangpetch, P.; Woodman, O.L. Relaxation to flavones and flavonols in rat isolated thoracic aorta: Mechanism of action and structure-activity relationships. *J. Cardiovasc. Pharmacol.* **2000**, *35*, 326–333. [CrossRef]

38. Calderone, V.; Chericoni, S.; Martinelli, C.; Testai, L.; Nardi, A.; Morelli, I.; Breschi, M.C.; Martinotti, E. Vasorelaxing effects of flavonoids: Investigation on the possible involvement of potassium channels. *Naunyn-Schmiedeberg's Arch. Pharmacol.* **2004**, *370*, 290–298. [CrossRef]

39. Je, H.D.; Kim, H.-D.; La, H.-O. The inhibitory effect of apigenin on the agonist-induced regulation of vascular contractility via calcium desensitization-related pathways. *Biomol. Ther.* **2014**, *22*, 100–105. [CrossRef]

40. Jiang, H.; Xia, Q.; Wang, X.; Song, J.; Bruce, I.C. Luteolin induces vasorelaxion in rat thoracic aorta via calcium and potassium channels. *Die Pharm.* **2005**, *60*, 444–447.

41. Si, H.; Wyeth, R.P.; Liu, D. The flavonoid luteolin induces nitric oxide production and arterial relaxation. *Eur. J. Nutr.* **2014**, *53*, 269–275. [CrossRef] [PubMed]

42. Saesong, T.; Temkitthawon, P.; Nangngam, P.; Ingkaninan, K. Pharmacognostic and physico-chemical investigations of the aerial part of *Bacopa monnieri* (L.) Wettst. *SJST* **2019**, *41*, 397–404.

43. Wu, H.; Jiang, H.; Wang, L.; Hu, Y. Relationship between vasorelaxation of flavonoids and their retention index in RP-HPLC. *Die Pharm.* **2006**, *61*, 667–669.

44. Jin, B.H.; Qian, L.B.; Chen, S.; Li, J.; Wang, H.P.; Bruce, I.C.; Lin, J.; Xia, Q. Apigenin protects endothelium-dependent relaxation of rat aorta against oxidative stress. *Eur. J. Pharmacol.* **2009**, *616*, 200–205. [CrossRef] [PubMed]

45. Ko, E.A.; Han, J.; Jung, I.D.; Park, W.S. Physiological roles of K^+ channels in vascular smooth muscle cells. *J. Smooth Muscle Res.* **2008**, *44*, 65–81. [CrossRef] [PubMed]

46. Ma, X.; Li, Y.F.; Gao, Q.; Ye, Z.G.; Lu, X.J.; Wang, H.P.; Jiang, H.D.; Bruce, I.C.; Xia, Q. Inhibition of superoxide anion-mediated impairment of endothelium by treatment with luteolin and apigenin in rat mesenteric artery. *Life Sci.* **2008**, *83*, 110–117. [CrossRef]

47. Qian, L.B.; Wang, H.P.; Chen, Y.; Chen, F.X.; Ma, Y.Y.; Bruce, I.C.; Xia, Q. Luteolin reduces high glucose-mediated impairment of endothelium-dependent relaxation in rat aorta by reducing oxidative stress. *Pharmacol. Res.* **2010**, *61*, 281–287. [CrossRef]

48. El-Bassossy, H.M.; Abo-Warda, S.M.; Fahmy, A. Chrysin and luteolin attenuate diabetes-induced impairment in endothelial-dependent relaxation: Effect on lipid profile, AGEs and NO generation. *Phytother. Res. PTR* **2013**, *27*, 1678–1684. [CrossRef]

49. Wisutthathum, S.; Kamkaew, N.; Inchan, A.; Chatturong, U.; Paracha, T.U.; Ingkaninan, K.; Wongwad, E.; Chootip, K. Extract of *Aquilaria crassna* leaves and mangiferin are vasodilators while showing no cytotoxicity. *J. Tradit. Complement. Med.* **2018**, in press. [CrossRef]

50. Wisutthathum, S.; Demougeot, C.; Totoson, P.; Adthapanyawanich, K.; Ingkaninan, K.; Temkitthawon, P.; Chootip, K. *Eulophia macrobulbon* extract relaxes rat isolated pulmonary artery and protects against monocrotaline-induced pulmonary arterial hypertension. *Phytomedicine* **2018**, *50*, 157–165. [CrossRef]

51. Wisutthathum, S.; Chootip, K.; Martin, H.; Ingkaninan, K.; Temkitthawon, P.; Totoson, P.; Demougeot, C. Vasorelaxant and hypotensive effects of an ethanolic extract of *Eulophia macrobulbon* and Its main compound 1-(4′-hydroxybenzyl)-4,8-dimethoxyphenanthrene-2,7-diol. *Front. Pharmacol.* **2018**, *9*, 484. [CrossRef] [PubMed]

Sample Availability: Samples of the compounds, luteolin, apigenin, bacoside A and bacopaside I are not available from the authors.

 *molecules* 

Article

The Anti-Aging Potential of Neohesperidin and Its Synergistic Effects with Other Citrus Flavonoids in Extending Chronological Lifespan of *Saccharomyces Cerevisiae* BY4742

Chunxia Guo [1,†], Hua Zhang [2,3,†], Xin Guan [1,*] and Zhiqin Zhou [1,4,5,*]

1   College of Horticulture and Landscape Architecture, Southwest University, Chongqing 400716, China; guochunxia@xdf.cn
2   College of Biology and Food Engineering, Chongqing Three Gorges University, Chongqing 404120, China; 20140055@sanxiau.edu.cn
3   Engineering Technology Research Center of Characteristic Biological Resources in Northeast of Chongqing, Three Gorges University, Chongqing 404120, China
4   Key Laboratory of Horticulture for Southern Mountainous Regions, Ministry of Education, Chongqing 400715, China
5   The Southwest Institute of Fruits Nutrition, Liang jiang New District, Chongqing 401121, China
*   Correspondence: xin.guan@swu.edu.cn (X.G.); zhouzhiqin@swu.edu.cn (Z.Z.); Tel.: +86-023-68251047 (Z.Z.)
†   These authors contribute equally to this work.

Academic Editor: H.P. Vasantha Rupasinghe
Received: 26 September 2019; Accepted: 8 November 2019; Published: 13 November 2019

Abstract: The anti-aging activity of many plant flavonoids, as well as their mechanisms of action, have been explored in the current literature. However, the studies on the synergistic effects between the different flavonoid compounds were quite limited in previous reports. In this study, by using a high throughput assay, we tested the synergistic effects between different citrus flavonoids throughout the yeast's chronological lifespan (CLS). We studied the effect of four flavonoid compounds including naringin, hesperedin, hesperitin, neohesperidin, as well as their different combinations on the CLS of the yeast strain BY4742. Their ROS scavenging ability, in vitro antioxidant activity and the influence on the extracellular pH were also tested. The results showed that neohesperidin extended the yeast's CLS in a concentration-dependent manner. Especially, we found that neohesperidin showed great potential in extending CLS of budding yeast individually or synergistically with hesperetin. The neohesperidin exhibited the strongest function in decreasing the reactive oxygen species (ROS) accumulation in yeast. These findings clearly indicated that neohesperidin is potentially an anti-aging citrus flavonoid, and its synergistic effect with other flavonoids on yeast's CLS will be an interesting subject for future research of the anti-aging function of citrus fruits.

Keywords: citrus flavonoids; neohesperidin; anti-aging activity; chronological lifespan; synergistic effect

---

1. Introduction

Aging, a complex and multifactorial biological process, can be defined as a gradual loss of physiological and psychological integrity, leading to gradual deterioration in almost all functions and the increased vulnerability to death [1]. A treatment that targets the multiple factors and/or pathways within the aging process is good candidate for study. At present, the trend of population aging is gradually increasing. Because of this trend in population, it is of great practical significance to find effective ways to slow aging or improve the healthy state of aging. The anti-aging activity of phytochemicals has been studied by the researchers from a wide variety of disciplines. Many different plant compounds have been suggested to have direct/potential anti-aging activity in the existing literature [2–5].

The budding yeast *Saccharomyces cerevisiae* has played a leading role as a model organism for studying evolutionarily conserved mechanisms, which are relevant to human aging and age-related diseases [6]. There are two types of lifespan in yeast, namely replicative lifespan (RLS) [7] and chronological lifespan (CLS) [8]. RLS is defined as the number of daughter cells a mother cell can produce before cell budding ceases [9], whereas CLS is the length of time budding yeast cells survive after undergoing a nutrient depletion-induced arrest of the cell cycle in stationary phase [10]. RLS and CLS can serve as models for proliferating and non-proliferating tissues in higher eukaryotes, respectively [6].

The longstanding and successful use of herbal drug combinations in traditional medicine inspired us to study the synergistic effect of phytochemicals that have healthy functions. Nowadays, synergy assessment has become a key area in medical research in order to enhance the efficiency of treatments and affect not only one single target, but several targets [11]. Previous investigations have shown that naringin, hesperidin, hesperetin and neohesperidin, widely distributed in citrus fruits, possess multiple biological activities relevant to anti-aging, and the detailed information of these phytochemicals are presented in literature [12–21]. However, the phytochemicals were less evaluated as combinations. At least, the ternary combinations of them were rarely reported. Meanwhile, the effect of neohesperidin on CLS of the budding yeast BY4742 was not revealed before.

In order to reveal the flavonoid combinations function on extending the CLS of the budding yeast BY4742, four flavonoid compounds, their binary and ternary combinations effect on CLS of yeast BY4742, and their ROS scavenging ability and in vitro antioxidant activity were also tested in the present work. Meanwhile, the extracellular pH values of yeast treated with the four compounds were detected since extracellular acidification of the culture medium might cause intracellular damage in the chronologically aging population [22].

2. Results

2.1. Neohesperidin Extended Yeast Lifespan in a Concentration-Dependent Manner

DMSO was universally used as a solvent for the water-insoluble drugs, and different concentrations of DMSO variably affected the growth of the yeast [23]. Therefore, we firstly screened the appropriate DMSO concentration. As shown in Figure 1 (A), there was no significant difference of yeast lifespan at 0.2% and 0.4% DMSO in the culture medium when compared with the control, while the lifespan at 0.6% and 0.8% DMSO were significantly decreased. To eliminate the influence of solvent in experiments, we set the final medium concentration of DMSO to 0.2% for carrying out the following experiments.

High-throughtput assays were used for rapid quantification of the CLS for the merit of the yeast chronological aging model [24,25]. Therefore, we employed this method to screen four citrus flavonoids (naringin, hesperidin, hesperetin and neohesperidin) for their anti-aging activity. We determined the longevity efficacy of the four phytochemicals in a range of doses, from 0 μM to 100 μM. From the result, it was found out that neohesperidin exhibited a potential to extend the CLS of BY4742 at 0.1 μM, while other three compounds could not increase the cell survival at the set concentrations (Figure 1B). For the convenience of combinatorial experiments, here the concentrations of A, B, C and D as 100, 0.1, 0.1 and 0.1 μM were chose, respectively. Namely, for the next combinational assays, A (100 μM naringin), B (0.1 μM hesperidin), C (0.1 μM hesperetin) and D (0.1 μM neohesperidin) were used.

The cumulative time of processing is thought to be an important factor for influencing cell growth. In accordance with this hypothesis, our data implied that the four flavonoid compounds did not influence the cell growth instantly after adding in 24 h at different processing times (Figure 2). Because the growth curves of yeast treating with A, B, C and D were almost the same as the control ones. This also showed that the four compounds did not inhibit the yeast cell growth. However, when the processing was started at day 2, the growth survival rates/CLS could change during long-lasting period (Figure 3).

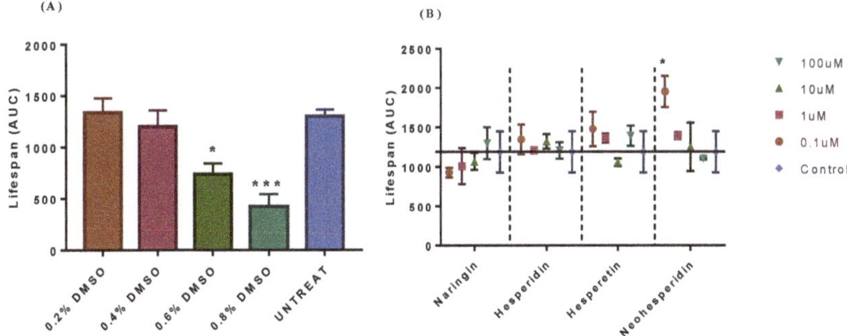

Figure 1. (**A**) Effect of different concentrations of dimethyl sulfoxide (DMSO) on lifespan of yeast BY4742. (**B**) Effect of different concentrations of the four flavonoid compounds on lifespan of yeast BY4742. AUC means area under the curve. The data was expressed as the mean values ± standard error of mean (SEM), $n = 3$. One-way ANOVA's Sidak's multiple comparisons test by GraphPad Prism 7.00 was used. (* $p < 0.05$, *** $p < 0.001$).

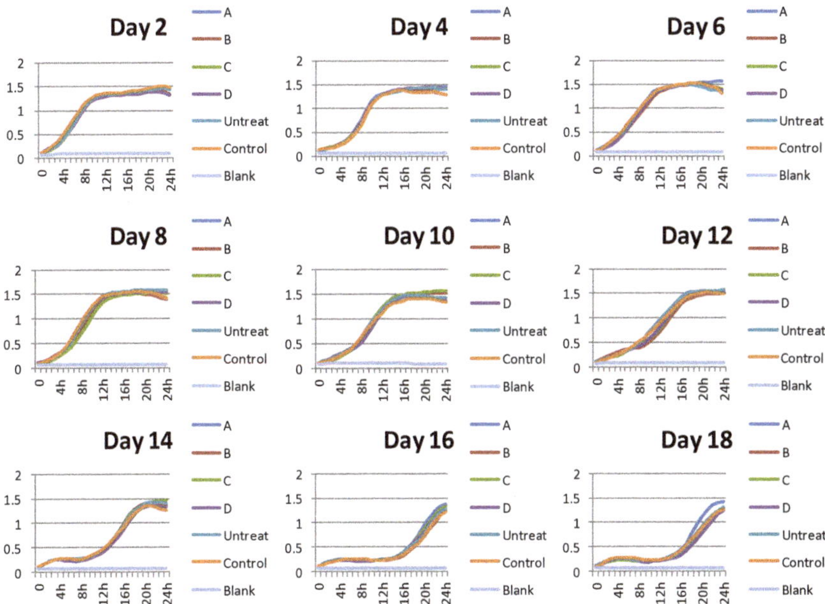

Figure 2. The growth curves of yeast BY4742 when treating with the four flavonoid compounds at different time at selected concentrations (100 µM A, 0.1 µM B, 0.1 µM C, and 0.1 µM D). The number of the day means the time of culturing in aging medium. And the growth curves were detected instantly after adding the compounds. The experiment was repeated three times. (A: naringin; B: hesperidin; C: hesperetin; D: neohesperidin).

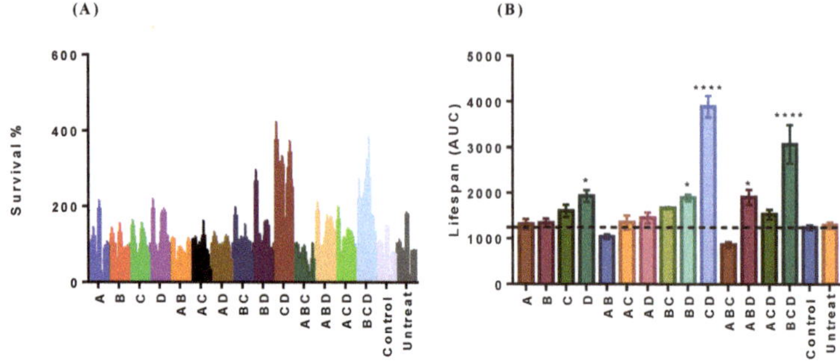

Figure 3. (**A**) The survival rates of yeast BY4742 under the treatment of different compounds and their combinations from day 2 to day 20, they were measured every two days. The survival rate was calculated as follows. Where tOD= 0.3, 2day is the time that OD value of day 2 age-point reaches 0.3 in the outgrowth curves. The initial age-point (day 2) is defined to be 100% viability and the relative survival percent of each successive age-point can be calculated as follows: $Vn = \frac{1}{2^{\frac{\Delta tn}{Dt_n}}} \times 100 n = days$, $\overline{Dt_n}$ represent the average doubling time. The survival integral (SI) for each well is defined as the area under the survival curve (AUC) and can be estimated by the formula: $\text{SI} = \sum\limits_{2}^{n} (\frac{V_{n-1}+V_n}{2})(day_n - day_{n-1})$, where day_n is the age point, such as days 2, 4, 6, 8, 10, 12, 14, 16, 18 and 20. (**B**) The effect of the four flavonoid compounds, their binary and ternary combinations on chronological lifespan in yeast BY4742 at their beneficial concentrations. AUC means area under the curve. One-way ANOVA multiple comparisons. The data was expressed as the mean values ± standard error of mean (SEM), $n = 4$. * $p < 0.05$, **** $p < 0.0001$. A (100 µM naringin), B (0.1 µM hesperidin), C (0.1 µM hesperetin), D (0.1 µM neohesperidin).

2.2. Neohesperidin Positively Interacted with Hesperetin for Extending the CLS of Yeast BY4742

The combination therapy was demonstrated to be a new and highly effective therapeutic strategy to manage many diseases, such as diabetes [26], cancer [27], cardiovascular disorders [28], obesity and osteoporosis [29]. Here, we studied the effect of binary and ternary combinations of naringin, hesperidin, hesperetin and neohesperidin on the CLS of budding yeast BY4742. Figure 3A showed the survival rates of yeast under different flavonoid treatments. For each treatment, 10 aging points were detected. It showed that the survival rates of some treatment were higher when compared to the counterparts of control. From Figure 3B, the result clearly showed us that the treatment of D, BD and ABD showed the significant differences ($p < 0.05$), just weaker than CD and BCD (**** $p < 0.0001$). We can see that D (neohesperidin) showed important function in the individual or synergistical treatments. Therefore, we could conclude that neohesperidin had great potential in increasing CLS of budding yeast BY4742 individually or synergistically with hesperetin.

2.3. Neohesperidin Significantly Reduced Intracellular Reactive Oxygen Species (ROS) Content

According to Harman, cellular component damage caused by ROS generated in mitochondria is the main force accelerating the aging process of the organism [30,31]. As shown in Figure 4A, the four flavonoids all exhibited a remarkable ROS scavenging capacity. Among four single treatments, neohesperidin had the most prominent effect. Other treatments combined, all the ternary combination, and BD as well, didn't decrease intracellular ROS. As for the binary combinations, such as AB, AC, AD, and BC, they showed higher ROS scavenging activities than their corresponding single substances. The ROS scavenging capability of the binary combination BD was between B and D, while the function of CD was almost the same as D.

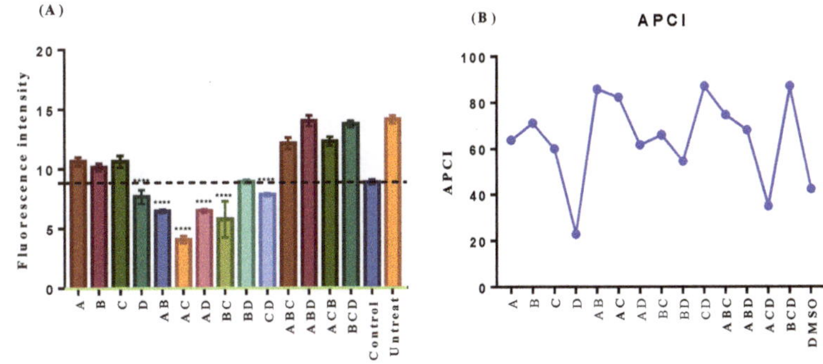

Figure 4. (**A**) Effect of A (100 µM naringin), B (0.1 µM hesperidin), C (0.1 µM hesperetin), D (0.1 µM neohesperidin), their binary and ternary combinations on intracellular ROS levels of yeast (BY4742) grown in standard SD medium after treating for 2 days (n = 12). The ROS probe H2DCFDA was used. Dichlorodihydrofluorescein (DCF). One-way ANOVA multiple comparisons. The data was expressed as the mean values ± standard error of mean (SEM), n = 12. **** $p < 0.0001$; (**B**) The antioxidant capacity (ӱ M trolox equivalents/µM phytochemical) of A (naringin), B (hesperidin), C (hesperetin), D (neohesperidin) and their binary and ternary combinations evaluated by 1,1-diphenyl-2-picrylhydrazyl (DPPH), FRAP and ABTS assays. APCI (antioxidant potency composite index) = Σ(the sample data of each method/the highest sample data of every method)/the number of methods •100. The higher the APCI is the lower the rank number is.

2.4. In Vitro Antioxidant Activity of Neohesperidin was Relatively Weak

Many methods are available for measuring the in vitro antioxidant capacity and most researchers apply one or more assays since each method measures different antioxidant characteristics of the compound [32]. In this study, we used three methods to determine antioxidant capacity include DPPH, ABTS, and FRAP assays. The antioxidant potency composite index (APCI) was defined to describe and evaluate the overall in vitro antioxidant capacity of the tested compounds and their combinations. The linear regression equations of the three assays are listed in Table 1. And all the related data were presented in Table 2. Meanwhile, the APCI was plotted in Figure 4B. From these results, we could see that neohesperidin had relatively weak in vitro antioxidant capacity; this implied the CLS extension function of neohesperidin was not depending on its in vitro antioxidant activity. However, with comparatively high in vitro antioxidant activity, CD and BCD extended the CLS of yeast BY4742. Overall, we cannot forecast a compound's CLS extending capacity just based on its antioxidant activity.

Table 1. The linear regression equations of the three assays.

Method	Equation	R^2	The Linear Range
DPPH	y = −0.0029x + 0.6716	0.9997	0–200 µM
ABTS	y = −0.0006x + 0.644	0.9962	0–1000 µM
FRAP	y = 0.0023x − 0.003	0.9992	0–560 µM

Table 2. The antioxidant capacities (µM trolox equivalents/µM phytochemicals) of the bioactive compounds analyzed in this study.

Phytochemicals	DPPH	ABTS	FRAP	APCI	Rank
A	11.33 ± 0.15 [j]	550.18 ± 1.21 [a]	350.74 ± 1.21 [b]	63.96	9
B	33.31 ± 1.16 [h]	533.57 ± 0.2 [a]	302.64 ± 0.82 [e]	71.32	6
C	41.39 ± 0.16 [f]	447.33 ± 0.59 [b]	160.01 ± 0.29 [h]	60.17	10
D	28.74 ± 0.17 [i]	119.33 ± 0.36 [c]	7.54 ± 0.21 [i]	23.02	15
AB	35.77 ± 0.06 [g]	571.97 ± 7.13 [a]	453.54 ± 0.86 [a]	85.97	3
AC	45.8 ± 0.24 [e]	548.94 ± 1.99 [a]	350.96 ± 0.46 [b]	82.51	4
AD	34.13 ± 0.10 [h]	500.91 ± 1.03 [b]	194.44 ± 0.62 [g]	61.90	10
BC	53.74 ± 0.23 [c]	432.93 ± 0.6 [b]	162.59 ± 0.71 [h]	66.19	8
BD	52.61 ± 0.05 [c]	444.8 ± 27.13 [b]	5.52 ± 0.22 [i]	54.73	12
CD	58.52 ± 0.06 [b]	546.17 ± 1.01 [a]	323.97 ± 0.04 [c]	87.23	1
ABC	58.61 ± 0.10 [b]	477.95 ± 3.37 [b]	208.58 ± 1.19 [f]	74.82	5
ABD	47.58 ± 0.12 [d]	489.66 ± 0.64 [b]	191.63 ± 1.50 [g]	68.30	7
ACD	61.75 ± 0.30 [a]	25.05 ± 0.74 [d]	4.03 ± 0.08 [i]	35.09	14
BCD	61.23 ± 0.11 [a]	538.31 ± 0.24 [a]	310.05 ± 0.52 [d]	87.21	2
DMSO	3.42 ± 0.18 [k]	455.56 ± 0.54 [b]	192.37 ± 0.26 [g]	42.53	13

A: naringin, B: hesperidin, C: hesperetin, D: neohesperidin, the concentration of A, B, C and D is 1µM. APCI = Σ (each sample value/the biggest sample value in that method)/the number of methods. Data were expressed as mean ± SEM (n = 9) and compared using one-way ANOVA's Sidak's multiple comparisons test at $p < 0.05$ by GraphPad Prism 7.00. Different letters (a, b, c, d, e, f, g, h, i, j, k) after data indicate values in the same column significant differences.

2.5. Neohesperidin Could Not Slow Down the Variation of Extracellular Acidification of Yeast BY4742

Important parameters include the composition of the growth medium as well as the pH value. The composition of the growth medium and pH value has been shown to have major impact on the CLS of S. cerevisiae [33]. The effects of pH on CLS of budding yeast were investigated by previous studies, and the results point to a mechanism of acetic acid toxicity related to the induction of growth signaling pathways and oxidative stress in yeast [34]. In order to know the effect of the four flavonoid compounds on the variation of extracellular acidification of yeast cultures, we detected the pH values every five minutes using a pH meter. In Figure 5, 10 µM naringin obviously slowed down the variation of extracellular acidification of budding yeast BY4742 at different aging states while the other three flavonoid compounds did not influence it significantly at the same concentration when compared to control groups.

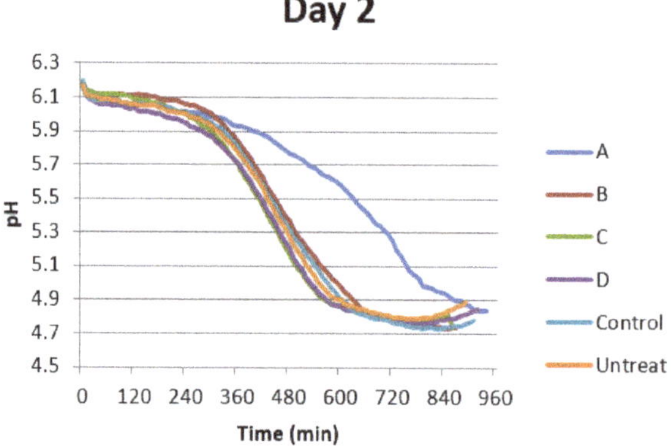

Figure 5. *Cont.*

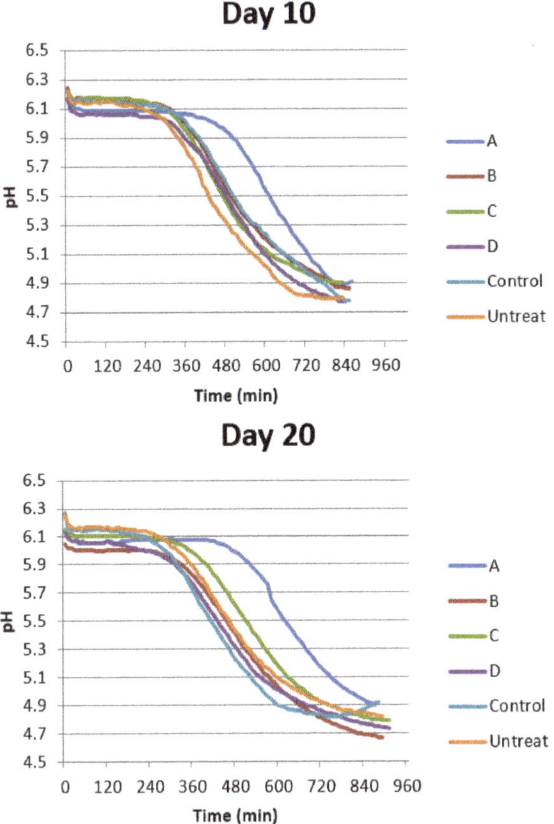

Figure 5. Variation of the pH of the culture mediums after the budding yeast BY4742 treated by the four flavoniod compounds at 10 μM. The experiment was performed at least in triplicate. A: naringin, B: hesperidin, C: hesperetin, D: neohesperidin.

3. Discussion

Former studies had reported that neohesperidin exhibited various anti-aging associated functions, such as the neuroprotective effect [15], ROS-scavenging and anti-inflammatory activities [35], attenuation of the decrease of mitochondrial membrane potential and the increase of caspase-3 activity evoked by H_2O_2 [16], and cellular apoptosis-inducing activity [21]. All these functions laid a good foundation for the result that neohesperidin increased the CLS of budding yeast BY4742 here. It is surprising that neohesperidin extended the CLS significantly only at the lowest concentration tested. In the report of Craker, et al. it also showed a lower concentration of auxin (10-6 IAA) promotes proton-extrusion. Proton-extrusion under a high concentration of auxin (10-4 IAA) is inhibited by auxin-induced ethylene [36]. The ROS-scavenging activity of neohesperidin was verified in our research. The in vitro antioxidant capacity of neohesperidin was relatively weak, which explained why the CLS extension function of neohesperidin did not depend on its in vitro antioxidant activity. However, the combinations CD and BCD had high in vitro antioxidant activity and increased the CLS of yeast (Figure 3B). The weak correlation of ROS and antioxidant activity may be caused by the method we used to analyze the antioxidant activity. The antioxidant activity method tried to reacted with a double bond at C 2–C 3 and/or a hydroxyl group at C 3 on the C ring of flavonoid. In Areias et al. the results strongly suggested that the higher antioxidant activity of the flavonoids is not correlated with the presence of a double bond at C 2–C 3 and/or a hydroxyl group at C 3 on the C ring, but rather may depend on the capacity to inhibit the production of reactive oxygen species to interact hydrophobically with membranes [37]. At the same times, a large number of studies have shown that some antioxidants do have the function of extending lifespan, their specific mechanisms of action are complex. Only some antioxidants have been shown to exhibit anti-aging effects related to the direct free radical and ROS clearance. But the life-extending effects of other antioxidants on model organisms were not limited to direct antioxidant function, but also include the regulation of stress-related genes expression and the induction of toxic stimulatory effects [38]. Therefore, it is impossible to predict the ability of a substance to extend the CLS of yeast based on its antioxidant activity. Though we did not test the result in other strains, it offered information for other researchers and scientists to validate the result in other strains and model organisms.

Many cellular processes and extrinsic factors negatively influence the yeast chronological lifespan, including medium acidification andoxidative stress [39]. One of the early changes that occurs in yeast cells grown in media containing 2% glucose (dextrose) is the production of acetic acid and acidification of the medium, which has been shown to influence chronological aging [38]. Buffering the medium to pH 6–7 prevents acidification and increases chronological life span [10,39–41]. Additionally, acetic acid can be utilized by *Saccharomyces cerevisiae* for growth and metabolism in spite of its potential toxicity [42]. 10 μM neohesperidin, hesperidin and hesperetin maintained the variation trend of extracellular pH values when compared with control (Figure 5). However, 10 μM naringin clearly slowed down the variation of extracellular acidification (Figure 5). At this concentration, the CLS of yeast BY4742 was treated with neohesperidin, hesperidin and hesperetin were almost the same as the control group.

Increased ROS scavenging has marked effects on CLS in yeast, but the reason remains an unresolved issue [33]. Aging and related diseases are the consequence of free radical-mediated damage to cellular macromolecules and their inability to counterbalance endogenous antioxidant defenses mechanisms [43]. However, data has indicated that ROS also can play a positive role in inducing stress response genes (hormesis) [43–46].

Moreover, recent findings suggest that also the type of ROS and the time they occur are important for lifespan extension in *S. cerevisiae* [47–49], which illustrates the complex role of ROS in yeast aging. Graziano et al. [47] reported that neohesperidin decreased ROS generation in human keratinocytes. Nohara, et al. recently revealed that nobiletin (one of flavonoids in citrus) fortifies mitochondrial respiration in skeletal muscle to promote healthy aging against metabolic challenge. ROS production was significantly suppressed by nobiletin treatment in a dose-dependent manner [50]. From Figures 3B and 4A, it showed that D and CD increased the CLS of yeast BY4742 and decreased the intracellular ROS content. But, for ABD and BCD, they prolonged the CLS while increased the intracellular ROS. This result was consistent with Wu's [51]. So, there was no certain positive or negative relationship between ROS scavenging activities of the compounds and their effects on lifespan, and this was also in line with the intricate role of ROS in yeast aging.

In Figure 3A, the survival rates of yeast BY4742 under the treatment of different compounds and their combinations from day 2 to day 20 were not gradually decreasing. They present double peaks. This can be also found in the result of Wu et al. (2014). For the first few days, the survival rates were relatively high. This can be explained by the enough nutrient and low survival pressure during this time. As time went on, valley points appeared for a very short time as the nutrition became less. Then, peaks appeared again. This phenomenon might attribute the success to the metabolism of another nutrient that alleviated the survival pressure.

Qi et al. [36] reported that the antioxidant activity of antioxidants mixture/compounds combination was more effective than a single compound. In our experiment, the combinations AB, AC, CD, ABC, and BCD had a stronger antioxidant capacity than any single substance corresponding to them. Based on our observations, it could be concluded that the flavonoids present in a mixture could interact, and their interactions could affect the total antioxidant capacity of a solution (Figure 4B). Although we demonstrated that the four flavonoid interactions trigger synergistic or antagonistic effects for the antioxidant power, there are other flavonoid combinations that require a more detailed study in order to better understand the mechanisms involved in these interactions. Lutchman et al. [52] had reported plant extracts that increased the yeast 's CLS. And the autophagy promoted by decreased TORC1 signaling is critically important for a long CLS [10]. By referring to these studies, we can explore the way by which the compounds execute their effect in future studies.

4. Materials and Methods

4.1. Materials

The wild-type strain *S. cerevisiae* BY4742 (ATCC®, 201389™) (*MATα his3Δ1 leu2Δ0 lys2Δ0 ura3Δ0*) was obtained from American Type Culture Collection (Manassas, VA, USA). The culture of yeast reference strain was aliquoted into 10 µL and stored at −80 °C. All L-amino acids, yeast nitrogen base w/o amino acids (YNB), ammonium sulfate, peptone, agar and yeast extract, H_2DCFDA, 2,4,6-tripyridyl-s-triazine (TPTZ), 2,2′-azino-bis (3-ethylbenzothiazoline-6-sulfonic acid) (ABTS), 1,1-diphenyl-2-picrylhydrazyl (DPPH), dimethyl sulfoxide (DMSO), neohesperidin, naringin, hesperidin and hesperetin were bought from Sigma-Aldrich (Shanghai, China). YPD Broth, YPD and other chemicals were from Solebo biotech Co., Ltd. (Beijing, China). The 96-well polystyrene microplates with flat bottom were purchased from Corning incorporated (Kennebunk, ME, USA).

4.2. Lifespan and Yeast Cell Growth Assay

The determination of chronological lifespan (CLS) of yeast was carried out according to the method of Wu et al. [25] with a moderate modification as follows. In brief, the yeast cells were prepared by transferring a streaked strain from frozen stocks onto YPD agar (0.5% yeast extract/1% peptone/2% dextrose/1.4% agar) plates. After incubating the cells at 30 °C for 2 days, a single colony was picked and inoculated into a 1.0-mL YPD liquid medium (1% yeast extract/2% peptone/2% dextrose) in a 10-mL sterilized centrifuge tube (round bottom) and cultured at 30 °C for 2 days in a flat incubator at 200 rpm. The 2-day YPD culture was diluted with autoclaved 18 mΩ Milli-Q grade water (1:10) and stored in a refrigerator at 4 °C for 2 days. After 2-day incubation at 4 °C, 5 µL ($\approx 1 \times 10^4$ cells) of the diluted culture was transferred to 993 microliter of synthetic-defined (SD) media (Supplementary Table S1, [51]) and maintained at 30 °C, 200 rpm for the entire experiment. Compounds in DMSO with several concentrations (2.0 µL) were added at the initial inoculation (0 h). Each experiment was performed at least in triplicate. Cell cultures were incubated at 30 °C without replacing the aging medium throughout the experiment. After 2 days of culture in an aging media, the cells reached stationary phase and the first age-point was then taken. Subsequent age-points were taken every 2–4 days. For each age point, 5.0 µL of the mixed culture was pipetted into each well of 96-well flat-bottom microplate. Ninety-five microliter YPD medium was then added to each well. The cell population was monitored with a microplate reader (Varioskan Flash; Thermo Scientific, Waltham, MA, USA) by recording OD660 every 10 min during 24 h.

The survival rate was calculated as follows [25]. Where tOD= 0.3, 2day is the time that OD value of day 2 age-point reaches 0.3 in the outgrowth curves. The initial age-point (day 2) is defined to be 100% viability and the relative survival percent of each successive age-point can be calculated as follows:

$$Vn = \frac{1}{2^{\frac{\Delta t_n}{\overline{Dt_n}}}} \times 100 n = days, \ \overline{Dt_n} \text{ represent the average doubling time.}$$

The survival integral (SI) for each well is defined as the area under the survival curve (AUC) and can be estimated by the formula:

$$\text{SI} = \sum_{2}^{n} \left(\frac{V_{n-1} + V_n}{2} \right) (day_n - day_{n-1})$$

where day_n is the age point, such as days 2, 4, 6, 8, 10, 12, 14, 16, 18 and 20.

4.3. Intracellular ROS Scavenging Ability Assays

To quantify the intracellular reactive oxygen species (ROS) level of yeast cells grown in standard SD medium with/without the four compounds, the method described in Wu et al. [51] had been referenced. Namely, 2 μL ROS probe H_2DCFDA from a fresh 5-mM stock solution in DMSO was added into 1.0 mL yeast aging culture (at day 2) at 30 °C for 1 h. The culture was then washed twice in sterile distilled water and suspended in 1.0 mL of 50 mM Tris/Cl buffer (pH 7.5). Twenty microliter of chloroform and 10 μL of 0.1% (*w/v*) sodium dodecyl sulfate (SDS) were added, and the cells were incubated at 200 rpm for 30 min to allow the dye to diffuse. The culture was centrifuged at 5000 rpm for 5 min, and the fluorescence of the supernatant was measured using a microplate reader with excitation at 480 nm and emission at 520 nm.

4.4. Antioxidant Activity Assays

In this study, DPPH, FRAP and ABTS assays were used for in vitro antioxidant capacity evaluation. The DPPH assay was performed according to the method described by Barreca et al. [53]. An aliquot of each sample (0.5 mL) was mixed with 75 μM (3.5 mL) of DPPH in methanol to a final volume of 4.0 mL. After reacting for 30 min without light, the absorption of the mixture was detected at a wavelength of 517 nm. The inhibition percentage of radical scavenging activity was the DPPH value. The FRAP assay was carried out according to Hungder et al. [54] with some modifications. A 0.2 mL aliquot of the sample was mixed with 3.8 mL of FRAP reagent (0.3 mol/L acetate buffer (pH 3.6), 10 mmol/L TPTZ solution, and 20 mmol/L ferric chloride ($FeCl_3$) were mixed (10:1:1, volume ratio)). After 30 min, the absorbance was detected at wavelength of 593 nm. And the ABTS assay was followed the method of Mnb et al. [55] with few modifications. The ABTS radical cation (ABTS •+) was generated by reaction of 176 μL of potassium persulfate solution (140 mM) and 10 mL aqueous ABTS solution (7 mM) under the condition of no light for 12–16 h. Then it was diluted with methanol to an absorption value of 0.7 ± 0.02 units at 734 nm. 0.1 mL sample was added to 4.9 mL ABTS reagent. The absorbance was measured at a wavelength of 734 nm after 10 min reaction. All absorbance values were determined by using the UV–VIS spectrophotometer (PerkinElmer Lambda 25 UV/VIS, Waltham, MA, USA). Antioxidant values were calculated by standard curve method and expressed as trolox equivalents (TE mg/g DW).

4.5. Extracellular pH Detection

The culture process of yeast in the early stage was almost the same as the method described in the part of "Lifespan and yeast cell growth assay". The specific operation was as follows. The yeast cells were prepared by transferring the yeast strain from frozen stocks onto YPD agar plates. After incubating the cells at 30 °C for 2 days, a single colony was picked and inoculated into a 1.0-mL YPD liquid medium in a 10-mL sterilized centrifuge tube and cultured at 30 °C for 2 days in a flat incubator at 200 rpm. The 2-day YPD culture was diluted with autoclaved 18 mΩ Milli-Q grade water (1:10) and stored in a refrigerator at 4 °C for 2 days. After 2-day incubation at 4 °C, 10 μL ($\approx 2 \times 10^4$ cells) of the diluted culture was transferred to 1986 μL of SD media and maintained at 30 °C, 200 rpm for 2/10/20 days. Various compounds in DMSO (4.0 μL) were add to the medium to a final concentration of 10 μM at initial inoculation (0 h). Then 1mL of the 2/10/20-day SD culture was added to 19 mL fresh YPD liquid medium. The pH was tested every five minutes using a pH meter while the yeast was cultured at 30 °C in a shaker at 200 rpm for the whole detection. Each experiment was performed at least in triplicate.

4.6. Data Analysis

The raw data from the microplate reader were exported to Microsoft Excel 2007 (Redmond, WA, USA). From the growth curves, the viability of the yeast can be obtained according to previous report [25]. Survival integral [56] of each aging culture was defined as the area under the survival curves [25,57]. The data were analyzed by one-way analysis of variance (one-way ANOVA), and were expressed as the mean values ± standard error of mean (SEM). The significance of difference (* $p < 0.05$; ** $p < 0.01$; *** $p < 0.001$; **** $p < 0.0001$) was determined using Sidak's multiple comparisons test. GraphPad Prism 7 (GraphPad Software, Inc., La Jolla, CA, USA) was used for the analysis.

5. Conclusions

In conclusion, neohesperidin, with relatively high capability of removing intracellular ROS, showed great potential in extending the CLS of budding yeast BY4742 individually or synergistically with hesperetin. This might lead to new choices for the treatment of aging problems and since there was a limited relationship between the CLS-extension of yeast and the tested indexes (e.g., ROS scavenging ability, in vitro antioxidant activity and the extracellular pH). Further studies to discover the molecular mechanisms of this phenomenon will be extremely beneficial to prevent aging. Because of this, the importance of choosing the best combination of flavonoids, should be borne in mind when designing new dietary supplements or functional foods.

Supplementary Materials: The following are available online at http://www.mdpi.com/1420-3049/24/22/4093/s1, Table S1: Composition of synthetic-defined (SD) medium used for yeast chronological lifespan analysis.

Author Contributions: Conceptualization, C.G., H.Z., X.G. and Z.Z.; Data curation, C.G. and X.G.; Formal analysis, C.G. and X.G.; Funding acquisition, H.Z., C.G. and Z.Z.; Methodology, Z.Z., C.G., H.Z. and X.G.; Project administration, Z.Z. and X.G.; Software, X.G.; Supervision, X.G. and Z.Z.; Validation, C.G. and H.Z.; Writing – original draft, Z.Z., X.G. and C.G.; Writing–review & editing, Z.Z., X.G., C.G. and H.Z.

Funding: This research was funded by the National Natural Science Foundation (31772260) and Chongqing Graduate Student Innovation Project for Chunxia Guo (CYS16081), the Program for Chongqing Municipal Education Commission (KJQN201901227), and the Foundation of Engineering Technology Research Center of Characteristic Biological Resources in Northeast of Chongqing.

Acknowledgments: We appreciated the technical support of Chun Tan in College of Horticulture and Landscape Architecture, Southwest University, China and Charles Wilson in Chongqing Three Gorges University, China.

Conflicts of Interest: The authors declare no conflict of interest.

References

1. López-Otín, C.; Blasco, M.A.; Partridge, L.; Serrano, M.; Kroemer, G. The hallmarks of aging. *Cell* **2013**, *153*, 1194–1217. [CrossRef]
2. Argyropoulou, A.; Aligiannis, N.; Trougakos, I.P.; Skaltsounis, A.L. Natural compounds with anti-ageing activity. *Nat. Prod. Rep.* **2013**, *30*, 1412–1437. [CrossRef]
3. Ding, A.J.; Zheng, S.Q.; Huang, X.B.; Xing, T.K.; Wu, G.S.; Sun, H.Y.; Qi, S.H.; Luo, H.R. Current perspective in the discovery of anti-aging agents from natural products. *Nat. Prod. Biol.* **2017**, *7*, 1–70. [CrossRef]
4. Pan, M.H.; Lai, C.S.; Tsai, M.L.; Wu, J.C.; Ho, C.T. Molecular mechanisms for anti-aging by natural dietary compounds. *Mol. Nutr. Food Res.* **2012**, *56*, 88–115. [CrossRef] [PubMed]
5. Si, H.; Liu, D. Dietary antiaging phytochemicals and mechanisms associated with prolonged survival. *J. Nutr. Biochem.* **2014**, *25*, 581–591. [CrossRef] [PubMed]
6. Kaeberlein, M. Lessons on longevity from budding yeast. *Nature* **2010**, *464*, 513. [CrossRef] [PubMed]
7. Prasain, J.K.; Carlson, S.H.; Wyss, J.M. Flavonoids and age-related disease, risk; benefits and critical windows. *Maturitas* **2010**, *66*, 163–171. [CrossRef] [PubMed]
8. Longo, V.D.; Shadel, G.S.; Kaeberlein, M.; Kennedy, B. Replicative and Chronological Aging in Saccharomyces cerevisiae. *Cell Metab.* **2012**, *16*, 18–31. [CrossRef] [PubMed]
9. Steinkraus, K.A.; Kaeberlein, M.; Kennedy, B.K. Replicative aging in yeast, the means to the end. *Annu. Rev. Cell Dev. Biol.* **2008**, *24*, 29. [CrossRef]

10. Burtner, C.R.; Murakami, C.J.; Kennedy, B.K.; Kaeberlein, M. A molecular mechanism of chronological aging in yeast. *Cell Cycle* **2009**, *8*, 1256–1270. [CrossRef]

11. Wagner, H. Synergy research, approaching a new generation of phytopharmaceuticals. *Fitoterapia* **2011**, *82*, 34–37. [CrossRef] [PubMed]

12. Benavente-Garcia, O.; Castillo, J. Update on uses and properties of citrus flavonoids, new findings in anticancer; cardiovascular; and anti-inflammatory activity. *J. Agric. Food chem.* **2008**, *56*, 6185–6205. [CrossRef] [PubMed]

13. Choi, E.J.; Ahn, W.S. Neuroprotective effects of chronic hesperetin administration in mice. *Arch. Pharm. Res.* **2008**, *31*, 1457. [CrossRef] [PubMed]

14. Elavarasan, J.; Velusamy, P.; Ganesan, T.; Ramakrishnan, S.K.; Rajasekaran, D.; Periandavan, K. Hesperidin-mediated expression of Nrf2 and upregulation of antioxidant status in senescent rat heart. *J. Pharm. Pharmacol.* **2012**, *64*, 1472–1482. [CrossRef] [PubMed]

15. Ho, S.L.; Poon, C.Y.; Lin, C.Y.; Yan, T.; Kwong, D.W.; Yung, K.K.; Wong, M.S.; Bian, Z.X.; Li, H.W. Inhibition of β-amyloid aggregation by albiflorin; aloeemodin and neohesperidin and their neuroprotective effect on primary hippocampal cells against β-amyloid induced toxicity. *Curr. Alzheimer Res.* **2015**, *12*, 424–433. [CrossRef]

16. Hwang, S.L.; Yen, G.C. Neuroprotective effects of the citrus flavanones against H_2O_2-induced cytotoxicity in PC12 cells. *J. Agric. Food Chem.* **2008**, *56*, 859–864. [CrossRef]

17. Kandhare, A.D.; Raygude, K.S.; Ghosh, P.; Ghule, A.E.; Bodhankar, S.L. Neuroprotective effect of naringin by modulation of endogenous biomarkers in streptozotocin induced painful diabetic neuropathy. *Fitoterapia* **2012**, *83*, 650–659. [CrossRef]

18. Kumar, A.; Dogra, S.; Prakash, A. Protective effect of naringin; a citrus flavonoid; against colchicine-induced cognitive dysfunction and oxidative damage in rats. *J. Med. Food* **2010**, *13*, 976–984. [CrossRef]

19. Parhiz, H.; Roohbakhsh, A.; Soltani, F.; Rezaee, R.; Iranshahi, M. Antioxidant and anti-inflammatory properties of the citrus flavonoids hesperidin and hesperetin, an updated review of their molecular mechanisms and experimental models. *Phytother. Res.* **2015**, *29*, 323–331. [CrossRef]

20. Sun, X.L.; Ishihara, S.; Matsuura, A.; Sakagami, Y.; Qi, J. Anti-aging effects of hesperidin on Saccharomyces cerevisiae via inhibition of reactive oxygen species and UTH1 gene expression. *Biosci. Biotechnol. Biochem.* **2012**, *76*, 640–645. [CrossRef]

21. Xu, F.; Zang, J.; Chen, D.; Zhang, T.; Zhan, H.Y.; Lu, M.D.; Zhuge, H.X. Neohesperidin induces cellular apoptosis in human breast adenocarcinoma MDA-MB-231 cells via activating the Bcl-2/Bax-mediated signaling pathway. *Nat. Prod. Commun.* **2012**, *7*, 1475. [CrossRef] [PubMed]

22. Murakami, C.; Delaney, J.R.; Chou, A.; Carr, D.; Schleit, J.; Sutphin, G.L.; An, E.H.; Castanza, A.S.; Fletcher, M.; Goswami, S. pH neutralization protects against reduction in replicative lifespan following chronological aging in yeast. *Cell Cycle* **2012**, *11*, 3087–3096. [CrossRef] [PubMed]

23. Hazen, K.C. Influence of DMSO on antifungal activity during susceptibility testing in vitro. *Diagn. Microbiol. Infect. Dis.* **2013**, *75*, 60–63. [CrossRef]

24. Murakami, C.J.; Burtner, C.R.; Kennedy, B.K.; Kaeberlein, M. A method for high-throughput quantitative analysis of yeast chronological life span. *J. Gerontol.* **2008**, *63*, 113. [CrossRef] [PubMed]

25. Wu, Z.; Song, L.; Liu, S.Q.; Huang, D. A high throughput screening assay for determination of chronological lifespan of yeast. *Exp. Gerontol.* **2011**, *46*, 915–922. [CrossRef]

26. Prabhakar, P.K.; Kumar, A.; Doble, M. Combination therapy: A new strategy to manage diabetes and its complications. *Phytomedicine Int. J. Phytother. Phytopharm.* **2014**, *21*, 123. [CrossRef]

27. Liu, R.H. Potential synergy of phytochemicals in cancer prevention: Mechanism of action. *J. Nutr.* **2004**, *134*, 3479S–3485S. [CrossRef]

28. Shen, J.Z.; Tlj, N.; Ho, W.S. Therapeutic potential of phytochemicals in combination with drugs for cardiovascular disorders. *Curr. Pharm. Des.* **2017**, *23*, 961–966. [CrossRef]

29. Rayalam, S.; Della-Fera, M.A.; Baile, C.A. Synergism between resveratrol and other phytochemicals: Implications for obesity and osteoporosis. *Mol. Nutr. Food Res.* **2011**, *55*, 1177–1185. [CrossRef]

30. Harman, D. Aging: A theory based on free radical and radiation chemistry. *J. Gerontol.* **1956**, *11*, 298. [CrossRef]

31. Moradas-Ferreira, P.; Costa, V.; Piper, P.; Mager, W. The molecular defences against reactive oxygen species in yeast. *Mol. Microbiol.* **1996**, *19*, 651–658. [CrossRef] [PubMed]

32. Hayes, J.E.; Allen, P.; Brunton, N.; O'Grady, M.N.; Kerry, J.P. Phenolic composition and in vitro antioxidant capacity of four commercial phytochemical products: Olive leaf extract (*Olea europaea* L.), lutein, sesamol and ellagic acid. *Food Chem.* **2011**, *126*, 948–955. [CrossRef]

33. Piper, P.W. Maximising the yeast chronological lifespan. *Sub-Cell. Biochem.* **2012**, *57*, 145.

34. Burhans, W.C.; Weinberger, M. Acetic acid effects on aging in budding yeast: Are they relevant to aging in higher eukaryotes? *Cell Cycle* **2009**, *8*, 2300–2302. [CrossRef] [PubMed]

35. Hwang, S.L.; Shih, P.H.; Yen, G.C. Neuroprotective effects of citrus flavonoids. *J. Agric. Food Chem.* **2012**, *60*, 877. [CrossRef]

36. Craker, L.E.; Cookson, C.; Osborne, D.J. Control of proton extrusion and cell elongation by ethylene and auxin in the water-plant Ranunculus sceleratus. *Plant. Sci. Lett.* **1978**, *12*, 379–385. [CrossRef]

37. Areias, F.M.; Rego, A.C.; Oliveira, C.R.; Seabra, R.M. Antioxidant effect of flavonoids after ascorbate/Fe^{2+}-induced oxidative stress in cultured retinal cells. *Biochem. Pharmacol.* **2001**, *62*, 111–118. [CrossRef]

38. Qi, W.; Zhu, C.; Fan, X.; Yang, D.; Yang, M. Research progress on the correlation between antioxidants and life extension. *Chin. J. Pharmacol. Toxicol.* **2016**, *3*, 588–597. (In Chinese)

39. Kumar, S.; Lefevre, S.D.; Veenhuis, M.; Klei, I.J.V.D. Extension of yeast chronological lifespan by methylamine. *PLoS ONE* **2012**, *7*, e48982. [CrossRef]

40. Murakami, C.J.; Wall, V.; Basisty, N.; Kaeberlein, M. Composition and acidification of the culture medium influences chronological aging similarly in vineyard and laboratory yeast. *PLoS ONE* **2011**, *6*, e24530. [CrossRef]

41. Morgunova, G.V.; Klebanov, A.A.; Marotta, F.; Khokhlov, A.N. Culture medium pH and stationary phase/chronological aging of different cells. *Mosc. Univ. Biol. Sci. Bull.* **2017**, *72*, 58–62. [CrossRef]

42. Gilvarg, C.; Bloch, K. The utilization of acetic acid for amino acid synthesis in yeast. *J. Biol. Chem.* **1951**, *193*, 339. [PubMed]

43. Yang, Y.; Zhao, X.J.; Pan, Y.; Zhou, Z. Identification of the chemical compositions of Ponkan peel by ultra performance liquid chromatography coupled with quadrupole time-of-flight mass spectrometry. *Anal. Methods* **2015**, *8*, 893–903. [CrossRef]

44. Cypser, J.R.; Johnson, T.E. Multiple stressors in caenorhabditis elegans induce stress hormesis and extended longevity. *J. Gerontol. Ser. A* **2002**, *57*, B109–B114. [CrossRef] [PubMed]

45. Ludovico, P.; Burhans, W.C. Reactive oxygen species, ageing and the hormesis police. *FEMS Yeast Res.* **2014**, *14*, 33–39. [CrossRef] [PubMed]

46. Pan, Y. Mitochondria, reactive oxygen species, and chronological aging: A message from yeast. *Exp. Gerontol.* **2011**, *46*, 847–852. [CrossRef]

47. Mesquita, A.; Weinberger, M.; Silva, A.; Sampaiomarques, B.; Almeida, B.; Leão, C.; Costa, V.; Rodrigues, F.; Burhans, W.C.; Ludovico, P. Caloric restriction or catalase inactivation extends yeast chronological lifespan by inducing H$_2$O$_2$ and superoxide dismutase activity. *Proc. Natl. Acad. Sci. USA* **2010**, *107*, 15123–15128. [CrossRef]

48. Pan, Y.; Schroeder, E.A.; Ocampo, A.; Barrientos, A.; Shadel, G.S. Regulation of yeast chronological life span by TORC1 via adaptive mitochondrial ROS signaling. *Cell Metab.* **2011**, *13*, 668–678. [CrossRef]

49. Graziano, A.C.; Cardile, V.; Crascì, L.; Caggia, S.; Dugo, P.; Bonina, F.; Panico, A. Protective effects of an extract from Citrus bergamia against inflammatory injury in interferon-γ and histamine exposed human keratinocytes. *Life Sci.* **2012**, *90*, 968–974. [CrossRef]

50. Nohara, K.; Mallampalli, V.; Nemkov, T.; Wirianto, M.; Yang, J.; Ye, Y.; Sun, Y.; Han, L.; Esser, K.A.; Mileykovskaya, E.; et al. Nobiletin fortifies mitochondrial respiration in skeletal muscle to promote healthy aging against metabolic challenge. *Nat. Commun.* **2019**, *10*, 1–15. [CrossRef]

51. Wu, Z.; Song, L.; Liu, S.Q.; Huang, D. Tanshinones extend chronological lifespan in budding yeast Saccharomyces cerevisiae. *Appl. Microbiol. Biotechnol.* **2014**, *98*, 8617–8628. [CrossRef] [PubMed]

52. Lutchman, V.; Medkour, Y.; Samson, E.; Arlia-Ciommo, A.; Dakik, P.; Cortes, B.; Rukundo, B. Discovery of plant extracts that greatly delay yeast chronological aging and have different effects on longevity-defining cellular processes. *Oncotarget* **2016**, *7*, 16542–16566. [CrossRef] [PubMed]

53. Barreca, D.; Bellocco, E.; Caristi, C.; Leuzzi, U.; Gattuso, G. Kumquat (*Fortunella japonica* Swingle) juice: Flavonoid distribution and antioxidant properties. *Food Res. Int.* **2011**, *44*, 2190–2197. [CrossRef]

54. Hungder, J.; Chang, K.S.; Chang, T.C.; Chuanliang, H. Antioxidant potentials of buntan pumelo (*Citrus grandis* Osbeck) and its ethanolic and acetified fermentation products. *Food Chem.* **2010**, *118*, 554–558.
55. Almeida, M.M.B.; de Sousa, P.H.M.; Arriaga, Â.M.C.; do Prado, G.M.; de Carvalho Magalhães, C.E.; Maia, G.A.; de Lemos, T.L.G. Bioactive compounds and antioxidant activity of fresh exotic fruits from northeastern Brazil. *Food Res. Int.* **2011**, *44*, 2155–2159. [CrossRef]
56. Park, S.J.; Ahmad, F.; Philp, A.; Baar, K.; Williams, T.; Luo, H.; Ke, H.; Rehmann, H.; Taussig, R.; Brown, A.L. Resveratrol ameliorates aging-related metabolic phenotypes by inhibiting cAMP phosphodiesterases. *Cell* **2012**, *148*, 421–433. [CrossRef]
57. Bottino, D.A.; Lopes, F.G.; Oliveira, F.J.D.; Mecenas, A.D.S.; Clapauch, R.; Bouskela, E. Relationship between biomarkers of inflammation, oxidative stress and endothelial/microcirculatory function in successful aging versus healthy youth: A transversal study. *BMC Geriatr.* **2015**, *15*, 41. [CrossRef]

Sample Availability: Samples are not available from the authors.

 *molecules* 

Review

The Targeted Impact of Flavones on Obesity-Induced Inflammation and the Potential Synergistic Role in Cancer and the Gut Microbiota

Meenakshi Sudhakaran [1,2] and Andrea I. Doseff [2,3,*]

[1]   Physiology Graduate Program, Michigan State University, East Lansing, MI 48824, USA; sudhaka7@msu.edu
[2]   Department of Physiology, Michigan State University, East Lansing, MI 48824, USA
[3]   Department of Pharmacology and Toxicology, Michigan State University, East Lansing, MI 48824, USA
*   Correspondence: doseffan@msu.edu

Academic Editors: H.P. Vasantha Rupasinghe and Celestino Santos-Buelga
Received: 26 March 2020; Accepted: 23 May 2020; Published: 27 May 2020

Abstract: Obesity is an inflammatory disease that is approaching pandemic levels, affecting nearly 30% of the world's total population. Obesity increases the risk of diabetes, cardiovascular disorders, and cancer, consequentially impacting the quality of life and imposing a serious socioeconomic burden. Hence, reducing obesity and related life-threatening conditions has become a paramount health challenge. The chronic systemic inflammation characteristic of obesity promotes adipose tissue remodeling and metabolic changes. Macrophages, the major culprits in obesity-induced inflammation, contribute to sustaining a dysregulated immune function, which creates a vicious adipocyte–macrophage crosstalk, leading to insulin resistance and metabolic disorders. Therefore, targeting regulatory inflammatory pathways has attracted great attention to overcome obesity and its related conditions. However, the lack of clinical efficacy and the undesirable side-effects of available therapeutic options for obesity provide compelling reasons for the need to identify additional approaches for the prevention and treatment of obesity-induced inflammation. Plant-based active metabolites or nutraceuticals and diets with an increased content of these compounds are emerging as subjects of intense scientific investigation, due to their ability to ameliorate inflammatory conditions and offer safe and cost-effective opportunities to improve health. Flavones are a class of flavonoids with anti-obesogenic, anti-inflammatory and anti-carcinogenic properties. Preclinical studies have laid foundations by establishing the potential role of flavones in suppressing adipogenesis, inducing browning, modulating immune responses in the adipose tissues, and hindering obesity-induced inflammation. Nonetheless, the understanding of the molecular mechanisms responsible for the anti-obesogenic activity of flavones remains scarce and requires further investigations. This review recapitulates the molecular aspects of obesity-induced inflammation and the crosstalk between adipocytes and macrophages, while focusing on the current evidence on the health benefits of flavones against obesity and chronic inflammation, which has been positively correlated with an enhanced cancer incidence. We conclude the review by highlighting the areas of research warranting a deeper investigation, with an emphasis on flavones and their potential impact on the crosstalk between adipocytes, the immune system, the gut microbiome, and their role in the regulation of obesity.

Keywords: flavones; inflammation; obesity; cancer; microbiome; molecular mechanisms; gene and protein regulatory networks; macrophages; NF-κB; IKKβ, inflammatory cytokines; apoptosis; apigenin; foods for health

---

1. Introduction

The incidence of obesity has ascended steadily in the last ~35 years and is reaching epidemic levels worldwide, inflicting life-threatening conditions and great socioeconomic burden. It is an alarming fact that almost half of the world's population is obese or overweight. Obesity is a major health concern often correlated with deteriorating life expectancy and increasing risks of several comorbid disorders, such as cardiovascular diseases, hypertension, type-2 diabetes mellitus, non-alcoholic fatty liver disease, steatohepatitis, osteoarthritis, and cancer [1]. The significant increase in obesity within the world's population prompted the need for identifying novel cost-effective interventions that are capable of controlling obesity with minimal harmful side effects. Obesity and obesity-linked diseases are associated with systemic chronic inflammation that leads to altered adipocyte functions [2]. Aberrant accumulation of macrophages (referred to as adipose tissue associated macrophages or ATMs) in adipose depots and other immune cells are vital contributors to obesity-induced inflammation [3,4]. The discovery that immune cell infiltration increases in adipose tissues of obese individuals has opened a new aspect in the research field and emphasizes the interest of using strategies that target immune cells to overcome the adversities associated with obesity. Thus, elucidating the mechanisms underlying obesity-linked inflammation has been suggested as a potential approach in preventing and battling obesity. High fat diets (HFD) induced harmful changes in the gut microbiome, leading to inflammation and systemic metabolic dysregulation [5]. Therefore, regulating the gut microbiome through the use of healthier diets could impact prevention and treatment of obesity.

Flavonoids are a large class of bioactive dietary nutraceuticals derived from the phenolic metabolism, which is widely distributed in plants and represents the important nutritional components of our diet [6]. Flavonoids, with more than 7000 identified, so far, have a myriad of health-promoting effects, owing to their potent antioxidant, anti-inflammatory, anti-carcinogenic, and immuno-modularity properties [7,8]. Due to these benefits, flavonoids are attracting great interest in the treatment and prevention of chronic inflammatory diseases. Emerging evidence suggests that the intake of flavonoid-rich diets exerts an inverse correlation with obesity and related inflammation [9,10]. Interestingly, recent studies showed that flavonoids can alter the gut microbiota ecosystem, reducing systemic inflammation [11]. Here, we reviewed the mechanistic aspects of obesity-induced inflammation, as well as the current knowledge on the role of dietary flavones, a subclass of flavonoids, and the molecular mechanisms that are involved in regulating obesity-induced inflammation and related diseases, such as cancer. We also highlight the potential beneficial effects of flavones on the relation between gut microbiota, immune and adipocyte homeostasis, and their impact on controlling and treating obesity.

2. Obesity-Induced Inflammation and Its Impact on Health

2.1. Obesity

Obesity is defined as an increase in body mass fat, resulting in excessive calorie consumption associated with a high incidence in the development of cardiovascular disease, metabolic dysfunction, diabetes, liver damage, and even cancer. Obesity is characterized by the presence of low and systemic chronic inflammation, which leads to dysregulated adipocyte function, promoting hormonal changes that alter the regulation of food consumption. In mammals, adipose tissue is classified into white adipose tissue (WAT) and brown adipose tissue (BAT). WAT functions as a reservoir of triglycerides from which free fatty acids (FFAs) are released to fuel the energy demands. BAT is considered to be a lipid reserve for cold-induced adaptive thermogenesis and is characterized by an increased mitochondrial count, lipolysis, and expression of uncoupling protein-1 (UCP-1), a key protein involved in the regulation of energy expenditure and protection against oxidative stress [12]. WAT is divided into two main depots, the subcutaneous adipose tissue (SAT) found under the skin, and the visceral adipose tissue (VAT) located around the internal organs. Among these depots, two types of thermogenic adipocytes are known to exist—classical brown and beige, which have disparate developmental and anatomical characteristics. The classical brown adipocytes found in the BAT have an embryonic origin, whereas the inducible thermogenic beige adipocytes exist in WAT and are derived either through transdifferentiation of WAT or from beige adipocyte precursor cells expressing platelet-derived growth factor receptor (PDGFR)-α [13]. This occasional development of beige adipocytes is referred to as the browning of WAT and was first identified in rodents, however, recent findings suggest the presence of beige adipocytes in humans as well. Browning is associated with resistance to HFD-induced obesity. In obese individuals, the conversion of triglycerides into FFAs through lipolytic enzymes such as adipose triglyceride lipase (ATGL) and hormone-sensitive lipase (HSL), and the subsequent FFA β-oxidation are impaired, consequently affecting the browning of adipose tissues [14]. Adipose tissue is also recognized as an endocrine organ that regulates systemic energy homeostasis, releasing a repertoire of cytokines (referred to as adipokines), hormones, and lipokines [15,16]. Leptin, an adipokine secreted during food intake, plays a key role in maintaining metabolic balance by suppressing appetite. Leptin levels are highly upregulated in obesity, but obese individuals become "leptin resistant" by losing their ability to control food ingestion, despite the presence of high levels of leptin [17]. The adipokine adiponectin has anti-obesogenic functions and regulates glucose levels and lipid oxidation [18]. Interestingly, recent studies involving single cell sequencing, and metabolomic and proteomic analyses of human mesenchymal progenitors and WAT, identified adipocyte progenitors that developed into adipocyte subsets with distinct metabolic and endocrine functions [19,20]. These findings highlight the cellular heterogeneity of adipose tissues and the need to gain a better understanding of the adipocyte populations, its precursors and the regulatory mechanisms that define their role in obesity.

Adipogenesis is a multi-step process that involves the development of a multipotent mesenchymal stem cell into a precursor preadipocyte, which then further differentiates into a mature lipid-laden adipocyte [21]. In the mouse cell line 3T3-L1, a broadly used model of adipogenesis, mitotic clonal expansion (MEC) involving multiple cell divisions, precedes the terminal differentiation [22]. Adipogenesis is regulated by a cascade of transcription factors, including peroxisome proliferator-activated receptor gamma (PPARγ), CAAT/enhancer-binding proteins (C/EBP), and sterol regulatory element binding protein isoform (SREBP)-1c, which induce temporal changes of adipogenesis-regulatory genes [23–25]. A recent study using a single molecule 5′ cap analysis of gene expression (CAGE) revealed dynamic patterns of gene expression profiles during adipocyte differentiation, in which the early stage involved an increase in genes related to structural remodeling and cell division, whereas genes in the later differentiation state were involved in the regulation of lipid metabolism and energy homeostasis characteristic to WAT [26]. Human transcriptome analyses reported several adipocyte-specific genes, such as leptin, adiponectin, fatty acid binding protein (FABP)-4 and ATGL, to be highly expressed in mature adipocytes [27].

In obesity, an imbalance in energy homeostasis causes adipose tissue remodeling, including adipocyte enlargement (hypertrophy) and an increase in numbers (hyperplasia) [28]. The adipose tissue constitutes adipocytes and stroma, which includes endothelial cells, pericytes, adipose stem cells, and immune cells (Figure 1). In lean conditions, macrophages, the predominant immune cell population accounting for 10% of all cells in the adipose tissue, are found in an alternatively activated M2 state, characterized by expressing CD206$^+$ CD163$^+$ CD301$^+$ surface receptors and are sparsely distributed [3]. Loss of M2 macrophages resulted in increased weight gain in myeloid-specific PPAR delta (δ) ablated mice fed with HFD, reflecting on the relevance of M2 ATMs in mitigating obesity [29]. In obese conditions, there is a significant increase in the number of ATMs, which are mainly found as classically activated M1 phenotype expressing CD11c$^+$ CD86$^+$ surface proteins responsible for promoting inflammatory conditions [30,31]. Notably, the main mechanism contributing to the increase of M1 phenotype is the recruitment of inflammatory monocytes (characterized by the presence of CCR2^{++} CX3CR1^{low} Ly6C^{high} CD11b$^+$ surface proteins) from the circulation, which on entering the adipose tissue differentiate into M1 ATMs [32]. The adipocyte secreted chemokine monocyte chemoattractant protein (MCP)-1 is known to be a key player in recruiting inflammatory monocytes to the adipose tissue [3,33]. Transgenic mice overexpressing MCP-1 showed an increased number of infiltrated macrophages in the adipose tissue, supporting the relevance of MCP-1 [34]. The recruitment of monocytes also requires CD11b integrin, as demonstrated using CD11b-deficient HFD-fed mice [35]. Several other adipocyte-induced chemokines such as colony stimulating factor (CSF)-1, C-X3-C motif ligand (CX3CL)-1, leukotriene B4, and macrophage migration inhibitory factor (MIF) also promoted macrophage infiltration [3,36–38]. Additionally, it was suggested that the proliferation of resident M2 ATMs might also contribute to an increase in the ATM population at the early stages of obesity [39,40]. Recent findings revealed the presence of a distinct pool of proinflammatory metabolically activated macrophages (MMe), which were stimulated by palmitate and participated in the trafficking and lysosomal metabolism of lipids, yet, failed to express the typical M1 markers [41,42]. These initial findings suggest a higher ATM heterogeneity than expected and would require further investigation. Other immune cells also contribute to maintaining the metabolic homeostasis in adipose tissue. Tregs and eosinophils aid in the polarization of ATMs into an M2 phenotype, by releasing cytokines such as interleukin (IL)-4, IL-13, and IL-10 [43–46]. In addition, innate lymphoid cells (ILC)-2 seem to induce adipocyte browning and ameliorate obesity through IL-33 dependent upregulation of UCP-1 [47]. Recent metabolomic and lipidomic profiling studies reported differential metabolite and lipid signatures in human and animal models, revealing a significant increase in glycerol 1-phosphate, glycolic acid, uric acid, polysaturated fatty acids, and fatty acyl chains in obese groups, thereby engendering fatty livers [48–50]. These studies highlight the need for the identification of molecules that can be used as an early diagnostic and prognostic marker in obesity-induced inflammation.

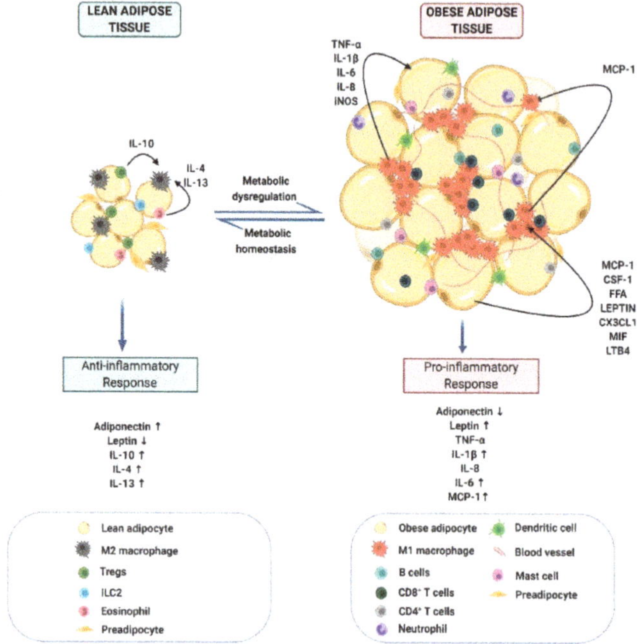

Figure 1. Schematic representation of the cellular dynamics of adipose tissue associated with obesity. As obesity develops, hypertrophic adipocytes and changes in immune cell populations contribute to the development of a chronic inflammatory adipose microenvironment that leads to metabolic dysregulation.

2.2. Inflammation and Its Link with Obesity

Chronic inflammation is a prolonged and progressive response that is accompanied by an altered immune function, ultimately leading to tissue dysfunction. It plays a key role in the initiation and progression of the pathophysiological alterations that are characteristic of obesity. Although there is an indisputable link between inflammation and obesity, there are still unresolved questions pertaining to its trigger and causative factors. It was hypothesized that inflammation initially originates as a consequence of homeostatic stress due to energy imbalance in the adipocytes [51]. Obesity-induced increase in gut permeability can give rise to circulating intestinal stemmed Gram-negative bacterial lipopolysaccharide (LPS) levels, which provoke inflammatory responses by interacting with toll-like receptor (TLR)-4 in adipocytes and macrophages [52,53]. Additionally, dietary or adipose tissue-derived FFAs binding to TLR2 and TLR4 can trigger the inflammatory signaling pathways [54]. Hypertrophic adipocytes can stimulate local induction of the transcription factor hypoxia-inducing factor (HIF)-1α, as a result of excessive oxygen depletion and decreased perfusion, leading to the upregulation of proinflammatory genes, the FFAs plasma levels, and macrophage infiltration [55,56]. FFAs released from dying adipocytes also exacerbate inflammation by binding to macrophage TLR2/4 receptors triggering the activation of nuclear factor-kappaB (NF-κB) and NOD-like receptor (NLR)P3, through the damage-associated molecular proteins (DAMPs), with subsequent production of inflammatory cytokines [57,58]. Often obese adipose tissues are characterized by the elevated generation of mitochondrial reactive oxygen species (ROS), leading to mitochondrial deregulation, oxidative stress, and inflammation [59]. In response to these stimuli, metabolically dysregulated adipocytes secrete proinflammatory cytokines, such as tumor necrosis factor (TNF)-α, which stimulate the neighboring adipocytes and the endothelial cells to secrete NF-κB regulated MCP-1 and other chemokines, thereby recruiting M1 ATMs [33]. High levels of MCP-1 were also implicated in mediating resident macrophage proliferation [60]. Transgenic and chemically induced mice models lacking macrophages conclusively supported the role of ATMs in promoting obesity-induced inflammation. Ablation of macrophages using clodronate liposomes (CL) treatment resulted in the reduction of HFD-induced adipose tissue inflammation [61]. Moreover, a significant decrease in inflammation and insulin resistance was observed in granulocyte macrophage colony stimulating factor (GM-CSF) knock-out mice fed with HFD, owing to the abatement of C-C motif chemokine receptor (CCR)-2-specific macrophage infiltration in adipose tissues, but had no effect on body weight [62]. Adipocyte-induced netrin-1 promoted the retention of inflammatory macrophages in obese adipose tissues by interacting with specific macrophage receptors [63,64]. The infiltrated M1 ATMs produce significantly large amounts of proinflammatory cytokines such as TNF-α, IL-6, IL-8, and IL-1β and attract more macrophages through MCP-1 secretion, creating a deleterious adipose microenvironment [31,34,65]. M1 macrophages can group around the dying adipocytes to form crown-like structures (CLS), which are indicative of an increase in hypoxia, hypertrophy, and stress [66]. This aberrant inflammatory environment creates further adipocyte dysregulation through increased secretion of leptin and lipolytic genes, and enhanced insulin resistance, eventually leading to adipocyte death.

Additional immune cells including mast cells, neutrophils, dendritic cells, T cells, B cells, and ILCs in the adipose tissue stroma participate in the onset of obesity (Figure 1). Nonetheless, how the interaction between these immune cells initiate and sustain adipose tissue inflammation remains elusive. In obesity, prior to ATM infiltration, there is a rise in the number of lymphocytes in the obese adipose tissues, which affects the sustenance of obesity-induced inflammation but is dispensable for the onset of obesity [4]. CD8$^+$ T cells, activated by hypertrophic adipocytes expressing major histocompatibility complex (MHC)-II, enhance ATM filtration and adipose tissue inflammation in HFD-fed mice, as confirmed by the CD8$^+$ T cells loss and gain of function in vivo studies [67–69]. CD8$^+$ T cells were found around CLS along with M1 ATMs in epididymal adipose tissues (EAT). Interestingly, CD4$^+$ T cells were identified to exhibit long-lasting obesity memory and induction of body mass regain in a weight gain–loss–regain C57BL/6J model, suggesting the potential role of an immune cell stimulated inflammatory condition in promoting obesity relapse [70]. B cells also stimulated secretion of proinflammatory cytokines, adipocyte hypertrophy and insulin resistance in obese mice. A decrease in systemic inflammation in obese B-cell-deficient mice was correlated with a significant reduction in Tregs, indicating the ability of B cells to regulate the T cell function in hypertrophic adipose tissue [71]. Additionally, B cell filtration in adipose tissues seems to precede other immune cells. B cells were infiltrated into the adipose tissues after 3 weeks of HFD, followed by T cell at week 6, and macrophage infiltration at week 12. Importantly, there was a significant rise in ATMs and natural killer cells (NK) in lymphocyte-deficient Rag2$^{-/-}$ mice, suggesting that T cells and B cells are not essential for the initiation of obesity [72]. Additionally, it remains debatable whether the early lymphocyte accumulation is a protective response rather than a stimulus of inflammatory conditions. Loss and gain of function in HFD-fed and leptin-mutated genetically obese mice revealed that CD4$^+$ Foxp3$^+$ Tregs secreted anti-inflammatory cytokines like IL-10 and influenced the insulin sensitivity of adipose tissues [44]. Interestingly, single cell RNA-sequencing of SAT, identified crosstalk induced between the adipocytes and the IL-10 secreting immune cells, wherein beige-like metabolically active adipocytes exhibited an enhanced expression of IL-10Rα responsive thermogenic genes [73]. A potential role of mast cells in mediating systemic thermogenesis, macrophage recruitment, and insulin resistance in high cholesterol and HFD-fed mice was reported [74]. Eosinophils-induced IL-4 in WAT was found to promote M2 macrophage polarization [75,76]. Neutrophils promote inflammation in HFD-induced obese mice, through the secretion of proteases, such as elastase [77]. CD11c$^+$ CD64$^-$ expressing dendritic cells accumulate in the SAT of obese mice and in humans, which promotes the activation of CD4$^+$ T-cell polarization and proliferation through Th17-type responses, to trigger inflammation [78,79]. The role of ILCs in obesity remains controversial. ILC2 depletion was reported to decline eosinophils, M2 ATMs, and anti-inflammatory cytokines, such as IL-13 and IL-5 in the VAT, thereby inducing adiposity [80]. On the contrary, recent studies identified that the loss of ILC2 and ILC3 resulted in decreased weight-gain in HFD-fed mice, which was reversed through the adoptive transfer of small intestine ILC2 [81]. The discrepancies in these studies suggest that further investigations to evaluate the role of different immune cells are necessary. Collectively, these findings support the complex roles of innate and adaptive immune cells during the early stages of obesity-induced inflammation, contributing to adipose tissue remodeling.

Several molecular pathways induce inflammation in obesity (Figure 2). TLR-mediated polarization of macrophages into an M1 inflammatory phenotype involves different transcription factors, such as NF-κB, PU.1, C/EBP-α, activator protein-1 (AP-1), STAT1, and interferon regulatory factor (IRF)-5 [82]. FFAs, TNF-α, or LPS can activate TLRs to stimulate c-Jun N-terminal kinases (JNK) or NF-κB mediated inflammation, resulting in enhanced innate immunity, activation of NLRP3 inflammasomes, and production of proinflammatory cytokines [83–85]. IKKβ and TLR4 deficiency in macrophages protected from insulin resistance in mice when exposed to HFD, and also inhibited FFA induced upregulation of TNF-α and IL-6 [54,86]. Mice, with a deficiency in TGF-β–activated kinase 1 (TAK1), an upstream modulator of NF-κB, in adipocytes, displayed an increased M2 ATM count in WAT, along with an enhanced resistance to HFD or leptin-deficiency-induced obesity [87]. Thus, targeting the IKKβ/NF-κB pathway has become an appealing approach to ameliorate the devastating effects of inflammation induced by obesity [88]. The association of obesity with increased insulin resistance has been extensively studied. Several adipokines such as leptin and retinol-binding protein (RBP)-4 were increased in obese insulin-resistant mice [33,89]. JNK and NF-κB pathways in adipocytes and macrophages, activated in response to obesity-induced stimuli, directly inhibit insulin response [63,90]. FFAs trigger diacylglycerol (DAG) and fatty acyl-CoA in the adipocytes, leading to protein kinase C (PKC) activation, which further phosphorylates insulin receptor substrate (IRS)-1 to inhibit AKT and GLUT-4, causing impairment in the ability of the liver to take up glucose and consequentially increased the circulating glucose levels (Figure 2) [91,92]. Obesity-induced inflammation and related dysregulated metabolic homeostasis often impact the liver, leading to nonalcoholic fatty liver diseases and steatosis (fat accumulation in the liver). Obesity enhances the supply of FFAs to the liver from the adipocytes, causing upregulation of PPARγ-dependent fatty acid translocase protein CD36. This increase in lipid storage elevates the activation of inflammatory responses in resident Kupffer cells and the recruitment of inflammatory myeloid cells to the liver, in a CCR2/MCP1-dependent manner, thereby elevating the severity of hepatic damage [93]. Evidence on the close overlap between the functional roles of adipocytes and macrophages imply inflammation to be the linking hub in obesity [94]. Therefore, defining, cell-specific regulators of obesity-induced inflammation can be promising in identifying therapeutic targets that can ameliorate the complications associated with obesity.

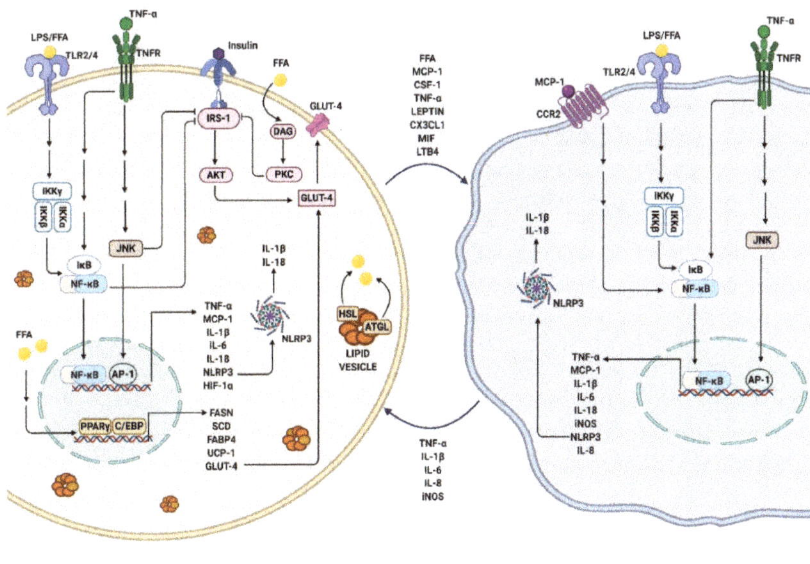

Figure 2. Adipocyte–macrophage crosstalk plays a key role in the induction and maintenance of obesity. Hypertrophic adipocytes release chemoattractants, promoting macrophage infiltration. Adipose-induced adipokines and free fatty acids (FFAs) stimulate adipose tissue macrophages (ATMs) into an M1 inflammatory stage to trigger JNK, NF-κB, and NLRP3-mediated pathways and inflammatory cytokines, which further induce adipocyte responses, including PPARγ and C/EBPs-regulated expressions of adipogenic, thermogenic, lipolytic, and lipogenic genes.

2.3. Obesity-Link Adipocyte and Macrophage Crosstalk

Adipocyte–macrophage crosstalk plays a central role in the induction and maintenance of obesity [95,96]. Macrophages constitute 40–60% of total cells in the adipose tissue depots of obese individuals, with an increase of ~5 fold, compared to lean [3,31]. Obese adipocyte-derived MCP-1, TNF-α and lipids, stimulate inflammatory monocyte infiltration and increase ATMs in adipose tissues [65,97]. Increased ATMs lead to higher levels of proinflammatory cytokines that cause further adipocyte dysregulation. ATM-derived TNF-α affects adipocytes by disrupting lipid and adipokine homeostasis, resulting in an increase of FFAs, among others. Adipocyte-induced FFAs further stimulate macrophages to express inflammatory cytokines. Hence, a disrupted macrophage–adipocyte crosstalk results in a harmful paracrine loop that exacerbates inflammation-mediated responses in the obese adipose tissue (Figure 2) [98–100]. Infiltrated ATMs in CLS are predominantly M1 phenotype, with just 10% in the outer rim of CLS accounting for M2 phenotype [39,43]. Initial studies suggested that adipocytes in CLS were highly necrotic in obese mice [101], while recent reports provide strong evidence that adipocytes undergo caspase-induced apoptosis [102,103]. Similar levels of adipocytes undergoing cell death were found in macrophage-depleted and control mice fed with HFD, suggesting that ATMs are not required for adipocyte cell death, rather the process is a response to elevated lipid dysregulation [104]. The ATMs in CLS function by taking up lipids, as well as removing the dead adipocytes through phagocytosis, and eventually form foam cells or become inactivated [105]. Further investigations on the understanding of the mechanistic nature of the macrophage–adipocytes crosstalk are much needed and guarantee to reveal vital knowledge, to help control obesity-induced inflammation.

3. Flavones and Their Impact on Obesity-Induced Inflammation

Obesity increases the incidence of heart disease by 30%, leads to diabetes, and is associated with cancer risk, as suggested by several meta-analyses [106]. Chronic inflammatory conditions represent one-third of the total $1.1 trillion US health care expenditure, representing approximately 20% of the annual national GDP. The currently used medications for obesity, such as orlistat, (a pancreatic lipase inhibitor), and liraglutide (an incretin mimetic), have severe side effects that accrue to other undesirable symptoms [107]. Hence, identifying additional approaches that lack unwanted effects is necessary.

Flavonoids are a large class of plant phenolic secondary metabolites with anti-obesogenic, anti-inflammatory, and immune-modulating activities. Higher consumption of flavonoid-rich diet has been linked to reduced energy consumption, food intake, and weight loss [108]. Thus, flavonoids might offer an economically favorable approach, with minimal, if any, side effects for the prevention and treatment of obesity. Flavonoids are structurally characterized by two benzene rings and a heterocyclic pyrone ring. Based on the oxidation and saturation status of the heterocyclic ring, flavonoids are categorized into different sub-groups, such as flavones, flavonols, flavanones, flavanonols, flavanols, isoflavones, and anthocyanidins [6]. Flavonoids are potent antioxidant agents and the molecular mechanisms by which they mitigate free-radical-derived oxidative stress have been extensively reported elsewhere [8,109]. The health beneficial effects of flavonoids are mediated primarily through their ability to modulate multiple gene/protein signaling networks. However, the basic mechanisms of action are not completely understood. Several studies support the beneficial role of flavonoids in obesity. The flavonol quercetin, perhaps one of the most studied in the context of obesity, increases adiponectin and downregulates MCP-1, TNF-α, and IL-6 expressions in adipocyte macrophage co-cultures and HFD mice models, via the inhibition of the NF-κB, AP-1, and mitogen-activated protein kinase (MAPK) pathways [110]. Resveratrol, a flavonoid found in red wine, decreased insulin resistance, inflammation, and CCR2-driven macrophage infiltration in SAT and VAT in HFD-fed mice [111]. Several studies have reported that soy isoflavones promote lipid homeostasis and fatty acid metabolism, and inhibit macrophage–adipocyte crosstalk both in vitro and in vivo [112,113]. Investigations on flavonoids as potential agents for treating obesity-linked cancers and obesity-associated modulation of gut microbiota are gaining interest. Resveratrol and naringenin suppressed inflammation and breast tumor growth by inhibiting adipocyte hypertrophy and tumor associated macrophages (TAM) in obese mice [114,115]. Consumption of anthocyanin containing foods can protect against diet-induced obesity and systemic inflammation, by modifying the gut microbial population in mice [116]. Striking associations of the dietary flavonoid intake with decreased obesity were found in numerous meta-analyses [117,118]. These findings established flavonoids as prospective arsenals in fighting obesity and reinforced the significance of their use in our daily diets and in clinical trials. Flavones, a sub-class of flavonoids, are highly efficacious as anti-inflammatory and anti-obesogenic agents. Here, we focus this review on the role of flavones in the prevention and treatment of obesity and its related disorders.

3.1. Flavone Sources and Structure

Flavones are gaining immense interest due to their diverse bioactivity in plants and animals. They differ in structure from the other flavonoids in terms of the presence of a double-bond between C2 and C3 in the flavonoid core skeleton, a ketone at C4, and the absence of any modifications in the C3 position (Figure 3) [119]. The flavone core is subjected to substitutional conjugations, such as hydroxylations (addition of OH groups), glycosylations (bound to sugar moieties), or methoxylations (addition of methyl groups) at different positions, accounting for its expansive range of health beneficial activities [8]. Flavones are naturally found in plants as glucosides, conjugated either through hydroxyl groups (*O*-glycosides) or directly linked through the carbon (*C*-glycosides) groups. Functional activity, absorption, and bioavailability of these flavones can largely be dependent upon the structure, linkage, and the number of sugar moieties [120,121]. Most of the studies investigating the beneficial effects of flavones use them in their sugar-free form (aglycone). However, studies using whole foods with a high content of these phytochemical components in their naturally occurring form, remain scarce.

We showed that aglycones are more easily absorbed than their glycosides, findings that are directly linked to their bioavailability and immunoregulatory functions [120]. Nevertheless, the poor solubility of aglycones imposes a great impediment for their clinical application in human health. We have overcome this gap in the field by developing foods from celery that increase the absorption and deliver bioactive concentrations of apigenin aglycone in vivo [120].

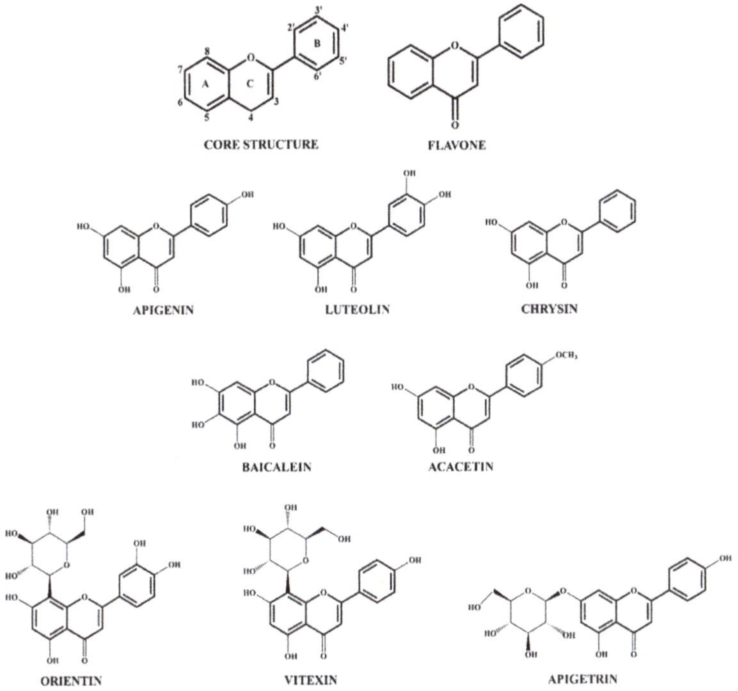

Figure 3. Structure of flavonoid core and different flavones.

The levels of flavonoids can significantly vary between plants and tissues. The common food sources of flavones include citrus fruits, vegetables, herbs, and grains. Albeit flavones represent only a small fraction of the total flavonoid intake, it is estimated to range between 0.7 to 9.0 mg/day [122]. Rich, natural flavone sources are parsley, celery, peppermint, and sage, which predominantly contain apigenin and luteolin in their *O*-glucoside forms [123]. In maize, maysin and apimaysin are common flavones modified in –C groups. Another group with a wide array of physiological effects is the methoxylated flavones, such as acacetin, diosmetin, and chrysoeriol, which are commonly found in the citrus family [124]. Despite the available and procuring knowledge on the bioactivity of pure flavones, further investigations on the effect of whole foods containing a high flavone content need to be adopted, with rigorous consideration on the estimation of consumption quantity. This is vital for overcoming hurdles in accurately interpreting the association between flavone intake and health outcomes at clinical levels.

3.2. Role of Flavones in Obesity-Induced Inflammation

Several studies have suggested promising effects of flavones on the prevention and treatment of obesity-induced inflammation, based on their ability to modulate adipocyte, as well as their immune cell function. Flavones inhibit different stages of adipogenesis by suppressing lipid accumulation in adipocytes, through the reduction of lipogenesis and lipolysis (Table 1).

Table 1. Flavones and their functional roles in obesity and its associated inflammation.

Flavone	Experimental Model	Concentration	Function	Reference
Apigenin	Mouse 3T3-L1 cells	10–50 μM	↓ adipogenesis: C/EBPβ and PPARγ ↓ lipolysis: HSL, MSL ↑ fatty acid oxidation: AMPK ↓ MCE, G0/G1 arrest	[125,126]
	Human mature adipocytes	25 μM	↓ lipogenesis: FASN No effect adipogenesis	[127]
	HFD-fed obese C57BL/6J mice	15–50 mg/kg/day	↓ adiposity ↓ lipogenesis: FASN ↑ lipolysis: ATGL, HSL ↑ fatty acid oxidation: AMPK and ACC ↓ inflammation: MAPK, NF-κB, TNF-α, IL-6 and MCP-1 ↓ ATM infiltration and M1 polarization ↑ thermogenesis: UCP-1 ↓ STAT3/CD36 ↓ liver steatosis and hepatic inflammation ↓ NLRP3 ↑ insulin sensitivity ↓ oxidative stress: XO and ROS ↑ Nrf2 activity	[128–132]
Luteolin	3T3-L1 cells	10–50 μM	↓ adipogenesis: C/EBPα and PPARγ ↓ lipogenesis	[133]
	HFD-fed obese C57BL/6J mice	5 mg/kg/day	↓ adiposity ↓ inflammation: IL-1β and IL-6 ↓ ATM infiltration and M1 polarization ↓ insulin resistance ↓ hepatic steatosis	[134–136]
Baicalein	3T3-L1 cells	12.5 μM	↓ adipogenesis: C/EBPα, C/EBPβ, FABP4 and PPARγ ↓ lipogenesis ↓ MCE, G0/G1 arrest	[137,138]
	Diet-induced obese C57BL/6J mice	20 mg/kg/day	↑ thermogenesis: UCP-1 ↑ insulin sensitivity: GLUT4	[139]
Orientin	3T3-L1 cells	50 μM	↓ adipogenesis: C/EBPα, C/EBPδ, PPARγ, FABP4 and GLUT4 ↓ lipogenesis: FASN, SCD, ACC ↓ lipolysis: HSL, MSL, ATGL ↓PI3K/Akt-FOXO1	[140]
Chrysin	3T3-L1 cells	50 μM	↑ adipogenesis: C/EBPα, C/EBPβ and PPARγ ↑ lipogenesis: ACC ↑ lipolysis: HSL, MSL, ↑ thermogenesis: UCP-1 ↑ AMPK	[141]
	Diet-induced obese C57BL/6J mice	20–30 mg/kg/day	↓ adiposity ↑ PPARγ ↓ inflammation: TNF-α, IL-6 and IL-1β ↓ ATM infiltration and M1 polarization	[142]
Apigetrin	3T3-L1 cells	100 μM	↓ adipogenesis: C/EBPα, PPARγ, and SREBP-1c ↓ lipogenesis: FASN ↓ inflammation: TNF-α and IL-6	[143]
Vitexin	3T3-L1 cells	25–100 μM	↓ adipogenesis: PPARγ ↓ lipogenesis ↑ ERK1/2 ↓ Akt	[144]
	HFD-fed obese C57BL/6 mice	5 mg/kg/day	↓ adiposity ↓ adipogenesis: C/EBPα and lipogenesis: FASN ↑ AMPK	[145]
Wogonin (*Scutellaria baicalensis*)	HFD-fed obese C57BL/6 mice	500 mg/kg/day	↓ insulin resistance ↓ inflammation: TNF-α and IFN-γ	[146]
Baicalin	HFD-fed obese C57BL/6 mice	5 mg/kg/day	↑ insulin sensitivity ↓ inflammation: TNF-α, MCP-1 and IL-1β ↓ oxidative stress ↑ Nrf2 activity ↑ CPT1A activity	[147,148]

Apigenin and luteolin inhibit adipogenesis at 10–50 μM by attenuating the accumulation of intracellular triglycerides and the expression of adipogenic transcriptional factors, such as C/EBP and PPARγ, in differentiated 3T3-L1 and primary adipocytes, through the upregulation of 5′-adenosine monophosphate-activated protein kinase (AMPK) activity [125,133]. Apigenin and baicalein suppress proliferation and differentiation of preadipocytes, by inducing cell cycle arrest at G0/G1, and inhibiting MCE during the early stages of differentiation [126,137]. Lipid accumulation in mature human adipocytes and differentiated 3T3-L1 adipocytes was suppressed by apigenin, orientin (luteolin-8-C-glucoside), and baicalein. This suppression occurs through the reduction of lipolytic genes ATGL, HSL and monoacyl glyceride lipase (MGL), and lipogenic genes like fatty acid synthase (FASN), acetyl-CoA carboxylase (ACC) and stearoyl-CoA desaturase (SCD), which reflect on the anti-lipolytic and anti-lipogenic role of flavones [127,138,140]. However, apigenin had no impact on adipogenesis or the expression of any of adipogenic genes, including SREBP-1c in mature human mesenchymal stem cell-derived adipocytes, while chrysin enhanced lipolysis, adipogenesis, and lipogenesis in differentiated 3T3-L1 cells [127,141]. Differences in these findings could be attributed to the etiology, the stage of differentiation of the cells used and the treatment dose, thus, highlighting the need for additional studies and the use of models that fully capture the complex balance of physiological whole-body metabolism. Reduction in transcriptional and translational levels of other adipogenic genes, such as FABP4 and GLUT4, were observed in the differentiated 3T3-L1 adipocytes treated with orientin and baicalein [138,140]. Chrysin enhances WAT thermogenesis by upregulating the expression of UCP-1 [141]. Although flavone glycosides apigetrin (apigenin 7-O-glucoside) and vitexin (apigenin 8-C-glucoside) suppressed adipocyte differentiation in 3T3-L1 cells, through the activation of the ERK/MAPK pathway, the flavone concentrations used were significantly high and unreachable in vivo in mice or human clinical trials, thereby possibly masking the key underlying mechanisms [143,144].

Anti-obesity responses of flavones were further corroborated using HFD obese mice models (Table 1). Baicalin (20 mg/kg/day), luteolin (5 mg/kg/day), apigenin (50 mg/kg/day), and vitexin (5 mg/kg/day) repress the expression of transcription factors associated with adipose differentiation, attenuate adiposity, and mitigate lipogenesis in WAT of HFD-fed C57BL/6J mice [128,134,139,145]. Apigenin (15–30 mg/kg) inhibit preadipocyte differentiation and visceral obesity in HFD-fed mice, by directly interacting with STAT3, hence inhibiting STAT3 transcriptional activity and reducing the expression of CD36 and PPARγ [129]. These results further support that flavones can exert their biological activities through direct binding to proteins. We previously demonstrated that apigenin binds with different affinities to 160 proteins, by screening a human peptide phage display library coupled with next generation sequencing (PD-Seq) [149]. Among these targets, cathepsin D (CTSD), was implicated as a pivot mediator in adipogenesis, lipid metabolism in mouse hepatic steatosis, mitochondrial dysfunction, cell death, and macrophage infiltration, in hypertrophic adipose tissues of genetically and HFD-induced obese mouse models [150,151]. Future mechanistic studies are vital to reveal the mechanisms of action of flavones in obesity-induced inflammation to facilitate its prevention and therapeutics.

Flavones inhibit obesity-induced inflammation by reducing macrophage numbers in adipose tissues, thereby diminishing a proinflammatory adipose environment (Table 1). Apigenin and chrysin reduce the levels of proinflammatory cytokines IL-12, TNF-α, IL-6, and MCP-1 in adipose tissue in obese C57BL/6J mice [132,142]. This effect seems to be due to the ability of apigenin to switch macrophage phenotype from M1 to M2 by binding to PPARγ, thereby suppressing the interaction between PPARγ and NF-κB. Luteolin also decreases the infiltration of ATMs in the EAT of HFD-fed mice, by reversing the polarization of obesity-associated M1 and MMe ATMs through the activation of the AMPKα1 pathway [135]. Interesting studies using adipocyte-RAW 264.7 macrophages co-cultures and cell-specific conditioned media revealed that luteolin reduces inflammation by suppressing macrophage-stimulated inflammatory cytokines, but has no effect on adipocytes-stimulated adipokines, suggesting that luteolin specifically targets macrophages [152]. An alternative explanation of these results could probably owe to the use of hypertrophic adipocytes, as it was previously reported that only preadipocytes and adipocyte progenitors release chemokines such as MCP-1, to stimulate macrophage accumulation in adipose tissues, while their expression levels are low in mature adipocytes [153,154]. *Scutellaria baicalensis* roots rich in baicalin, wogonin, and luteolin alleviate HFD-induced insulin-resistance in obese mice, through modulation of inflammation, by promoting M2 phenotype skewing and reducing the TLR5 signaling pathway [136,146]. Flavones prevent non-alcoholic fatty liver disease (NAFLD) and steatosis. This effect seems to be due to the ability of flavones to increase liver fatty acid oxidation and reduce oxidative stress. Apigenin (50 mg/kg/day) reduces HFD/FFA-induced hepatic steatosis, lipid peroxidation, and lipid accumulation in the liver, by downregulating the lipogenic genes and inhibiting the overexpression of inflammatory markers and Kupffer macrophage infiltration [131]. These hepatic protective effects of apigenin were ascribed to its xanthine oxidase (a purine nucleotide degradation and ROS generator) inhibitor role, which hence inhibited the NLRP3 inflammasome assembly, ROS generation, and the release of inflammatory cytokines IL-1β and IL-18, illustrating a combined anti-oxidant and anti-inflammatory mechanism of action [131]. Oxidative stress was found to be mitigated by apigenin (30 mg/kg/day) and baicalin (50 mg/kg/day), by inhibiting the mitochondrial dysfunction through Nrf2 activation in adipocytes and macrophages in HFD/FFA-fed NAFLD mice [130,147]. Interestingly, induced activation of Nrf2 negatively regulated the PPARγ function in the NAFLD model, probably through direct interaction with Nrf2 [130]. Luteolin (5 mg/kg/day) decreases liver lipotoxicity by inducing FFAs flux to WAT and attenuates liver fibrosis by reducing cathepsin and extracellular matrix accumulation [136]. A quantitative proteomic study identified baicalin as an allosteric activator of carnitine palmitoyltransferase 1 (CPT1), the rate-limiting enzyme of fatty acid β-oxidation, wherein it significantly improved hepatic steatosis and decreased diet-induced obesity, by directly binding with CPT1 to facilitate accelerated lipid influx into the mitochondria for β-oxidation and FFA degradation [148]. These findings underscore the efficacious nature of flavones in tackling obesity-induced inflammation, by actively affecting both the inflammatory macrophages and the adipocytes in the adipose depots, and also their crosstalk. Despite a large number of encouraging studies suggesting the health beneficial impacts of flavones, it warrants further investigation of the different upstream molecular mechanisms of their roles in modulating obesity-induced inflammation, using foods rich in flavonoids, at feasible treatment doses.

3.3. Controlling Obesity-Associated Cancer Using Flavones

Obesity was positively correlated with cancer morbidity and mortality in both men and women [155]. In addition, preclinical and epidemiological studies implicated obesity as a major risk factor for the development of cancer [156–158]. This is especially significant in the case of breast cancer, where the adipose tissue is a predominant component of the stroma in the mammary tissue [159]. Cancer cells spread to stromal compartments that possess abundant adipose tissue, while adipocytes along with ATMs serve as a tumor-favoring niche with endocrine resources to nurture and mold the tumor microenvironment, contributing to tumor progression and metastasis [160,161]. The insulin–insulin growth factor (IGF)-1 axis, sex hormones, and adipokines are key mediators between obesity and cancer, each of which are tightly linked to the endocrine and paracrine dysregulation of adipose tissue in obese individuals [162]. For instance, adipose-derived stem cells secrete chemokine adipsin and promote breast cancer growth [163,164]. Leptin released from adipose tissue is shown to induce vascular endothelial growth factor (VEGF) overexpression and enhanced cancer stem cell-like properties in breast cancer [165,166]. VAT-derived fibroblast growth factor (FGF)-2 was reported to stimulate cell transformation through FGF receptor-1 in melanoma and breast cancer [167].

Hypertrophic expansion of adipose tissues in obese individuals shares many common aspects with tumor growth. Both obesity and cancer progression are closely associated with energy intake and nutrient availability. Hypoxia, often linked to obesity, stimulates enhanced angiogenesis, creating a microenvironment that provides a tumor permissive niche for the transformed or infiltrating cells [168,169]. Certain fibrotic factors such as adipose-derived collagen-IV and endotrophin are the key mediators linking obesity and tumor growth [170–172]. While adipocyte-released mediators are the predominant regulators of tumor progression, cancer cells can also induce metabolic differences and condition adipocytes in a pro-tumorigenic fashion, to form cancer-associated adipocytes (CAA) [173]. Paracrine signals from CAAs induce lipid degradation resulting in the release of FFAs, which is used as an energy source during metastasis by facilitating β-oxidation in cancer cells [174]. This reciprocal crosstalk between the cancer cells and adipocytes within the microenvironment is crucial for creating a tumor-permissive niche. Most importantly, both are characterized by chronic inflammation. Inflamed adipose tissues induce an aggravated expression of proinflammatory mediators, increased aromatase levels, and elevated estrogen receptor-α (ER-α)-dependent gene expression, which are also involved in tumor growth and metastasis [175,176]. Tumor cells also release MCP-1 to trigger macrophage infiltration, which are key contributions for tumor maintenance. While adipocytes recruit M1 phenotype macrophages, cancer cells skew macrophages towards an M2 phenotype [177]. Interestingly, adipocyte-cancer cell crosstalk was shown to influence chemotherapy efficacy and outcome in obese patients. The efficiency of tamoxifen to inhibit the proliferation of breast cancer cell line MCF-7 was significantly reduced in the presence of matured adipocytes derived from adipocyte stem cells of obese women. This effect was attributed to the increased presence of inflammatory adipokines, such as leptin, IL-6, and TNF-α, in the co-cultures of MCF-7 and adipocytes [178]. It is noteworthy that some of the key molecular players involved in obesity were also strikingly critical in cancer progression, such as NF-κB, CCL2/CCR2, JNK, and HIF/VEGF.

The anti-cancer efficacy of flavones relied on their ability to regulate key molecular pathways related to cancer cell proliferation and immune cell function, thereby halting tumor growth and metastasis [8]. Flavones inhibit cell growth and promote cell death in various cancer types. Importantly, we found that apigenin induces cell death of numerous cancer cell types but had no effect on the proliferation of non-cancer cells in leukemia [179]. We previously showed that apigenin induces cell cycle arrest by inducing DNA damage through the phosphorylation of ataxia-telangiectasia mutated kinase (ATM) and H2A histone family member X (H2AX) [180]. Luteolin induces apoptosis of colon cancer cells through its interaction with p53 and upregulation of Nrf2 [181]. In the human xenograft prostate cancer model, apigenin, through the inhibition of IGF/IGFR-1, reduced tumor growth [182]. Flavones can also suppress stem-like properties in aggressive cancers [183]. Baicalein inhibits the expression of stem cell markers $CD44^{high}CD24^{low}$ and octamer-binding transcription factors (OCT)-3 and 4 in triple negative breast cancer cell lines, through the inhibition of interferon-induced protein with tetratricopeptide repeats 2 (IFIT2) [184]. In addition, flavones are also potent immunoregulators. Apigenin and a celery-based apigenin-rich (CEBAR) food, a diet developed by our team that delivers in vivo effective doses of apigenin [120,185], reduce inflammation through the inhibition of NF-κB and decreased proinflammatory TNF-α in vivo [186,187]. Apigenin and luteolin suppress MCP-1 and IL-6 release, inhibiting TAM infiltration and migration of cancer cells [188–190]. These findings support the potent role of flavones in the prevention and treatment of obesity-induced cancers, and in enhancing the efficacy of chemotherapeutic drugs. However, further investigations on the effects of flavones on HFD-induced mammary tumorigenesis in preclinical PyMT mouse models are required, which can be significantly informative for clinical studies.

3.4. Flavones as Emerging Mediators of Gut Microbiota and Its Link with Obesity-Induced Inflammation

The gastrointestinal tract (GI) is inhabited by a broad repertoire of microorganisms, generically referred to as the gut microbiota. While the intestine provides a nutrient-rich, protected environment in which microbiota thrive to create a diverse and stable ecosystem, the microbiome provides nutrients to human host cells and prevent the entry of potential pathogens [191]. Microbes play an essential role in vitamin production, the modification of food components, energy homeostasis, intestinal mucosa formation, and the development of immunity. The gut microbiota interacts with the host cells through molecular communication, using small molecules and other metabolites [192]. Metagenomic analysis revealed that bacteria in the intestine belong to mainly three phyla, Bacteroidetes, Firmicutes, and Actinobacteria. Diet plays a drastic role in maintaining gut microbiota diversity [193]. An imbalance in the composition, richness, and the metabolic activity of gut microbiota, known as dysbiosis, can give rise to dramatic changes in the symbiotic relationship between the bacteria consortium and the host, leading to a variety of chronic disease conditions, including obesity [194]. Obese individuals have an altered gut microbiota diversity with a reduction in barrier-protecting microbes, such as Lactobacillus and Bifidobacterium, and promotion of opportunistic pathogenic bacterial abundances like the Enterobacteriaceae, Desulfovibrionaceae, and Streptococcaceae families [195,196]. Compelling evidence demonstrating the role of the gut microbiota in obesity was provided by germ-free (GF) mice fed with HFD, showing a lesser weight-gain than non-GF mice, observations that were a result of enhanced fatty acid metabolism in GF mice [197]. Dysbiosis was found in both HFD and genetically induced obese mice, as evidenced by a 50% increase in the *Firmicutes* species and a 50% decrease in the *Bacteroidetes* species in obese conditions [198,199]. Additionally, transplantation of gut microbiome from genetically obese donor mice into GF mice increased adiposity, as compared to GF mice that received gut microbes from lean mice [200]. Similarly, dysbiosis and a significant reduction in bacterial diversity characterized by a higher *Firmicutes* to *Bacteroidetes* ratio were also observed in obese humans [201]. In HFD-induced obesity, enhanced growth of Enterobacteriaceae was correlated with an increase in intestinal endotoxin production, conditions that are known to contribute to an inflammatory intestinal microenvironment [202]. Intestinal bacteria inhibit fasting-induced adipocyte factor, which affects lipase activity and enhance triglyceride deposition in adipocytes. Furthermore, the obesity-altered gut microbiota is potent at harvesting energy from food by secreting enzymes that break down nutrients more efficaciously [200]. Strong links between diet, inflammation, and microbial dysbiosis were found. Increased fat intake causes a rise in Gram-negative bacteria, augmenting circulatory LPS levels and weakening the intestinal gut endothelium junctions that lead to enhanced intestinal permeability. Higher levels of IFN-γ and IL-1β increased gut epithelial permeability by suppressing the expression of tight junction proteins like occludin [203]. The innate immune system plays a critical role in regulating the crosstalk between the host and the microbiota during obesity-induced inflammation. An increase of macrophage infiltration into the intestinal lamina propia was observed in obese conditions, resulting from similar molecular mechanisms responsible for ATM infiltration [204,205]. LPS-induced TLR and NLR mediated the JNK and NF-κB pathways in intestinal epithelial cells and macrophages and stimulated the production of proinflammatory cytokines, which further impaired intestinal permeability. An abundance of *Bacteroidetes* and *Akkermansia muciniphila* increased in TLR4 and NLRP6 transgenic mice, thereby altering the microbiota profile and reducing inflammation [206,207]. HFD induced intestinal NF-κB and TNF-α expression and enhanced adiposity, which was resisted by the GF mice. Interestingly, intestinal changes induced by the HFD and microbiota-derived inflammatory changes seem to precede the onset of obesity [208]. Changes in the gut microbiota composition of genetically obese mice was associated with decreased MCP-1 levels [209]. HFD-induced alterations in gut microbiota spectrum hampered gut barrier function and enhanced macrophage infiltration and inflammation in mesenteric fat, suggesting a link between microbiota and inflammation [5]. These studies confirm that the microbiota is the main hub controlling the inflammatory responses in the intestine. There is growing evidence that establishes the role of microbiota in stimulating obesity-induced cancers. Fecal transfer from HFD-fed mice with aggressive intestinal tumor to healthy K-ras^{G12Dint} mice led to a microbial community shift and enhanced tumor progression [210]. Transferring the microbiota from HFD-fed mice into female GF mice was associated with progressive hepatic cancer in the offspring [211]. Interestingly, *Akkermansia muciniphila* was identified to be associated with a favorable outcome in lung and renal cancer patients undergoing PD (programmed cell death protein)-1 immune checkpoint inhibitor chemotherapy, implicating the potential role of gut microbes in modulating host response to therapy [212].

The interplay between the gut microbiome and flavonoid metabolism is emerging as an important player in health [213,214]. The gut microbiota plays a key role in modulating the chemistry, bioavailability, and absorption of flavonoids. Intestinal microbial glycohydrolases, glucosidase, demethylation, dihydroxylation, and decarboxylation, modified flavonoids and the resulting metabolites were more efficiently absorbed in the intestine. This was evident from their increased enterohepatic and plasma levels, and the elevated biological functions, as compared to their precursors [215]. Glycoside forms of flavonoids are often converted to their aglycones when metabolized by the gut microbiota [216]. Quercetin produced from the microbiota-mediated transformation of quercitrin (quercetin-3-*O*-rhamnoside) exhibited higher anti-inflammatory responses through the inhibition of the NF-κB pathway [217]. On the other hand, flavonoids can induce changes in gut microbiome composition, after the consumption of foods with a high content of polyphenols, predominantly via inhibition of pathogenic microbes and stimulation of commensal microbes [218,219]. Flavonoids might stimulate commensal bacteria *Lactobacillus* and *Bifidobacterium* in the gut microbiota, while hampering the colonization of the pathogenic strain *Clostridium*, thereby, reducing the gut microbiota dysbiosis [220]. Therefore, the reciprocal mutual effects involving the transformation of flavonoids by the gut microbiota and the modulation of microbiota by flavonoid and its metabolites can profoundly impact the flavonoid bioavailability, biological effects, and ultimately human health.

So far, studies on the effect of flavones on gut microbiota are lacking. However, studies reporting the beneficial role of other flavonoids as potent gut microbiota modifiers are emerging. For example, quercetin reduced the microbiota composition including the *Firmicutes/Bacteroidetes* ratio and the growth of species associated with diet-induced obesity like *Erysipelotrichaceae*, *Bacillus*, and *Eubacterium cylindroides*, while restoring the barrier integrity in obese NAFLD model [221,222]. In HFD/high sucrose-induced obesity models, polyphenol-rich cranberry extract diet and concord grape anthocyanins inhibited insulin resistance and inflammation by mediating an increase in *Akkermansia muciniphila* in the gut microbiota [116,223]. Anthocyanins from plum and peach juices decreased fecal short-chain fatty acids (SCFA), a subset of key gut microbiota metabolites, including acetate, propionate, and butyrate, and modified the bacterial composition of the microbiota, by increasing the population of *Faecalibacterium*, *Lactobacillus*, and *Bacteroidetes* [224,225]. Several flavonoids are shown to enrich beneficial bacterial abundance while reducing potential detrimental microbes in the human gut. Quercetin and resveratrol decreased the Enterobacteriaceae family and reduced the *Firmicutes/Bacteroidetes* ratio in the human gut, as shown through the 16S rRNA sequencing of fecal samples [226,227]. Diet supplementation with soy bars significantly enhances the abundance of beneficial *Bifidobacterium* bacteria in postmenopausal women, which results in increased lipid catabolism [228]. The ability of other flavonoids to regulate dysbiosis establishes a promising platform, urging the investigation of the potential of flavones in the regulation of the gut microbiota.

Studies related to the effect of flavone apigenin on the gut microbiota are gaining interest. Apigenin suppressed colonic inflammation by reducing IL-1β and IL-6 and immune cell infiltration [9]. Using NLRP6$^{-/-}$ mice and 16S rRNA gene sequencing of fecal samples, anti-inflammatory and anti-proliferative activity of apigenin was correlated to the apigenin-induced changes in the gut microbial composition, which was dependent on NLRP6 inflammasome. Notably, cohousing with apigenin-treated mice protected other mice against colitis, suggesting that the protective effects of apigenin were transmitted [229]. Apigenin inhibited *Enterococcus caccae* by upregulating genes pertaining to protein synthesis, DNA damage responses, and SCFA production, as identified by the 16S rRNA gene sequencing of apigenin-treated human fecal homogenates [230]. These findings suggest the potential role of apigenin as an active ingredient in modulating gut microbiota and hence mitigating obesity. To the best of our knowledge, studies pertaining to the effects of flavone intake on gut dysbiosis in humans remain to be reported. More studies considering the efficacy of various flavones in obese mouse models, as well as amongst humans within and between different regions, ethnicity, exercise regimes, and diets in regulating the intricate association between the gut microbiome and the immune system are necessary. Hence, an exhaustive understanding of the role of foods with a high content of flavones in the crosstalk between diet, gut microbiome, and immune system can provide a breakthrough in reducing obesity-induced inflammation.

4. Conclusions

Obesity and associated comorbidities have reached pandemic levels and require the identification of additional therapeutic and preventive approaches that lack adverse side effects and are cost-effective. Chronic inflammation has a crucial role in the initiation and maintenance of obesity, promoting metabolic dysregulation, microbiome dysbiosis, and increasing cancer incidence. A vicious crosstalk between adipocytes and the infiltrated immune cells led to a dramatic remodeling of the gene, protein, and lipid metabolic networks. Flavones, active plant metabolites or nutraceuticals, provide potential opportunities targeting numerous pathways that are central to obesity. The established evidence demonstrates that flavones ameliorate macrophage-mediated inflammation and reduce cancer progression and obesity. These recent findings underscore the efficacious nature of flavones in tackling obesity as a "sword of two edges", targeting macrophages and adipocytes, thus, reestablishing homeostasis. Studies on the health beneficial impacts of flavones through the modulation of adipogenic and immunogenic regulators warrant further investigation and future exciting discoveries.

Author Contributions: A.I.D. conceptualized the content; M.S. and A.I.D. contributed to the design and the writing of the manuscript. A.I.D. supervised the manuscript. A.I.D. acquired funding. M.S. is a predoctoral T32 fellow. All authors have read and agreed to the published version of the manuscript.

Funding: Work in Dr. A.I. Doseff's lab was supported by grant USDA-AFRI-2018-03994, NSF-IOS-1733633, USDA-AFRI-2020-67017-30838, and MSU general funds. Meenakshi Sudhakaran was supported by a graduate research fellowship support from the Plant Biotechnology for Health and Sustainability Training Program Project NIH T32-GM110523.

Acknowledgments: We apologize to the colleagues who made important contributions, but which were omitted due to space limitations.

Conflicts of Interest: The authors declare no conflict of interest.

References

1. Bluher, M. Obesity: Global epidemiology and pathogenesis. *Nat. Rev. Endocrinol.* **2019**, *15*, 288–298. [CrossRef] [PubMed]
2. Reilly, S.M.; Saltiel, A.R. Adapting to obesity with adipose tissue inflammation. *Nat. Rev. Endocrinol.* **2017**, *13*, 633–643. [CrossRef] [PubMed]
3. Weisberg, S.P.; McCann, D.; Desai, M.; Rosenbaum, M.; Leibel, R.L.; Ferrante, A.W., Jr. Obesity is associated with macrophage accumulation in adipose tissue. *J. Clin. Investig.* **2003**, *112*, 1796–1808. [CrossRef]
4. Liu, R.; Nikolajczyk, B.S. Tissue Immune Cells Fuel Obesity-Associated Inflammation in Adipose Tissue and Beyond. *Front. Immunol.* **2019**, *10*, 1587. [CrossRef] [PubMed]
5. Lam, Y.Y.; Ha, C.W.; Campbell, C.R.; Mitchell, A.J.; Dinudom, A.; Oscarsson, J.; Cook, D.I.; Hunt, N.H.; Caterson, I.D.; Holmes, A.J.; et al. Increased gut permeability and microbiota change associate with mesenteric fat inflammation and metabolic dysfunction in diet-induced obese mice. *PLoS ONE* **2012**, *7*, e34233. [CrossRef] [PubMed]
6. Jiang, N.; Doseff, A.I.; Grotewold, E. Flavones: From Biosynthesis to Health Benefits. *Plants* **2016**, *5*, 27. [CrossRef] [PubMed]
7. Panche, A.N.; Diwan, A.D.; Chandra, S.R. Flavonoids: An overview. *J. Nutr. Sci.* **2016**, *5*, e47. [CrossRef]
8. Sudhakaran, M.; Sardesai, S.; Doseff, A.I. Flavonoids: New Frontier for Immuno-Regulation and Breast Cancer Control. *Antioxidants* **2019**, *8*, 103. [CrossRef]
9. Gentile, D.; Fornai, M.; Colucci, R.; Pellegrini, C.; Tirotta, E.; Benvenuti, L.; Segnani, C.; Ippolito, C.; Duranti, E.; Virdis, A.; et al. The flavonoid compound apigenin prevents colonic inflammation and motor dysfunctions associated with high fat diet-induced obesity. *PLoS ONE* **2018**, *13*, e0195502. [CrossRef]
10. Vernarelli, J.A.; Lambert, J.D. Flavonoid intake is inversely associated with obesity and C-reactive protein, a marker for inflammation, in US adults. *Nutr. Diabetes* **2017**, *7*, e276. [CrossRef]
11. Gil-Cardoso, K.; Gines, I.; Pinent, M.; Ardevol, A.; Blay, M.; Terra, X. Effects of flavonoids on intestinal inflammation, barrier integrity and changes in gut microbiota during diet-induced obesity. *Nutr. Res. Rev.* **2016**, *29*, 234–248. [CrossRef] [PubMed]

12. Choe, S.S.; Huh, J.Y.; Hwang, I.J.; Kim, J.I.; Kim, J.B. Adipose Tissue Remodeling: Its Role in Energy Metabolism and Metabolic Disorders. *Front. Endocrinol. (Lausanne)* **2016**, *7*, 30. [CrossRef]

13. Shinoda, K.; Luijten, I.H.; Hasegawa, Y.; Hong, H.; Sonne, S.B.; Kim, M.; Xue, R.; Chondronikola, M.; Cypess, A.M.; Tseng, Y.H.; et al. Genetic and functional characterization of clonally derived adult human brown adipocytes. *Nat. Med.* **2015**, *21*, 389–394. [CrossRef] [PubMed]

14. Schweiger, M.; Schreiber, R.; Haemmerle, G.; Lass, A.; Fledelius, C.; Jacobsen, P.; Tornqvist, H.; Zechner, R.; Zimmermann, R. Adipose triglyceride lipase and hormone-sensitive lipase are the major enzymes in adipose tissue triacylglycerol catabolism. *J. Biol. Chem.* **2006**, *281*, 40236–40241. [CrossRef] [PubMed]

15. Scherer, P.E. Adipose tissue: From lipid storage compartment to endocrine organ. *Diabetes* **2006**, *55*, 1537–1545. [CrossRef]

16. Cao, H.; Gerhold, K.; Mayers, J.R.; Wiest, M.M.; Watkins, S.M.; Hotamisligil, G.S. Identification of a lipokine, a lipid hormone linking adipose tissue to systemic metabolism. *Cell* **2008**, *134*, 933–944. [CrossRef]

17. Scarpace, P.J.; Zhang, Y. Leptin resistance: A prediposing factor for diet-induced obesity. *Am. J. Physiol. Regul. Integr. Comp. Physiol.* **2009**, *296*, R493–R500. [CrossRef]

18. Yamauchi, T.; Kamon, J.; Minokoshi, Y.; Ito, Y.; Waki, H.; Uchida, S.; Yamashita, S.; Noda, M.; Kita, S.; Ueki, K.; et al. Adiponectin stimulates glucose utilization and fatty-acid oxidation by activating AMP-activated protein kinase. *Nat. Med.* **2002**, *8*, 1288–1295. [CrossRef]

19. Min, S.Y.; Desai, A.; Yang, Z.; Sharma, A.; DeSouza, T.; Genga, R.M.J.; Kucukural, A.; Lifshitz, L.M.; Nielsen, S.; Scheele, C.; et al. Diverse repertoire of human adipocyte subtypes develops from transcriptionally distinct mesenchymal progenitor cells. *Proc. Natl. Acad. Sci. USA* **2019**, *116*, 17970–17979. [CrossRef]

20. Raajendiran, A.; Ooi, G.; Bayliss, J.; O'Brien, P.E.; Schittenhelm, R.B.; Clark, A.K.; Taylor, R.A.; Rodeheffer, M.S.; Burton, P.R.; Watt, M.J. Identification of Metabolically Distinct Adipocyte Progenitor Cells in Human Adipose Tissues. *Cell Rep.* **2019**, *27*, 1528–1540.e1527. [CrossRef]

21. Rosen, E.D.; MacDougald, O.A. Adipocyte differentiation from the inside out. *Nat. Rev. Mol. Cell Biol.* **2006**, *7*, 885–896. [CrossRef] [PubMed]

22. Otto, T.C.; Lane, M.D. Adipose development: From stem cell to adipocyte. *Crit. Rev. Biochem. Mol. Biol.* **2005**, *40*, 229–242. [CrossRef] [PubMed]

23. Farmer, S.R. Transcriptional control of adipocyte formation. *Cell Metab.* **2006**, *4*, 263–273. [CrossRef] [PubMed]

24. Rosen, E.D.; Walkey, C.J.; Puigserver, P.; Spiegelman, B.M. Transcriptional regulation of adipogenesis. *Genes Dev.* **2000**, *14*, 1293–1307. [PubMed]

25. Tontonoz, P.; Hu, E.; Spiegelman, B.M. Stimulation of adipogenesis in fibroblasts by PPAR gamma 2, a lipid-activated transcription factor. *Cell* **1994**, *79*, 1147–1156. [CrossRef]

26. Ehrlund, A.; Mejhert, N.; Bjork, C.; Andersson, R.; Kulyte, A.; Astrom, G.; Itoh, M.; Kawaji, H.; Lassmann, T.; Daub, C.O.; et al. Transcriptional Dynamics during Human Adipogenesis and Its Link to Adipose Morphology and Distribution. *Diabetes* **2017**, *66*, 218–230. [CrossRef] [PubMed]

27. Ahn, J.; Wu, H.; Lee, K. Integrative Analysis Revealing Human Adipose-Specific Genes and Consolidating Obesity Loci. *Sci. Rep.* **2019**, *9*, 3087. [CrossRef]

28. Sun, K.; Kusminski, C.M.; Scherer, P.E. Adipose tissue remodeling and obesity. *J. Clin. Investig.* **2011**, *121*, 2094–2101. [CrossRef]

29. Kang, K.; Reilly, S.M.; Karabacak, V.; Gangl, M.R.; Fitzgerald, K.; Hatano, B.; Lee, C.H. Adipocyte-derived Th2 cytokines and myeloid PPARdelta regulate macrophage polarization and insulin sensitivity. *Cell Metab.* **2008**, *7*, 485–495. [CrossRef]

30. Fujisaka, S.; Usui, I.; Bukhari, A.; Ikutani, M.; Oya, T.; Kanatani, Y.; Tsuneyama, K.; Nagai, Y.; Takatsu, K.; Urakaze, M.; et al. Regulatory mechanisms for adipose tissue M1 and M2 macrophages in diet-induced obese mice. *Diabetes* **2009**, *58*, 2574–2582. [CrossRef]

31. Oh, D.Y.; Morinaga, H.; Talukdar, S.; Bae, E.J.; Olefsky, J.M. Increased macrophage migration into adipose tissue in obese mice. *Diabetes* **2012**, *61*, 346–354. [CrossRef]

32. Lumeng, C.N.; Deyoung, S.M.; Bodzin, J.L.; Saltiel, A.R. Increased inflammatory properties of adipose tissue macrophages recruited during diet-induced obesity. *Diabetes* **2007**, *56*, 16–23. [CrossRef] [PubMed]

33. Xu, H.; Barnes, G.T.; Yang, Q.; Tan, G.; Yang, D.; Chou, C.J.; Sole, J.; Nichols, A.; Ross, J.S.; Tartaglia, L.A.; et al. Chronic inflammation in fat plays a crucial role in the development of obesity-related insulin resistance. *J. Clin. Investig.* **2003**, *112*, 1821–1830. [CrossRef] [PubMed]

34. Kanda, H.; Tateya, S.; Tamori, Y.; Kotani, K.; Hiasa, K.; Kitazawa, R.; Kitazawa, S.; Miyachi, H.; Maeda, S.; Egashira, K.; et al. MCP-1 contributes to macrophage infiltration into adipose tissue, insulin resistance, and hepatic steatosis in obesity. *J. Clin. Investig.* **2006**, *116*, 1494–1505. [CrossRef] [PubMed]

35. Zheng, C.; Yang, Q.; Xu, C.; Shou, P.; Cao, J.; Jiang, M.; Chen, Q.; Cao, G.; Han, Y.; Li, F.; et al. CD11b regulates obesity-induced insulin resistance via limiting alternative activation and proliferation of adipose tissue macrophages. *Proc. Natl. Acad. Sci. USA* **2015**, *112*, E7239–E7248. [CrossRef] [PubMed]

36. Shah, R.; Hinkle, C.C.; Ferguson, J.F.; Mehta, N.N.; Li, M.; Qu, L.; Lu, Y.; Putt, M.E.; Ahima, R.S.; Reilly, M.P. Fractalkine is a novel human adipochemokine associated with type 2 diabetes. *Diabetes* **2011**, *60*, 1512–1518. [CrossRef]

37. Spite, M.; Hellmann, J.; Tang, Y.; Mathis, S.P.; Kosuri, M.; Bhatnagar, A.; Jala, V.R.; Haribabu, B. Deficiency of the leukotriene B4 receptor, BLT-1, protects against systemic insulin resistance in diet-induced obesity. *J. Immunol.* **2011**, *187*, 1942–1949. [CrossRef]

38. Finucane, O.M.; Reynolds, C.M.; McGillicuddy, F.C.; Harford, K.A.; Morrison, M.; Baugh, J.; Roche, H.M. Macrophage migration inhibitory factor deficiency ameliorates high-fat diet induced insulin resistance in mice with reduced adipose inflammation and hepatic steatosis. *PLoS ONE* **2014**, *9*, e113369. [CrossRef]

39. Haase, J.; Weyer, U.; Immig, K.; Kloting, N.; Bluher, M.; Eilers, J.; Bechmann, I.; Gericke, M. Local proliferation of macrophages in adipose tissue during obesity-induced inflammation. *Diabetologia* **2014**, *57*, 562–571. [CrossRef]

40. Zheng, C.; Yang, Q.; Cao, J.; Xie, N.; Liu, K.; Shou, P.; Qian, F.; Wang, Y.; Shi, Y. Local proliferation initiates macrophage accumulation in adipose tissue during obesity. *Cell Death Dis.* **2016**, *7*, e2167. [CrossRef]

41. Kratz, M.; Coats, B.R.; Hisert, K.B.; Hagman, D.; Mutskov, V.; Peris, E.; Schoenfelt, K.Q.; Kuzma, J.N.; Larson, I.; Billing, P.S.; et al. Metabolic dysfunction drives a mechanistically distinct proinflammatory phenotype in adipose tissue macrophages. *Cell Metab.* **2014**, *20*, 614–625. [CrossRef] [PubMed]

42. Xu, X.; Grijalva, A.; Skowronski, A.; van Eijk, M.; Serlie, M.J.; Ferrante, A.W., Jr. Obesity activates a program of lysosomal-dependent lipid metabolism in adipose tissue macrophages independently of classic activation. *Cell Metab.* **2013**, *18*, 816–830. [CrossRef] [PubMed]

43. Lumeng, C.N.; DelProposto, J.B.; Westcott, D.J.; Saltiel, A.R. Phenotypic switching of adipose tissue macrophages with obesity is generated by spatiotemporal differences in macrophage subtypes. *Diabetes* **2008**, *57*, 3239–3246. [CrossRef] [PubMed]

44. Feuerer, M.; Herrero, L.; Cipolletta, D.; Naaz, A.; Wong, J.; Nayer, A.; Lee, J.; Goldfine, A.B.; Benoist, C.; Shoelson, S.; et al. Lean, but not obese, fat is enriched for a unique population of regulatory T cells that affect metabolic parameters. *Nat. Med.* **2009**, *15*, 930–939. [CrossRef]

45. Liu, G.; Ma, H.; Qiu, L.; Li, L.; Cao, Y.; Ma, J.; Zhao, Y. Phenotypic and functional switch of macrophages induced by regulatory CD4+CD25+ T cells in mice. *Immunol. Cell Biol.* **2011**, *89*, 130–142. [CrossRef] [PubMed]

46. Qiu, Y.; Nguyen, K.D.; Odegaard, J.I.; Cui, X.; Tian, X.; Locksley, R.M.; Palmiter, R.D.; Chawla, A. Eosinophils and type 2 cytokine signaling in macrophages orchestrate development of functional beige fat. *Cell* **2014**, *157*, 1292–1308. [CrossRef]

47. Brestoff, J.R.; Kim, B.S.; Saenz, S.A.; Stine, R.R.; Monticelli, L.A.; Sonnenberg, G.F.; Thome, J.J.; Farber, D.L.; Lutfy, K.; Seale, P.; et al. Group 2 innate lymphoid cells promote beiging of white adipose tissue and limit obesity. *Nature* **2015**, *519*, 242–246. [CrossRef] [PubMed]

48. Chen, H.H.; Tseng, Y.J.; Wang, S.Y.; Tsai, Y.S.; Chang, C.S.; Kuo, T.C.; Yao, W.J.; Shieh, C.C.; Wu, C.H.; Kuo, P.H. The metabolome profiling and pathway analysis in metabolic healthy and abnormal obesity. *Int. J. Obes. (Lond.)* **2015**, *39*, 1241–1248. [CrossRef] [PubMed]

49. Eisinger, K.; Krautbauer, S.; Hebel, T.; Schmitz, G.; Aslanidis, C.; Liebisch, G.; Buechler, C. Lipidomic analysis of the liver from high-fat diet induced obese mice identifies changes in multiple lipid classes. *Exp. Mol. Pathol.* **2014**, *97*, 37–43. [CrossRef]

50. Hayakawa, J.; Wang, M.; Wang, C.; Han, R.H.; Jiang, Z.Y.; Han, X. Lipidomic analysis reveals significant lipogenesis and accumulation of lipotoxic components in ob/ob mouse organs. *Prostaglandins Leukot. Essent. Fat. Acids* **2018**, *136*, 161–169. [CrossRef]

51. Wang, H.; Ye, J. Regulation of energy balance by inflammation: Common theme in physiology and pathology. *Rev. Endocr. Metab. Disord.* **2015**, *16*, 47–54. [CrossRef] [PubMed]

52. Cani, P.D.; Amar, J.; Iglesias, M.A.; Poggi, M.; Knauf, C.; Bastelica, D.; Neyrinck, A.M.; Fava, F.; Tuohy, K.M.; Chabo, C.; et al. Metabolic endotoxemia initiates obesity and insulin resistance. *Diabetes* **2007**, *56*, 1761–1772. [CrossRef]

53. Cani, P.D.; Jordan, B.F. Gut microbiota-mediated inflammation in obesity: A link with gastrointestinal cancer. *Nat. Rev. Gastroenterol. Hepatol.* **2018**, *15*, 671–682. [CrossRef] [PubMed]

54. Shi, H.; Kokoeva, M.V.; Inouye, K.; Tzameli, I.; Yin, H.; Flier, J.S. TLR4 links innate immunity and fatty acid-induced insulin resistance. *J. Clin. Investig.* **2006**, *116*, 3015–3025. [CrossRef] [PubMed]

55. Jernas, M.; Palming, J.; Sjoholm, K.; Jennische, E.; Svensson, P.A.; Gabrielsson, B.G.; Levin, M.; Sjogren, A.; Rudemo, M.; Lystig, T.C.; et al. Separation of human adipocytes by size: Hypertrophic fat cells display distinct gene expression. *FASEB J.* **2006**, *20*, 1540–1542. [CrossRef]

56. Lee, Y.S.; Kim, J.W.; Osborne, O.; Oh, D.Y.; Sasik, R.; Schenk, S.; Chen, A.; Chung, H.; Murphy, A.; Watkins, S.M.; et al. Increased adipocyte O2 consumption triggers HIF-1alpha, causing inflammation and insulin resistance in obesity. *Cell* **2014**, *157*, 1339–1352. [CrossRef]

57. Strissel, K.J.; Stancheva, Z.; Miyoshi, H.; Perfield, J.W., 2nd; DeFuria, J.; Jick, Z.; Greenberg, A.S.; Obin, M.S. Adipocyte death, adipose tissue remodeling, and obesity complications. *Diabetes* **2007**, *56*, 2910–2918. [CrossRef]

58. Vandanmagsar, B.; Youm, Y.H.; Ravussin, A.; Galgani, J.E.; Stadler, K.; Mynatt, R.L.; Ravussin, E.; Stephens, J.M.; Dixit, V.D. The NLRP3 inflammasome instigates obesity-induced inflammation and insulin resistance. *Nat. Med.* **2011**, *17*, 179–188. [CrossRef]

59. Furukawa, S.; Fujita, T.; Shimabukuro, M.; Iwaki, M.; Yamada, Y.; Nakajima, Y.; Nakayama, O.; Makishima, M.; Matsuda, M.; Shimomura, I. Increased oxidative stress in obesity and its impact on metabolic syndrome. *J. Clin. Investig.* **2004**, *114*, 1752–1761. [CrossRef]

60. Amano, S.U.; Cohen, J.L.; Vangala, P.; Tencerova, M.; Nicoloro, S.M.; Yawe, J.C.; Shen, Y.; Czech, M.P.; Aouadi, M. Local proliferation of macrophages contributes to obesity-associated adipose tissue inflammation. *Cell Metab.* **2014**, *19*, 162–171. [CrossRef]

61. Feng, B.; Jiao, P.; Nie, Y.; Kim, T.; Jun, D.; van Rooijen, N.; Yang, Z.; Xu, H. Clodronate liposomes improve metabolic profile and reduce visceral adipose macrophage content in diet-induced obese mice. *PLoS ONE* **2011**, *6*, e24358. [CrossRef] [PubMed]

62. Kim, D.H.; Sandoval, D.; Reed, J.A.; Matter, E.K.; Tolod, E.G.; Woods, S.C.; Seeley, R.J. The role of GM-CSF in adipose tissue inflammation. *Am. J. Physiol. Endocrinol. Metab.* **2008**, *295*, E1038–E1046. [CrossRef] [PubMed]

63. Ramkhelawon, B.; Hennessy, E.J.; Menager, M.; Ray, T.D.; Sheedy, F.J.; Hutchison, S.; Wanschel, A.; Oldebeken, S.; Geoffrion, M.; Spiro, W.; et al. Netrin-1 promotes adipose tissue macrophage retention and insulin resistance in obesity. *Nat. Med.* **2014**, *20*, 377–384. [CrossRef] [PubMed]

64. McLaughlin, T.; Ackerman, S.E.; Shen, L.; Engleman, E. Role of innate and adaptive immunity in obesity-associated metabolic disease. *J. Clin. Investig.* **2017**, *127*, 5–13. [CrossRef]

65. Lee, C.H.; Lam, K.S. Obesity-induced insulin resistance and macrophage infiltration of the adipose tissue: A vicious cycle. *J. Diabetes Investig.* **2019**, *10*, 29–31. [CrossRef]

66. Murano, I.; Barbatelli, G.; Parisani, V.; Latini, C.; Muzzonigro, G.; Castellucci, M.; Cinti, S. Dead adipocytes, detected as crown-like structures, are prevalent in visceral fat depots of genetically obese mice. *J. Lipid Res.* **2008**, *49*, 1562–1568. [CrossRef]

67. McLaughlin, T.; Liu, L.F.; Lamendola, C.; Shen, L.; Morton, J.; Rivas, H.; Winer, D.; Tolentino, L.; Choi, O.; Zhang, H.; et al. T-cell profile in adipose tissue is associated with insulin resistance and systemic inflammation in humans. *Arterioscler. Thromb. Vasc. Biol.* **2014**, *34*, 2637–2643. [CrossRef]

68. Nishimura, S.; Manabe, I.; Nagasaki, M.; Eto, K.; Yamashita, H.; Ohsugi, M.; Otsu, M.; Hara, K.; Ueki, K.; Sugiura, S.; et al. CD8+ effector T cells contribute to macrophage recruitment and adipose tissue inflammation in obesity. *Nat. Med.* **2009**, *15*, 914–920. [CrossRef]

69. Deng, T.; Lyon, C.J.; Minze, L.J.; Lin, J.; Zou, J.; Liu, J.Z.; Ren, Y.; Yin, Z.; Hamilton, D.J.; Reardon, P.R.; et al. Class II major histocompatibility complex plays an essential role in obesity-induced adipose inflammation. *Cell Metab.* **2013**, *17*, 411–422. [CrossRef]

70. Zou, J.; Lai, B.; Zheng, M.; Chen, Q.; Jiang, S.; Song, A.; Huang, Z.; Shi, P.; Tu, X.; Wang, D.; et al. CD4+ T cells memorize obesity and promote weight regain. *Cell. Mol. Immunol.* **2018**, *15*, 630–639. [CrossRef]

71. DeFuria, J.; Belkina, A.C.; Jagannathan-Bogdan, M.; Snyder-Cappione, J.; Carr, J.D.; Nersesova, Y.R.; Markham, D.; Strissel, K.J.; Watkins, A.A.; Zhu, M.; et al. B cells promote inflammation in obesity and type 2 diabetes through regulation of T-cell function and an inflammatory cytokine profile. *Proc. Natl. Acad. Sci. USA* **2013**, *110*, 5133–5138. [CrossRef]

72. Duffaut, C.; Galitzky, J.; Lafontan, M.; Bouloumie, A. Unexpected trafficking of immune cells within the adipose tissue during the onset of obesity. *Biochem. Biophys. Res. Commun.* **2009**, *384*, 482–485. [CrossRef]

73. Rajbhandari, P.; Arneson, D.; Hart, S.K.; Ahn, I.S.; Diamante, G.; Santos, L.C.; Zaghari, N.; Feng, A.C.; Thomas, B.J.; Vergnes, L.; et al. Single cell analysis reveals immune cell-adipocyte crosstalk regulating the transcription of thermogenic adipocytes. *Elife* **2019**, *8*. [CrossRef]

74. Zhang, X.; Wang, X.; Yin, H.; Zhang, L.; Feng, A.; Zhang, Q.X.; Lin, Y.; Bao, B.; Hernandez, L.L.; Shi, G.P.; et al. Functional Inactivation of Mast Cells Enhances Subcutaneous Adipose Tissue Browning in Mice. *Cell Rep.* **2019**, *28*, 792–803.e794. [CrossRef] [PubMed]

75. Lee, E.H.; Itan, M.; Jang, J.; Gu, H.J.; Rozenberg, P.; Mingler, M.K.; Wen, T.; Yoon, J.; Park, S.Y.; Roh, J.Y.; et al. Eosinophils support adipocyte maturation and promote glucose tolerance in obesity. *Sci. Rep.* **2018**, *8*, 9894. [CrossRef]

76. Wu, D.; Molofsky, A.B.; Liang, H.E.; Ricardo-Gonzalez, R.R.; Jouihan, H.A.; Bando, J.K.; Chawla, A.; Locksley, R.M. Eosinophils sustain adipose alternatively activated macrophages associated with glucose homeostasis. *Science* **2011**, *332*, 243–247. [CrossRef] [PubMed]

77. Talukdar, S.; Oh, D.Y.; Bandyopadhyay, G.; Li, D.; Xu, J.; McNelis, J.; Lu, M.; Li, P.; Yan, Q.; Zhu, Y.; et al. Neutrophils mediate insulin resistance in mice fed a high-fat diet through secreted elastase. *Nat. Med.* **2012**, *18*, 1407–1412. [CrossRef] [PubMed]

78. Bertola, A.; Ciucci, T.; Rousseau, D.; Bourlier, V.; Duffaut, C.; Bonnafous, S.; Blin-Wakkach, C.; Anty, R.; Iannelli, A.; Gugenheim, J.; et al. Identification of adipose tissue dendritic cells correlated with obesity-associated insulin-resistance and inducing Th17 responses in mice and patients. *Diabetes* **2012**, *61*, 2238–2247. [CrossRef] [PubMed]

79. Cho, K.W.; Zamarron, B.F.; Muir, L.A.; Singer, K.; Porsche, C.E.; DelProposto, J.B.; Geletka, L.; Meyer, K.A.; O'Rourke, R.W.; Lumeng, C.N. Adipose Tissue Dendritic Cells Are Independent Contributors to Obesity-Induced Inflammation and Insulin Resistance. *J. Immunol.* **2016**, *197*, 3650–3661. [CrossRef]

80. Molofsky, A.B.; Nussbaum, J.C.; Liang, H.E.; Van Dyken, S.J.; Cheng, L.E.; Mohapatra, A.; Chawla, A.; Locksley, R.M. Innate lymphoid type 2 cells sustain visceral adipose tissue eosinophils and alternatively activated macrophages. *J. Exp. Med.* **2013**, *210*, 535–549. [CrossRef]

81. Sasaki, T.; Moro, K.; Kubota, T.; Kubota, N.; Kato, T.; Ohno, H.; Nakae, S.; Saito, H.; Koyasu, S. Innate Lymphoid Cells in the Induction of Obesity. *Cell Rep.* **2019**, *28*, 202–217.e207. [CrossRef] [PubMed]

82. Juhas, U.; Ryba-Stanislawowska, M.; Szargiej, P.; Mysliwska, J. Different pathways of macrophage activation and polarization. *Postępy Hig. Med. Dośw. (Online)* **2015**, *69*, 496–502. [CrossRef] [PubMed]

83. Suganami, T.; Tanimoto-Koyama, K.; Nishida, J.; Itoh, M.; Yuan, X.; Mizuarai, S.; Kotani, H.; Yamaoka, S.; Miyake, K.; Aoe, S.; et al. Role of the Toll-like receptor 4/NF-kappaB pathway in saturated fatty acid-induced inflammatory changes in the interaction between adipocytes and macrophages. *Arterioscler. Thromb. Vasc. Biol.* **2007**, *27*, 84–91. [CrossRef] [PubMed]

84. Pal, D.; Dasgupta, S.; Kundu, R.; Maitra, S.; Das, G.; Mukhopadhyay, S.; Ray, S.; Majumdar, S.; Bhattacharya, S. Fetuin-A acts as an endogenous ligand of TLR4 to promote lipid-induced insulin resistance. *Nat. Med.* **2012**, *18*, 1279–1285. [CrossRef] [PubMed]

85. Wen, H.; Gris, D.; Lei, Y.; Jha, S.; Zhang, L.; Huang, M.T.; Brickey, W.J.; Ting, J.P. Fatty acid-induced NLRP3-ASC inflammasome activation interferes with insulin signaling. *Nat. Immunol.* **2011**, *12*, 408–415. [CrossRef]

86. Arkan, M.C.; Hevener, A.L.; Greten, F.R.; Maeda, S.; Li, Z.W.; Long, J.M.; Wynshaw-Boris, A.; Poli, G.; Olefsky, J.; Karin, M. IKK-beta links inflammation to obesity-induced insulin resistance. *Nat. Med.* **2005**, *11*, 191–198. [CrossRef]

87. Sassmann-Schweda, A.; Singh, P.; Tang, C.; Wietelmann, A.; Wettschureck, N.; Offermanns, S. Increased apoptosis and browning of TAK1-deficient adipocytes protects against obesity. *JCI Insight* **2016**, *1*, e81175. [CrossRef]

88. Gaestel, M.; Kotlyarov, A.; Kracht, M. Targeting innate immunity protein kinase signalling in inflammation. *Nat. Rev. Drug Discov.* **2009**, *8*, 480–499. [CrossRef]

89. Yang, Q.; Graham, T.E.; Mody, N.; Preitner, F.; Peroni, O.D.; Zabolotny, J.M.; Kotani, K.; Quadro, L.; Kahn, B.B. Serum retinol binding protein 4 contributes to insulin resistance in obesity and type 2 diabetes. *Nature* **2005**, *436*, 356–362. [CrossRef] [PubMed]

90. Odegaard, J.I.; Ricardo-Gonzalez, R.R.; Goforth, M.H.; Morel, C.R.; Subramanian, V.; Mukundan, L.; Red Eagle, A.; Vats, D.; Brombacher, F.; Ferrante, A.W.; et al. Macrophage-specific PPARgamma controls alternative activation and improves insulin resistance. *Nature* **2007**, *447*, 1116–1120. [CrossRef] [PubMed]

91. Aguirre, V.; Uchida, T.; Yenush, L.; Davis, R.; White, M.F. The c-Jun NH(2)-terminal kinase promotes insulin resistance during association with insulin receptor substrate-1 and phosphorylation of Ser(307). *J. Biol. Chem.* **2000**, *275*, 9047–9054. [CrossRef] [PubMed]

92. Gao, Z.; Hwang, D.; Bataille, F.; Lefevre, M.; York, D.; Quon, M.J.; Ye, J. Serine phosphorylation of insulin receptor substrate 1 by inhibitor kappa B kinase complex. *J. Biol. Chem.* **2002**, *277*, 48115–48121. [CrossRef] [PubMed]

93. Obstfeld, A.E.; Sugaru, E.; Thearle, M.; Francisco, A.M.; Gayet, C.; Ginsberg, H.N.; Ables, E.V.; Ferrante, A.W., Jr. C-C chemokine receptor 2 (CCR2) regulates the hepatic recruitment of myeloid cells that promote obesity-induced hepatic steatosis. *Diabetes* **2010**, *59*, 916–925. [CrossRef] [PubMed]

94. Charriere, G.; Cousin, B.; Arnaud, E.; Andre, M.; Bacou, F.; Penicaud, L.; Casteilla, L. Preadipocyte conversion to macrophage. Evidence of plasticity. *J. Biol. Chem.* **2003**, *278*, 9850–9855. [CrossRef]

95. Russo, L.; Lumeng, C.N. Properties and functions of adipose tissue macrophages in obesity. *Immunology* **2018**, *155*, 407–417. [CrossRef]

96. Huh, J.Y.; Park, Y.J.; Ham, M.; Kim, J.B. Crosstalk between adipocytes and immune cells in adipose tissue inflammation and metabolic dysregulation in obesity. *Mol. Cells* **2014**, *37*, 365–371. [CrossRef]

97. Kamei, N.; Tobe, K.; Suzuki, R.; Ohsugi, M.; Watanabe, T.; Kubota, N.; Ohtsuka-Kowatari, N.; Kumagai, K.; Sakamoto, K.; Kobayashi, M.; et al. Overexpression of monocyte chemoattractant protein-1 in adipose tissues causes macrophage recruitment and insulin resistance. *J. Biol. Chem.* **2006**, *281*, 26602–26614. [CrossRef]

98. Wellen, K.E.; Hotamisligil, G.S. Obesity-induced inflammatory changes in adipose tissue. *J. Clin. Investig.* **2003**, *112*, 1785–1788. [CrossRef]

99. Suganami, T.; Nishida, J.; Ogawa, Y. A paracrine loop between adipocytes and macrophages aggravates inflammatory changes: Role of free fatty acids and tumor necrosis factor α. *Arterioscler. Thromb. Vasc. Biol.* **2005**, *25*, 2062–2068. [CrossRef]

100. Saltiel, A.R.; Olefsky, J.M. Inflammatory mechanisms linking obesity and metabolic disease. *J. Clin. Investig.* **2017**, *127*, 1–4. [CrossRef]

101. Cinti, S.; Mitchell, G.; Barbatelli, G.; Murano, I.; Ceresi, E.; Faloia, E.; Wang, S.; Fortier, M.; Greenberg, A.S.; Obin, M.S. Adipocyte death defines macrophage localization and function in adipose tissue of obese mice and humans. *J. Lipid Res.* **2005**, *46*, 2347–2355. [CrossRef] [PubMed]

102. Alkhouri, N.; Gornicka, A.; Berk, M.P.; Thapaliya, S.; Dixon, L.J.; Kashyap, S.; Schauer, P.R.; Feldstein, A.E. Adipocyte apoptosis, a link between obesity, insulin resistance, and hepatic steatosis. *J. Biol. Chem.* **2010**, *285*, 3428–3438. [CrossRef] [PubMed]

103. Keuper, M.; Bluher, M.; Schon, M.R.; Moller, P.; Dzyakanchuk, A.; Amrein, K.; Debatin, K.M.; Wabitsch, M.; Fischer-Posovszky, P. An inflammatory micro-environment promotes human adipocyte apoptosis. *Mol. Cell. Endocrinol.* **2011**, *339*, 105–113. [CrossRef] [PubMed]

104. Feng, D.; Tang, Y.; Kwon, H.; Zong, H.; Hawkins, M.; Kitsis, R.N.; Pessin, J.E. High-fat diet-induced adipocyte cell death occurs through a cyclophilin D intrinsic signaling pathway independent of adipose tissue inflammation. *Diabetes* **2011**, *60*, 2134–2143. [CrossRef]

105. Shapiro, H.; Pecht, T.; Shaco-Levy, R.; Harman-Boehm, I.; Kirshtein, B.; Kuperman, Y.; Chen, A.; Bluher, M.; Shai, I.; Rudich, A. Adipose tissue foam cells are present in human obesity. *J. Clin. Endocrinol. Metab.* **2013**, *98*, 1173–1181. [CrossRef]

106. Bogers, R.P.; Bemelmans, W.J.E.; Hoogenveen, R.T.; Boshuizen, H.C.; Woodward, M.; Knekt, P.; van Dam, R.M.; Hu, F.B.; Visscher, T.L.S.; Menotti, A.; et al. Association of Overweight with Increased Risk of Coronary Heart Disease Partly Independent of Blood Pressure and Cholesterol Levels: A Meta-analysis of 21 Cohort Studies Including More Than 300,000 Persons. *Arch. Intern. Med.* **2007**, *167*, 1720–1728. [CrossRef]

107. Li, M.F.; Cheung, B.M. Rise and fall of anti-obesity drugs. *World J. Diabetes* **2011**, *2*, 19–23. [CrossRef]

108. Bertoia, M.L.; Rimm, E.B.; Mukamal, K.J.; Hu, F.B.; Willett, W.C.; Cassidy, A. Dietary flavonoid intake and weight maintenance: three prospective cohorts of 124,086 US men and women followed for up to 24 years. *BMJ* **2016**, *352*, i17. [CrossRef]

109. Pietta, P.-G. Flavonoids as Antioxidants. *J. Nat. Prod.* **2000**, *63*, 1035–1042. [CrossRef]

110. Seo, M.J.; Lee, Y.J.; Hwang, J.H.; Kim, K.J.; Lee, B.Y. The inhibitory effects of quercetin on obesity and obesity-induced inflammation by regulation of MAPK signaling. *J. Nutr. Biochem.* **2015**, *26*, 1308–1316. [CrossRef]

111. Ding, S.; Jiang, J.; Wang, Z.; Zhang, G.; Yin, J.; Wang, X.; Wang, S.; Yu, Z. Resveratrol reduces the inflammatory response in adipose tissue and improves adipose insulin signaling in high-fat diet-fed mice. *PeerJ* **2018**, *6*, e5173. [CrossRef]

112. Sakamoto, Y.; Kanatsu, J.; Toh, M.; Naka, A.; Kondo, K.; Iida, K. The Dietary Isoflavone Daidzein Reduces Expression of Pro-Inflammatory Genes through PPARalpha/gamma and JNK Pathways in Adipocyte and Macrophage Co-Cultures. *PLoS ONE* **2016**, *11*, e0149676. [CrossRef]

113. Tan, J.; Huang, C.; Luo, Q.; Liu, W.; Cheng, D.; Li, Y.; Xia, Y.; Li, C.; Tang, L.; Fang, J.; et al. Soy Isoflavones Ameliorate Fatty Acid Metabolism of Visceral Adipose Tissue by Increasing the AMPK Activity in Male Rats with Diet-Induced Obesity (DIO). *Molecules* **2019**, *24*, 2809. [CrossRef] [PubMed]

114. Ke, J.Y.; Banh, T.; Hsiao, Y.H.; Cole, R.M.; Straka, S.R.; Yee, L.D.; Belury, M.A. Citrus flavonoid naringenin reduces mammary tumor cell viability, adipose mass, and adipose inflammation in obese ovariectomized mice. *Mol. Nutr. Food Res.* **2017**, *61*. [CrossRef] [PubMed]

115. Rossi, E.L.; Khatib, S.A.; Doerstling, S.S.; Bowers, L.W.; Pruski, M.; Ford, N.A.; Glickman, R.D.; Niu, M.; Yang, P.; Cui, Z.; et al. Resveratrol inhibits obesity-associated adipose tissue dysfunction and tumor growth in a mouse model of postmenopausal claudin-low breast cancer. *Mol. Carcinog.* **2018**, *57*, 393–407. [CrossRef] [PubMed]

116. Roopchand, D.E.; Carmody, R.N.; Kuhn, P.; Moskal, K.; Rojas-Silva, P.; Turnbaugh, P.J.; Raskin, I. Dietary Polyphenols Promote Growth of the Gut Bacterium Akkermansia muciniphila and Attenuate High-Fat Diet-Induced Metabolic Syndrome. *Diabetes* **2015**, *64*, 2847–2858. [CrossRef] [PubMed]

117. Akhlaghi, M.; Ghobadi, S.; Mohammad Hosseini, M.; Gholami, Z.; Mohammadian, F. Flavanols are potential anti-obesity agents, a systematic review and meta-analysis of controlled clinical trials. *Nutr. Metab. Cardiovasc. Dis.* **2018**, *28*, 675–690. [CrossRef]

118. Marranzano, M.; Ray, S.; Godos, J.; Galvano, F. Association between dietary flavonoids intake and obesity in a cohort of adults living in the Mediterranean area. *Int. J. Food Sci. Nutr.* **2018**, *69*, 1020–1029. [CrossRef]

119. Hostetler, G.L.; Ralston, R.A.; Schwartz, S.J. Flavones: Food Sources, Bioavailability, Metabolism, and Bioactivity. *Adv. Nutr.* **2017**, *8*, 423–435. [CrossRef] [PubMed]

120. Hostetler, G.; Riedl, K.; Cardenas, H.; Diosa-Toro, M.; Arango, D.; Schwartz, S.; Doseff, A.I. Flavone deglycosylation increases their anti-inflammatory activity and absorption. *Mol. Nutr. Food Res.* **2012**, *56*, 558–569. [CrossRef]

121. Xiao, J. Dietary flavonoid aglycones and their glycosides: Which show better biological significance? *Crit. Rev. Food Sci. Nutr.* **2017**, *57*, 1874–1905. [CrossRef]

122. Rodriguez-Garcia, C.; Sanchez-Quesada, C.; Gaforio, J.J. Dietary Flavonoids as Cancer Chemopreventive Agents: An Updated Review of Human Studies. *Antioxidants* **2019**, *8*, 137. [CrossRef] [PubMed]

123. Bhagwat, S.; Haytowitz, D.B.; Holden, J.M. *USDA Database for the Flavonoid Content of Selected Foods*; Release 3.1; US Department of Agriculture: Beltsville, MD, USA, 2014.

124. Berim, A.; Gang, D.R. Methoxylated flavones: Occurrence, importance, biosynthesis. *Phytochem. Rev.* **2016**, *15*, 363–390. [CrossRef]

125. Ono, M.; Fujimori, K. Antiadipogenic effect of dietary apigenin through activation of AMPK in 3T3-L1 cells. *J. Agric. Food Chem.* **2011**, *59*, 13346–13352. [CrossRef] [PubMed]

126. Kim, M.A.; Kang, K.; Lee, H.J.; Kim, M.; Kim, C.Y.; Nho, C.W. Apigenin isolated from Daphne genkwa Siebold et Zucc. inhibits 3T3-L1 preadipocyte differentiation through a modulation of mitotic clonal expansion. *Life Sci.* **2014**, *101*, 64–72. [CrossRef]

127. Gomez-Zorita, S.; Lasa, A.; Abendano, N.; Fernandez-Quintela, A.; Mosqueda-Solis, A.; Garcia-Sobreviela, M.P.; Arbones-Mainar, J.M.; Portillo, M.P. Phenolic compounds apigenin, hesperidin and kaempferol reduce in vitro lipid accumulation in human adipocytes. *J. Transl. Med.* **2017**, *15*, 237. [CrossRef]

128. Sun, Y.S.; Qu, W. Dietary Apigenin promotes lipid catabolism, thermogenesis, and browning in adipose tissues of HFD-Fed mice. *Food Chem. Toxicol.* **2019**, *133*, 110780. [CrossRef]

129. Su, T.; Huang, C.; Yang, C.; Jiang, T.; Su, J.; Chen, M.; Fatima, S.; Gong, R.; Hu, X.; Bian, Z.; et al. Apigenin inhibits STAT3/CD36 signaling axis and reduces visceral obesity. *Pharmacol. Res.* **2019**, *152*, 104586. [CrossRef]

130. Feng, X.; Yu, W.; Li, X.; Zhou, F.; Zhang, W.; Shen, Q.; Li, J.; Zhang, C.; Shen, P. Apigenin, a modulator of PPARgamma, attenuates HFD-induced NAFLD by regulating hepatocyte lipid metabolism and oxidative stress via Nrf2 activation. *Biochem. Pharmacol.* **2017**, *136*, 136–149. [CrossRef]

131. Lv, Y.; Gao, X.; Luo, Y.; Fan, W.; Shen, T.; Ding, C.; Yao, M.; Song, S.; Yan, L. Apigenin ameliorates HFD-induced NAFLD through regulation of the XO/NLRP3 pathways. *J. Nutr. Biochem.* **2019**, *71*, 110–121. [CrossRef]

132. Feng, X.; Weng, D.; Zhou, F.; Owen, Y.D.; Qin, H.; Zhao, J.; Wen, Y.; Huang, Y.; Chen, J.; Fu, H.; et al. Activation of PPARgamma by a Natural Flavonoid Modulator, Apigenin Ameliorates Obesity-Related Inflammation Via Regulation of Macrophage Polarization. *EBioMedicine* **2016**, *9*, 61–76. [CrossRef] [PubMed]

133. Park, H.S.; Kim, S.H.; Kim, Y.S.; Ryu, S.Y.; Hwang, J.T.; Yang, H.J.; Kim, G.H.; Kwon, D.Y.; Kim, M.S. Luteolin inhibits adipogenic differentiation by regulating PPARgamma activation. *Biofactors* **2009**, *35*, 373–379. [CrossRef]

134. Kwon, E.Y.; Kim, S.Y.; Choi, M.S. Luteolin-Enriched Artichoke Leaf Extract Alleviates the Metabolic Syndrome in Mice with High-Fat Diet-Induced Obesity. *Nutrients* **2018**, *10*, 979. [CrossRef] [PubMed]

135. Zhang, L.; Han, Y.J.; Zhang, X.; Wang, X.; Bao, B.; Qu, W.; Liu, J. Luteolin reduces obesity-associated insulin resistance in mice by activating AMPKα1 signalling in adipose tissue macrophages. *Diabetologia* **2016**, *59*, 2219–2228. [CrossRef] [PubMed]

136. Kwon, E.Y.; Choi, M.S. Luteolin Targets the Toll-Like Receptor Signaling Pathway in Prevention of Hepatic and Adipocyte Fibrosis and Insulin Resistance in Diet-Induced Obese Mice. *Nutrients* **2018**, *10*, 1415. [CrossRef] [PubMed]

137. Seo, M.-J.; Choi, H.-S.; Lee, O.-H.; Lee, B.-Y. *Baicalein Inhibits Lipid Accumulation through Regulation of MCE and Cell Cycle during 3T3-L1 Adipocyte Differentiation*; Federation of American Societies for Experimental Biology: Bethesda, MD, USA, 2013.

138. Nakao, Y.; Yoshihara, H.; Fujimori, K. Suppression of Very Early Stage of Adipogenesis by Baicalein, a Plant-Derived Flavonoid through Reduced Akt-C/EBPalpha-GLUT4 Signaling-Mediated Glucose Uptake in 3T3-L1 Adipocytes. *PLoS ONE* **2016**, *11*, e0163640. [CrossRef]

139. Min, W.; Wu, M.; Fang, P.; Yu, M.; Shi, M.; Zhang, Z.; Bo, P. Effect of Baicalein on GLUT4 Translocation in Adipocytes of Diet-Induced Obese Mice. *Cell. Physiol. Biochem.* **2018**, *50*, 426–436. [CrossRef]

140. Nagai, S.; Matsumoto, C.; Shibano, M.; Fujimori, K. Suppression of Fatty Acid and Triglyceride Synthesis by the Flavonoid Orientin through Decrease of C/EBPdelta Expression and Inhibition of PI3K/Akt-FOXO1 Signaling in Adipocytes. *Nutrients* **2018**, *10*, 130. [CrossRef]

141. Choi, J.H.; Yun, J.W. Chrysin induces brown fat-like phenotype and enhances lipid metabolism in 3T3-L1 adipocytes. *Nutrition* **2016**, *32*, 1002–1010. [CrossRef]

142. Feng, X.; Qin, H.; Shi, Q.; Zhang, Y.; Zhou, F.; Wu, H.; Ding, S.; Niu, Z.; Lu, Y.; Shen, P. Chrysin attenuates inflammation by regulating M1/M2 status via activating PPARgamma. *Biochem. Pharmacol.* **2014**, *89*, 503–514. [CrossRef]

143. Hadrich, F.; Sayadi, S. Apigetrin inhibits adipogenesis in 3T3-L1 cells by downregulating PPARgamma and CEBP-alpha. *Lipids Health Dis.* **2018**, *17*, 95. [CrossRef] [PubMed]

144. Lee, Y.-H.; Yang, S.-H.; Chen, S.-L.; Pan, Y.-F.; Liu, C.-M.; Li, M.-W.; Chou, S.-S.; Chou, M.-Y.; Youn, S.-C. The anti-adipogenic effect of vitexin is via ERK 1/2 MAPK signaling in 3T3-L1 adipocytes. *Int. J. Phytomed.* **2014**, *6*, 206.

145. Peng, Y.; Sun, Q.; Xu, W.; He, Y.; Jin, W.; Yuan, L.; Gao, R. Vitexin ameliorates high fat diet-induced obesity in male C57BL/6J mice via the AMPKalpha-mediated pathway. *Food Funct.* **2019**, *10*, 1940–1947. [CrossRef] [PubMed]

146. Na, H.Y.; Lee, B.C. Scutellaria baicalensis Alleviates Insulin Resistance in Diet-Induced Obese Mice by Modulating Inflammation. *Int. J. Mol. Sci.* **2019**, *20*, 727. [CrossRef]

147. Shen, K.; Feng, X.; Pan, H.; Zhang, F.; Xie, H.; Zheng, S. Baicalin Ameliorates Experimental Liver Cholestasis in Mice by Modulation of Oxidative Stress, Inflammation, and NRF2 Transcription Factor. *Oxidative Med. Cell. Longev.* **2017**, *2017*, 6169128. [CrossRef]

148. Dai, J.; Liang, K.; Zhao, S.; Jia, W.; Liu, Y.; Wu, H.; Lv, J.; Cao, C.; Chen, T.; Zhuang, S.; et al. Chemoproteomics reveals baicalin activates hepatic CPT1 to ameliorate diet-induced obesity and hepatic steatosis. *Proc. Natl. Acad. Sci. USA* **2018**, *115*, E5896–E5905. [CrossRef]

149. Arango, D.; Morohashi, K.; Yilmaz, A.; Kuramochi, K.; Parihar, A.; Brahimaj, B.; Grotewold, E.; Doseff, A.I. Molecular basis for the action of a dietary flavonoid revealed by the comprehensive identification of apigenin human targets. *Proc. Natl. Acad. Sci. USA* **2013**, *110*, E2153–E2162. [CrossRef]

150. Masson, O.; Prebois, C.; Derocq, D.; Meulle, A.; Dray, C.; Daviaud, D.; Quilliot, D.; Valet, P.; Muller, C.; Liaudet-Coopman, E. Cathepsin-D, a key protease in breast cancer, is up-regulated in obese mouse and human adipose tissue, and controls adipogenesis. *PLoS ONE* **2011**, *6*, e16452. [CrossRef]

151. Eguchi, A.; Feldstein, A.E. Lysosomal Cathepsin D contributes to cell death during adipocyte hypertrophy. *Adipocyte* **2013**, *2*, 170–175. [CrossRef]

152. Ando, C.; Takahashi, N.; Hirai, S.; Nishimura, K.; Lin, S.; Uemura, T.; Goto, T.; Yu, R.; Nakagami, J.; Murakami, S.; et al. Luteolin, a food-derived flavonoid, suppresses adipocyte-dependent activation of macrophages by inhibiting JNK activation. *FEBS Lett.* **2009**, *583*, 3649–3654. [CrossRef]

153. Gerhardt, C.C.; Romero, I.A.; Cancello, R.; Camoin, L.; Strosberg, A.D. Chemokines control fat accumulation and leptin secretion by cultured human adipocytes. *Mol. Cell. Endocrinol.* **2001**, *175*, 81–92. [CrossRef]

154. Kaplan, J.L.; Marshall, M.A.; McSkimming, C.C.; Harmon, D.B.; Garmey, J.C.; Oldham, S.N.; Hallowell, P.; McNamara, C.A. Adipocyte progenitor cells initiate monocyte chemoattractant protein-1-mediated macrophage accumulation in visceral adipose tissue. *Mol. Metab.* **2015**, *4*, 779–794. [CrossRef] [PubMed]

155. Parekh, N.; Chandran, U.; Bandera, E.V. Obesity in cancer survival. *Annu. Rev. Nutr.* **2012**, *32*, 311–342. [CrossRef] [PubMed]

156. Makowski, L.; Zhou, C.; Zhong, Y.; Kuan, P.F.; Fan, C.; Sampey, B.P.; Difurio, M.; Bae-Jump, V.L. Obesity increases tumor aggressiveness in a genetically engineered mouse model of serous ovarian cancer. *Gynecol. Oncol.* **2014**, *133*, 90–97. [CrossRef] [PubMed]

157. Dai, Z.; Xu, Y.C.; Niu, L. Obesity and colorectal cancer risk: A meta-analysis of cohort studies. *World J. Gastroenterol.* **2007**, *13*, 4199–4206. [CrossRef] [PubMed]

158. Nieman, K.M.; Romero, I.L.; Van Houten, B.; Lengyel, E. Adipose tissue and adipocytes support tumorigenesis and metastasis. *Biochim. Biophys. Acta* **2013**, *1831*, 1533–1541. [CrossRef]

159. Wang, Y.Y.; Attane, C.; Milhas, D.; Dirat, B.; Dauvillier, S.; Guerard, A.; Gilhodes, J.; Lazar, I.; Alet, N.; Laurent, V.; et al. Mammary adipocytes stimulate breast cancer invasion through metabolic remodeling of tumor cells. *JCI Insight* **2017**, *2*, e87489. [CrossRef]

160. Iyengar, P.; Combs, T.P.; Shah, S.J.; Gouon-Evans, V.; Pollard, J.W.; Albanese, C.; Flanagan, L.; Tenniswood, M.P.; Guha, C.; Lisanti, M.P.; et al. Adipocyte-secreted factors synergistically promote mammary tumorigenesis through induction of anti-apoptotic transcriptional programs and proto-oncogene stabilization. *Oncogene* **2003**, *22*, 6408–6423. [CrossRef]

161. Park, J.; Morley, T.S.; Kim, M.; Clegg, D.J.; Scherer, P.E. Obesity and cancer–mechanisms underlying tumour progression and recurrence. *Nat. Rev. Endocrinol.* **2014**, *10*, 455–465. [CrossRef]

162. Park, J.; Euhus, D.M.; Scherer, P.E. Paracrine and endocrine effects of adipose tissue on cancer development and progression. *Endocr. Rev.* **2011**, *32*, 550–570. [CrossRef]

163. Chen, Y.; He, Y.; Wang, X.; Lu, F.; Gao, J. Adiposederived mesenchymal stem cells exhibit tumor tropism and promote tumorsphere formation of breast cancer cells. *Oncol. Rep.* **2019**, *41*, 2126–2136. [CrossRef] [PubMed]

164. Goto, H.; Shimono, Y.; Funakoshi, Y.; Imamura, Y.; Toyoda, M.; Kiyota, N.; Kono, S.; Takao, S.; Mukohara, T.; Minami, H. Adipose-derived stem cells enhance human breast cancer growth and cancer stem cell-like properties through adipsin. *Oncogene* **2019**, *38*, 767–779. [CrossRef] [PubMed]

165. Gonzalez-Perez, R.R.; Xu, Y.; Guo, S.; Watters, A.; Zhou, W.; Leibovich, S.J. Leptin upregulates VEGF in breast cancer via canonic and non-canonical signalling pathways and NFkappaB/HIF-1alpha activation. *Cell Signal.* **2010**, *22*, 1350–1362. [CrossRef] [PubMed]

166. Zheng, Q.; Banaszak, L.; Fracci, S.; Basali, D.; Dunlap, S.M.; Hursting, S.D.; Rich, J.N.; Hjlemeland, A.B.; Vasanji, A.; Berger, N.A.; et al. Leptin receptor maintains cancer stem-like properties in triple negative breast cancer cells. *Endocr. Relat. Cancer* **2013**, *20*, 797–808. [CrossRef] [PubMed]

167. Chakraborty, D.; Benham, V.; Bullard, B.; Kearney, T.; Hsia, H.C.; Gibbon, D.; Demireva, E.Y.; Lunt, S.Y.; Bernard, J.J. Fibroblast growth factor receptor is a mechanistic link between visceral adiposity and cancer. *Oncogene* **2017**, *36*, 6668–6679. [CrossRef] [PubMed]

168. Harris, A.L. Hypoxia—A key regulatory factor in tumour growth. *Nat. Rev. Cancer* **2002**, *2*, 38–47. [CrossRef] [PubMed]

169. Cao, Y. Adipose tissue angiogenesis as a therapeutic target for obesity and metabolic diseases. *Nat. Rev. Drug Discov.* **2010**, *9*, 107–115. [CrossRef]

170. Bu, D.; Crewe, C.; Kusminski, C.M.; Gordillo, R.; Ghaben, A.L.; Kim, M.; Park, J.; Deng, H.; Xiong, W.; Liu, X.Z.; et al. Human endotrophin as a driver of malignant tumor growth. *JCI Insight* **2019**, *5*, e125094. [CrossRef]

171. Iyengar, P.; Espina, V.; Williams, T.W.; Lin, Y.; Berry, D.; Jelicks, L.A.; Lee, H.; Temple, K.; Graves, R.; Pollard, J.; et al. Adipocyte-derived collagen VI affects early mammary tumor progression in vivo, demonstrating a critical interaction in the tumor/stroma microenvironment. *J. Clin. Investig.* **2005**, *115*, 1163–1176. [CrossRef]

172. Park, J.; Scherer, P.E. Endotrophin—A novel factor linking obesity with aggressive tumor growth. *Oncotarget* **2012**, *3*, 1487–1488. [CrossRef]

173. Dirat, B.; Bochet, L.; Dabek, M.; Daviaud, D.; Dauvillier, S.; Majed, B.; Wang, Y.Y.; Meulle, A.; Salles, B.; Le Gonidec, S.; et al. Cancer-associated adipocytes exhibit an activated phenotype and contribute to breast cancer invasion. *Cancer Res.* **2011**, *71*, 2455–2465. [CrossRef] [PubMed]

174. Madak-Erdogan, Z.; Band, S.; Zhao, Y.C.; Smith, B.P.; Kulkoyluoglu-Cotul, E.; Zuo, Q.; Santaliz Casiano, A.; Wrobel, K.; Rossi, G.; Smith, R.L.; et al. Free Fatty Acids Rewire Cancer Metabolism in Obesity-Associated Breast Cancer via Estrogen Receptor and mTOR Signaling. *Cancer Res.* **2019**, *79*, 2494–2510. [CrossRef] [PubMed]

175. Park, E.J.; Lee, J.H.; Yu, G.Y.; He, G.; Ali, S.R.; Holzer, R.G.; Osterreicher, C.H.; Takahashi, H.; Karin, M. Dietary and genetic obesity promote liver inflammation and tumorigenesis by enhancing IL-6 and TNF expression. *Cell* **2010**, *140*, 197–208. [CrossRef] [PubMed]

176. Subbaramaiah, K.; Howe, L.R.; Bhardwaj, P.; Du, B.; Gravaghi, C.; Yantiss, R.K.; Zhou, X.K.; Blaho, V.A.; Hla, T.; Yang, P.; et al. Obesity is associated with inflammation and elevated aromatase expression in the mouse mammary gland. *Cancer Prev. Res. (Phila)* **2011**, *4*, 329–346. [CrossRef] [PubMed]

177. Pollard, J.W. Tumour-educated macrophages promote tumour progression and metastasis. *Nat. Rev. Cancer* **2004**, *4*, 71. [CrossRef] [PubMed]

178. Bougaret, L.; Delort, L.; Billard, H.; Le Huede, C.; Boby, C.; De la Foye, A.; Rossary, A.; Mojallal, A.; Damour, O.; Auxenfans, C.; et al. Adipocyte/breast cancer cell crosstalk in obesity interferes with the anti-proliferative efficacy of tamoxifen. *PLoS ONE* **2018**, *13*, e0191571. [CrossRef] [PubMed]

179. Vargo, M.A.; Voss, O.H.; Poustka, F.; Cardounel, A.J.; Grotewold, E.; Doseff, A.I. Apigenin-induced-apoptosis is mediated by the activation of PKCdelta and caspases in leukemia cells. *Biochem. Pharmacol.* **2006**, *72*, 681–692. [CrossRef]

180. Arango, D.; Parihar, A.; Villamena, F.A.; Wang, L.; Freitas, M.A.; Grotewold, E.; Doseff, A.I. Apigenin induces DNA damage through the PKCdelta-dependent activation of ATM and H2AX causing down-regulation of genes involved in cell cycle control and DNA repair. *Biochem. Pharmacol.* **2012**, *84*, 1571–1580. [CrossRef]

181. Kang, K.A.; Piao, M.J.; Hyun, Y.J.; Zhen, A.X.; Cho, S.J.; Ahn, M.J.; Yi, J.M.; Hyun, J.W. Luteolin promotes apoptotic cell death via upregulation of Nrf2 expression by DNA demethylase and the interaction of Nrf2 with p53 in human colon cancer cells. *Exp. Mol. Med.* **2019**, *51*, 40. [CrossRef]

182. Shukla, S.; Gupta, S. Apigenin suppresses insulin-like growth factor I receptor signaling in human prostate cancer: An in vitro and in vivo study. *Mol. Carcinog.* **2009**, *48*, 243–252. [CrossRef]

183. Li, Y.W.; Xu, J.; Zhu, G.Y.; Huang, Z.J.; Lu, Y.; Li, X.Q.; Wang, N.; Zhang, F.X. Apigenin suppresses the stem cell-like properties of triple-negative breast cancer cells by inhibiting YAP/TAZ activity. *Cell Death Discov.* **2018**, *4*, 105. [CrossRef]

184. Koh, S.Y.; Moon, J.Y.; Unno, T.; Cho, S.K. Baicalein Suppresses Stem Cell-Like Characteristics in Radio- and Chemoresistant MDA-MB-231 Human Breast Cancer Cells through Up-Regulation of IFIT2. *Nutrients* **2019**, *11*, 624. [CrossRef]

185. Arango, D.; Diosa-Toro, M.; Rojas-Hernandez, L.S.; Cooperstone, J.L.; Schwartz, S.J.; Mo, X.; Jiang, J.; Schmittgen, T.D.; Doseff, A.I. Dietary apigenin reduces LPS-induced expression of miR-155 restoring immune balance during inflammation. *Mol. Nutr. Food Res.* **2015**, *59*, 763–772. [CrossRef]

186. Cardenas, H.; Arango, D.; Nicholas, C.; Duarte, S.; Nuovo, G.J.; He, W.; Voss, O.H.; Gonzalez-Mejia, M.E.; Guttridge, D.C.; Grotewold, E.; et al. Dietary Apigenin Exerts Immune-Regulatory Activity in Vivo by Reducing NF-kappaB Activity, Halting Leukocyte Infiltration and Restoring Normal Metabolic Function. *Int. J. Mol. Sci.* **2016**, *17*, 323. [CrossRef] [PubMed]

187. Nicholas, C.; Batra, S.; Vargo, M.A.; Voss, O.H.; Gavrilin, M.A.; Wewers, M.D.; Guttridge, D.C.; Grotewold, E.; Doseff, A.I. Apigenin blocks lipopolysaccharide-induced lethality in vivo and proinflammatory cytokines expression by inactivating NF-κB through the suppression of p65 phosphorylation. *J. Immunol.* **2007**, *179*, 7121–7127. [CrossRef] [PubMed]

188. Bauer, D.; Redmon, N.; Mazzio, E.; Soliman, K.F. Apigenin inhibits TNFalpha/IL-1alpha-induced CCL2 release through IKBK-epsilon signaling in MDA-MB-231 human breast cancer cells. *PLoS ONE* **2017**, *12*, e0175558. [CrossRef] [PubMed]

189. Choi, H.J.; Choi, H.J.; Chung, T.W.; Ha, K.T. Luteolin inhibits recruitment of monocytes and migration of Lewis lung carcinoma cells by suppressing chemokine (C-C motif) ligand 2 expression in tumor-associated macrophage. *Biochem. Biophys. Res. Commun.* **2016**, *470*, 101–106. [CrossRef]

190. Fang, B.; Chen, X.; Wu, M.; Kong, H.; Chu, G.; Zhou, Z.; Zhang, C.; Chen, B. Luteolin inhibits angiogenesis of the M2like TAMs via the downregulation of hypoxia inducible factor1alpha and the STAT3 signalling pathway under hypoxia. *Mol. Med. Rep.* **2018**, *18*, 2914–2922. [CrossRef]

191. Brown, E.M.; Sadarangani, M.; Finlay, B.B. The role of the immune system in governing host-microbe interactions in the intestine. *Nat. Immunol.* **2013**, *14*, 660–667. [CrossRef]

192. Zierer, J.; Jackson, M.A.; Kastenmuller, G.; Mangino, M.; Long, T.; Telenti, A.; Mohney, R.P.; Small, K.S.; Bell, J.T.; Steves, C.J.; et al. The fecal metabolome as a functional readout of the gut microbiome. *Nat. Genet.* **2018**, *50*, 790–795. [CrossRef]

193. David, L.A.; Maurice, C.F.; Carmody, R.N.; Gootenberg, D.B.; Button, J.E.; Wolfe, B.E.; Ling, A.V.; Devlin, A.S.; Varma, Y.; Fischbach, M.A.; et al. Diet rapidly and reproducibly alters the human gut microbiome. *Nature* **2014**, *505*, 559–563. [CrossRef] [PubMed]

194. Turnbaugh, P.J.; Hamady, M.; Yatsunenko, T.; Cantarel, B.L.; Duncan, A.; Ley, R.E.; Sogin, M.L.; Jones, W.J.; Roe, B.A.; Affourtit, J.P.; et al. A core gut microbiome in obese and lean twins. *Nature* **2009**, *457*, 480–484. [CrossRef] [PubMed]

195. Shen, W.; Gaskins, H.R.; McIntosh, M.K. Influence of dietary fat on intestinal microbes, inflammation, barrier function and metabolic outcomes. *J. Nutr. Biochem.* **2014**, *25*, 270–280. [CrossRef] [PubMed]

196. Xiao, S.; Zhao, L. Gut microbiota-based translational biomarkers to prevent metabolic syndrome via nutritional modulation. *FEMS Microbiol. Ecol.* **2014**, *87*, 303–314. [CrossRef] [PubMed]

197. Backhed, F.; Manchester, J.K.; Semenkovich, C.F.; Gordon, J.I. Mechanisms underlying the resistance to diet-induced obesity in germ-free mice. *Proc. Natl. Acad. Sci. USA* **2007**, *104*, 979–984. [CrossRef] [PubMed]

198. Ley, R.E.; Backhed, F.; Turnbaugh, P.; Lozupone, C.A.; Knight, R.D.; Gordon, J.I. Obesity alters gut microbial ecology. *Proc. Natl. Acad. Sci. USA* **2005**, *102*, 11070–11075. [CrossRef]

199. Hildebrandt, M.A.; Hoffmann, C.; Sherrill-Mix, S.A.; Keilbaugh, S.A.; Hamady, M.; Chen, Y.Y.; Knight, R.; Ahima, R.S.; Bushman, F.; Wu, G.D. High-fat diet determines the composition of the murine gut microbiome independently of obesity. *Gastroenterology* **2009**, *137*, 1716–1724.e2. [CrossRef]

200. Turnbaugh, P.J.; Ley, R.E.; Mahowald, M.A.; Magrini, V.; Mardis, E.R.; Gordon, J.I. An obesity-associated gut microbiome with increased capacity for energy harvest. *Nature* **2006**, *444*, 1027–1031. [CrossRef]

201. Ley, R.E.; Turnbaugh, P.J.; Klein, S.; Gordon, J.I. Human gut microbes associated with obesity. *Nature* **2006**, *444*, 1022–1023. [CrossRef]

202. Kim, K.A.; Gu, W.; Lee, I.A.; Joh, E.H.; Kim, D.H. High fat diet-induced gut microbiota exacerbates inflammation and obesity in mice via the TLR4 signaling pathway. *PLoS ONE* **2012**, *7*, e47713. [CrossRef]

203. Luck, H.; Tsai, S.; Chung, J.; Clemente-Casares, X.; Ghazarian, M.; Revelo, X.S.; Lei, H.; Luk, C.T.; Shi, S.Y.; Surendra, A.; et al. Regulation of obesity-related insulin resistance with gut anti-inflammatory agents. *Cell Metab.* **2015**, *21*, 527–542. [CrossRef]

204. Winer, D.A.; Luck, H.; Tsai, S.; Winer, S. The Intestinal Immune System in Obesity and Insulin Resistance. *Cell Metab.* **2016**, *23*, 413–426. [CrossRef] [PubMed]

205. Kawano, Y.; Nakae, J.; Watanabe, N.; Kikuchi, T.; Tateya, S.; Tamori, Y.; Kaneko, M.; Abe, T.; Onodera, M.; Itoh, H. Colonic Pro-inflammatory Macrophages Cause Insulin Resistance in an Intestinal Ccl2/Ccr2-Dependent Manner. *Cell Metab.* **2016**, *24*, 295–310. [CrossRef] [PubMed]

206. Seregin, S.S.; Golovchenko, N.; Schaf, B.; Chen, J.; Pudlo, N.A.; Mitchell, J.; Baxter, N.T.; Zhao, L.; Schloss, P.D.; Martens, E.C.; et al. NLRP6 Protects Il10(-/-) Mice from Colitis by Limiting Colonization of Akkermansia muciniphila. *Cell Rep.* **2017**, *19*, 733–745. [CrossRef] [PubMed]

207. Xiao, L.; Chen, B.; Feng, D.; Yang, T.; Li, T.; Chen, J. TLR4 May Be Involved in the Regulation of Colonic Mucosal Microbiota by Vitamin A. *Front. Microbiol.* **2019**, *10*, 268. [CrossRef]

208. Ding, S.; Chi, M.M.; Scull, B.P.; Rigby, R.; Schwerbrock, N.M.; Magness, S.; Jobin, C.; Lund, P.K. High-fat diet: Bacteria interactions promote intestinal inflammation which precedes and correlates with obesity and insulin resistance in mouse. *PLoS ONE* **2010**, *5*, e12191. [CrossRef]

209. Cani, P.D.; Possemiers, S.; Van de Wiele, T.; Guiot, Y.; Everard, A.; Rottier, O.; Geurts, L.; Naslain, D.; Neyrinck, A.; Lambert, D.M.; et al. Changes in gut microbiota control inflammation in obese mice through a mechanism involving GLP-2-driven improvement of gut permeability. *Gut* **2009**, *58*, 1091–1103. [CrossRef]

210. Schulz, M.D.; Atay, C.; Heringer, J.; Romrig, F.K.; Schwitalla, S.; Aydin, B.; Ziegler, P.K.; Varga, J.; Reindl, W.; Pommerenke, C.; et al. High-fat-diet-mediated dysbiosis promotes intestinal carcinogenesis independently of obesity. *Nature* **2014**, *514*, 508–512. [CrossRef]

211. Poutahidis, T.; Varian, B.J.; Levkovich, T.; Lakritz, J.R.; Mirabal, S.; Kwok, C.; Ibrahim, Y.M.; Kearney, S.M.; Chatzigiagkos, A.; Alm, E.J.; et al. Dietary microbes modulate transgenerational cancer risk. *Cancer Res.* **2015**, *75*, 1197–1204. [CrossRef]

212. Routy, B.; Le Chatelier, E.; Derosa, L.; Duong, C.P.M.; Alou, M.T.; Daillere, R.; Fluckiger, A.; Messaoudene, M.; Rauber, C.; Roberti, M.P.; et al. Gut microbiome influences efficacy of PD-1-based immunotherapy against epithelial tumors. *Science* **2018**, *359*, 91–97. [CrossRef]

213. Laparra, J.M.; Sanz, Y. Interactions of gut microbiota with functional food components and nutraceuticals. *Pharmacol. Res.* **2010**, *61*, 219–225. [CrossRef] [PubMed]

214. Ahn-Jarvis, J.H.; Parihar, A.; Doseff, A.I. Dietary Flavonoids for Immunoregulation and Cancer: Food Design for Targeting Disease. *Antioxidants* **2019**, *8*, 202. [CrossRef] [PubMed]

215. Aura, A.-M. Microbial metabolism of dietary phenolic compounds in the colon. *Phytochem. Rev.* **2008**, *7*, 407–429. [CrossRef]

216. Kawabata, K.; Yoshioka, Y.; Terao, J. Role of Intestinal Microbiota in the Bioavailability and Physiological Functions of Dietary Polyphenols. *Molecules* **2019**, *24*, 370. [CrossRef] [PubMed]

217. Comalada, M.; Camuesco, D.; Sierra, S.; Ballester, I.; Xaus, J.; Gálvez, J.; Zarzuelo, A. In vivo quercitrin anti-inflammatory effect involves release of quercetin, which inhibits inflammation through down-regulation of the NF-κB pathway. *Eur. J. Immunol.* **2005**, *35*, 584–592. [CrossRef] [PubMed]

218. Lee, H.C.; Jenner, A.M.; Low, C.S.; Lee, Y.K. Effect of tea phenolics and their aromatic fecal bacterial metabolites on intestinal microbiota. *Res. Microbiol.* **2006**, *157*, 876–884. [CrossRef]

219. Collins, B.; Hoffman, J.; Martinez, K.; Grace, M.; Lila, M.A.; Cockrell, C.; Nadimpalli, A.; Chang, E.; Chuang, C.C.; Zhong, W.; et al. A polyphenol-rich fraction obtained from table grapes decreases adiposity, insulin resistance and markers of inflammation and impacts gut microbiota in high-fat-fed mice. *J. Nutr. Biochem.* **2016**, *31*, 150–165. [CrossRef]

220. Zhao, L.; Zhang, Q.; Ma, W.; Tian, F.; Shen, H.; Zhou, M. A combination of quercetin and resveratrol reduces obesity in high-fat diet-fed rats by modulation of gut microbiota. *Food Funct.* **2017**, *8*, 4644–4656. [CrossRef]

221. Etxeberria, U.; Arias, N.; Boque, N.; Macarulla, M.T.; Portillo, M.P.; Martinez, J.A.; Milagro, F.I. Reshaping faecal gut microbiota composition by the intake of trans-resveratrol and quercetin in high-fat sucrose diet-fed rats. *J. Nutr. Biochem.* **2015**, *26*, 651–660. [CrossRef]

222. Porras, D.; Nistal, E.; Martinez-Florez, S.; Pisonero-Vaquero, S.; Olcoz, J.L.; Jover, R.; Gonzalez-Gallego, J.; Garcia-Mediavilla, M.V.; Sanchez-Campos, S. Protective effect of quercetin on high-fat diet-induced non-alcoholic fatty liver disease in mice is mediated by modulating intestinal microbiota imbalance and related gut-liver axis activation. *Free Radic. Biol. Med.* **2017**, *102*, 188–202. [CrossRef]

223. Anhe, F.F.; Roy, D.; Pilon, G.; Dudonne, S.; Matamoros, S.; Varin, T.V.; Garofalo, C.; Moine, Q.; Desjardins, Y.; Levy, E.; et al. A polyphenol-rich cranberry extract protects from diet-induced obesity, insulin resistance and intestinal inflammation in association with increased Akkermansia spp. population in the gut microbiota of mice. *Gut* **2015**, *64*, 872–883. [CrossRef] [PubMed]

224. den Besten, G.; Bleeker, A.; Gerding, A.; van Eunen, K.; Havinga, R.; van Dijk, T.H.; Oosterveer, M.H.; Jonker, J.W.; Groen, A.K.; Reijngoud, D.J.; et al. Short-Chain Fatty Acids Protect Against High-Fat Diet-Induced Obesity via a PPARgamma-Dependent Switch from Lipogenesis to Fat Oxidation. *Diabetes* **2015**, *64*, 2398–2408. [CrossRef]

225. Noratto, G.D.; Garcia-Mazcorro, J.F.; Markel, M.; Martino, H.S.; Minamoto, Y.; Steiner, J.M.; Byrne, D.; Suchodolski, J.S.; Mertens-Talcott, S.U. Carbohydrate-Free Peach (*Prunus persica*) and Plum (*Prunus salicina*) [corrected] Juice Affects Fecal Microbial Ecology in an Obese Animal Model. *PLoS ONE* **2014**, *9*, e101723. [CrossRef] [PubMed]

226. Ghimire, S.; Wongkuna, S.; Sankaranarayanan, R.; Ryan, E.P.; Bhat, G.J.; Scaria, J. Rice Bran and Quercetin Produce a Positive Synergistic Effect on Human Gut Microbiota, Elevate the Level of Propionate, and Reduce the Population of *Enterobacteriaceae* family when Determined using a Bioreactor Model. *bioRxiv* **2020**. [CrossRef]

227. Jaimes, J.D.; Jarosova, V.; Vesely, O.; Mekadim, C.; Mrazek, J.; Marsik, P.; Killer, J.; Smejkal, K.; Kloucek, P.; Havlik, J. Effect of Selected Stilbenoids on Human Fecal Microbiota. *Molecules* **2019**, *24*, 744. [CrossRef]

228. Nakatsu, C.H.; Armstrong, A.; Clavijo, A.P.; Martin, B.R.; Barnes, S.; Weaver, C.M. Fecal bacterial community changes associated with isoflavone metabolites in postmenopausal women after soy bar consumption. *PLoS ONE* **2014**, *9*, e108924. [CrossRef]

229. Radulovic, K.; Normand, S.; Rehman, A.; Delanoye-Crespin, A.; Chatagnon, J.; Delacre, M.; Waldschmitt, N.; Poulin, L.F.; Iovanna, J.; Ryffel, B.; et al. A dietary flavone confers communicable protection against colitis through NLRP6 signaling independently of inflammasome activation. *Mucosal Immunol.* **2018**, *11*, 811–819. [CrossRef]

230. Wang, M.; Firrman, J.; Zhang, L.; Arango-Argoty, G.; Tomasula, P.; Liu, L.; Xiao, W.; Yam, K. Apigenin Impacts the Growth of the Gut Microbiota and Alters the Gene Expression of Enterococcus. *Molecules* **2017**, *22*, 1292. [CrossRef] [PubMed]

Review

Therapeutic Potential of Flavonoids in Pain and Inflammation: Mechanisms of Action, Pre-Clinical and Clinical Data, and Pharmaceutical Development

Camila R. Ferraz [1], Thacyana T. Carvalho [1], Marília F. Manchope [1], Nayara A. Artero [1], Fernanda S. Rasquel-Oliveira [1], Victor Fattori [1], Rubia Casagrande [2,*] and Waldiceu A. Verri Jr. [1,*]

[1] Departament of Pathology, Center of Biological Sciences, Londrina State University, 86057–970 Londrina, Paraná, Brazil; camila_ferraz96@hotmail.com (C.R.F.); thacy_thacy@yahoo.com.br (T.T.C.); marilia_manchope@hotmail.com (M.F.M.); naayanitelli@hotmail.com (N.A.A.); fernandarasquel@gmail.com (F.S.R.-O.); vfattori@outlook.com (V.F.)

[2] Departament of Pharmaceutical Sciences, Center of Health Sciences, Londrina State University, 86057–970 Londrina, Paraná, Brazil

* Correspondence: rubiacasa@yahoo.com.br (R.C.); waldiceujr@yahoo.com.br (W.A.V.J.); Tel.: +55-43-3371-2475 (R.C.); +55-43-3371-4979 (W.A.V.J.); Fax: +55-43-3371-4387 (W.A.V.J.)

Academic Editor: Derek J. McPhee

Received: 26 December 2019; Accepted: 7 February 2020; Published: 10 February 2020

Abstract: Pathological pain can be initiated after inflammation and/or peripheral nerve injury. It is a consequence of the pathological functioning of the nervous system rather than only a symptom. In fact, pain is a significant social, health, and economic burden worldwide. Flavonoids are plant derivative compounds easily found in several fruits and vegetables and consumed in the daily food intake. Flavonoids vary in terms of classes, and while structurally unique, they share a basic structure formed by three rings, known as the flavan nucleus. Structural differences can be found in the pattern of substitution in one of these rings. The hydroxyl group (–OH) position in one of the rings determines the mechanisms of action of the flavonoids and reveals a complex multifunctional activity. Flavonoids have been widely used for their antioxidant, analgesic, and anti-inflammatory effects along with safe preclinical and clinical profiles. In this review, we discuss the preclinical and clinical evidence on the analgesic and anti-inflammatory proprieties of flavonoids. We also focus on how the development of formulations containing flavonoids, along with the understanding of their structure-activity relationship, can be harnessed to identify novel flavonoid-based therapies to treat pathological pain and inflammation.

Keywords: clinical trials; natural products; hyperalgesia; allodynia; analgesia; flavonoid; hypersensitivity; inflammation; cytokines; NF-kB

---

1. Introduction

Inflammatory response induced by micro-organisms or tissue damage trigger the release of pathogen-associated molecular patterns (PAMPs) and damage-associated molecular patterns (DAMPs), respectively [1,2]. Tissue-resident immune cells such as macrophages and dendritic cells (DC) recognize these molecules though receptors namely pattern recognition receptors (PRRs). Once activated, these immune cells produce chemoattractant molecules, which are mainly governed by the transcription factor NF-κB [1,2]. The transcription factor NF-κB regulates the expression of inflammatory enzymes, such as COX-2 [3] and pro-inflammatory cytokines [4–6], which makes it one of the most important transcription factors during the inflammatory process and pain. Cytokines and chemokines released by these immune cells along with formyl-peptide (fMLP) released by dying cells activate vascular endothelial cells and provide a gradient of signals that precisely guide neutrophils to the inflamed tissue following a spatial, temporal and hierarchic cascade of mediators [7,8]. Specifically, neutrophils rapidly migrated away from high concentrations of CXCR2 ligands to follow fMLP signal, indicating that the necrotactic stimulus hierarchically override CXCR2 signaling. Accordingly, the lack of fMLP receptor, but not CXCR2, impairs the chemotaxis of neutrophils to the necrotic foci in the context of sterile inflammation [8]. In addition to follow a spatial, temporal and hierarchic cascade of mediators, the recruitment of neutrophils is also context dependent. Using *E. coli* as stimulus, it has been demonstrated that the adhesion of neutrophils in sinusoids depends on CD44 rather than Mac-1 (required for sterile inflammation), revealing, therefore, different adhesion molecules for neutrophil recruitment in sterile vs non-sterile inflammation [8]. In the inflammatory foci, neutrophils release reactive oxygen species (ROS), pro-inflammatory cytokines, and pro-inflammatory lipid mediators, which ultimately contribute to inflammatory pain. Prostaglandin (PG) E_2, for instance, is produced by either COX-1 or COX-2 and during inflammation, increased production of PGE_2 by COX-2 contributes to neuronal sensitization leading to pain [9]. Nociceptive pain has a protective function and occurs upon the detection of noxious stimuli by nociceptor. For the detection of these stimuli, nociceptors express ion channels such as the TRP (TRPA1, TRPM8, TRPV1 etc), Nav (Nav1.7, Nav1.8, Nav1.9 etc), and ASIC (ASIC1 to 4) families in their peripheral ends [9]. Pathological pain, on the other hand, is characterized by an amplified response to a noxious stimulus (hyperalgesia) or by a response to a normally innocuous stimulus (allodynia) [10]. Nociceptive pain, therefore, involves the activation of high threshold nociceptors while in inflammatory pain, non-noxious stimuli can now generate mechanical and thermal hypersensitivity [9,11,12]. This phenomenon is described as nociceptor sensitization and occurs due to a shift from a high threshold to a low threshold type of pain [9,11,12]. Current knowledge on persistent pathological pain includes the understanding on how peripheral and spinal cord sensitization of nociceptors occur and changes in the immune cell phenotypes influences pain perception. In the periphery, innate immune cells such as macrophages, neutrophils, and mast cell release mediators that act on the peripheral nerve endings [9,11,12]. PGE_2, histamine, and cytokines, for example, are the main molecules responsible for lowering neuronal activation threshold and producing peripheral neuronal sensitization [9,11,12]. In the spinal cord, this sensitization is mediated by cytokines, chemokines, and growth factors released by tissue resident cells known as microglia, astrocytes, and oligodendrocytes [13–16]. All of these mechanisms ultimately contribute to pathological pain by promoting plastic changes in the periphery and central nervous system (CNS) that modify the neuronal phenotype and function.

Pain management is a worldwide challenge due to side effects induced by classical treatments. Acetaminophen and NSAIDs are effective for the management pain. While preclinical data demonstrate that COX-2 selective inhibitors are effective, clinical data show that they induce several side effects such as kidney and heart diseases [17], and non-selective COX inhibitors also induce gastro-intestinal ulcers and kidney injury [18,19]. Acetaminophen is widely known to induce liver injury both in mouse and human [20,21]. This means that there is need of drugs with lessened side effects or different side effects allowing to choose the best option considering the patient's comorbidities. Depending on the intensity of the pain, opioids are one of the drugs used for relief. However, millions of patients cope with side effects that include constipation, drowsiness, risk of addiction, and sometimes even respiratory failure and death [22]. Even upon opioid therapy, neuropathic pain, for instance, remains challenging to treat, with only half of the treated population typically report a significant reduction in pain and complete resolution of symptoms is rarely achieved [23]. Paradoxically, while recognized by their potent analgesic effects, opioids can induce pain [24]. The release of DAMPs, such as HMGB-1 and biglycan, induced by morphine is of the main mechanisms related to opioid-induced pain [25]. These molecules increase the production of spinal cord IL-1β by microglia in a NLRP3-dependent manner [26]. Furthermore, morphine increases IL-1β mRNA expression via TLR4 and increases IL-1β release via P2X4 receptor [27]. Altogether, this cascade of events in microglia potentiates morphine-induced hyperalgesia by stimulating IL-1β production in the spinal cord and explains this paradoxical effect [25–27]. Corticosteroids and immunobiological agents are other group of molecules used to treat pain. While the first cause hormonal changes (leading to suppression of adrenal function and bone loss) and Cushing syndrome [28], the later induces heart failure [29]. Moreover, corticosteroids, immunobiological agents, and opioids are linked with immunosuppression and consequently pathogen spread [30–32]. Thus, the search for natural compounds with lower adverse effects on patients has been growing, and over the past years higher attention has been given to the flavonoids.

Flavonoids are an essential group of polyphenolic compounds, and their flavan nucleus is the main structural characteristic. Figure 1 shows the structures of the flavonoids discussed in this review.

Group of flavonoids				
Flavones	**Flavonols**	**Flavanones**	**Chalcones**	**Flavan-3-ols**

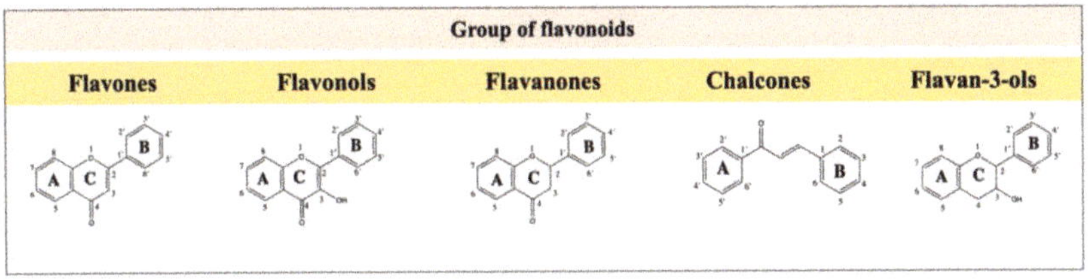

Figure 1. The chemical structures of the flavonoid groups discussed in this review.

They are one of the most widely found classes of compounds in vegetables and fruits. The chemical structure of flavonoids is based on a fifteen-carbon skeleton consisting of two benzene rings connected through a heterocyclic pyrane ring. The flavonoids can be divided into an assortment of classes, for example, flavones (e.g., flavone, apigenin, and luteolin), flavonols (e.g., quercetin, kaempferol, myricetin, and fisetin), and flavanones (e.g., flavanone, hesperetin, and naringenin). While individual compounds within a class differ in the pattern of substitution of the A and B rings, the classes itself vary in the degree of oxidation and pattern of substitution of the C ring. Flavonoids are known to have analgesic, anti-inflammatory, and antioxidant properties. These effects are related to the inhibition of NF-κB-dependent pro-inflammatory cytokines [33], VEGF, ICAM-1, STAT3 [34], and activation of antioxidant transcription factor Nrf2 [33]. Many flavonoids, as apigenin and vitexin, were reported to be safe natural alternative therapeutic against neuroinflammatory diseases as such Multiple Sclerosis (MS) [35,36]. Therefore, flavonoids are multi-target drugs, which explain their wide array of actions. Further, as the activity of flavonoids does not depend on abolishing a single mechanism, but rather reducing varied mechanisms, it is likely that physiological function is maintained, which reduces the incidence of side effects compared to single target drugs.

2. Pre-Clinical Evidence of Flavonoids for Pain Control

Many mediators are produced by tissue immune resident cells and immune cells that undergo recruitment to the injured tissue during an inflammatory response. Among these mediators, PGE2 [37], along with bradykinin, sympathetic amines, and pro-inflammatory cytokines (e.g., TNF-α, IL-6, and IL-1β) act to promote either sensitization or activation of nociceptors and then evoke pain [37,38]. Flavonoids are plant derivative compounds, and their potential usage as medicinal natural products is growing exponentially over the years (Figure 2).

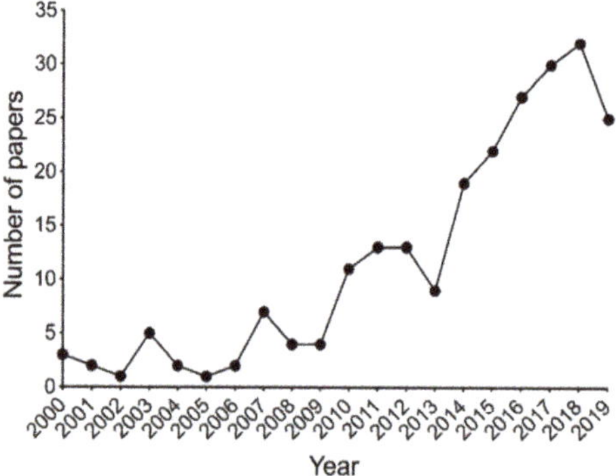

Figure 2. The number of manuscripts published on flavonoids, pain, and inflammation during the last 20 years at PubMed. The keywords search at PubMed was "flavonoids and pain and inflammation", and only original research papers were considered.

Studies about flavonoids' effects on inflammatory diseases and pain have been increasing in the last decade as several groups are demonstrating the involvement of these phenolic compounds as anti-inflammatory, analgesic, and antioxidant molecules. The search for new therapeutic drugs with less or no side effects is the major reason leading to this growing interest in natural products for the treatment of inflammatory and painful conditions. The flavonoids discussed in this section were selected based on the ones that also present clinical data (Section 4: Clinical Studies and Safety), to align pre-clinical data with clinical data. We discuss their targets in the disease context, which explains the anti-inflammatory and analgesic actions of flavonoids (Figures 3 and 4). Figure 3 shows flavonoids whose anti-inflammatory and analgesic effects do not involve, at least at the present date, the direct interaction and modulation of ion channels.

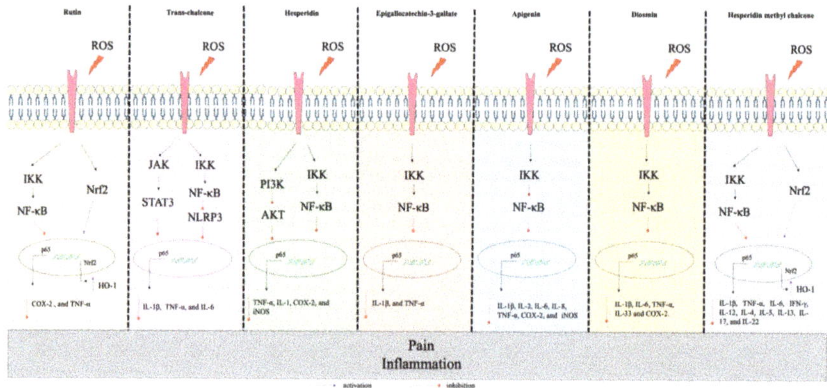

Figure 3. The anti-inflammatory and analgesic effects of flavonoids. Intracellular targets of **Rutin**: NF-κB [39,40] and Nrf2 [40], **Trans-chalcone**: NF-κB [41] and STAT3 [41] and NLRP3 [42], **Hesperidin**: PI3K/ AKT [43] and NF-κB [44], **Epigallocatechin-3-gallate**: NF-κB [45], **Apigerin**: NF-κB [46], **Diosmin**: NF-κB [47], and **Hesperidin methyl chalcone**: NF-κB [48–50] and Nrf2 [49,50]. ROS and inflammatory stimuli that activate specific receptors trigger intracellular signaling that will result in pain and inflammation. The **blue** arrows indicate endogenous pathways that are stimulated by flavonoids resulting in the reduction of pain and inflammation. The **red** arrows represent endogenous pathways that are inhibited by flavonoids resulting in reduced pain and inflammation.

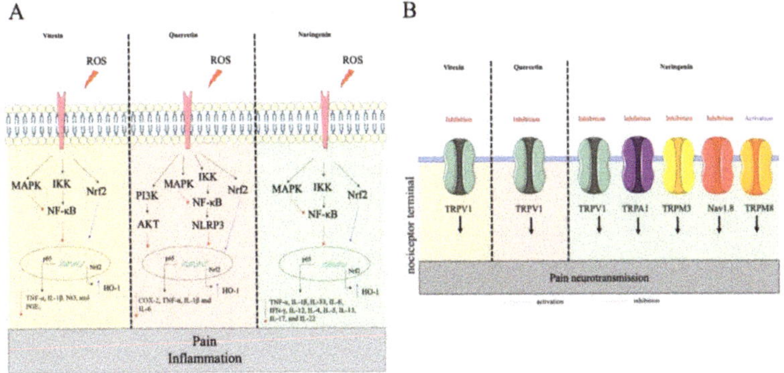

Figure 4. The anti-inflammatory and analgesic effects of Vitexin, Quercetin, and Naringenin. (**A**) Intracellular targets of **Vitexin**: MAPK [51], NF-κB [51] and Nrf2 [52], **Quercetin**: MAPK [53], NF-κB [53–55], AKT [56], Nrf2 [33,54], and NLRP3 [57] and **Naringenin:** NF-κB [58–60] and Nrf2 [59, 61,62]. (**B**) Ion channels expressed by neurons that are targeted by Vitexin, Quercetin, and Naringenin to reduce pain. **Vitexin:** TRPV1 [38], **Quercetin:** TRPV1 [63], and **Naringenin:** TRPV1 [58], TRPA1 [58], TRPM3 [64], Nav 1.8 [65], and TRPM8 [64]. In panel (**A**), ROS and inflammatory stimuli that activate specific receptors trigger intracellular signaling that will result in pain and inflammation. The **blue** arrows indicate endogenous pathways that are stimulated by flavonoids, and the **red** arrows represent endogenous pathways that are inhibited by flavonoids resulting in reduced pain and inflammation.

2.1. Flavonols

Quercetin (3,3′,4′,5,7-pentahydroxyflavone) is a flavonol found in different types of fruits, vegetables, and plants, including: berries, apples, tomatoes, cocoa, onions, and medicinal plants [66–68]. It is an antioxidant flavonoid that presents anti-inflammatory activity by inhibiting pro-inflammatory cytokines in mouse models of colitis [69], periodontitis [70], cancer pain [71], and acute [54] and chronic arthritis [33]. That effect is mainly related to the ability of quercetin in reducing recruitment of neutrophils (myeloperoxidase [MPO] activity [33,69,72]), oxidative stress [33,54,69,73,74], COX-2 in the knee joint (arthritis model) [33], NLRP3 inflammasome activation (inhibition of ASC speck formation and ASC oligomerization) in macrophages [57], p65 NF-κB activation [53,54] and MAP kinases signaling in macrophages [53], p50 NF-κB activation in primary human keratinocytes [55], and TNF-α, IL-1β and IL-6 production in RAW 264.7 cells stimulated by LPS [75]. Recent studies demonstrated that quercetin is able to modulate the neutrophils actin polymerization [76], pro-inflammatory cytokines expression by mast cells [77,78] and monocytes [79,80], dendritic cell activation [81] and maturation [82], and the phenotype M1 to M2 in macrophages [83–86]. In the past few years, increasing attention has been paid to the analgesic effect of quercetin [33,54,71,73,74,87]. In addition to the inhibition of the aforementioned pro-inflammatory signaling pathways, part of that analgesic effect is through the activation of Nrf2/HO-1 pathway [33,54]. In fact, co-treatment with a HO-1 inducer potentiates the analgesic effect of opioids and cannabinoids through the activation of cGMP/PKG/ATP-sensitive potassium channels pathway in a model of CFA-induced pain [88], indicating that stimulating HO-1 expression contributes to analgesia. This is a relevant mechanism because it has been demonstrated that morphine activates PI3Kγ/AKT pathway that, in turn, stimulates NO production in an nNOS-dependent manner in the DRG neurons [89,90]. NO also indirectly acts through stimulation of cGMP/PKG and causes the up-regulation of ATP-sensitive potassium channels to promote the hyperpolarization of primary nociceptive neurons [89,90]. Moreover, when phosphorylated, p65 subunit competes with Nrf2 for the adaptor protein CREB binding protein (CBP) [91,92]. Consequently, there is a hypoacetylation that blocks chromatin condensation, suppressing Nrf2/ARE gene expression. That mechanism dampens the Nrf2 signalling pathway activation and consequently antioxidant response [91,92]. Quercetin can also suppress the expression levels of PKCε and TRPV1 in the spinal cords and DRGs as demonstrated in a model of paclitaxel-induced peripheral neuropathy in rats and mice [63]. Therefore, drugs such as quercetin with the ability of activating cGMP/PKG/ATP-sensitive potassium channels pathway in neurons without the important side effects of opioids might be promising candidates for pain treatment.

Rutin (3,3′,4′,5,7-pentahydroxyflavone-3-rhamnoglucoside), a glycoside of quercetin, is also a flavonol that can be found in several plants, such as buckwheat, white mulberry, passion flower, tomatoes and apples [93,94]. It possesses anti-inflammatory and anti-hyperalgesic effects through the inhibition of oxidative stress and neuroinflammation [39]. Specifically, rutin inhibits NF-κB activation [39,40], activates Nrf2/HO-1 and NO/cGMP/PKG/ATP-sensitive potassium channels channel signaling pathways and inhibits pro-inflammatory cytokine production (IL-1β and TNF-α) [40]. Rutin promotes M2 polarization of Th1 primed RAW264.7 and CD11b+ primary macrophages [95] and modulates the production of mediators by neutrophils [96], mast cells [77], and monocytes [97]. Furthermore, by testing the same doses of rutin and quercetin (80 mg/kg), it was demonstrated that rutin was more efficient than quercetin on reducing edema in a model of experimental arthritis of adjuvant-carrageenan-induced inflammation (ACII) in rats likely because better absorption due to higher hydrophilicity than quercetin [98]. Altogether, these data indicate that both quercetin and rutin block pain and inflammation by activating cGMP/PKG/ATP-sensitive potassium channels pathway in neurons and inhibiting NF-κB and inducing Nrf2 activation in immune cells.

2.2. Flavones

Apigenin (4′,5,6-trihydroxyflavone) is abundantly present in vegetables, fruits, and medicinal herbs, including parsley, apples, grape, and *Matricaria chamomilla* [99,100]. It was demonstrated that this flavone possesses anti-inflammatory effects by blocking NF-κB activation in a model of LPS-induced apoptosis. In this model, apigenin decreases the neutrophil infiltration to the lungs by reducing the Macrophage Inflammatory Protein-2 (MIP-2) that is a neutrophil chemoattractant molecule [46]. Furthermore, previous studies showed apigenin has anti-inflammatory activities on LPS-induced inflammation through the inhibition of the expression of iNOS, COX-2, cytokines (IL-1β, IL-2, IL-6, IL-8, and TNF-α), and AP-1 proteins in human lung epithelial cells [101]. In addition, apigenin modulates macrophages polarization [102], promotes down-regulation of Mcl-1 in neutrophils [103], suppresses the inflammatory activity of DCs [35,104], regulates the production of inflammatory mediators by monocytes [105] and mast cells [106]. Among the antioxidant effects, apigenin scavenges hydroxyl (•OH) radicals generated by UV photolysis of hydrogen peroxide [107] and chelates iron [108]. Further, apigenin reduces the levels of MPO and malondialdehyde activity (MDA) in acetic acid-induced ulcerative colitis and these effects were comparable to the corticoid, prednisolone [109].

Vitexin, a flavone C-glycoside (5,7,4-trihydroxyflavone-8-glucoside) found in medicinal and other plants, presents anti-inflammatory activities by preventing the oxidative stress, inhibiting pro-inflammatory cytokines production, and increasing the anti-inflammatory cytokine IL-10 [38]. Vitexin inhibited the production of TNF-α, IL-6, nitric oxide (NO), and PGE$_2$ in human osteoarthritis chondrocytes [110]. Furthermore, vitexin prevents mast cells degranulation [111], reduces the production of pro-inflammatory mediators by neutrophils [112,113] and macrophages [113]. Vitexin attenuated RANKL-induced activation of MAPK signal pathways and NF-κB signaling pathways [51]. Also, the activation of Nrf2/HO-1 pathway is one of the mechanisms of vitexin to limit inflammation in a model of LPS-induced acute lung injury [52]. It was demonstrated by Chen et al., [114] in isoflurane-treated PC12 cells that vitexin is neuroprotective by downregulating the expression of TRPV1 and glutamate ionotropic receptor NMDA type subunit 2B (NR2B) protein expression. In corroboration, in vivo data show that vitexin reduces capsaicin-induced pain behaviors, indicating that part of its analgesic effect is through inhibition of TRPV1 activation [38]. Moreover, vitexin showed analgesic activities by reducing the writhing response induced by acetic acid [38,115] and phenyl-*P*-benzoquinone (PBQ) [38], and the thermal and mechanical hyperalgesia [38,115]. The analgesic effects of vitexin in the model of hind-paw incisional surgery seem to be mediated by opioid-related mechanisms since delta, mu, and κ-opioid receptor antagonists reversed the analgesic effects of this flavonoid [115].

Diosmin (diosmetin-7-*O*-rutinoside) is a flavonoid abundant in citrus fruits [116]. This flavonoid presents anti-inflammatory and anti-hyperalgesic mechanisms described in different models. Some of these effects have been attributed to the reduction of NF-κB activation [47], TNF-α in PC12 cell induced by LPS [117], *Il-1β, Tnf-α, and Il-33/St2* mRNA expression in the spinal cord induced by CCI [118], TNF-α, IL-1β and IL-6 in the spinal cord and sciatic nerve tissues [119], NO, PGE$_2$, IL-6, IL-12, TNF-α production by macrophages [120], COX-2 and MPO activity [121,122]. In addition, diosmin inhibited LTB$_4$ synthesis [121]. The antioxidant effects of diosmin were demonstrated in colitis induced by acetic acid or trinitrobenzenesulfonic acid (TNBS). Diosmin reduced the levels of MDA and prevented the consumption of GSH [121,122]. Regarding the analgesic effects of diosmin, this flavonoid reduces neuropathic pain in the sciatic nerve chronic constriction injury (CCI) model by activating the analgesic NO/cGMP/PKG/ATP-sensitive potassium channels channel signaling pathway and blocking central sensitization [118]. Also, in a model of CCI in rats, diosmin acts at central level through opioid and dopaminergic receptors to inhibit mechanical and thermal hyperalgesia [119]. Unpublished data of the Verri laboratory also show that diosmin treats LPS-induced peritonitis and inflammatory pain by blocking NF-κB activation in leukocytes. Therefore, diosmin might be a promising drug to treat chronic and non-sterile inflammatory pain.

2.3. Flavanones

Naringenin (4′,5,7-trihydroxyflavanone) is found in citric fruits such as lemon, grapefruit, orange, and tangerine [123]. Its effects on inflammation have been demonstrated in UVB irradiation [61,124], inflammatory pain [58,62,125], and arthritis [59]. The mechanisms by which naringenin reduces inflammation and pain are related to inhibition of oxidative stress [61,62,124,126], leukocyte recruitment [59,60], MPO activity [124,127], NF-κB activation [58–60], mRNA expression of inflammasome components [59], pro-inflammatory cytokine production (TNF-α, IL-1β, IL-33, IL-6, IFN-γ, IL-12, IL-4, IL-5, IL-13, IL-17, and IL-22) [58–60,62,124,125,127], and MAPK signaling activation [128]. In addition, naringenin modulates macrophages activation [129] and microbicidal activity of neutrophils [130], and reduces DC maturation [131]. Further, naringenin acts by the inducing Nrf2/HO-1 [59,61,62] and activation of the analgesic signaling pathway NO/cGMP/PKG/ATP-sensitive potassium channels [62]. Naringenin also reduces formalin- and capsaicin-induced pain behaviors, demonstrating that part of its mechanism is through inhibition of TRPA1 and TRPV1 activation, respectively [58]. In addition, naringenin induces analgesia by blocking TRPM3 and activating TRPM8 [64], which might contribute to its analgesic effect. Naringenin also blocks sodium influx in rat DRG neurons by selecting inhibiting Nav1.8 (and not Nav1.7 or Nav1.9) [65]. Collectively, naringenin targets different channels expressed by neurons and immune cell signaling pathways (decreases NFkB activation and stimulates Nrf2) to reduce pain and inflammation.

Hesperidin (hesperetin-7-rhamnoglucoside) is a flavonoid that belongs to the flavanones class mainly found in citrus fruits [132]. Pharmacological effects of hesperidin have been reported such as anti-inflammatory [133,134], analgesic [134–136], antioxidant [133,137], antiallergic [138], and antidiabetic [139]. Hesperidin reduces TNF-α, IL-1β, ICAM-1, VEGF, the neutrophil ability to generate superoxide radical [140], and advanced glycosylation end products levels [133]. Also, hesperidin modulates polarization of macrophages to M1 through suppression of the PI3K/AKT pathway [43], and inhibits TNF-α and IL-1β production by mast cells [141]. Further, it was demonstrated that hesperidin reduces MDA levels [44,133] and increases the antioxidant enzyme superoxide dismutase (SOD) activity [133] and GSH [44]. Moreover, hesperidin inhibits neuroinflammation in mice as observed by decreased levels of GFAP (an astrocyte activation marker), NF-κB, iNOS, and COX-2 in the coronal brain induced by an intracerebroventricular infusion of streptozotocin (STZ) [44]. Treatment with hesperidin also ameliorates mechanical hyperalgesia in STZ-induced diabetic rats [136] and CCI experimental model of neuropathic pain [134]. Therefore, both hesperidin and naringenin show neuroprotective and antioxidant properties that contribute to their analgesic effects.

2.4. Chalcone

Under alkaline conditions, the methylation of hesperidin produces hesperidin methyl chalcone (HMC) [142]. HMC is an antioxidant flavonoid with anti-inflammatory and analgesic effects in experimental models. For instance, HMC inhibits oxidative stress by preventing the decrease of the antioxidant capacity (FRAP, ABTS, and GSH) and inhibiting the superoxide anion production, lipid peroxidation and gp91phox mRNA expression. Inducing the Nrf2/HO-1 pathway [49,50] contributes to the in vivo antioxidant effects of HMC [49,50,143,144]. HMC also inhibits the production of pro-inflammatory cytokines, such as TNF-α, IL-1β, and IL-6 induced by UVB irradiation in mouse skin [50]. Further, HMC inhibited edema, neutrophil recruitment (MPO activity), and MMP-9 activity [143]. HMC showed anti-inflammatory effects by inhibiting the NF-κB-dependent pro-inflammatory cytokine production (IL-1β, TNF-α, IL-6, IFN-γ, IL-12, IL-4, IL-5, IL-13, IL-17, IL-22) [49,50,143,144] and leukocyte recruitment [49,143]. In addition, targeting the TRPV1 receptor activation is another important mechanism by which hesperidin exerts its analgesic effects [144]. We recently demonstrated that HMC interacts with NFκB (Ser276 residue), reduces *gp91phox* and increases *Ho-1* mRNA expression in knee joint tissue after stimulus with zymosan. HMC also reduces the cytokines IL-33, TNF-α, and IL-6 levels in zymosan-stimulated RAW264.7 macrophages [48].

Trans-chalcone (1,3-diphenyl-2-propen-1-one) is a flavonoid precursor that can be obtained from plants such as *Piper methysticum* (Kava-Kava), *Aniba riparia* and *Didymocarpus corchorijolia* [145]. Trans-chalcone presents in vivo antioxidant effects [42,145] and anti-hyperalgesic mechanisms through the inhibition of leukocyte recruitment, IL-1β, TNF-α, and IL-6 production, and NLRP3 inflammasome activation in macrophages stimulated with MSU crystals [42]. It also inhibits VEGF and ICAM-1 expression and activation of STAT3 and NF-κB [41], indicating that part of its analgesic mechanism is related to the blockade of NF-κB-dependent pro-inflammatory mediators.

2.5. Flavanols, Flavan-3-ols or Catechins

Epigallocatechin-3-gallate (EGCG) is a polyphenol very abundant in green tea (*Camellia sinensis*) and its main active component [146]. This flavonoid has several biological activities, including anti-inflammatory, anti-hyperalgesic, and antioxidant. EGCG inhibited the COX-2 mRNA expression and PGE$_2$ production induced by IL-1β in human osteoarthritis chondrocytes via up-regulation of hsa-miR-199a-3p expression [146]. Another mechanism by which EGCG inhibits inflammation is via down-regulation of NF-κB and consequently inhibition of iNOS stimulated by LPS in mouse macrophages [45], and reduced NO, prostaglandin PGE$_2$ and COX-2 production in macrophages [147], dendritic cell differentiation and maturation [148], and inhibited mast cell degranulation [149] and neutrophil chemotaxis [150]. Peripheral analgesic effects of EGCG are related to the suppression of CCR2, IL-1β, and TNF-α mRNA expression in the DRG in a model of OA [151] and through down-regulation of CX3CL1 protein in the spinal cord in a CCI-induced neuropathic pain [146]. In terms of spinal effects, EGCG decreases TNF-α expression in the spinal cord in mouse models of bone cancer-caused pain [152] and reduces glial cell activation via inhibition of fatty acid synthase (FASN), Ras homologue gene family member A (RhoA), and TNF-α in a model of spinal cord injury (SCI) [153].

Summing up, the findings from these studies suggest that treatment with flavonoids effectively alleviate inflammation and pain in varied preclinical models by inhibiting leukocyte recruitment, pro-inflammatory/pro-hyperalgesic mediator production, oxidative stress, activation of signaling pathways (Figures 3 and 4) and the modulation of channels to inhibit inflammation and pain (Figure 4). The data in Table 1 summarize some effects of flavonoids on different cell lines.

Table 1. Pre-clinical studies analyzing the effects of different flavonoids on cell lines.

Flavonoids Groups	Flavonoid	Cell Line		Effects	Refs
Flavonols	Quercetin	macrophages	RAW 264.7	Reduce TNF-α, IL-1β and IL-6 production	[75]
			BMDM	Inhibit ASC speck formation and ASC oligomerization	[57]
			BMDM	Modulate M1 and M2	[85]
			RAW 264.7		[83]
			J 774		[84]
		neutrophils	Human neutrophils	Modulate actin polymerization	[76]
		dendritic cell	BMDC	Activation and Maturation	[81, 82]
		mast cells	HMC-1	Reduce TNF-α, IL-1β, IL-8 and IL-6 production	[77]
			hCBMCs	Reduce histamine, leukotrienes and PGD2	[78]
		monocytes	Human THP-1 monocytic cells THP-1	Reduce TNF-α, and IL-1β production	[79, 80]
	Rutin	macrophages	RAW 264.7	promote M2 polarization	[95]
			CD11b+ primary macrophages		
		neutrophils	Human peripheral blood neutrophils	Reduce NO and TNF-α production	[96]
		mast cells	HMC-1	Reduce TNF-α, IL-1β, IL-8 and IL-6 production	[77]
		monocytes	Human THP-1	Inhibit adhesion	[97]
Flavones	Apigenin	macrophages	ANA-1	Modulate macrophages polarization	[102]
			RAW264.7		
			RAW 264.7	Reduce NO production and COX-2 expression	[105]
		neutrophils	Human peripheral blood neutrophils	Down-regulation of Mcl-1	[103]
		dendritic cell	BMDC	Inhibit maturation and migration	[35, 104]
		mast cells	HMC-1	Inhibit TNF-α, IL-8, IL-6, GM-CSF, and COX-2 expression and NF-kB activation	[106]
		monocytes	monocytes to HUVEC	Reduce TNF-α, production	[105]
	Vitexin	macrophages	RAW 264.7	Inhibit TNF-α, IL-1β, NO, PGE2 and increase in IL-10 release	[113]
		neutrophils	Human peripheral blood neutrophils	Reduce NO, TNF-α, and MPO production	[112]
		mast cells	RBL-2H3	Prevent degranulation	[111]
	Diosmin	macrophages	RAW264.7	Reduce NO, PGE2, IL-6, IL-12, TNF-α production	[120]

Table 1. *Cont.*

Flavonoids Groups	Flavonoid	Cell Line		Effects	Refs
Flavanones	Naringenin	macrophages	U937	Regulate activation	[129]
		neutrophils	Human peripheral blood neutrophils	Regulate microbicidal activity	[130]
		dendritic cell	BMDC	Reduce maturation	[131]
	Hesperidin	macrophages	RAW264.7	Modulate M1 polarization	[70]
		neutrophils	Human peripheral blood neutrophils	Reduce generate superoxide radical	[140]
		mast cells	HMC-1	Reduce TNF-α and IL-1β production	[141]
Chalcone	Trans-chalcone	macrophages	BMDM	Reduce IL-1β production	[42]
	Hesperidin methyl chalcone	macrophages	RAW264.7	Reduce IL-33, TNF-α, and IL-6 levels	[141]
Flavan-3-ols	Epigallocatechin-3-gallate	macrophages	RAW 264.7	Reduce NO, prostaglandin PGE2 and COX-2 production	[147]
		neutrophils	Murine peritoneal neutrophils	Reduce chemotaxis	[150]
		dendritic cell	Human dendritic cell	Differentiation and maturation	[148]
		mast cells	RBL-2H3	Inhibit degranulation	[149]

BMDM: Bone Marrow-derived Macrophage; BMDC: Bone marrow-derived dendritic cells; hCBMCs: Human umbilical cord blood-derived mast cells.

3. Structure-Activity Relationship (SAR)

Flavonoids are divided into different classes with a basic structure of 3 rings. Figure 5 shows the basic structure of the flavonoids discussed in this section.

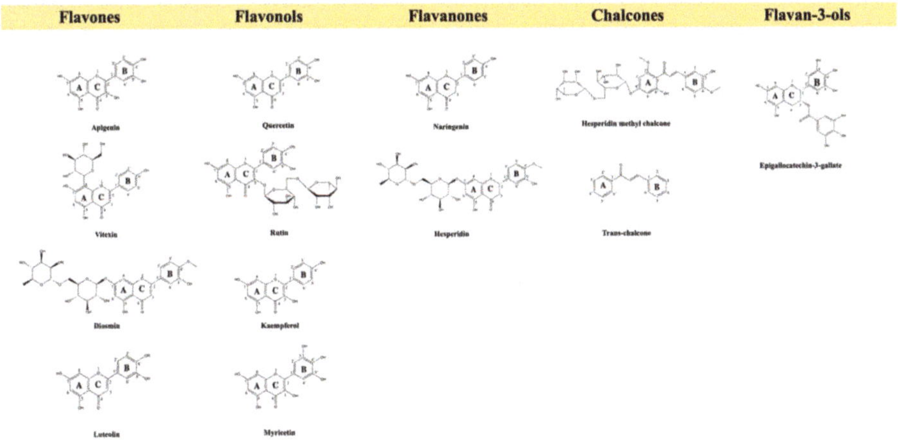

Figure 5. The chemical structures of the flavonoids discussed in this review.

These classes differ on where the C-ring carbon is attached to the B ring, and the C-ring saturation and oxidation degree [154]. Therefore, the structure of the flavonoids is fundamental to the understanding of their activity. Relevant to flavonoids' biological effect are: (i) B ring containing O-dihydroxy confers stability after hydrogen donation, and phenoxyl radical formation which participates in the electron delocalization; (ii) C ring allows electron dislocation from phenoxyl radical from B ring when a 2,3-double-bound bond is in conjugation with 4-oxo group on C ring; (iii) electron dislocation is favorable in combination of 2,3-double bond and 3-hydroxyl and 5-hydroxyl groups by increasing the resonance stability [155]. The amount of hydroxyl groups is less important than their position at flavonoid basic structure. For example, the isovitexin has 3 hydroxyls groups and baicalin has 2 hydroxyls groups differing into hydroxyls amount, positions and scavenging DPPH activity. Isovitexin (apigenin-6-C-glucoside) does not scavenge DPPH radicals ($IC_{50} > 1000$ μM) while baicalin (Baicalein 7-O-glucuronide) scavenges DPPH radicals ($IC_{50} = 15.5$ μM). On the other hand, the monomer epicatechin scavenges DPPH radicals at same level as its dimer Procyanidin B-2 ($IC_{50} = 11.7$ μM) [156]. While the class of flavonoid is not determinant for their activity (for instance, flavonols [kaempferol, quercetin, and myricetin] and flavones [chrysin, flavone, apigenin, baicalein, baicalin, and luteolin] block TNFα-induced ICAM-1 expression on alveolar epithelial cells A549), the hydroxyls at positions 5 and 7 on the A ring and at position 4 on the B ring are important for activity [157]. Specifically, hydroxyls at position 3 on B ring reduce flavonoid activity and at position 5 position abolish its activity [157]. Thus, changes into basic flavonoid structure could increase, decrease, or even not alter flavonoids antioxidant activity. In addition, flavonoids such as trans-chalcone that does not present antioxidant chemical groups presents anti-inflammatory and analgesic effects in vivo and reduce oxidative stress in vivo likely due to inhibiting inflammation since no antioxidant effect was observed in vitro in cell-free systems [42,158]. Thus, defining whether a flavonoid has therapeutic potential solely by its structure and chemical groups with antioxidant potential is not adequate to take full advantage of plant flavonoids. Further, there is more detailed understanding on the structure activity relationship regarding antioxidant activity without clear conclusions on anti-inflammatory and analgesic mechanisms. In this section, we discuss how flavonoid basic structure and their substitutions correlate with their activity.

Flavonoids can be found on nature as their glycoside or aglycone forms. Flavonoid glycosides present increased solubility and stability in water, which, however, interferes with their activity. Given glycosylation occurs in hydroxyl groups, it changes structural key elements for their radical scavenging activity. Specifically, glycosylation changes the double bond in conjugation with the 4-OXO group in the C-ring at C2, C3 position, the *O*-dihydroxy (catechol group) at the B-ring, and the presence of hydroxyl groups in positions C-3 (C-ring), C-5 and C-7 (A-ring) [159]. For instance, the aglycone quercetin and its glycoside form quercitrin both inhibit leukocyte recruitment and LTB_4 production in carrageenan-induced pleurisy in rats, but not at the same level. Quercitrin showed lower activity compared to its aglycon form quercetin [160]. Also, aglycones quercetin and hesperetin (75 mg/kg) inhibit carrageenan-induced paw edema in mice but not their glycosides form rutin and hesperidin (150 mg/kg, i.p.) [161]. Regarding glycosylation effect on antioxidant activity, quercetin scavenged more DPPH and peroxynitrite ($[ONOO^-]$, IC_{50}= 5.5±0.1 and 48.8 ±0.5 μM) than rutin (IC_{50}= 7.4 ±0.2 and 92.4±1.0 μM) [162]. Thus, while glycosylation increases flavonoid solubility, it decreases anti-inflammatory and antioxidant activities because it changes key features in the flavonoid structure.

$ONOO^-$ is produced by the reaction between superoxide anion and nitric oxide. Both mediators are produced at high amounts during inflammatory process by NADPH oxidase and inducible nitric oxide synthase [163]. These mediators induce pain by increasing hyperalgesic mediators through NFkB activation [164], such as TNFα acting via TNFR1 [165] and COX-2/PGE2 axis [74,166], or by directly inducing neuronal depolarization. For the scavenge ability of $ONOO^-$ in vitro, 3-hydroxyl moiety at B ring demonstrated to be important for flavonoids. For instance, quercetin a flavonoid that possess the 3-hydroxyl group has a higher scavenger ability (IC_{50} = 0.93 ±0.12 μM) when compared to the flavonols galangin (IC_{50} = 3.37 ±0.99 μM) and kaempferol (IC_{50} = 4.35 ±0.27 μM) which have the group 4-hydroxyl [167]. Similarly, the *O*-dihydroxy (catechol group) is determinant to superoxide anion scavenging activity of flavones and flavanones. For instance, the presence of the following elements increases scavenging activity: no hydroxyl groups at the B-ring, a 4′-hydroxyl substitution, and *O*-dihydroxy (catechol group), where the last feature shows higher antioxidant capacity [168]. Therefore, the hydroxyl group at B ring seems to be important for $ONOO^-$ and superoxide anion scavenging activity and its reasonable that this scavenging activity accounts for flavonoids analgesic and anti-inflammatory activity. Flavonoids also present antioxidant effects by indirect mechanisms through activation of Nrf2 signaling pathway, for example [169]. Chalcones are more potent than other types of flavonoids, where the double bond at C2-C3 position of their structure are particularly important for Nrf2 induction. In fact, reduction of that double bond impairs Nrf2 activation. Chemical addition of sugar moiety to the flavonoid basic structure or naturally flavonoid glycosides present less activation of this important signaling pathway [170].

The analgesic and anti-inflammatory activities of flavonoids are related, at least in part, to their NF-κB inhibitory effects [40,54,58,60,62]. Shin et al. screened 30 flavonoid derivatives against their inhibitory activity on TNFα-induced NF-κB activation in HCT116 human colon cancer cells. Moiety 3′,5′-dimethoxy at ring B seems to be important to inhibit TNFα-induced NF-κB activation in vitro [171]. Specifically, compounds 2,3′,5′-Trimethoxychalcone, 3,3′,5′-Trimethoxychalcone, 3,3′,5,5′-Tetramethoxychalcone, and 2-Hydroxy-3′,4,5′-trimethoxychalcone had better inhibitory effect than 3,3′,5-Trimethoxychalcone, 2′,3,5-Trimethoxychalcone and 2-Hydroxy-4,4′-dimethoxychalcone. Compound 3,3′,5,5′-Tetramethoxychalcone showed the best inhibitory activity against TNFα-induced NF-κB activation decreasing IkB phosphorylation at Ser^{32} and RelA phosphorylation at Ser^{536}. Thus, 3′,3″,5′,5″-Tetramethoxychalcone seems to interfere with IKK complex responsible for IkB and RelA phosphorylation [172]. Moreover, apigenin, a flavone, interferes with IKK complex by decreasing RelA phosphorylation at Ser^{536} in LPS-induced NF-κB activation in human primary monocytes [172]. Comparing 5-hydroxyl against 5-methoxyl substitution at flavone A-ring, hydroxyl moiety displays higher effect than methoxyl in TNFα-induced NF-κB activation in HCT116 cells [171]. Regarding flavonoids, the inhibition of NF-κB activation in a context of TNF-α-induced ICAM-1 expression by luteolin and apigenin is dependent on the presence of a double bond at position C2-C3 of the C ring with OXO function at position 4, along with the presence of hydroxyl groups at position 4′ of the B ring. In fact, chrysin that lacks hydroxyl group at position 4′ of the B ring has lower inhibitory effect over NF-κB activation when compared to apigenin and luteolin [157]. Thus, there are different features in each ring that alters NF-κB inhibitory activity. (i) At B ring 3′,5′-dimethoxy moiety and hydroxyl group at position 5 in ring A, and (ii) at ring C a double bond at position C2-C3 with OXO function at position 4 and combined with hydroxyl group at position 4′ at ring B increased inhibitory activity in TNFα-induced NF-κB activation. More studies are needed to determine more important SAR for flavonoids′ NF-κB inhibitory activity.

As mentioned, flavonoids are drugs with safe pre-clinical profile without the common side effect of standard NSAIDs. For instance, luteolin inhibits PGI_2 produced by COX-2 without the usual side effects produced by NSAIDs [173,174]. Ribeiro et al. screened at whole blood flavonoids inhibitory activity against COX-1 and COX-2. A common feature of those flavonoids which inhibited COX-2 activity (2-(3,4-dihydroxyphenyl)-4*H*-chromen-4-one; 5,3′,4′-Trihydroxyflavone, 7,3′,4′-Trihydroxyflavone; and luteolin) is the catechol group in the B-ring [175]. Docking analysis shows three active sites between COX-2 and luteolin with B-ring oriented onto COX-2 hydrophobic pocket (Tyr^{385}, Trp^{387}, Phe^{518}, Ala^{201}, Tyr^{248}) and luteolin 3,4-dihydroxy groups formed H-bond between Tyr^{385} and Ser^{530} [176]. Regarding on flavonoids and COX-1 inhibitory activity, flavonoids flavone, 5-hydroxyflavone, and 7-hydroxyflavone have none or just one hydroxyl group at B ring. Fewer substitutions in flavonoids backbone is important for COX-1 inhibitory activity because COX-1 pocket is small and the interaction between less substituted flavonoids may occur easier than more substituted molecules [175]. It is reasonable that this inhibitory effect of flavonoids on COX-2 and COX-1 is responsible, at least in part, for their analgesic effect. In addition, because flavonoids are multitarget drugs physiological systems are less affected compared to single target drugs that almost abolish a unique mechanism involved in disease and physiological functions. Because of this, despite the inhibition of COX, flavonoids do not present the common side effects of NSAIDs. On the other hand, flavonoids reduce the side effects caused by NSAIDs. For instance, hypericum perforatum inhibited acetaminophen-induced hepatotoxicity and lethality in mice which is mainly constituted by flavonoids as quercetin and rutin [177,178].

IL-1β is a pro-inflammatory and hyperalgesic cytokine matured by NOD-like receptor (NLR) family, pyrin domain-containing 3 (NLRP3) inflammasome [179]. Lim et al. [180] screened 56 flavonoids towards their SAR in monosodium urate (MSU) crystals-induced IL-1β release by THP-1 cells [180]. Among those, 56 molecules such as flavone, 2′,4′-dihhydroxyflavone, 3′,4′-dichloroflavone, 4′,5,7-trihydroxyflavone (apigenin), 3,4′,5,7-tetrahydroxyflavone (kaempferol), and 3,3′,4′,5,7-pentahydroxyflavone (quercetin) inhibit MSU-induced IL-1β production [180]. Specifically, quercetin interferes with ASC oligomerization, thus, inhibiting inflammasome activation [57], The 4-carbonyl group, 2,3- double bond, and 3-hydroxyl moieties at ring C are important in MSU-induced IL-1β maturation since catechin did not show an inhibitory effect when compared to flavone and quercetin. Also, flavanones naringenin (10 µM) compared to apigenin (10 µM) did not inhibit IL-1β released by MSU [179]. Nevertheless, data in murine bone marrow-derived macrophages show that naringenin (300 µM) reduces the release of IL-1β [181]. Therefore, the lack of activity of naringenin on THP-1 cells might be cell-specific and dependent on concentration and experimental conditions. In addition, 4-hydroxyl substitution at B-ring, such as in apigenin and kaempferol, enhances inhibitory activity of IL-1β production compared to chrysin and galangin. Moreover, hydroxyl substitution at position 4 at B ring combined with 5,7-dihydroxyl at A ring further increase the inhibitory activity in apigenin and kaempferol compared to 4-hydroxyflavone. On the other hand, methylated, methoxylated, and glycosides derivates were unable to inhibit MSU-induced IL-1β release by THP-1 [180]. Thus, the presence of 5,7-dihydroxyl groups at A ring, 4-hydroxyl or 3,4-dichloro substitutions at B ring are important for the inhibition of IL-1β release by MSU. Therefore, flavonoids have important structures that confer their activity such as, 2,3-double-bound bond in conjugation with 4-oxo group, or hydroxyl groups in a specific position as 5,7-dihydroxyl groups at A ring, 4-hydroxyl or 3,4-dichloro substitutions at B ring. The structure of flavonoids is closely related to their activity and changes such as methylation, methoxylations, glycosylation, and polymerization can increase, decrease, or even not alter their activity, such as occur with the *O*-dihydroxy group at the B ring, the 2,3-double-bound bond in conjugation with the 4-oxo group at C ring, and the 3-hydroxyl and 5-hydroxyl groups at the C and A rings. Moreover, flavonoids derivation improves their activity, as in the case of the methoxy substitution in 3,3′,5,5′-Tetramethoxychalcone.

The PI3K/Akt pathway plays an essential role in the regulation of inflammatory responses [182–184]. The inhibition of PI3K protein by quercetin and myricetin was investigated ny crystallographic approach. The results demonstrate that the hydrogen bond between the 3′-OH (B ring) of quercetin and the side chain of Lys^{833} mimics the interaction made by the ketone moiety of LY294002 (PI3K inhibitor) and myricetin is recognized through B ring by Val^{882} residue of PI3K [185]. In this sense, the treatment of the T47D cells with epidermal growth factor (EGF) induced Akt phosphorylation at Ser^{473} and pretreatment the cells with quercetin (25 μM) suppressed the EGF-induced Akt phosphorylation at Ser^{473} [56]. These findings provide a molecular rationale for designing molecules based on the inhibition of PI3K/Akt pathway by quercetin and myricetin. More studies are needed to determine flavonoids SAR and their interaction with inflammatory targets aiming to develop flavonoids targeting selected pain and inflammation pathways.

4. Clinical Studies and Safety

The use of plants for therapeutic purposes is an ancient practice [186]. As pharmaceutical tools evolve, the discovery of new drugs from plants leads to the isolation of many compounds that are widely used clinically today [186], or originate prototypes for synthesis of new drugs [187,188]. It is estimated that around 48% of new chemical entities discovered between 1981 and 2002 are derived from natural sources, including plant-based ones [189].

Flavonoids represent one of the most studied classes of metabolites with diverse phenolic groups, with an estimated 10,000 different members, having pollinator attractants, antimicrobial, and UV-protective properties in plant kingdom [34]. The antioxidant potential is the most shared among different flavonoids, explaining part of their beneficial effects on human health [34]. As an unprecedented demand for better therapeutic approaches keeps growing worldwide, flavonoids are one of the natural group of compounds that have been extensively studied pre-clinically (Figures 3 and 4) and clinically (Table 1). Randomized controlled trials and other population-based studies suggest that many flavonoids contribute to cardiometabolic health. For instance, kaempferol, naringenin, and hesperetin reduce incidence of cerebrovascular disease, and anthocyanin decreases hypertension in adults [190,191]. Other studies state that anthocyanidins improve disease conditions involving cognitive function, such as Parkinson [192,193]. Incidence of breast, lung, and prostate cancers are affected by consumption of quercetin, myricetin and other flavonols [191,194], as well as other chronic conditions such as asthma and type 2 diabetes [191]. Therefore, in this section we highlight clinical studies addressing the effect of flavonoids or flavonoid-based drugs for the treatment of inflammatory diseases. Table 2 summarizes data discussed in this section.

Table 2. Clinical studies analyzing the effects of different flavonoids in diets or as supplements in patients.

Flavonoid/Flavonoid-Based Compound	n	Treatment	Duration	Outcomes	Refs
Flavanones, anthocyanins, flavan-3-ols, flavonols, flavones, and polymers	49,281 men in the HPFS and 80,336 women from the NSH	Food frequency questionnaire	20–22 years of follow-up	Intake of some flavonoids may reduce Parkinson disease risk, particularly in men	[193]
Hesperidin	24	500 mg, daily	3 weeks	Increased flow-mediated dilation and reduced concentrations of circulating inflammatory biomarkers	[195]
	100	Daflon 500 mg (3 tablets bid. the first 4 days and 2 tablets bid. the following 3 days)	7 days	Clinical severity, inflammation, congestion, edema, prolapse, duration, and severity of hemorrhoidal episode diminished	[196]
	120	Daflon 500 mg, 2 tablets, daily	2 months	The overall symptom score decreased when compared to placebo	[197]
	105	Daflon 500 mg, 2 tablets, daily	4 weeks + follow-up for 6 months	Improvement in pain, heaviness, bleeding, pruritus, and mucosal discharge from baseline	[198]
	56	379 mg of green tea extract	3 months	Improvements in blood pressure, insulin resistance, inflammation and oxidative stress, and lipid profile in patients with obesity-related hypertension	[199]
Quercetin	50	500 mg, daily	8 weeks	Improvements in clinical symptoms, disease activity, hs-TNFα, and health assessment questionnaire in women with RA	[200]
Apigenin	100	2 mL of an oleogel preparation of reformulated traditional chamomile oil	Topical application, once	Pain, nausea, vomiting, photophobia, and phonophobia significantly decreased in patients with migraine without aura	[201]
Silymarin (Livergol®, Goldaruo pharmaceutical, Iran)	44	420 mg, daily	90 days	Joint swelling, tenderness, and pain were reduced	[202]
Pycnogenol® (Horphag Research Ltd., UK, Geneve, Switzerland)	67	220 mg, daily	3 weeks	Patients with OA decreased C reactive protein levels and reduced use of painkillers and non-steroidal anti-inflammatory drugs	[203]
	100	150 mg/day	3 months	Patients with OA presented relief from daily pain, stiffness, and physical function	[204]
	37	150 mg/day	3 months	Alleviating OA symptoms and reducing the need for NSAIDs or COX-2 inhibitors administration	[205]
Alvocidib or Flavopiridol (Tolero Pharmaceuticals, Inc.)	10	30-min loading dose of 30 mg/m (2) followed by a 4-h infusion of 30 mg/m (2) once weekly	3 weeks every 5 weeks, twice	Reduction in tumor burden on chronic lymphocytic leukemia patients	[206]

NHS: Nurses' Health Study; HPFS: Health Professionals Follow-Up Study; hs-TNFα: high-sensitivity tumor necrosis factor-α; RA: rheumatoid arthritis; OA: osteoarthritis.

4.1. Hesperidin

Hesperidin ($C_{28}H_{34}O_{15}$), a flavonoid found in citrus fruits [207], has been linked to the improvement of human endothelial function. That is related to an improve on flow-mediated dilation and lower soluble E-selectin concentrations after 500 mg/day consumption by 24 volunteers, during a 3-week randomized, placebo-controlled, double-blind, crossover trial [195]. In another study, results show a lower diastolic blood pressure and improvement of endothelium-dependent microvascular reactivity in overweight individuals consuming 292 mg of hesperidin during a 4-week randomized, controlled, crossover study. Highe plasma levels of hesperidin correlated with these outcomes [208]. Hesperidin is associated with alterations in lipid profiles as well. A 3-week daily intake of 500 mg hesperidin by 25 individuals with metabolic syndrome reduced total cholesterol and increased HDL levels [209], and 100 mg or 500 mg/d for 6 weeks reduced serum triglyceride levels in hyperlipidemic subjects [210]. Moreover, in a study conducted by Rizza et al. [195], anti-inflammatory effects were observed after 500 mg daily hesperidin treatment, as circulating levels of inflammatory biomarkers (high-sensitivity C-reactive protein and serum amyloid A protein) decreased when compared to placebo-treated patients [195]. In combination with other compounds, hesperidin exerts beneficial properties. Clinical trials with Daflon 500 mg, a venotropic drug composed of diosmin (450 mg) and hesperidin (50 mg), observed vasoprotector effects in subjects suffering from chronic venous insufficiency. That includes improvement of venous capacitance, lessened edema, limitation of skin disorders, stimulation of ulcer healing, and improvements of clinical signs and symptoms (leg pain, heat sensation, heaviness, redness, etc.) [211–213]. Moreover, it also reduced the severity, frequency, and duration of the hemorrhoidal attacks. As consequence, patients experience reduced pain, bleeding, and discharge [196–198]. Rikkunshito, an herbal medicine of which hesperidin is one of main active ingredients, is used in the treatment of gastrointestinal symptoms of patients with functional dyspepsia [214,215]. In addition, it promotes secretion of ghrelin and, therefore, stimulates appetite in cancer patients suffering from anorexia during chemotherapy [216–218]. HMC ($C_{29}H_{36}O_{15}$), a chalcone methylated compound, derived from hesperidin, with higher solubility, has been described as effective and safe in the treatment of vascular diseases when combined with other compounds such as vitamin C and Ruscus aculeatus extract (Cyclo 3 Fort®, Laboratoires Pierre Fabre, Paris, France). Treatment with HMC reduces symptoms such as pain, heaviness, and paresthesia, as reviewed by Boyle et al. [219] and Kakkos and Allaert [220].

4.2. Catechins

Catechins are polyphenolic molecules present in various types of teas, commonly related to health benefits due to their antioxidant capacity and attenuation of NFκB activation [221–224]. Epigallocatechin-3-gallate (EGCG) is the major constituent among catechins present in tea leaves [224]. Regarding safety, an experimental study published in 2016 concluded that EGCG administration presents no toxicity at doses up to 500 mg/kg/day for 13 weeks in rats [225]. As reviewed previously [223,226,227], clinical studies analyzing the effects of green tea consumption linked catechins to cardioprotective and anti-inflammatory effects, such as lower blood pressure, lower levels of serum TNFα, C-reactive protein, cholesterol and triglycerides in patients [199]. Upon green tea consumption, individuals with prostate cancer presented a significant decrease in the disease progression [228–231] and risk of developing lung cancer was diminished [232,233]. Despite the strong evidence of cancer-preventive effect suggested by some studies, there are conflicting results as well, and several limitations should be considered – genetic variations between individuals and populations, confounding factors, and regarding experimental results, human translational difficulties such as equivalent dose [234]. Furthermore, in a double-blind controlled study, type 2 diabetes patients had improvements in obesity and glucose levels after 582.8 mg or 96.3 mg consumption of catechins for 12 weeks [235]. Bone health improvements and Alzheimer's disease prevention by catechins in pre-clinical and clinical studies were summarized somewhere else [236,237].

4.3. Quercetin

Quercetin ($C_{15}H_{10}O_7$) is the most abundant dietary flavonol [66]. One study published in 2009 aiming to test the effects of 150 mg quercetin/day for 6 weeks in subjects with high-cardiovascular disease risk observed a reduction in systolic blood pressure and plasma oxidized LDL [238]. Another randomized, double-blind, placebo-controlled, crossover study, aiming to test the efficacy of 730 mg quercetin/day for 4 weeks in hypertensive subjects observed reductions in systolic, diastolic and mean arterial pressures [239]. Similar blood-pressure regulation efficacy was observed in type-2 diabetes women taking 500 mg quercetin once daily, for 10 weeks [240]. A meta-analysis of clinical studies evaluating quercetin actions on cardiovascular protection concludes that quercetin supplementation should be considered as an add-on therapeutic approach [241]. Quercetin doses above 500 mg/day show effects on the C-reactive protein levels [242], and in women with polycystic ovary syndrome, 1g quercetin intake for 12 weeks improved adiponectin-mediated insulin resistance and hormonal profiles [242]. Moreover, in 50 rheumatoid arthritis (RA) patients, 500 mg quercetin supplementation during 8 weeks ameliorates clinical symptoms such as early morning stiffness and pain, disease activity, and TNFα plasma levels [200].

4.4. Apigenin

Apigenin ($C_{15}H_{10}O_5$) consumption is considered safe, although muscle relaxation and sedation can occur when 30–100 mg/kg body weight doses are administrated [243]. Anxiolytic effects were observed in patients with generalized anxiety disorder treated with chamomile extract for 8 weeks [244]. A crossover double-blind clinical trial, enrolling 100 patients diagnosed with migraine without aura, tested the efficacy of an oleogel with chamomile extracts. Results showed that pain, nausea, vomiting, photophobia, and phonophobia significantly decreased [201]. Topic application of apigenin-containing cream in 20 women reduces skin aging, besides increasing dermal density and elasticity [245]. In addition, treatment with a flavonoid mixture composed of 20 mg epigallocathechin-gallate plus 20 mg apigenin for 2–5 years reduced the recurrence rate of colon neoplasia in patients with resected colon cancer [246].

4.5. Flavonoid-Based Compounds

Plant extracts containing different classes of flavonoids have been studied in clinical trials and presents promising results. Silymarin (Livergol®, Goldaruo pharmaceutical, Iran), present in species derived from Silybum marianum, contains seven flavolignans and the flavonoid taxifolin ($C_{15}H_{12}O_7$) [247]. Parameters such as joint swelling, tenderness, and pain were reduced in RA patients taking 420 mg of silymarin daily for three months [202]. A meta-analysis involving 8 randomized controlled trials found that Silymarin intake has efficacy in the treatment of alcoholic fatty disease by reducing transaminases levels in patients [248]. Pycnogenol® (Horphag Research Ltd., UK, Geneve, Switzerland), extract of *Pinus maritima*, is a mixture of flavonoids, mainly procyanidins [249]. Sixty-seven subjects with osteoarthritis (OA) taking 220 mg of Pycnogenol® daily for 3 weeks presented a decrease in C reactive protein levels and reduced use of painkillers and non-steroidal anti-inflammatory drugs [203,250]. In other trials with OA patients taking 150 mg/day Pycnogenol®, relief from daily pain, stiffness, and physical function were observed [204,205]. Alvocidib, or Flavopiridol (Tolero Pharmaceuticals, Inc.) is a synthetic analog of a naturally occurring flavone extracted from Dysoxylum binectariferum. It has a CDK9 kinase inhibitor activity, arresting cell cycle at G1 phase, and, therefore, has been shown as a promising therapy for patients with acute myeloid leukemia and chronic lymphocytic leukemia [206,251]. Enzogenol®, a flavonoid-rich extract of Pinus radiate containing approximately 80% total proanthocyanidins and other water-soluble flavonoids, has been linked to neuroprotection properties [252,253] and beneficial vascular effects [254]. Furthermore, in a double-blind, placebo-controlled trial, women taking Colladeen® (Lamberts Healthcare Limited, UK) –

320 mg oligomeric procyanidins – improves leg health and reduces fluid retention in pre-menopausal phase [255].

So far, most clinical reports in the literature states flavonoids consumption as safe. Despite the main source of flavonoids is diet, it is important to highlight that supplement intake is growing, which leads to the concern of high-dosage toxicity and/or interactions with other dietary elements or medications, resulting in possible adverse effects [256]. Although there are a lot of patent requested to pharmaceutical compositions containing flavonoids and to use of flavonoids [257], to our knowledge, the Food and Drug Administration (FDA) has not yet endorsed any flavonoids as pharmaceutical drugs. Although antioxidant properties are commonly linked to flavonoid molecules, under specific conditions they can be prooxidant (hydroxyl radicals can be produced due to their iron and copper reducing activities [258]) and present side effects, as genotoxicity, nausea, and headache [259,260]. Supplementation of high doses may cause antithyroid and goitrogenic effects, in addition to lower bioavailability of vitamin C, trace elements, and vitamins such as folate [256]. While presenting safe pre-clinical and clinical profiles, further studies addressing toxicity, interactions, contraindications, and safety of flavonoids intake are needed to avoid indiscriminate consumption based on the idea that "natural molecules" do no harm.

5. Development of Pharmaceutical Formulations Containing Flavonoids

Different types of formulations have been applied to flavonoids, such as topical and oral formulations [261,262]. Topical formulations of flavonoids are a promising therapeutic option to provide a site-specific application of the drug with important pharmaceutical uses [261]. Several studies demonstrated that topical formulations containing quercetin, trans-chalcone, naringenin, and hesperidin methyl chalcone inhibit the UVB irradiation-induced inflammation and stress oxidative [50,61,72,263]. Therefore, their topical use may provide the required photochemical protection in addition to human sunscreens. Quercetin gel reduced CFA-induced inflammation such as increased paw edema, erythema, swelling, joint stiffness and disturb in the movement [264]. In this regard, treatments with flavonoids incorporated into topical formulations can decrease the local inflammatory process, pain, and stress oxidative.

The most challenging factor is the flavonoids' low water solubility, which leads to lower absorption and consequently lower bioavailability after oral administration [265]. In this sense, protective delivery systems of flavonoids may be a promising therapeutic option to significantly expand the water solubility, dissolution, absorption, thermal stability and bioavailability of flavonoid, thus the construction of flavonoid microcapsules, nanoparticles and nano-formulations is an effective approach to increase its bioavailability, including the use of liposomes, inclusion complex (cyclodextrin and phospholipid complexes), micelles, and solid dispersion (Table 3).

Microencapsulation (pectin/casein by complex coacervation) of quercetin ameliorates acetic acid-induced colitis by providing a controlled release of quercetin at the mouse colon and improves the anti-inflammatory and antioxidant effects of quercetin compared to the non-encapsulated drug [69]. Rutin-Loaded Microparticles (pectin/casein by complex coacervation as well) induce the inhibition of carrageenan-induced mechanical hyperalgesia better than nonmicroencapsulated rutin does [266].

Superparamagnetic iron oxide nanoparticle (SPION) drug delivery system enhanced the bioavailability of quercetin. The SPION increases blood circulation time of quercetin and increases the concentration of quercetin in the brain that leads to higher antioxidant activity and more efficient interactions of quercetin with RSK2, MSK1, CytC, Cdc42, Apaf1, FADD, CRK proteins [267]. Naringenin nanosuspensions formulated using polyvinylpyrrolidone displayed a higher dissolution amount (91 ± 4.4% during 60 min) compared to pure naringenin (42 ± 0.41%). The apparent and effective permeability of naringenin nanosuspension was increased as compared to the pure naringenin. The in vivo naringenin nanosuspension treatment showed maximum concentration and area under curve (0–24 h) values approximately 2-fold superior than the pure drug [268]. The flavonoid fisetin–loaded polymeric nanoparticles had protected and preserved the release of flavonoid fisetin in gastric and intestinal conditions. The ABTS scavenging capacity of flavonoid fisetin, as well as α-glucosidase inhibition activity, were enhanced about 20-fold compared to pure compounds [269]. The nanoparticles and microencapsulates provide controlled release of agent and can be further optimized to be used as an efficient flavonoids' delivery.

Table 3. Delivery systems containing flavonoids and key improvement effects.

Formulations Containing Flavonoids	Flavonoids	Biological Effects
Microparticles	Quercetin [69] Rutin [266]	
Nanoparticles	Quercetin [267] Naringenin [268] Fisetin [269]	
SNEDDS	Naringenin [270] Quercetin [271] Rutin [272,273]	↑ bioavailability ↑ absorption ↑ solubility ↑ stability ↑ efficacy
Liposomal	Quercetin [274,275] Rutin [274] Kaempferol [275] Luteolin [275]	
Inclusion Complex	Quercetin [276] Rutin [277] Naringenin [278]	
Micelles	Quercetin [279] Rutin [280]	
Solid Dispersion	Naringenin [281]	

Pre-clinical findings.

Self-Nanoemulsifying Drug Delivery System (SNEDDS) has been proved to improve the bioavailability of naringenin, quercetin, and rutin. The drug concentration time-curve of naringenin from SNEDDS revealed a significant increase in naringenin absorption compared to naringenin suspension [270]. The optimized SNEDDS formulation of rutin had a globule size in the nanometric range, which led to faster and better absorption in comparison to rutin suspension [272,273]. The quercetin SNEDDS significantly improved quercetin transport across a human colon cell monolayer and demonstrated rapid absorption within 40 min of oral ingestion [271]. SNEDDS increased absorption, optimum globule size and higher solubility as well as higher bioavailability. Thus, the SNEDDS could be used an effective approach for enhancing the solubility and bioavailability of flavonoids.

The liposomal encapsulation for four types of flavonoids demonstrated that the quercetin-loaded liposomes showed higher stability and more antioxidant capacity than those loading rutin, luteolin and kaempferol [274,275]. The water solubility of quercetin and rutin was improved by phospholipid complex [276,277]. There was no statistical difference between the quercetin or rutin complex and quercetin or rutin in the in vitro antioxidant activity, indicating that the process of complexation does not adversely affect the bioactivity of the active ingredient [276,277]. Naringenin-cyclodextrin complex presented a release of 98.0 - 100% at 60 min and enhanced the thermal stability of naringenin [278]. Polymeric micelles exhibited sustained release of quercetin or rutin compared to free quercetin or rutin, as demonstrated by in vitro assays. The solubility of quercetin and rutin was markedly improved compared to pure molecules [279,280]. Amorphous solid dispersion (ASD) of naringenin has been developed with various polymers like cellulose derivatives. Poly-vinylpyrrolidinone-based ASDs demonstrated drug release in gastric (1.2 pH buffer) and small intestine (6.8 pH buffer) conditions. The cellulosic polymer delivers naringenin selectively at a neutral pH at the site for its absorption, whereas it inhibits the drug degradation in the gastric environment environment [281]. At present, a wide range of protective drug delivery systems are available, showing promising results for flavonoids' delivery. In this sense, protective formulation strategies can be explored as per the pharmacological activity of flavonoids.

6. Conclusions

We have reviewed pre-clinical and clinical data about the anti-inflammatory and analgesic properties of flavonoids. Most studies have been demonstrating the potential therapeutic role and efficacy of flavonoids in cardiovascular diseases, osteoarthritis, Parkinson disease, colitis, cancer pain, arthritis, and neuropathic pain. The mechanisms of action of flavonoids are yet to be fully elucidated, but many studies have shown their relevance as a potent anti-inflammatory, analgesic, and antioxidative group of molecules. Indeed, flavonoids can block the expression and activation of many cellular regulatory proteins such as cytokines and transcription factors, resulting in diminished cellular inflammatory responses and pain. In conclusion, in view of the pharmacological activities of flavonoids, it could also be interesting to further develop protective delivery formulations containing flavonoids to treat inflammatory diseases and pain, since promising effects were already observed [69,266].

Flavonoids are multi-target molecules, and increasing attention has been given to these molecules due to their anti-inflammatory and analgesic properties. Diminishing the activity of varied pathways seems to present fewer side effects than abolishing the activity of one target, since, in general, the targets also have endogenous physiological roles. Flavonoids block the synthesis of inflammatory mediators such as IL-1β, TNF-α, NO, and COX-2, suppress VEGF and ICAM-1 expression, along with the activation of STAT3, NFkB, NLRP3 inflammasome, and MAP kinases pathways. Regardless of their broad pharmacological properties, flavonoids show a poor water solubility, inadequate permeability, and constrained bioavailability, potentially requiring high doses to show efficacy in [282]. Phase 2 metabolism is known to affect the bioavailability of flavonoids in humans [283]. Generally, most flavonoids experience sulfation, methylation, and glucuronidation in the small digestive system and liver and conjugated metabolites can be found in plasma after flavonoid ingestion [284]. Nevertheless, some metabolites of flavonoids are still active [285]. In such manner, various endeavors have been made to expand bioavailability, for example, improving the intestinal absorption by means of utilization of absorption enhancers, novel delivery systems; improving metabolic stability; changing the site of absorption from large intestine to small intestine.

Thus, flavonoid pharmacology, therapeutics, and pharmaceutical development remain a hot topic in inflammatory diseases and pain treatment.

Author Contributions: C.R.F., T.T.C., M.F.M., N.A.A., F.S.R.-O., and V.F. contributed to the design and to the writing of the manuscript. W.A.V.J. and R.C. supervised the manuscript and were in charge of overall direction. All authors have read and agreed to the published version of the manuscript.

Funding: Authors also acknowledge the PPSUS grant funded by Decit/SCTIE/MS intermediated by CNPq with support of Fundação Araucária and SESA-PR (agreement 041/2017); and PRONEX grant (Programa de Apoio a Grupos de Excelência) supported by SETI/Fundação Araucária and MCTI/CNPq, and Governo do Estado do Paraná (agreement 014/2017).

Acknowledgments: C.R.F. acknowledges Post-doc fellowship from Conselho Nacional de Desenvolvimento Científico e Tecnológico (CNPq) and T.T.C. from Coordenação de Aperfeiçoamento de Pessoal de Nível Superior (CAPES; finance code 001). M.F.M., N.A.A. and V.F. acknowledge PhD scholarship from CAPES and FRS-O acknowledges Master scholarship from CAPES (finance code 001). R.C. and W.A.V.J. thank the Research Fellowship from CNPq.

Conflicts of Interest: The authors have no conflicts of interest to declare.

References

1. Hayden, M.S.; Ghosh, S. Shared principles in NF-κB signaling. *Cell* **2008**, *132*, 344–362. [CrossRef]
2. Ghosh, S.; Hayden, M.S. New regulators of NF-κB in inflammation. *Nat. Rev Immunol* **2008**, *8*, 837–848. [CrossRef] [PubMed]

3. Lee, K.M.; Kang, B.S.; Lee, H.L.; Son, S.J.; Hwang, S.H.; Kim, D.S.; Park, J.S.; Cho, H.J. Spinal NF-kB activation induces COX-2 upregulation and contributes to inflammatory pain hypersensitivity. *Eur. J. Neurosci.* **2004**, *19*, 3375–3381. [CrossRef]

4. Liu, T.; Zhang, L.; Joo, D.; Sun, S.C. NF-κB signaling in inflammation. *Signal. Transduct Target.* **2017**, *2*, 1–9. [CrossRef] [PubMed]

5. Souza, G.R.; Cunha, T.M.; Silva, R.L.; Lotufo, C.M.; Verri, W.A., Jr.; Funez, M.I.; Villarreal, C.F.; Talbot, J.; Sousa, L.P.; Parada, C.A.; et al. Involvement of nuclear factor κ B in the maintenance of persistent inflammatory hypernociception. *Pharm. Biochem. Behav.* **2015**, *134*, 49–56. [CrossRef] [PubMed]

6. Ferraz, C.R.; Calixto-Campos, C.; Manchope, M.F.; Casagrande, R.; Clissa, P.B.; Baldo, C.; Verri, W.A., Jr. Jararhagin-induced mechanical hyperalgesia depends on TNF-alpha, IL-1beta and NFκB in mice. *Toxicon* **2015**, *103*, 119–128. [CrossRef]

7. Fattori, V.; Amaral, F.A.; Verri, W.A., Jr. Neutrophils and arthritis: Role in disease and pharmacological perspectives. *Pharm. Res.* **2016**, *112*, 84–98. [CrossRef]

8. McDonald, B.; Pittman, K.; Menezes, G.B.; Hirota, S.A.; Slaba, I.; Waterhouse, C.C.; Beck, P.L.; Muruve, D.A.; Kubes, P. Intravascular danger signals guide neutrophils to sites of sterile inflammation. *Science* **2010**, *330*, 362–366. [CrossRef]

9. Verri, W.A., Jr.; Cunha, T.M.; Parada, C.A.; Poole, S.; Cunha, F.Q.; Ferreira, S.H. Hypernociceptive role of cytokines and chemokines: Targets for analgesic drug development? *Pharmacol. Ther.* **2006**, *112*, 116–138. [CrossRef]

10. Woolf, C.J. What is this thing called pain? *J. Clin. Invest.* **2010**, *120*, 3742–3744. [CrossRef]

11. Woolf, C.J.; Salter, M.W. Neuronal plasticity: Increasing the gain in pain. *Science* **2000**, *288*, 1765–1769. [CrossRef] [PubMed]

12. Pinho-Ribeiro, F.A.; Verri, W.A., Jr.; Chiu, I.M. Nociceptor Sensory Neuron-Immune Interactions in Pain and Inflammation. *Trends Immunol.* **2017**, *38*, 5–19. [CrossRef] [PubMed]

13. Fattori, V.; Borghi, S.M.; Rossaneis, A.C.; Bertozzi, M.M.; Cunha, T.M.; Verri, W.A., Jr. Neuroimmune Regulation of Pain and Inflammation: Targeting Glial Cells and Nociceptor Sensory Neurons Interaction. In *Frontiers in CNS Drug Discovery*; Atta-ur, R., Choudhary, M.I., Eds.; Bentham: New York, NY, USA, 2017; Volume 3, pp. 146–200.

14. Zarpelon, A.C.; Rodrigues, F.C.; Lopes, A.H.; Souza, G.R.; Carvalho, T.T.; Pinto, L.G.; Xu, D.; Ferreira, S.H.; Alves-Filho, J.C.; McInnes, I.B.; et al. Spinal cord oligodendrocyte-derived alarmin IL-33 mediates neuropathic pain. *Faseb J.* **2016**, *30*, 54–65. [CrossRef] [PubMed]

15. Scholz, J.; Woolf, C.J. The neuropathic pain triad: Neurons, immune cells and glia. *Nat. Neurosci.* **2007**, *10*, 1361–1368. [CrossRef] [PubMed]

16. Fattori, V.; Pinho-Ribeiro, F.A.; Staurengo-Ferrari, L.; Borghi, S.M.; Rossaneis, A.C.; Casagrande, R.; Verri, W.A., Jr. The specialized pro-resolving lipid mediator Maresin-1 reduces inflammatory pain with a long-lasting analgesic effect. *Br. J. Pharm.* **2019**, *176*, 1728–1744. [CrossRef]

17. Rayar, A.M.; Lagarde, N.; Ferroud, C.; Zagury, J.F.; Montes, M.; Sylla-Iyarreta Veitia, M. Update on COX-2 selective inhibitors: Chemical classification, side effects and their use in cancers and neuronal diseases. *Curr. Top. Med. Chem.* **2017**, *17*, 2935–2956. [CrossRef]

18. Fattori, V.; Borghi, S.M.; Guazelli, C.F.; Giroldo, A.C.; Crespigio, J.; Bussmann, A.J.; Coelho-Silva, L.; Ludwig, N.G.; Mazzuco, T.L.; Casagrande, R.; et al. Vinpocetine reduces diclofenac-induced acute kidney injury through inhibition of oxidative stress, apoptosis, cytokine production, and NF-κB activation in mice. *Pharmacol. Res.* **2017**, *120*, 10–22. [CrossRef]

19. Ungprasert, P.; Srivali, N.; Thongprayoon, C. Nonsteroidal anti-inflammatory drugs and risk of incident heart failure: A systematic review and meta-analysis of observational studies. *Clin. Cardiol.* **2016**, *39*, 111–118. [CrossRef]

20. Marcondes-Alves, L.; Fattori, V.; Borghi, S.M.; Lourenco-Gonzalez, Y.; Bussmann, A.J.C.; Hirooka, E.Y.; Casagrande, R.; Verri, W.A., Jr.; Arakawa, N.S. Kaurenoic acid extracted from Sphagneticola trilobata reduces acetaminophen-induced hepatotoxicity through inhibition of oxidative stress and pro-inflammatory cytokine production in mice. *Nat. Prod. Res.* **2019**, *33*, 921–924. [CrossRef]

21. Larsen, F.S.; Wendon, J. Understanding paracetamol-induced liver failure. *Intensive Care Med.* **2014**, *40*, 888–890. [CrossRef]

22. Karp, J.F.; Shega, J.W.; Morone, N.E.; Weiner, D.K. Advances in understanding the mechanisms and management of persistent pain in older adults. *Br. J. Anaesth.* **2008**, *101*, 111–120. [CrossRef] [PubMed]

23. Schutz, S.G.; Robinson-Papp, J. HIV-related neuropathy: Current perspectives. *Hiv Aids (Auckl)* **2013**, *5*, 243–251.

24. Grace, P.M.; Maier, S.F.; Watkins, L.R. Opioid-induced central immune signaling: Implications for opioid analgesia. *Headache* **2015**, *55*, 475–489. [CrossRef]

25. Grace, P.M.; Strand, K.A.; Galer, E.L.; Rice, K.C.; Maier, S.F.; Watkins, L.R. Protraction of neuropathic pain by morphine is mediated by spinal damage associated molecular patterns (DAMPs) in male rats. *Brain Behav. Immun.* **2018**, *72*, 45–50. [CrossRef]

26. Grace, P.M.; Strand, K.A.; Galer, E.L.; Urban, D.J.; Wang, X.; Baratta, M.V.; Fabisiak, T.J.; Anderson, N.D.; Cheng, K.; Greene, L.I.; et al. Morphine paradoxically prolongs neuropathic pain in rats by amplifying spinal NLRP3 inflammasome activation. *Proc. Natl. Acad. Sci. USA* **2016**, *113*, E3441–E3450. [CrossRef]

27. Liang, Y.; Chu, H.; Jiang, Y.; Yuan, L. Morphine enhances IL-1beta release through toll-like receptor 4-mediated endocytic pathway in microglia. *Purinergic Signal.* **2016**, *12*, 637–645. [CrossRef] [PubMed]

28. Poetker, D.M.; Reh, D.D. A comprehensive review of the adverse effects of systemic corticosteroids. *Otolaryngol Clin. N. Am.* **2010**, *43*, 753–768. [CrossRef] [PubMed]

29. Slordal, L.; Spigset, O. Heart Failure Induced by Non-Cardiac Drugs. *Drug Saf.* **2006**, *29*, 567–586. [CrossRef] [PubMed]

30. Orlicka, K.; Barnes, E.; Culver, E.L. Prevention of infection caused by immunosuppressive drugs in gastroenterology. *Adv. Chronic Dis.* **2013**, *4*, 167–185. [CrossRef]

31. Cabral, V.P.; Andrade, C.A.; Passos, S.R.; Martins, M.F.; Hokerberg, Y.H. Severe infection in patients with rheumatoid arthritis taking anakinra, rituximab, or abatacept: A systematic review of observational studies. *Rev. Bras. Reum. Engl. Ed.* **2016**, *56*, 543–550. [CrossRef]

32. Plein, L.M.; Rittner, H.L. Opioids and the immune system - friend or foe. *Br. J. Pharm.* **2018**, *175*, 2717–2725. [CrossRef]

33. Borghi, S.M.; Mizokami, S.S.; Pinho-Ribeiro, F.A.; Fattori, V.; Crespigio, J.; Clemente-Napimoga, J.T.; Napimoga, M.H.; Pitol, D.L.; Issa, J.P.M.; Fukada, S.Y.; et al. The flavonoid quercetin inhibits titanium dioxide (TiO2)-induced chronic arthritis in mice. *J. Nutr. Biochem.* **2018**, *53*, 81–95. [CrossRef]

34. Verri, W.A., Jr.; Vicentini, F.T.; Baracat, M.M.; Georgetti, S.R.; Cardoso, R.D.; Cunha, T.M.; Ferreira, S.H.; Cunha, F.Q.; Fonseca, M.J.; Casagrande, R. Flavonoids as Anti-Inflammatory and Analgesic Drugs: Mechanisms of Action and Perspectives in the Development of Pharmaceutical Forms. In *Studies in Natural Products Chemistry*, 1st ed.; Rahman, A.U., Ed.; Elsevier: Amsterdam, The Netherlands, 2012; Volume 36, pp. 297–330.

35. Ginwala, R.; Bhavsar, R.; Chigbu, D.I.; Jain, P.; Khan, Z.K. Potential role of flavonoids in treating chronic inflammatory diseases with a special focus on the anti-inflammatory activity of apigenin. *Antioxidants* **2019**, *8*, 35. [CrossRef] [PubMed]

36. Anusha, C.; Sumathi, T.; Joseph, L.D. Protective role of apigenin on rotenone induced rat model of Parkinson's disease: Suppression of neuroinflammation and oxidative stress mediated apoptosis. *Chem. Biol. Interact.* **2017**, *269*, 67–79. [CrossRef] [PubMed]

37. Ronchetti, S.; Migliorati, G.; Delfino, D.V. Association of inflammatory mediators with pain perception. *Biomed. Pharm.* **2017**, *96*, 1445–1452. [CrossRef] [PubMed]

38. Borghi, S.M.; Carvalho, T.T.; Staurengo-Ferrari, L.; Hohmann, M.S.; Pinge-Filho, P.; Casagrande, R.; Verri, W.A., Jr. Vitexin inhibits inflammatory pain in mice by targeting TRPV1, oxidative stress, and cytokines. *J. Nat. Prod.* **2013**, *76*, 1141–1149. [CrossRef]

39. Tian, R.; Yang, W.; Xue, Q.; Gao, L.; Huo, J.; Ren, D.; Chen, X. Rutin ameliorates diabetic neuropathy by lowering plasma glucose and decreasing oxidative stress via Nrf2 signaling pathway in rats. *Eur. J. Pharm.* **2016**, *771*, 84–92. [CrossRef]

40. Carvalho, T.T.; Mizokami, S.S.; Ferraz, C.R.; Manchope, M.F.; Borghi, S.M.; Fattori, V.; Calixto-Campos, C.; Camilios-Neto, D.; Casagrande, R.; Verri, W.A., Jr. The granulopoietic cytokine granulocyte colony-stimulating factor (G-CSF) induces pain: Analgesia by rutin. *Inflammopharmacology* **2019**, *27*, 1285–1296. [CrossRef]

41. Lamoke, F.; Labazi, M.; Montemari, A.; Parisi, G.; Varano, M.; Bartoli, M. Trans-Chalcone prevents VEGF expression and retinal neovascularization in the ischemic retina. *Exp. Eye Res.* **2011**, *93*, 350–354. [CrossRef]

42. Staurengo-Ferrari, L.; Ruiz-Miyazawa, K.W.; Pinho-Ribeiro, F.A.; Fattori, V.; Zaninelli, T.H.; Badaro-Garcia, S.; Borghi, S.M.; Carvalho, T.T.; Alves-Filho, J.C.; Cunha, T.M.; et al. Trans-Chalcone Attenuates Pain and Inflammation in Experimental Acute Gout Arthritis in Mice. *Front. Pharm.* **2018**, *9*, 1123. [CrossRef]

43. Qi, W.; Lin, C.; Fan, K.; Chen, Z.; Liu, L.; Feng, X.; Zhang, H.; Shao, Y.; Fang, H.; Zhao, C.; et al. Hesperidin inhibits synovial cell inflammation and macrophage polarization through suppression of the PI3K/AKT pathway in complete Freund's adjuvant-induced arthritis in mice. *Chem Biol Interact.* **2019**, *306*, 19–28. [CrossRef] [PubMed]

44. Javed, H.; Vaibhav, K.; Ahmed, M.E.; Khan, A.; Tabassum, R.; Islam, F.; Safhi, M.M.; Islam, F. Effect of hesperidin on neurobehavioral, neuroinflammation, oxidative stress and lipid alteration in intracerebroventricular streptozotocin induced cognitive impairment in mice. *J. Neurol. Sci.* **2015**, *348*, 51–59. [CrossRef] [PubMed]

45. Lin, Y.L.; Lin, J.K. (−)-Epigallocatechin-3-gallate blocks the induction of nitric oxide synthase by down-regulating lipopolysaccharide-induced activity of transcription factor nuclear factor-κB. *Mol. Pharm.* **1997**, *52*, 465–472. [CrossRef]

46. Cardenas, H.; Arango, D.; Nicholas, C.; Duarte, S.; Nuovo, G.J.; He, W.; Voss, O.H.; Gonzalez-Mejia, M.E.; Guttridge, D.C.; Grotewold, E.; et al. Dietary Apigenin Exerts Immune-Regulatory Activity in Vivo by Reducing NF-κB Activity, Halting Leukocyte Infiltration and Restoring Normal Metabolic Function. *Int. J. Mol. Sci.* **2016**, *17*, 323. [CrossRef] [PubMed]

47. Tahir, M.; Rehman, M.U.; Lateef, A.; Khan, R.; Khan, A.Q.; Qamar, W.; Ali, F.; O'Hamiza, O.; Sultana, S. Diosmin protects against ethanol-induced hepatic injury via alleviation of inflammation and regulation of TNF-alpha and NF-κB activation. *Alcohol* **2013**, *47*, 131–139. [CrossRef]

48. Rasquel-Oliveira, F.S.; Manchope, M.F.; Staurengo-Ferrari, L.; Ferraz, C.R.; Santos, T.S.; Zaninelli, T.H.; Fattori, V.; Antero, N.A.; Badaro-Garcia, S.; Freitas, A.; et al. Hesperidin methyl chalcone interacts with NFκB Ser276 and inhibits zymosan-induced joint pain and inflammation, and RAW 264.7 macrophage activation. *Inflammopharmacology* **2020**. accepted.

49. Ruiz-Miyazawa, K.W.; Pinho-Ribeiro, F.A.; Borghi, S.M.; Staurengo-Ferrari, L.; Fattori, V.; Amaral, F.A.; Teixeira, M.M.; Alves-Filho, J.C.; Cunha, T.M.; Cunha, F.Q.; et al. Hesperidin Methylchalcone Suppresses Experimental Gout Arthritis in Mice by Inhibiting NF-κB Activation. *J. Agric. Food Chem.* **2018**, *66*, 6269–6280. [CrossRef]

50. Martinez, R.M.; Pinho-Ribeiro, F.A.; Steffen, V.S.; Caviglione, C.V.; Pala, D.; Baracat, M.M.; Georgetti, S.R.; Verri, W.A.; Casagrande, R. Topical formulation containing hesperidin methyl chalcone inhibits skin oxidative stress and inflammation induced by ultraviolet B irradiation. *Photochem. Photobiol Sci.* **2016**, *15*, 554–563. [CrossRef]

51. Jiang, J.; Jia, Y.; Lu, X.; Zhang, T.; Zhao, K.; Fu, Z.; Pang, C.; Qian, Y. Vitexin suppresses RANKL-induced osteoclastogenesis and prevents lipopolysaccharide (LPS)-induced osteolysis. *J. Cell Physiol.* **2019**, *234*, 17549–17560. [CrossRef]

52. Lu, Y.; Yu, T.; Liu, J.; Gu, L. Vitexin attenuates lipopolysaccharide-induced acute lung injury by controlling the Nrf2 pathway. *PLoS ONE* **2018**, *13*, e0196405. [CrossRef]

53. Lee, H.N.; Shin, S.A.; Choo, G.S.; Kim, H.J.; Park, Y.S.; Kim, B.S.; Kim, S.K.; Cho, S.D.; Nam, J.S.; Choi, C.S.; et al. Antiinflammatory effect of quercetin and galangin in LPSstimulated RAW264.7 macrophages and DNCBinduced atopic dermatitis animal models. *Int J. Mol. Med.* **2018**, *41*, 888–898. [PubMed]

54. Guazelli, C.F.S.; Staurengo-Ferrari, L.; Zarpelon, A.C.; Pinho-Ribeiro, F.A.; Ruiz-Miyazawa, K.W.; Vicentini, F.; Vignoli, J.A.; Camilios-Neto, D.; Georgetti, S.R.; Baracat, M.M.; et al. Quercetin attenuates zymosan-induced arthritis in mice. *Biomed. Pharm.* **2018**, *102*, 175–184. [CrossRef] [PubMed]

55. Vicentini, F.T.; He, T.; Shao, Y.; Fonseca, M.J.; Verri, W.A., Jr.; Fisher, G.J.; Xu, Y. Quercetin inhibits UV irradiation-induced inflammatory cytokine production in primary human keratinocytes by suppressing NF-κB pathway. *J. Derm. Sci.* **2011**, *61*, 162–168. [CrossRef] [PubMed]

56. Gulati, N.; Harter, D.; Desai, D.; Amin, S.; Murali, R.; Jhanwar-Uniyal, M. Quercetin inhibits the akt pathway, leading to suppression of survival and induction of apoptosis in cancer cells. *Cancer Res.* **2005**, *65*, 536.

57. Domiciano, T.P.; Wakita, D.; Jones, H.D.; Crother, T.R.; Verri, W.A., Jr.; Arditi, M.; Shimada, K. Quercetin Inhibits Inflammasome Activation by Interfering with ASC Oligomerization and Prevents Interleukin-1 Mediated Mouse Vasculitis. *Sci Rep.* **2017**, *7*, 41539. [CrossRef]

58. Pinho-Ribeiro, F.A.; Zarpelon, A.C.; Fattori, V.; Manchope, M.F.; Mizokami, S.S.; Casagrande, R.; Verri, W.A., Jr. Naringenin reduces inflammatory pain in mice. *Neuropharmacology* **2016**, *105*, 508–519. [CrossRef]

59. Bussmann, A.J.C.; Borghi, S.M.; Zaninelli, T.H.; Dos Santos, T.S.; Guazelli, C.F.S.; Fattori, V.; Domiciano, T.P.; Pinho-Ribeiro, F.A.; Ruiz-Miyazawa, K.W.; Casella, A.M.B.; et al. The citrus flavanone naringenin attenuates zymosan-induced mouse joint inflammation: Induction of Nrf2 expression in recruited CD45(+) hematopoietic cells. *Inflammopharmacology* **2019**, *27*, 1229–1242. [CrossRef]

60. Manchope, M.F.; Artero, N.A.; Fattori, V.; Mizokami, S.S.; Pitol, D.L.; Issa, J.P.M.; Fukada, S.Y.; Cunha, T.M.; Alves-Filho, J.C.; Cunha, F.Q.; et al. Naringenin mitigates titanium dioxide (TiO2)-induced chronic arthritis in mice: Role of oxidative stress, cytokines, and NFκB. *Inflamm Res.* **2018**, *67*, 997–1012. [CrossRef]

61. Martinez, R.M.; Pinho-Ribeiro, F.A.; Steffen, V.S.; Silva, T.C.; Caviglione, C.V.; Bottura, C.; Fonseca, M.J.; Vicentini, F.T.; Vignoli, J.A.; Baracat, M.M.; et al. Topical Formulation Containing Naringenin: Efficacy against Ultraviolet B Irradiation-Induced Skin Inflammation and Oxidative Stress in Mice. *PLoS ONE* **2016**, *11*, e0146296. [CrossRef]

62. Manchope, M.F.; Calixto-Campos, C.; Coelho-Silva, L.; Zarpelon, A.C.; Pinho-Ribeiro, F.A.; Georgetti, S.R.; Baracat, M.M.; Casagrande, R.; Verri, W.A., Jr. Naringenin Inhibits Superoxide Anion-Induced Inflammatory Pain: Role of Oxidative Stress, Cytokines, Nrf-2 and the NO-cGMP-PKG-KATPChannel Signaling Pathway. *PLoS ONE* **2016**, *11*, e0153015. [CrossRef]

63. Gao, W.; Zan, Y.; Wang, Z.J.; Hu, X.Y.; Huang, F. Quercetin ameliorates paclitaxel-induced neuropathic pain by stabilizing mast cells, and subsequently blocking PKCepsilon-dependent activation of TRPV1. *Acta Pharm. Sin.* **2016**, *37*, 1166–1177. [CrossRef] [PubMed]

64. Straub, I.; Mohr, F.; Stab, J.; Konrad, M.; Philipp, S.E.; Oberwinkler, J.; Schaefer, M. Citrus fruit and fabacea secondary metabolites potently and selectively block TRPM3. *Br. J. Pharmacol.* **2013**, *168*, 1835–1850. [CrossRef] [PubMed]

65. Zhou, Y.; Cai, S.; Moutal, A.; Yu, J.; Gomez, K.; Madura, C.L.; Shan, Z.; Pham, N.Y.N.; Serafini, M.J.; Dorame, A.; et al. The Natural Flavonoid Naringenin Elicits Analgesia through Inhibition of NaV1.8 Voltage-Gated Sodium Channels. *ACS Chem. Neurosci.* **2019**, *10*, 4834–4846. [CrossRef]

66. Li, Y.; Yao, J.; Han, C.; Yang, J.; Chaudhry, M.T.; Wang, S.; Liu, H.; Yin, Y. Quercetin, Inflammation and Immunity. *Nutrients* **2016**, *8*, 167. [CrossRef]

67. Pietta, P.G. Flavonoids as antioxidants. *J. Nat. Prod.* **2000**, *63*, 1035–1042. [CrossRef]

68. Ferraz, C.R.; Silva, D.B.; Prado, L.; Canabrava, H.A.N.; Bispo-da-Silva, L.B. Antidiarrhoeic effect and dereplication of the aqueous extract of Annona crassiflora (Annonaceae). *Nat. Prod. Res.* **2019**, *33*, 563–567. [CrossRef]

69. Guazelli, C.F.; Fattori, V.; Colombo, B.B.; Georgetti, S.R.; Vicentini, F.T.; Casagrande, R.; Baracat, M.M.; Verri, W.A., Jr. Quercetin-loaded microcapsules ameliorate experimental colitis in mice by anti-inflammatory and antioxidant mechanisms. *J. Nat. Prod.* **2013**, *76*, 200–208. [CrossRef]

70. Napimoga, M.H.; Clemente-Napimoga, J.T.; Macedo, C.G.; Freitas, F.F.; Stipp, R.N.; Pinho-Ribeiro, F.A.; Casagrande, R.; Verri, W.A., Jr. Quercetin inhibits inflammatory bone resorption in a mouse periodontitis model. *J. Nat. Prod.* **2013**, *76*, 2316–2321. [CrossRef]

71. Calixto-Campos, C.; Correa, M.P.; Carvalho, T.T.; Zarpelon, A.C.; Hohmann, M.S.; Rossaneis, A.C.; Coelho-Silva, L.; Pavanelli, W.R.; Pinge-Filho, P.; Crespigio, J.; et al. Quercetin Reduces Ehrlich Tumor-Induced Cancer Pain in Mice. *Anal. Cell Pathol.* **2015**, *2015*, 285708. [CrossRef]

72. Casagrande, R.; Georgetti, S.R.; Verri, W.A., Jr.; Dorta, D.J.; dos Santos, A.C.; Fonseca, M.J. Protective effect of topical formulations containing quercetin against UVB-induced oxidative stress in hairless mice. *J. Photochem. Photobiol. B* **2006**, *84*, 21–27. [CrossRef]

73. Valerio, D.A.; Georgetti, S.R.; Magro, D.A.; Casagrande, R.; Cunha, T.M.; Vicentini, F.T.; Vieira, S.M.; Fonseca, M.J.; Ferreira, S.H.; Cunha, F.Q.; et al. Quercetin reduces inflammatory pain: Inhibition of oxidative stress and cytokine production. *J. Nat. Prod.* **2009**, *72*, 1975–1979. [CrossRef] [PubMed]

74. Maioli, N.A.; Zarpelon, A.C.; Mizokami, S.S.; Calixto-Campos, C.; Guazelli, C.F.; Hohmann, M.S.; Pinho-Ribeiro, F.A.; Carvalho, T.T.; Manchope, M.F.; Ferraz, C.R.; et al. The superoxide anion donor, potassium superoxide, induces pain and inflammation in mice through production of reactive oxygen species and cyclooxygenase-2. *Braz. J. Med. Biol Res.* **2015**, *48*, 321–331. [CrossRef] [PubMed]

75. Cho, S.Y.; Park, S.J.; Kwon, M.J.; Jeong, T.S.; Bok, S.H.; Choi, W.Y.; Jeong, W.I.; Ryu, S.Y.; Do, S.H.; Lee, C.S.; et al. Quercetin suppresses proinflammatory cytokines production through MAP kinases andNF-κB pathway in lipopolysaccharide-stimulated macrophage. *Mol. Cell Biochem.* **2003**, *243*, 153–160. [CrossRef] [PubMed]

76. Souto, F.O.; Zarpelon, A.C.; Staurengo-Ferrari, L.; Fattori, V.; Casagrande, R.; Fonseca, M.J.; Cunha, T.M.; Ferreira, S.H.; Cunha, F.Q.; Verri, W.A., Jr. Quercetin reduces neutrophil recruitment induced by CXCL8, LTB4, and fMLP: Inhibition of actin polymerization. *J. Nat. Prod.* **2011**, *74*, 113–118. [CrossRef]

77. Park, H.H.; Lee, S.; Son, H.Y.; Park, S.B.; Kim, M.S.; Choi, E.J.; Singh, T.S.; Ha, J.H.; Lee, M.G.; Kim, J.E.; et al. Flavonoids inhibit histamine release and expression of proinflammatory cytokines in mast cells. *Arch. Pharm. Res.* **2008**, *31*, 1303–1311. [CrossRef] [PubMed]

78. Weng, Z.; Zhang, B.; Asadi, S.; Sismanopoulos, N.; Butcher, A.; Fu, X.; Katsarou-Katsari, A.; Antoniou, C.; Theoharides, T.C. Quercetin is more effective than cromolyn in blocking human mast cell cytokine release and inhibits contact dermatitis and photosensitivity in humans. *PLoS ONE* **2012**, *7*, e33805. [CrossRef]

79. Huang, S.M.; Wu, C.H.; Yen, G.C. Effects of flavonoids on the expression of the pro-inflammatory response in human monocytes induced by ligation of the receptor for AGEs. *Mol. Nutr. Food Res.* **2006**, *50*, 1129–1139. [CrossRef] [PubMed]

80. Boomgaarden, I.; Egert, S.; Rimbach, G.; Wolffram, S.; Muller, M.J.; Doring, F. Quercetin supplementation and its effect on human monocyte gene expression profiles in vivo. *Br. J. Nutr.* **2010**, *104*, 336–345. [CrossRef] [PubMed]

81. Huang, R.Y.; Yu, Y.L.; Cheng, W.C.; OuYang, C.N.; Fu, E.; Chu, C.L. Immunosuppressive effect of quercetin on dendritic cell activation and function. *J. Immunol.* **2010**, *184*, 6815–6821. [CrossRef]

82. Lin, W.; Wang, W.; Wang, D.; Ling, W. Quercetin protects against atherosclerosis by inhibiting dendritic cell activation. *Mol. Nutr. Food Res.* **2017**, *61*, 1700031. [CrossRef]

83. Kim, Y.J.; Park, W. Anti-Inflammatory Effect of Quercetin on RAW 264.7 Mouse Macrophages Induced with Polyinosinic-Polycytidylic Acid. *Molecules* **2016**, *21*, 450. [CrossRef] [PubMed]

84. Hamalainen, M.; Nieminen, R.; Asmawi, M.Z.; Vuorela, P.; Vapaatalo, H.; Moilanen, E. Effects of flavonoids on prostaglandin E2 production and on COX-2 and mPGES-1 expressions in activated macrophages. *Planta Med.* **2011**, *77*, 1504–1511. [CrossRef]

85. Lara-Guzman, O.J.; Tabares-Guevara, J.H.; Leon-Varela, Y.M.; Alvarez, R.M.; Roldan, M.; Sierra, J.A.; Londono-Londono, J.A.; Ramirez-Pineda, J.R. Proatherogenic macrophage activities are targeted by the flavonoid quercetin. *J. Pharm. Exp.* **2012**, *343*, 296–306. [CrossRef]

86. Dong, J.; Zhang, X.; Zhang, L.; Bian, H.X.; Xu, N.; Bao, B.; Liu, J. Quercetin reduces obesity-associated ATM infiltration and inflammation in mice: A mechanism including AMPKalpha1/SIRT1. *J. Lipid Res.* **2014**, *55*, 363–374. [CrossRef] [PubMed]

87. Borghi, S.M.; Pinho-Ribeiro, F.A.; Fattori, V.; Bussmann, A.J.; Vignoli, J.A.; Camilios-Neto, D.; Casagrande, R.; Verri, W.A., Jr. Quercetin Inhibits Peripheral and Spinal Cord Nociceptive Mechanisms to Reduce Intense Acute Swimming-Induced Muscle Pain in Mice. *PLoS ONE* **2016**, *11*, e0162267. [CrossRef] [PubMed]

88. Carcole, M.; Castany, S.; Leanez, S.; Pol, O. Treatment with a heme oxygenase 1 inducer enhances the antinociceptive effects of micro-opioid, delta-opioid, and cannabinoid 2 receptors during inflammatory pain. *J. Pharm. Exp.* **2014**, *351*, 224–232. [CrossRef]

89. Cunha, T.M.; Roman-Campos, D.; Lotufo, C.M.; Duarte, H.L.; Souza, G.R.; Verri, W.A., Jr.; Funez, M.I.; Dias, Q.M.; Schivo, I.R.; Domingues, A.C.; et al. Morphine peripheral analgesia depends on activation of the PI3Kgamma/AKT/nNOS/NO/KATP signaling pathway. *Proc. Natl Acad Sci. USA* **2010**, *107*, 4442–4447. [CrossRef]

90. Sachs, D.; Cunha, F.Q.; Ferreira, S.H. Peripheral analgesic blockade of hypernociception: Activation of arginine/NO/cGMP/protein kinase G/ATP-sensitive K+ channel pathway. *Proc. Natl. Acad. Sci. USA* **2004**, *101*, 3680–3685. [CrossRef]

91. Liu, G.H.; Qu, J.; Shen, X. NF-κB/p65 antagonizes Nrf2-ARE pathway by depriving CBP from Nrf2 and facilitating recruitment of HDAC3 to MafK. *Biochim. Biophys. Acta* **2008**, *1783*, 713–727. [CrossRef]

92. Yu, M.; Li, H.; Liu, Q.; Liu, F.; Tang, L.; Li, C.; Yuan, Y.; Zhan, Y.; Xu, W.; Li, W.; et al. Nuclear factor p65 interacts with Keap1 to repress the Nrf2-ARE pathway. *Cell Signal.* **2011**, *23*, 883–892. [CrossRef]

93. Wu, C.H.; Lin, M.C.; Wang, H.C.; Yang, M.Y.; Jou, M.J.; Wang, C.J. Rutin inhibits oleic acid induced lipid accumulation via reducing lipogenesis and oxidative stress in hepatocarcinoma cells. *J. Food Sci.* **2011**, *76*, T65–T72. [CrossRef] [PubMed]

94. Hosseinzadeh, H.; Nassiri-Asl, M. Review of the protective effects of rutin on the metabolic function as an important dietary flavonoid. *J. Endocrinol. Invest.* **2014**, *37*, 783–788. [CrossRef] [PubMed]

95. Nadella, V.; Ranjan, R.; Senthilkumaran, B.; Qadri, S.; Pothani, S.; Singh, A.K.; Gupta, M.L.; Prakash, H. Podophyllotoxin and Rutin Modulate M1 (iNOS+) Macrophages and Mitigate Lethal Radiation (LR) Induced Inflammatory Responses in Mice. *Front. Immunol.* **2019**, *10*, 106. [CrossRef] [PubMed]

96. Nikfarjam, B.A.; Adineh, M.; Hajiali, F.; Nassiri-Asl, M. Treatment with Rutin - A Therapeutic Strategy for Neutrophil-Mediated Inflammatory and Autoimmune Diseases: - Anti-inflammatory Effects of Rutin on Neutrophils. *J. Pharmacopunct.* **2017**, *20*, 52–56.

97. Lee, W.; Ku, S.-K.; Bae, J.-S. Barrier protective effects of rutin in LPS-induced inflammation in vitro and in vivo. *Food Chem. Toxicol.* **2012**, *50*, 3048–3055. [CrossRef]

98. Guardia, T.; Rotelli, A.E.; Juarez, A.O.; Pelzer, L.E. Anti-inflammatory properties of plant flavonoids. Effects of rutin, quercetin and hesperidin on adjuvant arthritis in rat. *Farmaco* **2001**, *56*, 683–687. [CrossRef]

99. Yan, X.; Qi, M.; Li, P.; Zhan, Y.; Shao, H. Apigenin in cancer therapy: Anti-cancer effects and mechanisms of action. *Cell Biosci.* **2017**, *7*, 50. [CrossRef]

100. Jiang, P.Y.; Zhu, X.J.; Zhang, Y.N.; Zhou, F.F.; Yang, X.F. Protective effects of apigenin on LPS-induced endometritis via activating Nrf2 signaling pathway. *Microb. Pathog.* **2018**, *123*, 139–143. [CrossRef]

101. Patil, R.H.; Babu, R.L.; Naveen Kumar, M.; Kiran Kumar, K.M.; Hegde, S.M.; Nagesh, R.; Ramesh, G.T.; Sharma, S.C. Anti-Inflammatory Effect of Apigenin on LPS-Induced Pro-Inflammatory Mediators and AP-1 Factors in Human Lung Epithelial Cells. *Inflammation* **2016**, *39*, 138–147. [CrossRef]

102. Feng, X.; Weng, D.; Zhou, F.; Owen, Y.D.; Qin, H.; Zhao, J.; Huang, Y.; Chen, J.; Fu, H.; Yang, N.; et al. Activation of PPARgamma by a Natural Flavonoid Modulator, Apigenin Ameliorates Obesity-Related Inflammation Via Regulation of Macrophage Polarization. *EBioMedicine* **2016**, *9*, 61–76. [CrossRef]

103. Lucas, C.D.; Allen, K.C.; Dorward, D.A.; Hoodless, L.J.; Melrose, L.A.; Marwick, J.A.; Tucker, C.S.; Haslett, C.; Duffin, R.; Rossi, A.G. Flavones induce neutrophil apoptosis by down-regulation of Mcl-1 via a proteasomal-dependent pathway. *FASEB J.* **2012**, *27*, 1084–1094. [CrossRef] [PubMed]

104. Li, X.; Han, Y.; Zhou, Q.; Jie, H.; He, Y.; Han, J.; He, J.; Jiang, Y.; Sun, E. Apigenin, a potent suppressor of dendritic cell maturation and migration, protects against collagen-induced arthritis. *J. Cell. Mol. Med.* **2016**, *20*, 170–180. [CrossRef] [PubMed]

105. Lee, J.H.; Zhou, H.Y.; Cho, S.Y.; Kim, Y.S.; Lee, Y.S.; Jeong, C.S. Anti-inflammatory mechanisms of apigenin: Inhibition of cyclooxygenase-2 expression, adhesion of monocytes to human umbilical vein endothelial cells, and expression of cellular adhesion molecules. *Arch. Pharm. Res.* **2007**, *30*, 1318–1327. [CrossRef] [PubMed]

106. Kang, O.H.; Lee, J.H.; Kwon, D.Y. Apigenin inhibits release of inflammatory mediators by blocking the NF-κB activation pathways in the HMC-1 cells. *Immunopharmacol. Immunotoxicol.* **2011**, *33*, 473–479. [CrossRef] [PubMed]

107. Rafat Husain, S.; Cillard, J.; Cillard, P. Hydroxyl radical scavenging activity of flavonoids. *Phytochemistry* **1987**, *26*, 2489–2491. [CrossRef]

108. Van Acker, S.A.; van Balen, G.P.; van den Berg, D.J.; Bast, A.; van der Vijgh, W.J. Influence of iron chelation on the antioxidant activity of flavonoids. *Biochem Pharm.* **1998**, *56*, 935–943. [CrossRef]

109. Ganjare, A.B.; Nirmal, S.A.; Patil, A.N. Use of apigenin from Cordia dichotoma in the treatment of colitis. *Fitoterapia* **2011**, *82*, 1052–1056. [CrossRef]

110. Yang, H.; Huang, J.; Mao, Y.; Wang, L.; Li, R.; Ha, C. Vitexin alleviates interleukin-1beta-induced inflammatory responses in chondrocytes from osteoarthritis patients: Involvement of HIF-1alpha pathway. *Scand. J. Immunol.* **2019**, *90*, e12773. [CrossRef]

111. Kim, H.J.; Nam, Y.R.; Kim, E.J.; Nam, J.H.; Kim, W.K. Spirodela polyrhiza and its Chemical Constituent Vitexin Exert Anti-Allergic Effect via ORAI1 Channel Inhibition. *Am. J. Chin. Med.* **2018**, *46*, 1243–1261. [CrossRef]

112. Nikfarjam, B.A.; Hajiali, F.; Adineh, M.; Nassiri-Asl, M. Anti-inflammatory Effects of Quercetin and Vitexin on Activated Human Peripheral Blood Neutrophils: - The effects of quercetin and vitexin on human neutrophils. *J. Pharmacopunct.* **2017**, *20*, 127–131.

113. Rosa, S.I.; Rios-Santos, F.; Balogun, S.O.; Martins, D.T. Vitexin reduces neutrophil migration to inflammatory focus by down-regulating pro-inflammatory mediators via inhibition of p38, ERK1/2 and JNK pathway. *Phytomedicine* **2016**, *23*, 9–17. [CrossRef] [PubMed]

114. Chen, L.; Zhang, B.; Shan, S.; Zhao, X. Neuroprotective effects of vitexin against isoflurane-induced neurotoxicity by targeting the TRPV1 and NR2B signaling pathways. *Mol. Med. Rep.* **2016**, *14*, 5607–5613. [CrossRef] [PubMed]

115. Demir Ozkay, U.; Can, O.D. Anti-nociceptive effect of vitexin mediated by the opioid system in mice. *Pharm. Biochem. Behav.* **2013**, *109*, 23–30. [CrossRef]

116. Nogata, Y.; Sakamoto, K.; Shiratsuchi, H.; Ishii, T.; Yano, M.; Ohta, H. Flavonoid composition of fruit tissues of citrus species. *Biosci. Biotechnol. Biochem.* **2006**, *70*, 178–192. [CrossRef] [PubMed]

117. Dholakiya, S.L.; Benzeroual, K.E. Protective effect of diosmin on LPS-induced apoptosis in PC12 cells and inhibition of TNF-alpha expression. *Toxicol. In Vitro* **2011**, *25*, 1039–1044. [CrossRef] [PubMed]

118. Bertozzi, M.M.; Rossaneis, A.C.; Fattori, V.; Longhi-Balbinot, D.T.; Freitas, A.; Cunha, F.Q.; Alves-Filho, J.C.; Cunha, T.M.; Casagrande, R.; Verri, W.A., Jr. Diosmin reduces chronic constriction injury-induced neuropathic pain in mice. *Chem. Biol. Interact.* **2017**, *273*, 180–189. [CrossRef]

119. Carballo-Villalobos, A.I.; Gonzalez-Trujano, M.E.; Pellicer, F.; Alvarado-Vasquez, N.; Lopez-Munoz, F.J. Central and peripheral anti-hyperalgesic effects of diosmin in a neuropathic pain model in rats. *Biomed. Pharm.* **2018**, *97*, 310–320. [CrossRef]

120. Berkoz, M. Diosmin suppresses the proinflammatory mediators in lipopolysaccharide-induced RAW264.7 macrophages via NF-κB and MAPKs signal pathways. *Gen. Physiol. Biophys.* **2019**, *38*, 315–324. [CrossRef]

121. Crespo, M.E.; Galvez, J.; Cruz, T.; Ocete, M.A.; Zarzuelo, A. Anti-inflammatory activity of diosmin and hesperidin in rat colitis induced by TNBS. *Planta Med.* **1999**, *65*, 651–653. [CrossRef]

122. Shalkami, A.S.; Hassan, M.; Bakr, A.G. Anti-inflammatory, antioxidant and anti-apoptotic activity of diosmin in acetic acid-induced ulcerative colitis. *Hum. Exp. Toxicol.* **2018**, *37*, 78–86. [CrossRef]

123. Lee, C.H.; Jeong, T.S.; Choi, Y.K.; Hyun, B.H.; Oh, G.T.; Kim, E.H.; Kim, J.R.; Han, J.I.; Bok, S.H. Anti-atherogenic effect of citrus flavonoids, naringin and naringenin, associated with hepatic ACAT and aortic VCAM-1 and MCP-1 in high cholesterol-fed rabbits. *Biochem. Biophys. Res. Commun.* **2001**, *284*, 681–688. [CrossRef] [PubMed]

124. Martinez, R.M.; Pinho-Ribeiro, F.A.; Steffen, V.S.; Caviglione, C.V.; Vignoli, J.A.; Barbosa, D.S.; Baracat, M.M.; Georgetti, S.R.; Verri, W.A., Jr.; Casagrande, R. Naringenin Inhibits UVB Irradiation-Induced Inflammation and Oxidative Stress in the Skin of Hairless Mice. *J. Nat. Prod.* **2015**, *78*, 1647–1655. [CrossRef] [PubMed]

125. Pinho-Ribeiro, F.A.; Zarpelon, A.C.; Mizokami, S.S.; Borghi, S.M.; Bordignon, J.; Silva, R.L.; Cunha, T.M.; Alves-Filho, J.C.; Cunha, F.Q.; Casagrande, R.; et al. The citrus flavonone naringenin reduces lipopolysaccharide-induced inflammatory pain and leukocyte recruitment by inhibiting NF-κB activation. *J. Nutr. Biochem.* **2016**, *33*, 8–14. [CrossRef] [PubMed]

126. Al-Rejaie, S.S.; Aleisa, A.M.; Abuohashish, H.M.; Parmar, M.Y.; Ola, M.S.; Al-Hosaini, A.A.; Ahmed, M.M. Naringenin neutralises oxidative stress and nerve growth factor discrepancy in experimental diabetic neuropathy. *Neurol. Res.* **2015**, *37*, 924–933. [CrossRef]

127. Oguido, A.; Hohmann, M.S.N.; Pinho-Ribeiro, F.A.; Crespigio, J.; Domiciano, T.P.; Verri, W.A., Jr.; Casella, A.M.B. Naringenin Eye Drops Inhibit Corneal Neovascularization by Anti-Inflammatory and Antioxidant Mechanisms. *Invest. Ophthalmol. Vis. Sci.* **2017**, *58*, 5764–5776. [CrossRef]

128. Zhang, B.; Wei, Y.Z.; Wang, G.Q.; Li, D.D.; Shi, J.S.; Zhang, F. Targeting MAPK Pathways by Naringenin Modulates Microglia M1/M2 Polarization in Lipopolysaccharide-Stimulated Cultures. *Front. Cell. Neurosci.* **2018**, *12*, 531. [CrossRef]

129. Bodet, C.; La, V.D.; Epifano, F.; Grenier, D. Naringenin has anti-inflammatory properties in macrophage and ex vivo human whole-blood models. *J. Periodontal. Res.* **2008**, *43*, 400–407. [CrossRef]

130. Nishimura Fde, C.; de Almeida, A.C.; Ratti, B.A.; Ueda-Nakamura, T.; Nakamura, C.V.; Ximenes, V.F.; Silva Sde, O. Antioxidant effects of quercetin and naringenin are associated with impaired neutrophil microbicidal activity. *Evid. Based Complement. Altern. Med.* **2013**, *2013*, 795916. [CrossRef]

131. Li, Y.R.; Chen, D.Y.; Chu, C.L.; Li, S.; Chen, Y.K.; Wu, C.L.; Lin, C.C. Naringenin inhibits dendritic cell maturation and has therapeutic effects in a murine model of collagen-induced arthritis. *J. Nutr. Biochem.* **2015**, *26*, 1467–1478. [CrossRef]

132. Manach, C.; Morand, C.; Gil-Izquierdo, A.; Bouteloup-Demange, C.; Remesy, C. Bioavailability in humans of the flavanones hesperidin and narirutin after the ingestion of two doses of orange juice. *Eur. J. Clin. Nutr.* **2003**, *57*, 235–242. [CrossRef]

133. Shi, X.; Liao, S.; Mi, H.; Guo, C.; Qi, D.; Li, F.; Zhang, C.; Yang, Z. Hesperidin prevents retinal and plasma abnormalities in streptozotocin-induced diabetic rats. *Molecules* **2012**, *17*, 12868–12881. [CrossRef] [PubMed]

134. Carballo-Villalobos, A.I.; Gonzalez-Trujano, M.E.; Alvarado-Vazquez, N.; Lopez-Munoz, F.J. Pro-inflammatory cytokines involvement in the hesperidin antihyperalgesic effects at peripheral and central levels in a neuropathic pain model. *Inflammopharmacology* **2017**, *25*, 265–269. [CrossRef] [PubMed]

135. Galati, E.M.; Monforte, M.T.; Kirjavainen, S.; Forestieri, A.M.; Trovato, A.; Tripodo, M.M. Biological effects of hesperidin, a citrus flavonoid. (Note I): Antiinflammatory and analgesic activity. *Farmaco* **1994**, *40*, 709–712. [PubMed]

136. Visnagri, A.; Kandhare, A.D.; Chakravarty, S.; Ghosh, P.; Bodhankar, S.L. Hesperidin, a flavanoglycone attenuates experimental diabetic neuropathy via modulation of cellular and biochemical marker to improve nerve functions. *Pharm. Biol.* **2014**, *52*, 814–828. [CrossRef] [PubMed]

137. Kaur, G.; Tirkey, N.; Chopra, K. Beneficial effect of hesperidin on lipopolysaccharide-induced hepatotoxicity. *Toxicology* **2006**, *226*, 152–160. [CrossRef]

138. Lee, N.K.; Choi, S.H.; Park, S.H.; Park, E.K.; Kim, D.H. Antiallergic activity of hesperidin is activated by intestinal microflora. *Pharmacology* **2004**, *71*, 174–180. [CrossRef]

139. Akiyama, S.; Katsumata, S.; Suzuki, K.; Ishimi, Y.; Wu, J.; Uehara, M. Dietary hesperidin exerts hypoglycemic and hypolipidemic effects in streptozotocin-induced marginal type 1 diabetic rats. *J. Clin. Biochem. Nutr.* **2010**, *46*, 87–92. [CrossRef]

140. Zielinska-Przyjemska, M.; Ignatowicz, E. Citrus fruit flavonoids influence on neutrophil apoptosis and oxidative metabolism. *Phytother. Res.* **2008**, *22*, 1557–1562. [CrossRef]

141. Choi, I.Y.; Kim, S.J.; Jeong, H.J.; Park, S.H.; Song, Y.S.; Lee, J.H.; Kang, T.H.; Park, J.H.; Hwang, G.S.; Lee, E.J.; et al. Hesperidin inhibits expression of hypoxia inducible factor-1 alpha and inflammatory cytokine production from mast cells. *Mol. Cell Biochem.* **2007**, *305*, 153–161. [CrossRef]

142. Gil-Izquierdo, A.; Gil, M.I.; Ferreres, F.; Tomas-Barberan, F.A. In vitro availability of flavonoids and other phenolics in orange juice. *J. Agric. Food Chem.* **2001**, *49*, 1035–1041. [CrossRef]

143. Martinez, R.M.; Pinho-Ribeiro, F.A.; Steffen, V.S.; Caviglione, C.V.; Vignoli, J.A.; Baracat, M.M.; Georgetti, S.R.; Verri, W.A., Jr.; Casagrande, R. Hesperidin methyl chalcone inhibits oxidative stress and inflammation in a mouse model of ultraviolet B irradiation-induced skin damage. *J. Photochem. Photobiol. B* **2015**, *148*, 145–153. [CrossRef] [PubMed]

144. Pinho-Ribeiro, F.A.; Hohmann, M.S.; Borghi, S.M.; Zarpelon, A.C.; Guazelli, C.F.; Manchope, M.F.; Casagrande, R.; Verri, W.A., Jr. Protective effects of the flavonoid hesperidin methyl chalcone in inflammation and pain in mice: Role of TRPV1, oxidative stress, cytokines and NF-κB. *Chem. Biol. Interact.* **2015**, *228*, 88–99. [CrossRef]

145. Singh, H.; Sidhu, S.; Chopra, K.; Khan, M.U. Hepatoprotective effect of trans-Chalcone on experimentally induced hepatic injury in rats: Inhibition of hepatic inflammation and fibrosis. *Can. J. Physiol. Pharm.* **2016**, *94*, 879–887. [CrossRef] [PubMed]

146. Bosch-Mola, M.; Homs, J.; Alvarez-Perez, B.; Puig, T.; Reina, F.; Verdu, E.; Boadas-Vaello, P. (−)-Epigallocatechin-3-Gallate Antihyperalgesic Effect Associates With Reduced CX3CL1 Chemokine Expression in Spinal Cord. *Phytother. Res.* **2017**, *31*, 340–344. [CrossRef] [PubMed]

147. Zhong, Y.; Chiou, Y.S.; Pan, M.H.; Shahidi, F. Anti-inflammatory activity of lipophilic epigallocatechin gallate (EGCG) derivatives in LPS-stimulated murine macrophages. *Food Chem.* **2012**, *134*, 742–748. [CrossRef]

148. Yoneyama, S.; Kawai, K.; Tsuno, N.H.; Okaji, Y.; Asakage, M.; Tsuchiya, T.; Yamada, J.; Sunami, E.; Osada, T.; Kitayama, J.; et al. Epigallocatechin gallate affects human dendritic cell differentiation and maturation. *J. Allergy Clin. Immunol.* **2008**, *121*, 209–214. [CrossRef] [PubMed]

149. Inoue, T.; Suzuki, Y.; Ra, C. Epigallocatechin-3-gallate inhibits mast cell degranulation, leukotriene C4 secretion, and calcium influx via mitochondrial calcium dysfunction. *Free Radic Biol Med.* **2010**, *49*, 632–640. [CrossRef]

150. Takano, K.; Nakaima, K.; Nitta, M.; Shibata, F.; Nakagawa, H. Inhibitory effect of (-)-epigallocatechin 3-gallate, a polyphenol of green tea, on neutrophil chemotaxis in vitro and in vivo. *J. Agric. Food Chem.* **2004**, *52*, 4571–4576. [CrossRef]

151. Leong, D.J.; Choudhury, M.; Hanstein, R.; Hirsh, D.M.; Kim, S.J.; Majeska, R.J.; Schaffler, M.B.; Hardin, J.A.; Spray, D.C.; Goldring, M.B.; et al. Green tea polyphenol treatment is chondroprotective, anti-inflammatory and palliative in a mouse post-traumatic osteoarthritis model. *Arthritis Res.* **2014**, *16*, 508. [CrossRef]

152. Li, Q.; Zhang, X. Epigallocatechin-3-gallate attenuates bone cancer pain involving decreasing spinal Tumor Necrosis Factor-alpha expression in a mouse model. *Int. Immunopharmacol.* **2015**, *29*, 818–823. [CrossRef]

153. Alvarez-Perez, B.; Homs, J.; Bosch-Mola, M.; Puig, T.; Reina, F.; Verdu, E.; Boadas-Vaello, P. Epigallocatechin-3-gallate treatment reduces thermal hyperalgesia after spinal cord injury by down-regulating RhoA expression in mice. *Eur. J. Pain* **2016**, *20*, 341–352. [CrossRef] [PubMed]

154. Panche, A.N.; Diwan, A.D.; Chandra, S.R. Flavonoids: An overview. *J. Nutr. Sci.* **2016**, *5*, e47. [CrossRef] [PubMed]

155. Bors, W.; Heller, W.; Michel, C.; Saran, M. Flavonoids as antioxidants: Determination of radical-scavenging efficiencies. *Methods Enzymol.* **1990**, *186*, 343–355. [PubMed]

156. Okawa, M.; Kinjo, J.; Nohara, T.; Ono, M. DPPH (1,1-diphenyl-2-picrylhydrazyl) radical scavenging activity of flavonoids obtained from some medicinal plants. *Biol. Pharm. Bull.* **2001**, *24*, 1202–1205. [CrossRef]

157. Chen, C.C.; Chow, M.P.; Huang, W.C.; Lin, Y.C.; Chang, Y.J. Flavonoids inhibit tumor necrosis factor-alpha-induced up-regulation of intercellular adhesion molecule-1 (ICAM-1) in respiratory epithelial cells through activator protein-1 and nuclear factor-κB: Structure-activity relationships. *Mol. Pharm.* **2004**, *66*, 683–693.

158. Martinez, R.M.; Pinho-Ribeiro, F.A.; Steffen, V.S.; Caviglione, C.V.; Fattori, V.; Bussmann, A.J.C.; Bottura, C.; Fonseca, M.J.V.; Vignoli, J.A.; Baracat, M.M.; et al. trans-Chalcone, a flavonoid precursor, inhibits UV-induced skin inflammation and oxidative stress in mice by targeting NADPH oxidase and cytokine production. *Photochem. Photobiol. Sci.* **2017**, *16*, 1162–1173. [CrossRef]

159. Heim, K.E.; Tagliaferro, A.R.; Bobilya, D.J. Flavonoid antioxidants: Chemistry, metabolism and structure-activity relationships. *J. Nutr. Biochem.* **2002**, *13*, 572–584. [CrossRef]

160. Mascolo, N.; Pinto, A.; Capasso, F. Flavonoids, leucocyte migration and eicosanoids. *J. Pharm. Pharm.* **1988**, *40*, 293–295. [CrossRef]

161. Rotelli, A.E.; Guardia, T.; Juarez, A.O.; de la Rocha, N.E.; Pelzer, L.E. Comparative study of flavonoids in experimental models of inflammation. *Pharm. Res.* **2003**, *48*, 601–606. [CrossRef]

162. Iacopini, P.; Baldi, M.; Storchi, P.; Sebastiani, L. Catechin, epicatechin, quercetin, rutin and resveratrol in red grape: Content, in vitro antioxidant activity and interactions. *J. Food Compos. Anal.* **2008**, *21*, 589–598. [CrossRef]

163. Szabo, C.; Ischiropoulos, H.; Radi, R. Peroxynitrite: Biochemistry, pathophysiology and development of therapeutics. *Nat. Rev. Drug Discov.* **2007**, *6*, 662–680. [CrossRef] [PubMed]

164. Pinho-Ribeiro, F.A.; Fattori, V.; Zarpelon, A.C.; Borghi, S.M.; Staurengo-Ferrari, L.; Carvalho, T.T.; Alves-Filho, J.C.; Cunha, F.Q.; Cunha, T.M.; Casagrande, R.; et al. Pyrrolidine dithiocarbamate inhibits superoxide anion-induced pain and inflammation in the paw skin and spinal cord by targeting NF-κB and oxidative stress. *Inflammopharmacology* **2016**, *24*, 97–107. [CrossRef] [PubMed]

165. Yamacita-Borin, F.Y.; Zarpelon, A.C.; Pinho-Ribeiro, F.A.; Fattori, V.; Alves-Filho, J.C.; Cunha, F.Q.; Cunha, T.M.; Casagrande, R.; Verri, W.A., Jr. Superoxide anion-induced pain and inflammation depends on TNFalpha/TNFR1 signaling in mice. *Neurosci. Lett.* **2015**, *605*, 53–58. [CrossRef] [PubMed]

166. Ndengele, M.M.; Cuzzocrea, S.; Esposito, E.; Mazzon, E.; Di Paola, R.; Matuschak, G.M.; Salvemini, D. Cyclooxygenases 1 and 2 contribute to peroxynitrite-mediated inflammatory pain hypersensitivity. *Faseb J.* **2008**, *22*, 3154–3164. [CrossRef]

167. Chen, J.W.; Zhu, Z.Q.; Hu, T.X.; Zhu, D.Y. Structure-activity relationship of natural flavonoids in hydroxyl radical-scavenging effects. *Acta Pharm. Sin.* **2002**, *23*, 667–672.

168. Cos, P.; Calomme, M.; Pieters, L.; Vlietinck, A.J.; Berghe, D.V. Structure-Activity Relationship of Flavonoids as Antioxidant and Pro-Oxidant Compounds. In *Studies in Natural Products Chemistry*; Atta ur, R., Ed.; Elsevier: Amsterdam, The Netherlands, 2000; Volume 22, pp. 307–341.

169. Staurengo-Ferrari, L.; Badaro-Garcia, S.; Hohmann, M.S.N.; Manchope, M.F.; Zaninelli, T.H.; Casagrande, R.; Verri, W.A., Jr. Contribution of Nrf2 Modulation to the Mechanism of Action of Analgesic and Anti-inflammatory Drugs in Pre-clinical and Clinical Stages. *Front. Pharm.* **2018**, *9*, 1536. [CrossRef]

170. Li, Y.-R.; Li, G.-H.; Zhou, M.-X.; Xiang, L.; Ren, D.-M.; Lou, H.-X.; Wang, X.-N.; Shen, T. Discovery of natural flavonoids as activators of Nrf2-mediated defense system: Structure-activity relationship and inhibition of intracellular oxidative insults. *Bioorg. Med. Chem.* **2018**, *26*, 5140–5150. [CrossRef]

171. Shin, S.Y.; Woo, Y.; Hyun, J.; Yong, Y.; Koh, D.; Lee, Y.H.; Lim, Y. Relationship between the structures of flavonoids and their NF-κB-dependent transcriptional activities. *Bioorg. Med. Chem. Lett.* **2011**, *21*, 6036–6041. [CrossRef]

172. Nicholas, C.; Batra, S.; Vargo, M.A.; Voss, O.H.; Gavrilin, M.A.; Wewers, M.D.; Guttridge, D.C.; Grotewold, E.; Doseff, A.I. Apigenin blocks lipopolysaccharide-induced lethality in vivo and proinflammatory cytokines expression by inactivating NF-κB through the suppression of p65 phosphorylation. *J. Immunol* **2007**, *179*, 7121–7127. [CrossRef]

173. Ziyan, L.; Yongmei, Z.; Nan, Z.; Ning, T.; Baolin, L. Evaluation of the anti-inflammatory activity of luteolin in experimental animal models. *Planta Med.* **2007**, *73*, 221–226. [CrossRef]

174. Lodhi, S.; Vadnere, G.P.; Patil, K.D.; Patil, T.P. Protective effects of luteolin on injury induced inflammation through reduction of tissue uric acid and pro-inflammatory cytokines in rats. *J. Tradit. Complement. Med.* **2020**, *10*, 60–69. [CrossRef] [PubMed]

175. Ribeiro, D.; Freitas, M.; Tome, S.M.; Silva, A.M.; Laufer, S.; Lima, J.L.; Fernandes, E. Flavonoids inhibit COX-1 and COX-2 enzymes and cytokine/chemokine production in human whole blood. *Inflammation* **2015**, *38*, 858–870. [CrossRef] [PubMed]

176. D'Mello, P.; Gadhwal, M.; Joshi, U.; Shetgiri, P.; Pharmacy, P. Mumbai, Modeling of COX-2 inhibotory activity of flavonoids. *Int. J. Pharm. Pharm. Sci.* **2011**, *3*, 33–40.

177. Hohmann, M.S.; Cardoso, R.D.; Fattori, V.; Arakawa, N.S.; Tomaz, J.C.; Lopes, N.P.; Casagrande, R.; Verri, W.A., Jr. Hypericum perforatum Reduces Paracetamol-Induced Hepatotoxicity and Lethality in Mice by Modulating Inflammation and Oxidative Stress. *Phytother Res.* **2015**, *29*, 1097–1101. [CrossRef]

178. Shanmugam, S.; Thangaraj, P.; Lima, B.D.S.; Chandran, R.; de Souza Araujo, A.A.; Narain, N.; Serafini, M.R.; Junior, L.J.Q. Effects of luteolin and quercetin 3-beta-d-glucoside identified from Passiflora subpeltata leaves against acetaminophen induced hepatotoxicity in rats. *Biomed. Pharm.* **2016**, *83*, 1278–1285. [CrossRef]

179. Ren, K.; Torres, R. Role of interleukin-1beta during pain and inflammation. *Brain Res. Rev.* **2009**, *60*, 57–64. [CrossRef]

180. Lim, H.; Min, D.S.; Park, H.; Kim, H.P. Flavonoids interfere with NLRP3 inflammasome activation. *Toxicol Appl. Pharm.* **2018**, *355*, 93–102. [CrossRef] [PubMed]

181. Ruiz-Miyazawa, K.W.; Borghi, S.M.; Pinho-Ribeiro, F.A.; Staurengo-Ferrari, L.; Fattori, V.; Fernandes, G.S.A.; Casella, A.M.; Alves-Filho, J.C.; Cunha, T.M.; Cunha, F.Q.; et al. The citrus flavanone naringenin reduces gout-induced joint pain and inflammation in mice by inhibiting the activation of NFκB and macrophage release of IL-1β. *J. Funct. Foods* **2018**, *48*, 106–116. [CrossRef]

182. Lu, H.; Yao, H.; Zou, R.; Chen, X.; Xu, H. Galangin Suppresses Renal Inflammation via the Inhibition of NF-κB, PI3K/AKT and NLRP3 in Uric Acid Treated NRK-52E Tubular Epithelial Cells. *Biomed. Res. Int.* **2019**, *2019*, 3018357. [CrossRef] [PubMed]

183. Vergadi, E.; Ieronymaki, E.; Lyroni, K.; Vaporidi, K.; Tsatsanis, C. Akt Signaling Pathway in Macrophage Activation and M1/M2 Polarization. *J. Immunol.* **2017**, *198*, 1006–1014. [CrossRef]

184. Schabbauer, G.; Tencati, M.; Pedersen, B.; Pawlinski, R.; Mackman, N. PI3K-Akt pathway suppresses coagulation and inflammation in endotoxemic mice. *Arter. Thromb Vasc. Biol.* **2004**, *24*, 1963–1969. [CrossRef] [PubMed]

185. Walker, E.H.; Pacold, M.E.; Perisic, O.; Stephens, L.; Hawkins, P.T.; Wymann, M.P.; Williams, R.L. Structural determinants of phosphoinositide 3-kinase inhibition by wortmannin, LY294002, quercetin, myricetin, and staurosporine. *Mol. Cell* **2000**, *6*, 909–919. [CrossRef]

186. Balunas, M.J.; Kinghorn, A.D. Drug discovery from medicinal plants. *Life Sci.* **2005**, *78*, 431–441. [CrossRef] [PubMed]

187. Fabricant, D.S.; Farnsworth, N.R. The value of plants used in traditional medicine for drug discovery. *Environ. Health Perspect.* **2001**, *109*, 69–75. [PubMed]

188. Viegas Jr, C.; Bolzani, V.D.S.; Barreiro, E.J. Os produtos naturais e a química medicinal moderna. *Química Nova* **2006**, *29*, 326–337. [CrossRef]

189. Newman, D.J.; Cragg, G.M.; Snader, K.M. Natural Products as Sources of New Drugs over the Period 1981–2002. *J. Nat. Prod.* **2003**, *66*, 1022–1037. [CrossRef]

190. Cassidy, A.; O'Reilly, É.J.; Kay, C.; Sampson, L.; Franz, M.; Forman, J.P.; Curhan, G.; Rimm, E.B. Habitual intake of flavonoid subclasses and incident hypertension in adults. *Am. J. Clin. Nutr.* **2011**, *93*, 338–347. [CrossRef]

191. Knekt, P.; Kumpulainen, J.; Järvinen, R.; Rissanen, H.; Heliövaara, M.; Reunanen, A.; Hakulinen, T.; Aromaa, A. Flavonoid intake and risk of chronic diseases. *Am. J. Clin. Nutr.* **2002**, *76*, 560–568. [CrossRef]

192. Devore, E.E.; Kang, J.H.; Breteler, M.M.B.; Grodstein, F. Dietary intakes of berries and flavonoids in relation to cognitive decline. *Ann. Neurol.* **2012**, *72*, 135–143. [CrossRef]

193. Gao, X.; Cassidy, A.; Schwarzschild, M.A.; Rimm, E.B.; Ascherio, A. Habitual intake of dietary flavonoids and risk of Parkinson disease. *Neurology* **2012**, *78*, 1138–1145. [CrossRef]

194. Hui, C.; Qi, X.; Qianyong, Z.; Xiaoli, P.; Jundong, Z.; Mantian, M. Flavonoids, Flavonoid Subclasses and Breast Cancer Risk: A Meta-Analysis of Epidemiologic Studies. *PLoS ONE* **2013**, *8*, e54318. [CrossRef] [PubMed]

195. Rizza, S.; Muniyappa, R.; Iantorno, M.; Kim, J.-A.; Chen, H.; Pullikotil, P.; Senese, N.; Tesauro, M.; Lauro, D.; Cardillo, C.; et al. Citrus polyphenol hesperidin stimulates production of nitric oxide in endothelial cells while improving endothelial function and reducing inflammatory markers in patients with metabolic syndrome. *J. Clin. Endocrinol. Metab.* **2011**, *96*, E782–E792. [CrossRef] [PubMed]

196. Cospite, M. Double-blind, placebo-controlled evaluation of clinical activity and safety of Daflon 500 mg in the treatment of acute hemorrhoids. *Angiology* **1994**, *45*, 566–573. [PubMed]

197. Godeberge, P. Daflon 500 mg in the treatment of hemorrhoidal disease: A demonstrated efficacy in comparison with placebo. *Angiology* **1994**, *45*, 574–578. [PubMed]

198. Meshikhes, A.-W.N. Efficacy of Daflon in the treatment of hemorrhoids. *Saudi Med. J.* **2002**, *23*, 1496–1498.

199. Bogdanski, P.; Suliburska, J.; Szulinska, M.; Stepien, M.; Pupek-Musialik, D.; Jablecka, A. Green tea extract reduces blood pressure, inflammatory biomarkers, and oxidative stress and improves parameters associated with insulin resistance in obese, hypertensive patients. *Nutr. Res.* **2012**, *32*, 421–427. [CrossRef]

200. Javadi, F.; Ahmadzadeh, A.; Eghtesadi, S.; Aryaeian, N.; Zabihiyeganeh, M.; Rahimi Foroushani, A.; Jazayeri, S. The Effect of Quercetin on Inflammatory Factors and Clinical Symptoms in Women with Rheumatoid Arthritis: A Double-Blind, Randomized Controlled Trial. *J. Am. Coll. Nutr.* **2017**, *36*, 9–15. [CrossRef]

201. Zargaran, A.; Borhani-Haghighi, A.; Salehi-Marzijarani, M.; Faridi, P.; Daneshamouz, S.; Azadi, A.; Sadeghpour, H.; Sakhteman, A.; Mohagheghzadeh, A. Evaluation of the effect of topical chamomile (Matricaria chamomilla L.) oleogel as pain relief in migraine without aura: A randomized, double-blind, placebo-controlled, crossover study. *Neurol. Sci.* **2018**, *39*, 1345–1353. [CrossRef]

202. Shavandi, M.; Moini, A.; Shakiba, Y.; Mashkorinia, A.; Dehghani, M.; Asar, S.; Kiani, A. Silymarin (Livergol®) Decreases Disease Activity Score in Patients with Rheumatoid Arthritis: A Non-randomized Single-arm Clinical Trial. *Iran. J. Allergyasthmaand Immunol.* **2017**, *16*, 99–106.

203. Feragalli, B.; Dugall, M.; Luzzi, R.; Ledda, A.; Hosoi, M.; Belcaro, G.; Cesarone, M.R. Pycnogenol®: Supplementary management of symptomatic osteoarthritis with a patch. An observational registry study. *Minerva Endocrinol.* **2018**, *44*, 97–101. [CrossRef]

204. Cisár, P.; Jány, R.; Waczulíková, I.; Sumegová, K.; Muchová, J.; Vojtaššák, J.; Ďuracková, Z.; Lisý, M.; Rohdewald, P. Effect of pine bark extract (Pycnogenol®) on symptoms of knee osteoarthritis. *Phytother. Res.* **2008**, *22*, 1087–1092. [CrossRef] [PubMed]

205. Farid, R.; Mirfeizi, Z.; Mirheidari, M.; Rezaieyazdi, Z.; Mansouri, H.; Esmaelli, H.; Zibadi, S.; Rohdewald, P.; Watson, R.R. Pycnogenol supplementation reduces pain and stiffness and improves physical function in adults with knee osteoarthritis. *Nutr. Res.* **2007**, *27*, 692–697. [CrossRef]

206. Awan, F.T.; Jones, J.A.; Maddocks, K.; Poi, M.; Grever, M.R.; Johnson, A.; Byrd, J.C.; Andritsos, L.A. A phase 1 clinical trial of flavopiridol consolidation in chronic lymphocytic leukemia patients following chemoimmunotherapy. *Ann. Hematol.* **2016**, *95*, 1137–1143. [CrossRef] [PubMed]

207. Hajialyani, M.; Hosein Farzaei, M.; Echeverria, J.; Nabavi, S.M.; Uriarte, E.; Sobarzo-Sanchez, E. Hesperidin as a Neuroprotective Agent: A Review of Animal and Clinical Evidence. *Molecules* **2019**, *24*, 648. [CrossRef]

208. Morand, C.; Dubray, C.; Milenkovic, D.; Lioger, D.; Martin, J.F.; Scalbert, A.; Mazur, A. Hesperidin contributes to the vascular protective effects of orange juice: A randomized crossover study in healthy volunteers. *Am. J. Clin. Nutr.* **2011**, *93*, 73–80. [CrossRef]

209. Kurowska, E.M.; Spence, J.D.; Jordan, J.; Wetmore, S.; Freeman, D.J.; Piché, L.A.; Serratore, P. HDL-cholesterol-raising effect of orange juice in subjects with hypercholesterolemia. *Am. J. Clin. Nutr.* **2000**, *72*, 1095–1100. [CrossRef]

210. Miwa, Y.; Yamada, M.; Sunayama, T.; Mitsuzumi, H.; Tsuzaki, Y.; Chaen, H.; Mishima, Y.; Kibata, M. Effects of Glucosyl Hesperidin on Serum Lipids in Hyperlipidemic Subjects: Preferential Reduction in Elevated Serum Triglyceride Level. *J. Nutr. Sci. Vitaminol.* **2004**, *50*, 211–218. [CrossRef]

211. Amato, C. Advantage of a micronized flavonoidic fraction (Daflon 500 mg) in comparison with a nonmicronized diosmin. *Angiology* **1994**, *45*, 531–536.

212. Laurent, R.; Gilly, R.; Frileux, C. Clinical evaluation of a venotropic drug in man. Example of Daflon 500 mg. *Int. Angiol.* **1988**, *7*, 39–43.

213. Geroulakos, G.; Nicolaides, A.N. Controlled studies of Daflon 500 mg in chronic venous insufficiency. *Angiology* **1994**, *45*, 549–553.

214. Kusunoki, H.; Haruma, K.; Hata, J.; Ishii, M.; Kamada, T.; Yamashita, N.; Honda, K.; Inoue, K.; Imamura, H.; Manabe, N.; et al. Efficacy of Rikkunshito, a Traditional Japanese Medicine (Kampo), in Treating Functional Dyspepsia. *Intern. Med.* **2010**, *49*, 2195–2202. [CrossRef] [PubMed]

215. Arai, M.; Matsumura, T.; Tsuchiya, N.; Sadakane, C.; Inami, R.; Suzuki, T.; Yoshikawa, M.; Imazeki, F.; Yokosuka, O. Rikkunshito Improves the Symptoms in Patients with Functional Dyspepsia, Accompanied by an Increase in the Level of Plasma Ghrelin. *Hepatogastroenterology* **2012**, *59*, 62–66. [CrossRef] [PubMed]

216. Fujitsuka, N.; Asakawa, A.; Uezono, Y.; Minami, K.; Yamaguchi, T.; Niijima, A.; Yada, T.; Maejima, Y.; Sedbazar, U.; Sakai, T.; et al. Potentiation of ghrelin signaling attenuates cancer anorexia–cachexia and prolongs survival. *Transl. Psychiatry* **2011**, *1*, e23. [CrossRef] [PubMed]

217. Ohno, T.; Yanai, M.; Ando, H.; Toyomasu, Y.; Ogawa, A.; Morita, H.; Ogata, K.; Mochiki, E.; Asao, T.; Kuwano, H. Rikkunshito, a traditional Japanese medicine, suppresses cisplatin-induced anorexia in humans. *Clin. Exp. Gastroenterol.* **2011**, *4*, 291. [CrossRef]

218. Takiguchi, S.; Hiura, Y.; Takahashi, T.; Kurokawa, Y.; Yamasaki, M.; Nakajima, K.; Miyata, H.; Mori, M.; Hosoda, H.; Kangawa, K.; et al. Effect of rikkunshito, a Japanese herbal medicine, on gastrointestinal symptoms and ghrelin levels in gastric cancer patients after gastrectomy. *Gastric Cancer* **2013**, *16*, 167–174. [CrossRef]

219. Boyle, P.; Diehm, C.; Robertson, C. Meta-analysis of clinical trials of Cyclo 3 Fort in the treatment of chronic venous insufficiency. *Int. Angiol.* **2003**, *22*, 250.

220. Kakkos, S.K.; Allaert, F.A. Efficacy of Ruscus extract, HMC and vitamin C, constituents of Cyclo 3 fort®, on improving individual venous symptoms and edema: A systematic review and meta-analysis of randomized double-blind placebo-controlled trials. *Int. Angiol.* **2017**, *36*, 93–106.

221. Matsui, T. Condensed catechins and their potential health-benefits. *Eur. J. Pharmacol.* **2015**, *765*, 495–502. [CrossRef]

222. Braicu, C.; Ladomery, M.R.; Chedea, V.S.; Irimie, A.; Berindan-Neagoe, I. The relationship between the structure and biological actions of green tea catechins. *Food Chem.* **2013**, *141*, 3282–3289. [CrossRef]

223. Chen, X.-Q.; Hu, T.; Han, Y.; Huang, W.; Yuan, H.-B.; Zhang, Y.-T.; Du, Y.; Jiang, Y.-W. Preventive Effects of Catechins on Cardiovascular Disease. *Molecules* **2016**, *21*, 1759. [CrossRef]

224. Mandel, S.; Weinreb, O.; Amit, T.; Youdim, M.B.H. Cell signaling pathways in the neuroprotective actions of the green tea polyphenol (-)-epigallocatechin-3-gallate: Implications for neurodegenerative diseases. *J. Neurochem.* **2004**, *88*, 1555–1569. [CrossRef]

225. Isbrucker, R.A.; Edwards, J.A.; Wolz, E.; Davidovich, A.; Bausch, J. Safety studies on epigallocatechin gallate (EGCG) preparations. Part 2: Dermal, acute and short-term toxicity studies. *Food Chem. Toxicol.* **2006**, *44*, 636–650. [CrossRef]

226. Babu, P.V.A.; Liu, D.; Liu, D. Green tea catechins and cardiovascular health: An update. *Curr. Med. Chem.* **2008**, *15*, 1840–1850. [CrossRef] [PubMed]

227. Ohishi, T.; Goto, S.; Monira, P.; Isemura, M.; Nakamura, Y. Anti-inflammatory Action of Green Tea. *Anti-Inflamm. Anti-Allergy Agents Med. Chem.* **2016**, *15*, 74–90. [CrossRef] [PubMed]

228. Bettuzzi, S.; Brausi, M.; Rizzi, F.; Castagnetti, G.; Peracchia, G.; Corti, A. Chemoprevention of Human Prostate Cancer by Oral Administration of Green Tea Catechins in Volunteers with High-Grade Prostate Intraepithelial Neoplasia: A Preliminary Report from a One-Year Proof-of-Principle Study. *Cancer Res.* **2006**, *66*, 1234–1240. [CrossRef] [PubMed]

229. Jian, L.; Xie, L.P.; Lee, A.H.; Binns, C.W. Protective effect of green tea against prostate cancer: A case-control study in southeast China. *Int. J. Cancer* **2004**, *108*, 130–135. [CrossRef]

230. Kurahashi, N.; Sasazuki, S.; Iwasaki, M.; Inoue, M.; Tsugane, S.; Group, J.S. Green Tea Consumption and Prostate Cancer Risk in Japanese Men: A Prospective Study. *Am. J. Epidemiol.* **2007**, *167*, 71–77. [CrossRef]

231. Brausi, M.; Rizzi, F.; Bettuzzi, S. Chemoprevention of Human Prostate Cancer by Green Tea Catechins: Two Years Later. A Follow-up Update. *Eur. Urol.* **2008**, *54*, 472–473. [CrossRef]

232. Bonner, M.R.; Rothman, N.; Mumford, J.L.; He, X.; Shen, M.; Welch, R.; Yeager, M.; Chanock, S.; Caporaso, N.; Lan, Q. Green tea consumption, genetic susceptibility, PAH-rich smoky coal, and the risk of lung cancer. *Mutat. Res. Toxicol. Environ. Mutagen.* **2005**, *582*, 53–60. [CrossRef]

233. Ohno, Y.; Wakai, K.; Genka, K.; Ohmine, K.; Kawamura, T.; Tamakoshi, A.; Aoki, R.; Senda, M.; Hayashi, Y.; Nagao, K.; et al. Tea Consumption and Lung Cancer Risk: A Case-Control Study in Okinawa, Japan. *Jpn. J. Cancer Res.* **1995**, *86*, 1027–1034. [CrossRef]

234. Yang, C.S.; Lambert, J.D.; Ju, J.; Lu, G.; Sang, S. Tea and cancer prevention: Molecular mechanisms and human relevance. *Toxicol. Appl. Pharmacol.* **2007**, *224*, 265–273. [CrossRef] [PubMed]

235. Nagao, T.; Meguro, S.; Hase, T.; Otsuka, K.; Komikado, M.; Tokimitsu, I.; Yamamoto, T.; Yamamoto, K. A Catechin-rich Beverage Improves Obesity and Blood Glucose Control in Patients With Type 2 Diabetes. *Obesity* **2009**, *17*, 310–317. [CrossRef] [PubMed]

236. Ide, K.; Matsuoka, N.; Yamada, H.; Furushima, D.; Kawakami, K. Effects of Tea Catechins on Alzheimer's Disease: Recent Updates and Perspectives. *Molecules* **2018**, *23*, 2357. [CrossRef] [PubMed]

237. Shen, C.-L.; Chyu, M.-C. Tea flavonoids for bone health: From animals to humans. *J. Investig. Med.* **2016**, *64*, 1151–1157. [CrossRef]

238. Egert, S.; Bosy-Westphal, A.; Seiberl, J.; Kürbitz, C.; Settler, U.; Plachta-Danielzik, S.; Wagner, A.E.; Frank, J.; Schrezenmeir, J.; Rimbach, G.; et al. Quercetin reduces systolic blood pressure and plasma oxidised low-density lipoprotein concentrations in overweight subjects with a high-cardiovascular disease risk phenotype: A double-blinded, placebo-controlled cross-over study. *Br. J. Nutr.* **2009**, *102*, 1065–1074. [CrossRef]

239. Edwards, R.L.; Lyon, T.; Litwin, S.E.; Rabovsky, A.; Symons, J.D.; Jalili, T. Quercetin Reduces Blood Pressure in Hypertensive Subjects. *J. Nutr.* **2007**, *137*, 2405–2411. [CrossRef]

240. Zahedi, M.; Ghiasvand, R.; Feizi, A.; Asgari, G.; Darvish, L. Does Quercetin Improve Cardiovascular Risk factors and Inflammatory Biomarkers in Women with Type 2 Diabetes: A Double-blind Randomized Controlled Clinical Trial. *Int. J. Prev. Med.* **2013**, *4*, 777–785.

241. Serban, M.; Sahebkar, A.; Zanchetti, A.; Mikhailidis, D.P.; Howard, G.; Antal, D.; Andrica, F.; Ahmed, A.; Aronow, W.S.; Muntner, P.; et al. Effects of Quercetin on Blood Pressure: A Systematic Review and Meta-Analysis of Randomized Controlled Trials. *J. Am. Hear. Assoc.* **2016**, *5*, e002713.

242. Mohammadi-Sartang, M.; Mazloom, Z.; Sherafatmanesh, S.; Ghorbani, M.; Firoozi, D. Effects of supplementation with quercetin on plasma C-reactive protein concentrations: A systematic review and meta-analysis of randomized controlled trials. *Eur. J. Clin. Nutr.* **2017**, *71*, 1033–1039. [CrossRef]

243. Ross, J.A.; Kasum, C.M. DIETARY FLAVONOIDS: Bioavailability, Metabolic Effects, and Safety. *Annu. Rev. Nutr.* **2002**, *22*, 19–34. [CrossRef]

244. Ross, S.M. Generalized Anxiety Disorder (GAD). *Holist. Nurs. Pract.* **2013**, *27*, 366–368. [CrossRef] [PubMed]

245. Choi, S.; Youn, J.; Kim, K.; Joo, D.H.; Shin, S.; Lee, J.; Lee, H.K.; An, I.-S.; Kwon, S.; Youn, H.J.; et al. Apigenin inhibits UVA-induced cytotoxicity in vitro and prevents signs of skin aging in vivo. *Int. J. Mol. Med.* **2016**, *38*, 627–634. [CrossRef]

246. Hoensch, H.; Groh, B.; Edler, L.; Kirch, W. Prospective cohort comparison of flavonoid treatment in patients with resected colorectal cancer to prevent recurrence. *World J. Gastroenterol.* **2008**, *14*, 2187–2193. [CrossRef]

247. Vargas-Mendoza, N.; Madrigal-Santillán, E.; Morales-González, Á.; Esquivel-Soto, J.; Esquivel-Chirino, C.; González-Rubio, M.G.-L.Y.; Gayosso-De-Lucio, J.A.; Morales-González, J.A. Hepatoprotective effect of silymarin. *World J. Hepatol.* **2014**, *6*, 144–149. [CrossRef] [PubMed]

248. Zhong, S.; Fan, Y.; Yan, Q.; Fan, X.; Wu, B.; Han, Y.; Zhang, Y.; Chen, Y.; Zhang, H.; Niu, J. The therapeutic effect of silymarin in the treatment of nonalcoholic fatty disease: A meta-analysis (PRISMA) of randomized control trials. *Medicine* **2017**, *96*, e9061. [CrossRef] [PubMed]

249. D'Andrea, G. Pycnogenol: A blend of procyanidins with multifaceted therapeutic applications? *Fitoterapia* **2010**, *81*, 724–736. [CrossRef] [PubMed]

250. Belcaro, G.; Cesarone, M.R.; Errichi, S.; Zulli, C.; Errichi, B.M.; Vinciguerra, G.; Ledda, A.; Di Renzo, A.; Stuard, S.; Dugall, M.; et al. Variations in C-reactive protein, plasma free radicals and fibrinogen values in patients with osteoarthritis treated with Pycnogenol®. *Redox Rep.* **2008**, *13*, 271–276. [CrossRef]

251. Zeidner, J.F.; Karp, J.E. Clinical activity of alvocidib (flavopiridol) in acute myeloid leukemia. *Leuk. Res.* **2015**, *39*, 1312–1318. [CrossRef]

252. Pipingas, A.; Silberstein, R.B.; Vitetta, L.; Rooy, C.V.; Harris, E.V.; Young, J.M.; Frampton, C.M.; Sali, A.; Nastasi, J. Improved cognitive performance after dietary supplementation with a *Pinus radiata* bark extract Formulation. *Phytother. Res.* **2008**, *22*, 1168–1174. [CrossRef]

253. Theadom, A.; Mahon, S.; Barker-Collo, S.; McPherson, K.; Rush, E.; Vandal, A.C.; Feigin, V.L. Enzogenol for cognitive functioning in traumatic brain injury: A pilot placebo-controlled RCT. *Eur. J. Neurol.* **2013**, *20*, 1135–1144. [CrossRef]

254. Shand, B.; Strey, C.; Scott, R.; Morrison, Z.; Gieseg, S. Pilot study on the clinical effects of dietary supplementation With Enzogenol®, a flavonoid extract of pine bark and vitamin C. *Phytother. Res.* **2003**, *17*, 490–494. [CrossRef] [PubMed]

255. Christie, S.; Walker, A.F.; Hicks, S.M.; Abeyasekera, S. Flavonoid supplement improves leg health and reduces fluid retention in pre-menopausal women in a double-blind, placebo-controlled study. *Phytomedicine* **2004**, *11*, 11–17. [CrossRef] [PubMed]

256. Egert, S.; Rimbach, G. Which Sources of Flavonoids: Complex Diets or Dietary Supplements? *Adv. Nutr.* **2011**, *2*, 8–14. [CrossRef] [PubMed]

257. Prochaska, H.J.S.; Scotto, K.W. Use of flavonoids to treat multidrug resistant cancer cells. U.S. Patent 5,336,685, 9 August 1994.

258. Mira, L.; Fernandez, M.T.; Santos, M.; Rocha, R.; Florêncio, M.H.; Jennings, K.R. Interactions of flavonoids with iron and copper ions: A mechanism for their antioxidant activity. *Free Radic. Res.* **2002**, *36*, 1199–1208. [CrossRef] [PubMed]

259. Calderon-Montano, J.M.; Burgos-Moron, E.; Perez-Guerrero, C.; Lopez-Lazaro, M. A review on the dietary flavonoid kaempferol. *Mini-Rev. Med. Chem.* **2011**, *11*, 298–344. [CrossRef] [PubMed]

260. Soleimani, V.; Delghandi, P.S.; Moallem, S.A.; Karimi, G. Safety and toxicity of silymarin, the major constituent of milk thistle extract: An updated review. *Phytother. Res.* **2019**, *33*, 1627–1638. [CrossRef]

261. Nagula, R.L.; Wairkar, S. Recent advances in topical delivery of flavonoids: A review. *J. Control. Release* **2019**, *296*, 190–201. [CrossRef]

262. Lauro, M.R.; Torre, M.L.; Maggi, L.; De Simone, F.; Conte, U.; Aquino, R.P. Fast- and slow-release tablets for oral administration of flavonoids: Rutin and quercetin. *Drug Dev. Ind. Pharm.* **2002**, *28*, 371–379. [CrossRef]

263. Martinez, R.M.; Pinho-Ribeiro, F.A.; Vale, D.L.; Steffen, V.S.; Vicentini, F.; Vignoli, J.A.; Baracat, M.M.; Georgetti, S.R.; Verri, W.A., Jr.; Casagrande, R. Trans-chalcone added in topical formulation inhibits skin inflammation and oxidative stress in a model of ultraviolet B radiation skin damage in hairless mice. *J. Photochem. Photobiol B* **2017**, *171*, 139–146. [CrossRef]

264. Gokhale, J.P.; Mahajan, H.S.; Surana, S.J. Quercetin loaded nanoemulsion-based gel for rheumatoid arthritis: In vivo and in vitro studies. *Biomed. Pharm.* **2019**, *112*, 108622. [CrossRef]

265. Muller, R.H.; Schmidt, S.; Buttle, I.; Akkar, A.; Schmitt, J.; Bromer, S. SolEmuls-novel technology for the formulation of i.v. emulsions with poorly soluble drugs. *Int. J. Pharm.* **2004**, *269*, 293–302. [CrossRef] [PubMed]

266. De Medeiros, D.C.; Mizokami, S.S.; Sfeir, N.; Georgetti, S.R.; Urbano, A.; Casagrande, R.; Verri, W.A.; Baracat, M.M. Preclinical Evaluation of Rutin-Loaded Microparticles with an Enhanced Analgesic Effect. *ACS Omega* **2019**, *4*, 1221–1227. [CrossRef]

267. Amanzadeh, E.; Esmaeili, A.; Abadi, R.E.N.; Kazemipour, N.; Pahlevanneshan, Z.; Beheshti, S. Quercetin conjugated with superparamagnetic iron oxide nanoparticles improves learning and memory better than free quercetin via interacting with proteins involved in LTP. *Sci. Rep.* **2019**, *9*, 6876. [CrossRef] [PubMed]

268. Gera, S.; Talluri, S.; Rangaraj, N.; Sampathi, S. Formulation and Evaluation of Naringenin Nanosuspensions for Bioavailability Enhancement. *AAPS Pharmscitech.* **2017**, *18*, 3151–3162. [CrossRef]

269. Sechi, M.; Syed, D.N.; Pala, N.; Mariani, A.; Marceddu, S.; Brunetti, A.; Mukhtar, H.; Sanna, V. Nanoencapsulation of dietary flavonoid fisetin: Formulation and in vitro antioxidant and alpha-glucosidase inhibition activities. *Mater. Sci Eng. C Mater. Biol. Appl.* **2016**, *68*, 594–602. [CrossRef]

270. Khan, A.W.; Kotta, S.; Ansari, S.H.; Sharma, R.K.; Ali, J. Self-nanoemulsifying drug delivery system (SNEDDS) of the poorly water-soluble grapefruit flavonoid Naringenin: Design, characterization, in vitro and in vivo evaluation. *Drug Deliv.* **2015**, *22*, 552–561. [CrossRef]

271. Tran, T.H.; Guo, Y.; Song, D.; Bruno, R.S.; Lu, X. Quercetin-containing self-nanoemulsifying drug delivery system for improving oral bioavailability. *J. Pharm. Sci.* **2014**, *103*, 840–852. [CrossRef]

272. Sharma, S.; Narang, J.K.; Ali, J.; Baboota, S. Synergistic antioxidant action of vitamin E and rutin SNEDDS in ameliorating oxidative stress in a Parkinson's disease model. *Nanotechnology* **2016**, *27*, 375101. [CrossRef]

273. Macedo, A.S.; Quelhas, S.; Silva, A.M.; Souto, E.B. Nanoemulsions for delivery of flavonoids: Formulation and in vitro release of rutin as model drug. *Pharm. Dev. Technol.* **2014**, *19*, 677–680. [CrossRef]

274. Bonechi, C.; Donati, A.; Tamasi, G.; Leone, G.; Consumi, M.; Rossi, C.; Lamponi, S.; Magnani, A. Protective effect of quercetin and rutin encapsulated liposomes on induced oxidative stress. *Biophys. Chem.* **2018**, *233*, 55–63. [CrossRef]

275. Huang, M.; Su, E.; Zheng, F.; Tan, C. Encapsulation of flavonoids in liposomal delivery systems: The case of quercetin, kaempferol and luteolin. *Food Funct.* **2017**, *8*, 3198–3208. [CrossRef]

276. Singh, D.; Rawat, M.S.; Semalty, A.; Semalty, M. Quercetin-phospholipid complex: An amorphous pharmaceutical system in herbal drug delivery. *Curr. Drug Discov. Technol.* **2012**, *9*, 17–24. [CrossRef]

277. Singh, D.; Rawat, M.S.; Semalty, A.; Semalty, M. Rutin-phospholipid complex: An innovative technique in novel drug delivery system- NDDS. *Curr. Drug Deliv.* **2012**, *9*, 305–314. [CrossRef]

278. Yang, L.J.; Ma, S.X.; Zhou, S.Y.; Chen, W.; Yuan, M.W.; Yin, Y.Q.; Yang, X.D. Preparation and characterization of inclusion complexes of naringenin with beta-cyclodextrin or its derivative. *Carbohydr. Polym.* **2013**, *98*, 861–869. [CrossRef]

279. Patra, A.; Satpathy, S.; Shenoy, A.K.; Bush, J.A.; Kazi, M.; Hussain, M.D. Formulation and evaluation of mixed polymeric micelles of quercetin for treatment of breast, ovarian, and multidrug resistant cancers. *Int. J. Nanomed.* **2018**, *13*, 2869–2881. [CrossRef] [PubMed]

280. Guo, R.; Wei, P. Studies on the antioxidant effect of rutin in the microenvironment of cationic micelles. *Microchim. Acta* **2008**, *161*, 233–239. [CrossRef]

281. Li, B.; Liu, H.; Amin, M.; Wegiel, L.A.; Taylor, L.S.; Edgar, K.J. Enhancement of naringenin solution concentration by solid dispersion in cellulose derivative matrices. *Cellulose* **2013**, *20*, 2137–2149. [CrossRef]

282. Kumar, S.; Pandey, A.K. Chemistry and Biological Activities of Flavonoids: An Overview. *Sci. World J.* **2013**, *2013*, 1–16. [CrossRef] [PubMed]

283. Manach, C.; Williamson, G.; Morand, C.; Scalbert, A.; Remesy, C. Bioavailability and bioefficacy of polyphenols in humans. I. Review of 97 bioavailability studies. *Am. J. Clin. Nutr.* **2005**, *81*, 230S–242S. [CrossRef]

284. Thilakarathna, S.H.; Rupasinghe, H.P. Flavonoid bioavailability and attempts for bioavailability enhancement. *Nutrients* **2013**, *5*, 3367–3387. [CrossRef]

285. Wang, S.; Yang, C.; Tu, H.; Zhou, J.; Liu, X.; Cheng, Y.; Luo, J.; Deng, X.; Zhang, H.; Xu, J. Characterization and Metabolic Diversity of Flavonoids in Citrus Species. *Sci. Rep.* **2017**, *7*, 10549. [CrossRef]

Article

Potential Therapeutic Anti-Inflammatory and Immunomodulatory Effects of Dihydroflavones, Flavones, and Flavonols

Cristina Zaragozá [1,*], Lucinda Villaescusa [1], Jorge Monserrat [2], Francisco Zaragozá [1] and Melchor Álvarez-Mon [2,3,4,5]

[1] Biomedical Sciences Department, Pharmacology Unit, University of Alcalá, Alcalá de Henares, 28871 Madrid, Spain; lucinda.villaescusa@uah.es (L.V.); francisco.zaragoza@uah.es (F.Z.)

[2] Laboratory of Immune System Diseases and Oncology, Department of Medicine and Medical Specialties, University of Alcalá, Alcala de Henares, 28871 Madrid, Spain; jorge.monserrat@uah.es (J.M.); mademons@gmail.com (M.Á.-M.)

[3] Immune System Diseases and Oncology Service, University Hospital "Príncipe de Asturias", Alcalá de Henares, 28871 Madrid, Spain

[4] Biomedical Research Networking Center in Hepatic and Digestive Diseases (CIBERehd), 28871 Madrid, Spain

[5] Ramón y Cajal Health Research Institute (IRYCIS), 28034 Madrid, Spain

* Correspondence: cristina.zaragoza@uah.es

Received: 21 January 2020; Accepted: 22 February 2020; Published: 24 February 2020

Abstract: Systemic inflammation, circulating immune cell activation, and endothelial cell damage play a critical role in vascular pathogenesis. Flavonoids have shown anti-inflammatory effects. In this study, we investigated the effects of different flavonoids on the production of pro-inflammatory interleukin (IL) 1β, 6, and 8, and tumor necrosis factor α (TNF-α), in peripheral blood cells. Methods: We studied the whole blood from 36 healthy donors. Lipopolysaccharide (LPS)-stimulated (0.5 µg/mL) whole-blood aliquots were incubated in the presence or absence of different concentrations of quercetin, rutin, naringenin, naringin, diosmetin, and diosmin for 6 h. Cultures were centrifuged and the supernatant was collected in order to measure IL-1β, TNF-α, IL-6, and IL-8 production using specific immunoassay techniques. This production was significantly inhibited by quercetin, naringenin, naringin, and diosmetin, but in no case by rutin or diosmin. Flavonoids exert different effects, maybe due to the differences between aglycons and glucosides present in their chemical structures. However, these studies suggest that quercetin, naringenin, naringin, and diosmetin could have a potential therapeutic effect in the inflammatory process of cardiovascular disease.

Keywords: flavonoids; aglycons; glycosides; IL-1β; TNF-α; IL-6; IL-8; pro-inflammatory cytokines

---

1. Introduction

Cardiovascular diseases are the number one cause of death in the world [1]. Nowadays, the important role of inflammatory events in cardiovascular diseases is widely known [2]. Systemic inflammation, circulating immune cell activation, and endothelial cell damage are critical events, along with arterial wall damage [3–6]. Furthermore, the relevance of inflammation in the pathogenesis of chronic venous disorder has also been shown [7,8]. Monocytes play a critical role in the inflammatory response [9]. Activated monocytes display relevant immunomodulatory activities, including the secretion of pivotal cytokines, such as pro-inflammatory cytokines interleukin (IL)-6, IL-1β, IL-8, and tumor necrosis factor α (TNF-α). Different mechanisms may be involved in the abnormal activation of monocytes in chronic diseases [10]. There is increasing evidence that gut dysbiosis; increased intestinal permeability, also known as "leaky gut"; and bacterial translocation, are key mechanisms in the induction of the systemic increase of lipopolysaccharide (LPS) and pro-inflammatory monocyte activation in several chronic diseases [11–13]. LPS is a potent signal for monocyte stimulation [14]. Monocytes have been demonstrated to be involved in the pathogenesis of several cardiovascular diseases, such as atherosclerosis [15].

Flavonoids are the largest group of naturally occurring polyphenolic compounds, present in almost all parts of flowering plants [16]. Their basic chemical structure is made up of two aromatic rings (rings A and B) connected by an oxygen-containing pyran ring (ring C). Flavonoids are classified into different subgroups, including flavones, isoflavones, flavonols, dihydroflavones, flavanes, chalcones, and anthocyanidins, according to the hydroxylation pattern of rings A and B, the oxidative degree of ring C, and the structure and position of the substitutions [17].

The structural diversity of flavonoids results in a wide range of biological effects; the different substitutions on the carbon atoms determine the biological effects of the flavonoids [18]. It has been reported that flavones show different pharmacological actions, such as antinociceptive [19], anti-inflammatory [20], antioxidant [21], antiulcerogenic [22], and anticarcinogenic [23] actions. The combination of multiple pharmacological properties in a single nucleus is quite interesting [16]. Several mechanisms have been shown to be involved in the anti-inflammatory effects of flavonoids. Flavonoids have the ability to modulate macrophages from pro- to anti-inflammatory phenotypes, potentially contributing to the resolution of pre-established inflammatory processes [24,25]. Furthermore, flavonoids have inhibitory effects in platelet activation, being critical cells for vascular inflammation [26].

The therapeutic activity of flavonoids has been suggested for inflammatory diseases such as cardiovascular diseases, obesity, diabetes, bone health, and asthma [27–31]. Furthermore, the potential use of these molecules in the adjuvant treatment of cancer has also been suggested [25,32].

Flavonoids are secondary metabolites in plants, occurring in virtually all plant parts; especially photosynthesizing plant cells [33]. Polyphenols represent the most abundant compounds among secondary metabolites produced by plants. Naringenin, diosmetin, and quercetin are the aglycons of different kinds of flavonoids found in their heteroside form, specifically naringin, diosmin, and rutin. Naringin and its aglycon naringenin are the most important dihydroflavones that have been isolated from citrus fruits [34]. Quercetin and rutin are two flavonols widely distributed among plants and commonly found in daily diets, predominantly in fruits and vegetables [35,36]. On the other hand, diosmin and diosmetin are two flavones found in various dietary sources, such as oregano spice; oregano leaves; citrus fruits; and extracts from specific medicinal herbs of Rosaceae, Asteraceae, Brassicaceae, and Caryophyllaceae [37].

A flavonoid's biological in vivo activity is very dependent on its bioavailability and this is determined by the chemical structure, mainly the type of sugar moiety. In general, the glycoside levels in plasma are low. Deglycosylation occurs both in the small intestine and in the large intestine, depending on the type of sugar moiety and the aglycon produced by the microbiota. Then, metabolites are absorbed via the large intestine and transported into the circulation [38].

It is possible to hypothesize that flavonoids might inhibit the production of pro-inflammatory cytokines through the activation of leukocytes and focusing on those that are mainly secreted by activated monocytes. The aim of this work was to evaluate the immunomodulatory effect of quercetin, rutin, naringenin, naringin, diosmetin, and diosmin on the production of pro-inflammatory cytokines TNF-α, IL-1β, IL-6, and IL-8 in whole-blood cells stimulated by LPS.

2. Results

2.1. Time Course Cytokine Production Curves

Firstly, the kinetic of cytokines produced by lipopolysaccharide (LPS)-stimulated whole blood was investigated (Figure 1). The culture medium concentration of IL-1β, TNF-α, IL-6, and IL-8 at basal conditions and after 4, 6, 8, and 24 h was quantified. The maximum cytokine concentration was found at 6h of culture and this time was chosen as the time condition for subsequent assays.

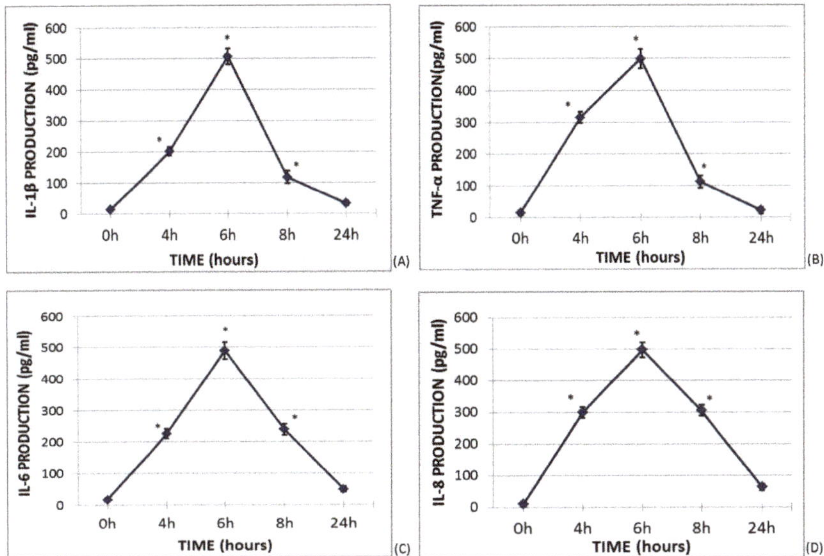

Figure 1. Cytokine production curves in lipopolysaccharide (LPS)-stimulated whole blood. One milliliter of whole blood was incubated in the presence of LPS (0.5 μg/mL) in darkness and continuously shaken at 37 °C. The supernatant concentrations of interleukin (IL)-1β, tumor necrosis factor α (TNF-α), IL-6, and IL-8 were measured by ELISA at 0, 4, 6, 8, and 24 h of culture. The diamonds and vertical segments represent the mean ± SEM of six different donors. * $p < 0.05$ represents a significant difference each time the measurement was compared to baseline production.

2.2. Study of the Effects of Flavonoids in Cytokine Production in LPS-Stimulated Whole Blood

The effects of quercetin, rutin, naringenin, naringin, diosmetin, and diosmin on IL-1β production in LPS-stimulated whole blood (Figure 2) were investigated. As a control, the inhibitor of IL-1β rhein (diacerhein-derived metabolite, inhibitor of IL-1β production) was used [39]. It was found that quercetin, diosmetin, and naringin significantly reduced IL-1β production in a dose-dependent manner. Naringenin also significantly inhibited IL-1β production, but in an inverse dose-dependent manner. In contrast, rutin and diosmin did not modify IL-1β production in LPS-stimulated whole blood.

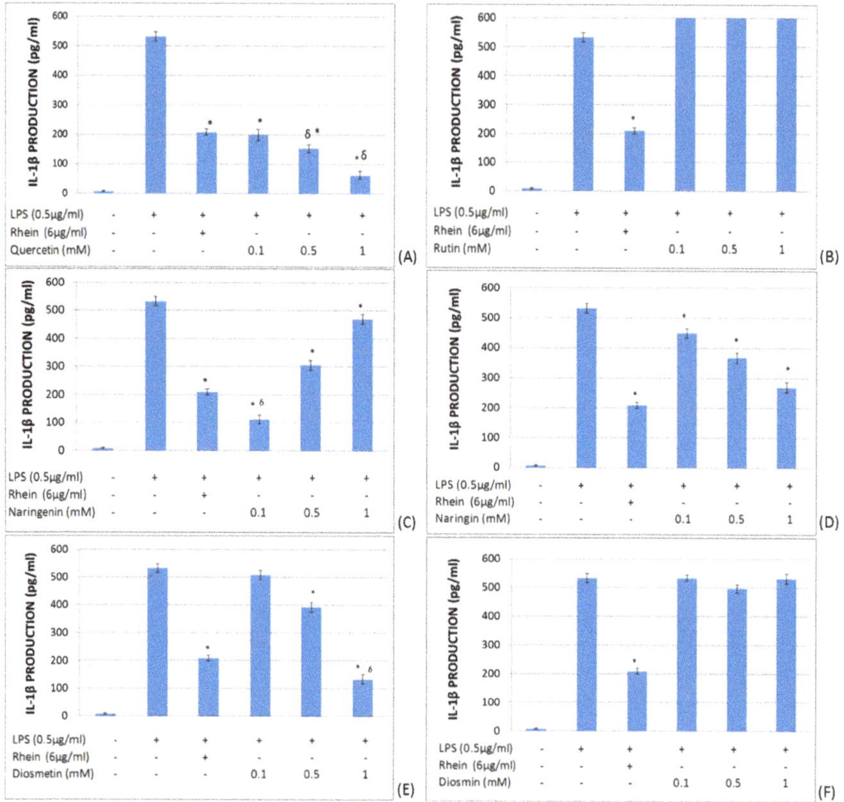

Figure 2. Effects of different flavonoids on the production of IL-1β in LPS-stimulated (0.5 μg/mL) whole blood after 6 h of culture. The different panels show the results of the effects of quercetin (**A**), rutin (**B**), naringenin (**C**), naringin (**D**), diosmetin (**E**), and diosmin (**F**). All data are expressed as the mean (top segment of the rectangles) ± SEM (vertical segment) of thirty independent experiments. * $p < 0.05$: significantly different when compared to the LPS control. $\delta p < 0.05$: significantly different when compared to the rhein control.

The effects of quercetin, rutin, naringenin, naringin, diosmetin, and diosmin on the TNF-α secretion in LPS-stimulated whole blood (Figure 3) were investigated. Quercetin showed a dramatic inhibitory effect on TNF-α production. Both naringenin and naringin showed a dose-dependent suppressor effect upon TNF-α production. Diosmetin significantly inhibited TNF-α production, but in an inverse dose-dependent manner. On the other hand, rutin and diosmin did not modify TNF-α production in LPS-stimulated whole blood.

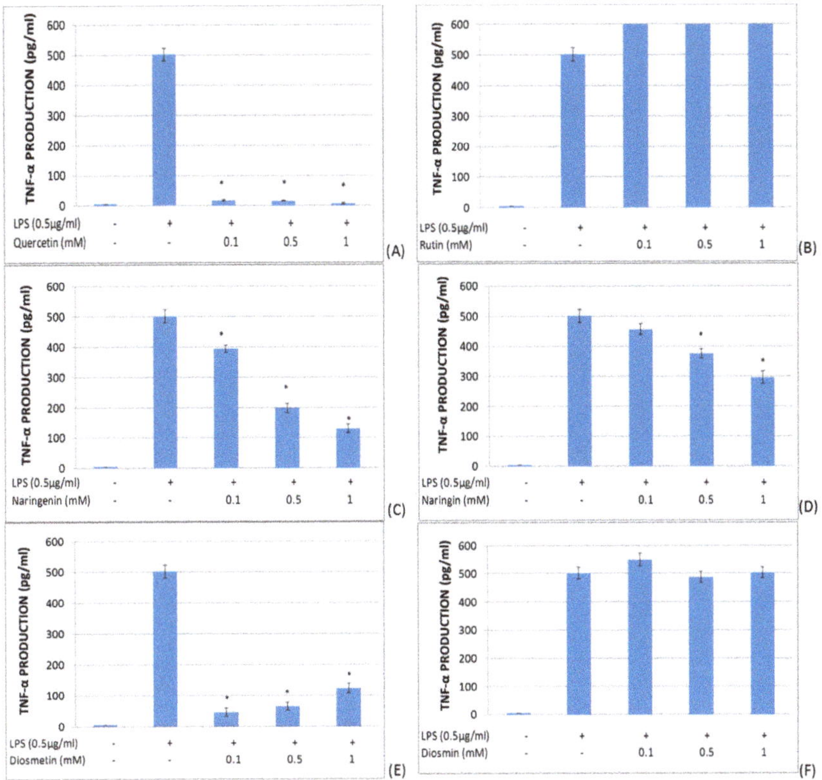

Figure 3. Effects of different flavonoids on the production of TNF-α in LPS-stimulated (0.5 μg/mL) whole blood after 6 h of culture. The different panels show the results of the effects of quercetin (**A**), rutin (**B**), naringenin (**C**), naringin (**D**), diosmetin (**E**), and diosmin (**F**). All data are expressed as the mean (top segment of the rectangles) ± SEM (vertical segment) of thirty independent experiments. * $p < 0.05$: significantly different when compared to the LPS control.

The effects of quercetin, rutin, naringenin, naringin, diosmetin, and diosmin on IL-6 secretion in LPS-stimulated whole blood were studied (Figure 4). It was found that quercetin, naringenin, diosmetin, and naringin significantly reduced IL-6 production in a dose-dependent manner. On the other hand, rutin and diosmin did not significantly modify the IL-6 production in LPS-stimulated whole blood.

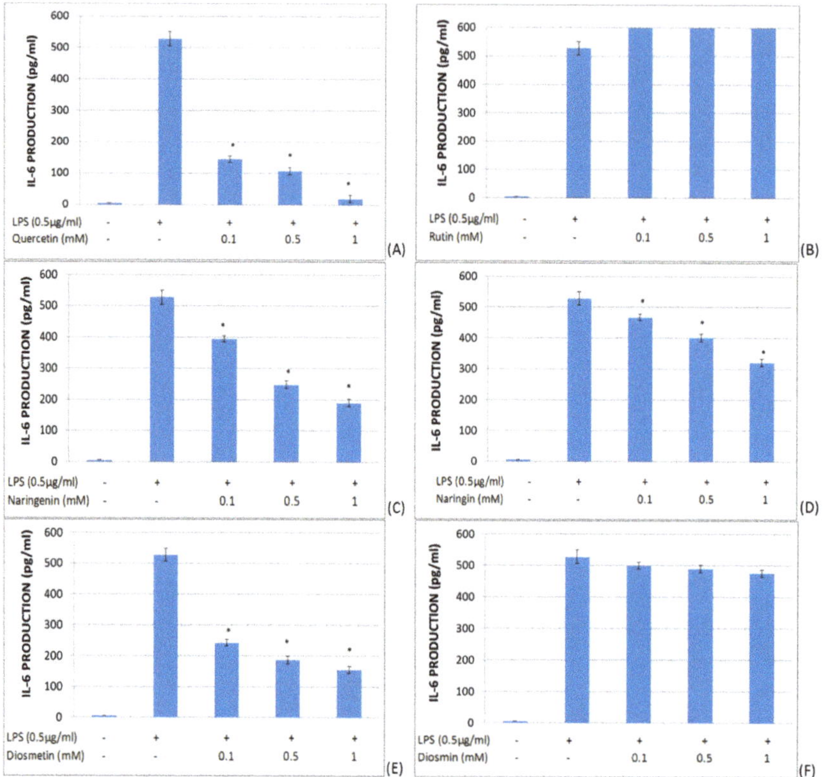

Figure 4. Effects of different flavonoids on the production of IL-6 in LPS-stimulated (0.5 µg/mL) whole blood after 6 h of culture. The different panels show the results of the effects of quercetin (**A**), rutin (**B**), naringenin (**C**), naringin (**D**), diosmetin (**E**), and diosmin (**F**). All data are expressed as the mean (top segment of the rectangles) ± SEM (vertical segment) of thirty independent experiments. * $p < 0.05$: significantly different when compared to the LPS control.

The effect of quercetin, rutin, naringenin, naringin, diosmetin, and diosmin on IL-8 secretion in LPS-stimulated whole blood was studied (Figure 5). In these assays, quercetin, naringenin, diosmetin, and naringin showed a significant decrease in IL-8 production in a dose-dependent manner. Furthermore, rutin and diosmin did not alter IL-8 production in LPS-stimulated whole blood.

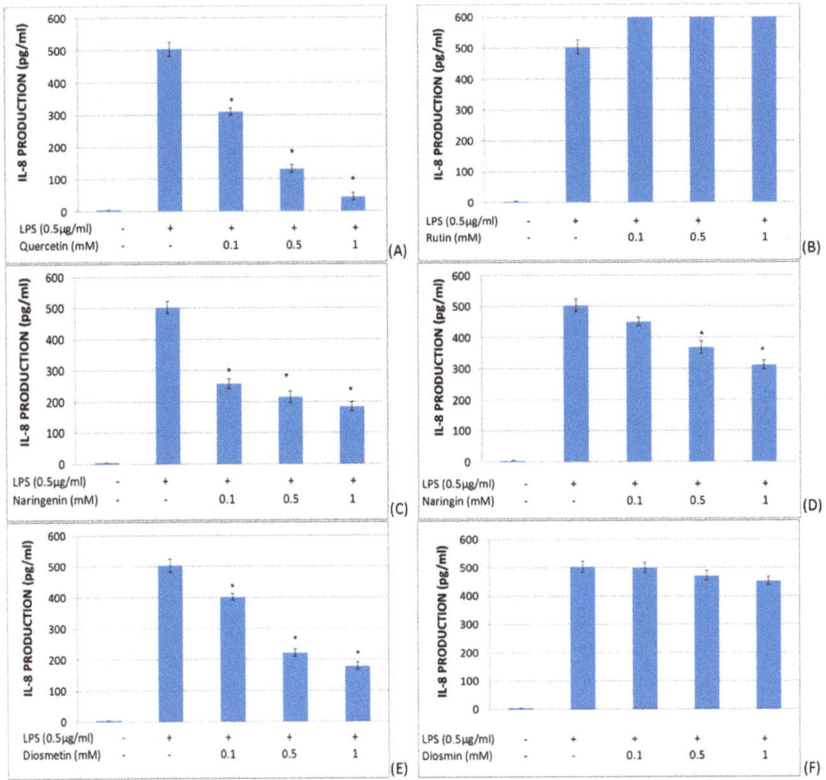

Figure 5. Effects of different flavonoids on the production of IL-8 in LPS-stimulated (0.5 µg/mL) whole blood after 6 h of culture. The different panels show the results of the effects of quercetin (**A**), rutin (**B**), naringenin (**C**), naringin (**D**), diosmetin (**E**), and diosmin (**F**). All data are expressed as the mean (top segment of the rectangles) ± SEM (vertical segment) of thirty independent experiments. * $p < 0.05$: significantly different when compared to the LPS control.

3. Discussion

In this work, we have demonstrated that certain members of the flavonoid family have an inhibitory effect on the production of IL-1β, TNF-α, IL-6, and IL-8 in LPS-stimulated whole blood. Quercetin, naringenin, naringin, and diosmetin have this immunosuppressor effect. However, rutin and diosmin lack this anti-inflammatory regulatory effect.

The inflammatory-immune system plays a critical role in the pathogenesis of atherosclerosis [40]. Large thrombogenic necrotic cores, along with lipids, intraplaque hemorrhage, a thinner fibrous cap, and inflammatory cell infiltration, pathologically characterize unstable plaque rupture, and these pathological features eventually result in platelet aggregation and thrombus formation [41]. Control of the molecular and cellular mechanisms involved in these processes is an area of intense research. Some flavonoids have shown the ability to inhibit platelet activation and aggregation [26]. In this study, we have further investigated the anti-inflammatory and immunoregulatory effects of flavonoids, focusing on the production of pro-inflammatory cytokines in LPS-stimulated whole blood. LPS is the principal component of the outer membrane of gram-negative bacteria, and one of the most potent inflammation inducers of pro-inflammatory cytokines [42]. LPS is a relevant monocyte activation signal with a marked effect on the production of inflammatory cytokines, including IL-1β, TNF-α, IL-6, and IL-8 [43]. The effect of flavonoids varies, depending on their chemical structures and functional groups [18]. The presence of C3–hydroxyl (OH) may result in an anti-inflammatory effect. Indeed, quercetin, which has a C3–OH group (the characteristic of flavonols), enhanced LPS-stimulated cytokine inhibition. The other aglycons studied, like naringenin and diosmetin, demonstrated a relevant cytokine suppressor effect. Diosmetin is a flavone which contains a carbon–carbon double bond (C2=C3) in ring C, and a C5–OH and C7–OH group in ring A (Figure 6). The dihydroflavone naringenin differs from diosmetin by the presence of a carbon–carbon single bond (C2–C3), and by the different substitutions on the carbon atoms of ring B (Figure 6). The glycoside naringin is a dihydroflavone which has glycosylation on C7–OH (Figure 6) and exerted a cytokine-suppressor effect, but one that was not as relevant as the aglycons. However, rutin and diosmin did not show any inhibitory capacity in the release of pro-inflammatory cytokines. Rutin, a flavonol with glycosylation of C3–OH (Figure 6), did not have any effect on the LPS-induced release of cytokines, similar to diosmin, which belongs to the flavones group, with the glycosylation of C7–OH (Figure 6).

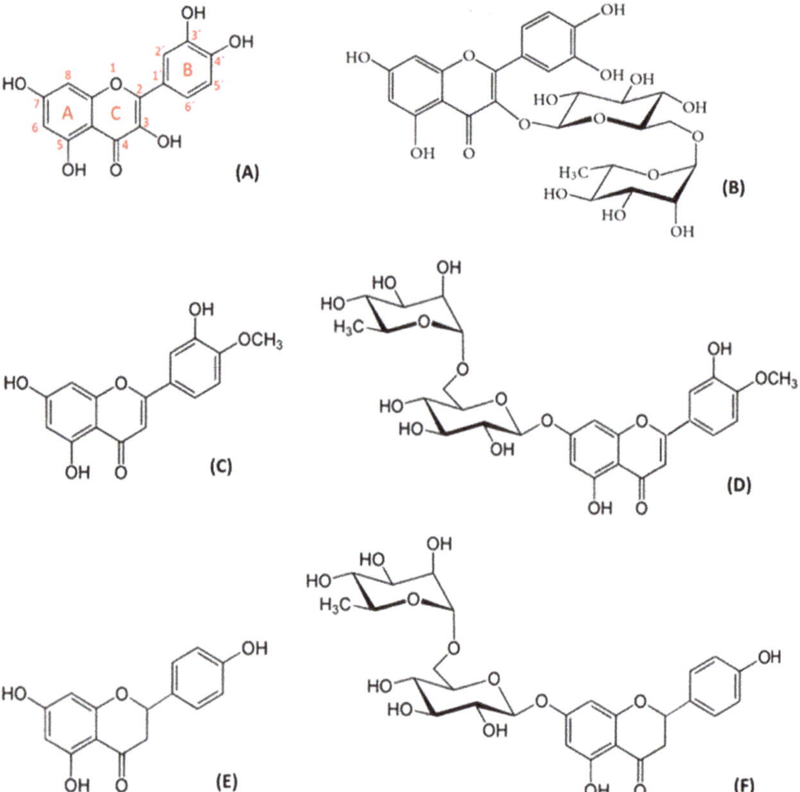

Figure 6. Chemical structure of flavonoids: (**A**) Quercetin, (**B**) rutin, (**C**) diosmetin, (**D**) diosmin, (**E**) naringenin, and (**F**) naringin.

It is well-known that disaccharide structures of glycosides block the anti-inflammatory activity. There are works on the metabolism of flavonoids and the major effect of their metabolites in the organism [44,45]. Aglycons may enter epithelial cells by passive diffusion because of their increased lipophilicity. Therefore, generally, flavonoid glycosides are cleaved either in the intestinal lumen or in the epithelium before absorption [46]. In vitro, according to our results, aglycons exert higher activity versus heterosides. In this sense, quercetin has five hydroxyl groups (Figure 6) capable of interacting with the receptor binding site, and one of them is in the most active position: the C3–OH on ring

C [47]. Due to these structural characteristics, quercetin has shown an intense and wide inhibitory effect on whole-blood production of TNF-α at all assayed concentrations. These inhibitory results of quercetin are also remarkable in the studied interleukin release. Likewise, diosmetin has three hydroxyl groups, which could justify its inhibitory effect, although this effect is lower than that of quercetin, since none of the OH is located at the 3 position. The inhibitory effect of diosmetin is exerted in a dose-dependent manner for IL-1β, IL-6, and IL-8 production; however, the release of TNF-α is inhibited in an inverse dose-dependent manner. Naringenin also contains three hydroxyl groups in its structure, and no C3–OH on ring C. This product inhibits, in a dose-dependent manner, IL-6, IL-8, and TNF-α, and in an inverse dose-dependent manner, IL-1β. Rutin and diosmin did not exert any activity in our in vitro assays, maybe due to their disaccharide structures, although naringin, despite containing disaccharides in its structure, showed some activity.

On the other hand, the basic structure of flavonoids and which type of sugar moiety is attached strongly affect their bioavailability. Bioavailability is a crucial factor determining their biological activity in vivo. Dietary flavonoids are mostly present in their glycoside forms. However, this is not the case in the plasma, where glycosides are scarce. As we discussed previously, deglycosylation occurs both in the small intestine and in the large intestine, depending on the type of sugar moiety. In the small intestine, two enzymes have been reported to act, such as β-glucosidases, against flavonoid monoglucosides.

In the case of non-monoglucosidic glycosides of flavonols, such as rutin, intestinal β-glucosidases cannot hydrolyze the sugar moiety. Therefore, the intestinal microbiota acts to yield absorbable aglycon in the cecum and in the large intestine. The aglycon produced by the microbiota is absorbed via the large intestine and transported into the circulation [38]. Regarding flavones, such as diosmin, the oral absorption is poor, reaching plasma concentrations that are typically < 1 µg/mL in humans. However, formulations of micronized diosmin have improved the bioavailability and they are widely used in the treatment of chronic venous disease, hemorrhoidal disease, and other indications [48]. Numerous studies have confirmed the superior efficacy of micronized diosmin (therapeutic dose = 1000 mg) compared to non-micronized diosmin (1000 mg), through the diosmetin plasma concentration as the main metabolic by-product of diosmin [49]. There were no significant adverse events in the study groups [48].

Regarding dihydroflavones, naringin in humans is metabolized in aglycon naringenin by naringinase found in the liver. Its solubility is limited in water and has a low oral bioavailability of around 5% [50]. Dihydroflavones are absorbed into enterocytes after oral intake; they are rapidly metabolized, in particular, into conjugates, sulfates, and glucuronides, which are the major circulating forms in plasma. A large fraction reaches the colon, where it is efficiently metabolized into small absorbable phenolics. The form (aglycon vs. glycoside) and species (e.g., human vs. rat) have important impacts [51].

These results support the anti-inflammatory and immunomodulatory effects of flavonoids, in addition to their role in platelet activation [26]. Nowadays, it is not possible to reach high concentrations in plasma due to the lack of flavonoid solubility, but these assays hint at clarifying the possible mechanism of action of these compounds. Further studies are required to define the different cellular targets of flavonoids in pro-inflammatory cytokines. It is possible to suggest in our experimental system that monocytes, as LPS responders, are potential targets of the flavonoid immunomodulatory activity. Additional studies may establish the potential therapeutic relevance of these flavonoid effects in vascular diseases.

4. Materials and Methods

4.1. Study Cohort, and Inclusion and Exclusion Criteria

Forty healthy non-smokers (age 21.8 ± 1.01 years (mean ± SEM); 11 men, 25 women) provided blood for the present study. Possible donors were excluded if they showed evidence of heart, kidney, lung, or autoimmune disease; had a history of tumors, any chronic or acute infection, diabetes mellitus, hypercholesterolemia, endocrine diseases, immunodeficiency, or thrombocytopathy; were undergoing immunosuppressant, immunomodulatory, cytostatic, or nonsteroidal anti-inflammatory drug (NSAID) treatment; or took any other medication within the three months prior to the study that might modify the cytokine response.

Written informed consent was obtained from each donor. The study was conducted according to the ethical guidelines of the 1975 Declaration of Helsinki, with the approval of the Biomedical Ethics Committee of the University of Alcalá de Henares.

4.2. Peripheral Blood Extraction

Peripheral blood was collected by antecubital puncture in sodium citrate-containing (3.8% wt/vol) Vacutainer® tubes (Dismadel, Spain), discarding the first 2 mL. All extractions were performed at the Dept. of Haematology of the *Principe de Asturias Hospital*, Alcala de Henares (Spain). Sodium citrate was selected as the anticoagulant instead of heparin, ethylenediaminetetraacetic acid (EDTA), or D-phenylalanyl-L-prolyl-L-arginine chloromethyl ketone (PPACK), given its lesser impact on complement activation pathways [52].

4.3. Selected Drugs

Quercetin (Sigma-Aldrich Chemical, Spain), rutin (Sigma-Aldrich Chemical, Spain), diosmetin (Zoster Ferrer, Spain), diosmin (Zoster Ferrer, Spain), naringenin (Zoster Ferrer, Spain), and naringin (Zoster Ferrer, Spain) (Figure 6) were dissolved in dimethyl sulphoxide (DMSO) (Dismadel, Spain) to a final plasmatic concentration of 0.5 mM. This concentration is close to the clinical dose used in the flavone diosmin (Daflon®) [53]. The difficulty in dissolving the compounds meant that no concentrations higher than 1mM could be tested. Then, further dilutions of 0.1, 0.5, and 1 mM were produced for use in later assays. The smallest volume of DMSO (2 μL), which allowed a perfect dissolution of compounds, was added to blood to avoid modifying or changing the cell structure.

4.4. Assay of IL-1β, TNF-α, Il-6, and IL-8 Production in LPS-Stimulated Whole Blood

Whole-blood aliquots of 1mL, extracted from donors, were incubated with flavonoids at 37 °C for 30 min in darkness and with continuous shaking. Afterwards, 0.5 μg/mL of lipopolysaccharide was added, to boost the production of different cytokines by monocytes [54], and incubated at 37 °C for 6 h in darkness and with continuous shaking.

Next, samples were introduced in dry ice and were gradually centrifuged (Centrifugal machine JOUAN B3.11 model) for 10 min at 4000 rpm (revolutions per minute), and then, the supernatant was collected. Every control and product was assayed in quadruplicate.

The levels of different cytokines were measured by using a specific enzyme immunoassay kit for each cytokine (IL-1β Biotrak ELISA kit, Amersham Biosciences, Little Chalfont, UK; TNF-α Biotrak ELISA kit, Amersham Biosciences, Little Chalfont, UK; Il-6 Biotrak ELISA kit, Amersham Biosciences, Little Chalfont, UK; IL_8 Biotrak ELISA kit, Amersham Biosciences, Little Chalfont, UK), in accordance with the manufacturer's instructions. Absorbance was measured with a spectrophotometer (ELx800 Absorbance Microplate Reader of Biotek, Wisconsin, USA) at 450 nm. The standard curve for cytokines covered the range from 7.8 to 500 pg/mL. The intra- and inter-assay coefficients of variation (CVs) were 7.6% and 10.3%, respectively. The assay sensitivity was 1.1 pg/mL.

Dimethyl sulfoxide (DMSO) and other reagents, unless specifically stated elsewhere, were purchased from Sigma-Aldrich (St. Louis, MO, USA). The final volume of DMSO in the reaction mixture was 0.5%.

4.5. Statistical Analysis

All results are expressed as the mean ± SEM (standard error mean) of values obtained in each experiment. Since most variables did not fulfil the normality hypothesis, a Wilcoxon test was used to analyse the variance of paired groups. The significance level was set at $p < 0.05$. Statistical analysis was performed using SPSS-19 software (SPSS-IBM, Armonk, NY, USA).

Author Contributions: Conceptualization: C.Z. and F.Z.; methodology: C.Z. and L.V.; formal analysis: C.Z., J.M. and M.Á.-M.; software: C.Z. and J.M.; supervision: C.Z. and M.Á.-M.; writing—original draft: C.Z. and L.V.; writing—review and editing: C.Z. and M.Á.-M.; funding acquisition: C.Z. and F.Z. All authors have read and agreed to the published version of the manuscript.

Funding: This research was funded by the University of Alcalá de Henares-Reig Jofré Art. 83 LOU, grant number (154/2018), entitled, Asesoramiento en materia de medicamentos y programas de investigación, desarrollo e innovación.

Acknowledgments: We thank the Haematology Service of the Principe de Asturias Hospital for their help with blood extractions. We appreciate the technical support.

Conflicts of Interest: The authors declare no conflict of interest.

References

1. Kaptoge, S.; Pennells, L.; De Bacquer, D.; Cooney, M.T.; Kavousi, M.; Stevens, G.; Riley, L.M.; Savin, S.; Khan, T.; Altay, S.; et al. World Health Organization cardiovascular disease risk charts: Revised models to estimate risk in 21 global regions. *Lancet Glob. Health* **2019**, *7*, e1332–e1345. [CrossRef]

2. Becker, R.C.; Owens, A.P.; Sadayappan, S. Tissue-level inflammation and ventricular remodeling in hypertrophic cardiomyopathy. *J. Thromb Thrombolysis* **2020**, *49*, 177–183. [CrossRef] [PubMed]

3. Melnikov, I.S.; Kozlov, S.G.; Saburova, O.S.; Avtaeva, N.Y.; Prokofieva, L.V.; Gabbasov, Z.A. Current position on the role of monomeric C-reactive protein in vascular pathology and atherothrombosis. *CPD* **2019**, *25*. [CrossRef]

4. Chiu, Y.-J.; Hsieh, Y.-H.; Lin, T.-H.; Lee, G.-C.; Hsieh-Li, H.M.; Sun, Y.-C.; Chen, C.-M.; Chang, K.-H.; Lee-Chen, G.-J. Novel compound VB-037 inhibits Aβ aggregation and promotes neurite outgrowth through enhancement of HSP27 and reduction of P38 and JNK-mediated inflammation in cell models for Alzheimer's disease. *Neurochem. Int.* **2019**, *125*, 175–186. [CrossRef] [PubMed]

5. Vogel, S.; Thein, S.L. Platelets at the crossroads of thrombosis, inflammation and haemolysis. *Br. J. Haematol* **2018**, *180*, 761–767. [CrossRef] [PubMed]

6. Mkhize, N.V.P.; Qulu, L.; Mabandla, M.V. The Effect of Quercetin on Pro- and Anti-Inflammatory Cytokines in a Prenatally Stressed Rat Model of Febrile Seizures. *J. Exp. Neurosci.* **2017**, *11*, 117906951770466. [CrossRef] [PubMed]

7. Ortega, M.A.; Asúnsolo, Á.; Romero, B.; Álvarez-Rocha, M.J.; Sainz, F.; Leal, J.; Álvarez-Mon, M.; Buján, J.; García-Honduvilla, N. Unravelling the Role of MAPKs (ERK1/2) in Venous Reflux in Patients with Chronic Venous Disorder. *Cells Tissues Organs* **2018**, *206*, 272–282. [CrossRef]

8. Ortega, M.A.; Asúnsolo, Á.; Leal, J.; Romero, B.; Alvarez-Rocha, M.J.; Sainz, F.; Álvarez-Mon, M.; Buján, J.; García-Honduvilla, N. Implication of the PI3K/Akt/mTOR Pathway in the Process of Incompetent Valves in Patients with Chronic Venous Insufficiency and the Relationship with Aging. *Oxidative Med. Cell. Longev.* **2018**, *2018*, 1–14. [CrossRef]

9. Colmorten, K.B.; Nexoe, A.B.; Sorensen, G.L. The Dual Role of Surfactant Protein-D in Vascular Inflammation and Development of Cardiovascular Disease. *Front. Immunol.* **2019**, *10*, 2264. [CrossRef]

10. Kuznetsova, T.; Prange, K.H.M.; Glass, C.K.; de Winther, M.P.J. Transcriptional and epigenetic regulation of macrophages in atherosclerosis. *Nat. Rev. Cardiol* **2019**. [CrossRef]

11. Alvarez-Mon, M.A.; Gómez, A.M.; Orozco, A.; Lahera, G.; Sosa, M.D.; Diaz, D.; Auba, E.; Albillos, A.; Monserrat, J.; Alvarez-Mon, M. Abnormal Distribution and Function of Circulating Monocytes and Enhanced Bacterial Translocation in Major Depressive Disorder. *Front. Psychiatry* **2019**, *10*, 812. [CrossRef] [PubMed]

12. Jayashree, B.; Bibin, Y.S.; Prabhu, D.; Shanthirani, C.S.; Gokulakrishnan, K.; Lakshmi, B.S.; Mohan, V.; Balasubramanyam, M. Increased circulatory levels of lipopolysaccharide (LPS) and zonulin signify novel biomarkers of proinflammation in patients with type 2 diabetes. *Mol. Cell. Biochem.* **2014**, *388*, 203–210. [CrossRef] [PubMed]

13. Sieve, I.; Ricke-Hoch, M.; Kasten, M.; Battmer, K.; Stapel, B.; Falk, C.S.; Leisegang, M.S.; Haverich, A.; Scherr, M.; Hilfiker-Kleiner, D. A positive feedback loop between IL-1β, LPS and NEU1 may promote atherosclerosis by enhancing a pro-inflammatory state in monocytes and macrophages. *Vasc. Pharmacol.* **2018**, *103–105*, 16–28. [CrossRef] [PubMed]

14. Murray, P.J. Immune regulation by monocytes. *Semin. Immunol.* **2018**, *35*, 12–18. [CrossRef]

15. Swirski, F.K.; Nahrendorf, M. Leukocyte Behavior in Atherosclerosis, Myocardial Infarction, and Heart Failure. *Science* **2013**, *339*, 161–166. [CrossRef]

16. Umamaheswari, S. Anti-Inflammatory Effect of Selected Dihydroxyflavones. *JCDR* **2015**, *9*, FF05. [CrossRef]

17. González, R.; Ballester, I.; López-Posadas, R.; Suárez, M.D.; Zarzuelo, A.; Martínez-Augustin, O.; Medina, F.S.D. Effects of Flavonoids and other Polyphenols on Inflammation. *Crit. Rev. Food Sci. Nutr.* **2011**, *51*, 331–362. [CrossRef]

18. Dymarska, M.; Janeczko, T.; Kostrzewa-Susłow, E. Glycosylation of Methoxylated Flavonoids in the Cultures of Isaria fumosorosea KCH J2. *Molecules* **2018**, *23*, 2578. [CrossRef]

19. Justino, A.B.; Costa, M.S.; Saraiva, A.L.; Silva, P.H.; Vieira, T.N.; Dias, P.; Linhares, C.R.B.; Dechichi, P.; de Melo Rodrigues Avila, V.; Espindola, F.S.; et al. Protective effects of a polyphenol-enriched fraction of the fruit peel of Annona crassiflora Mart. on acute and persistent inflammatory pain. *Inflammopharmacology* **2019**. [CrossRef]

20. Domingos, O.D.S.; Alcântara, B.G.V.; Santos, M.F.C.; Maiolini, T.C.S.; Dias, D.F.; Baldim, J.L.; Lago, J.H.G.; Soares, M.G.; Chagas-Paula, D.A. Anti-Inflammatory Derivatives with Dual Mechanism of Action from the Metabolomic Screening of Poincianella pluviosa. *Molecules* **2019**, *24*, 4375. [CrossRef]

21. Saha, S.; Panieri, E.; Suzen, S.; Saso, L. The Interaction of Flavonols with Membrane Components: Potential Effect on Antioxidant Activity. *J. Membr. Biol* **2020**, *253*, 57–71. [CrossRef] [PubMed]

22. Boligon, A.A.; de Freitas, R.B.; de Brum, T.F.; Waczuk, E.P.; Klimaczewski, C.V.; de Ávila, D.S.; Athayde, M.L.; de Freitas Bauermann, L. Antiulcerogenic activity of Scutia buxifolia on gastric ulcers induced by ethanol in rats. *Acta Pharm. Sin. B* **2014**, *4*, 358–367. [CrossRef] [PubMed]

23. Li, X.; Sdiri, M.; Peng, J.; Xie, Y.; Yang, B.B. Identification and characterization of chemical components in the bioactive fractions of *Cynomorium coccineum* that possess anticancer activity. *Int. J. Biol. Sci.* **2020**, *16*, 61–73. [CrossRef] [PubMed]

24. Mendes, L.F.; Gaspar, V.M.; Conde, T.A.; Mano, J.F.; Duarte, I.F. Flavonoid-mediated immunomodulation of human macrophages involves key metabolites and metabolic pathways. *Sci Rep.* **2019**, *9*, 14906. [CrossRef]

25. Magne Nde, C.B.; Zingue, S.; Winter, E.; Creczynski-Pasa, T.B.; Michel, T.; Fernandez, X.; Njamen, D.; Clyne, C. Flavonoids, Breast Cancer Chemopreventive and/or Chemotherapeutic Agents. *Curr. Med. Chem.* **2015**, *22*, 3434–3446.

26. Zaragozá, C.; Monserrat, J.; Mantecón, C.; Villaescusa, L.; Zaragozá, F.; Álvarez-Mon, M. Antiplatelet activity of flavonoid and coumarin drugs. *Vasc. Pharmacol.* **2016**, *87*, 139–149. [CrossRef]

27. Khalilpourfarshbafi, M.; Gholami, K.; Murugan, D.D.; Abdul Sattar, M.Z.; Abdullah, N.A. Differential effects of dietary flavonoids on adipogenesis. *Eur J. Nutr* **2019**, *58*, 5–25. [CrossRef]

28. Meng, H.; Shao, D.; Li, H.; Huang, X.; Yang, G.; Xu, B.; Niu, H. Resveratrol improves neurological outcome and neuroinflammation following spinal cord injury through enhancing autophagy involving the AMPK/mTOR pathway. *Mol Med. Rep.* **2018**, *18*, 2237–2244. [CrossRef]

29. Tanaka, T.; Takahashi, R. Flavonoids and Asthma. *Nutrients* **2013**, *5*, 2128–2143. [CrossRef]

30. Che, C.-T.; Wong, M.; Lam, C. Natural Products from Chinese Medicines with Potential Benefits to Bone Health. *Molecules* **2016**, *21*, 239. [CrossRef]

31. Gómez-Guzmán, M.; Rodríguez-Nogales, A.; Algieri, F.; Gálvez, J. Potential Role of Seaweed Polyphenols in Cardiovascular-Associated Disorders. *Mar. Drugs* **2018**, *16*, 250. [CrossRef] [PubMed]

32. Gonçalves, C.; de Freitas, M.; Ferreira, A. Flavonoids, Thyroid Iodide Uptake and Thyroid Cancer—A Review. *IJMS* **2017**, *18*, 1247. [CrossRef] [PubMed]

33. Kumar, S.; Pandey, A.K. Chemistry and Biological Activities of Flavonoids: An Overview. *Sci. World J.* **2013**, *2013*, 1–16. [CrossRef] [PubMed]

34. Tripoli, E.; Guardia, M.L.; Giammanco, S.; Majo, D.D.; Giammanco, M. Citrus flavonoids: Molecular structure, biological activity and nutritional properties: A review. *Food Chem.* **2007**, *104*, 466–479. [CrossRef]

35. Khan, H.; Ullah, H.; Aschner, M.; Cheang, W.S.; Akkol, E.K. Neuroprotective Effects of Quercetin in Alzheimer's Disease. *Biomolecules* **2019**, *10*, 59. [CrossRef]

36. Heim, K.E.; Tagliaferro, A.R.; Bobilya, D.J. Flavonoid antioxidants: Chemistry, metabolism and structure-activity relationships. *J. Nutr. Biochem.* **2002**, *13*, 572–584. [CrossRef]

37. Doostdar, H.; Burke, M.D.; Mayer, R.T. Bioflavonoids: Selective substrates and inhibitors for cytochrome P450 CYP1A and CYP1B1. *Toxicology* **2000**, *144*, 31–38. [CrossRef]

38. Murota, K.; Nakamura, Y.; Uehara, M. Flavonoid metabolism: The interaction of metabolites and gut microbiota. *Biosci. Biotechnol. Biochem.* **2018**, *82*, 600–610. [CrossRef]

39. De Isla, N.G.; Yang, J.W.; Huselstein, C.; Muller, S.; Stoltz, J.F. IL-1beta synthesis by chondrocyte analyzed by 3D microscopy and flow cytometry: Effect of Rhein. *Biorheology* **2006**, *43*, 595–601.

40. Pescetelli, I.; Zimarino, M.; Ghirarduzzi, A.; De Caterina, R. Localizing factors in atherosclerosis. *J. Cardiovasc. Med.* **2015**, *16*, 824–830. [CrossRef]

41. Sun, B.; Zhao, H.; Li, X.; Yao, H.; Liu, X.; Lu, Q.; Wan, J.; Xu, J. Angiotensin II-accelerated vulnerability of carotid plaque in a cholesterol-fed rabbit model-assessed with magnetic resonance imaging comparing to histopathology. *Saudi J. Biol. Sci.* **2017**, *24*, 495–503. [CrossRef] [PubMed]

42. Dholakiya, S.L.; Benzeroual, K.E. Protective effect of diosmin on LPS-induced apoptosis in PC12 cells and inhibition of TNF-α expression. *Toxicology in Vitro* **2011**, *25*, 1039–1044. [CrossRef] [PubMed]

43. Lee, S.-B.; Lee, W.S.; Shin, J.-S.; Jang, D.S.; Lee, K.T. Xanthotoxin suppresses LPS-induced expression of iNOS, COX-2, TNF-α, and IL-6 via AP-1, NF-κB, and JAK-STAT inactivation in RAW 264.7 macrophages. *Int. Immunopharmacol.* **2017**, *49*, 21–29. [CrossRef] [PubMed]

44. Serreli, G.; Deiana, M. In vivo formed metabolites of polyphenols and their biological efficacy. *Food Funct.* **2019**, *10*, 6999–7021. [CrossRef]

45. Olivares-Vicente, M.; Barrajon-Catalan, E.; Herranz-Lopez, M.; Segura-Carretero, A.; Joven, J.; Encinar, J.A.; Micol, V. Plant-Derived Polyphenols in Human Health: Biological Activity, Metabolites and Putative Molecular Targets. *Curr. Drug Metab.* **2018**, *19*, 351–369. [CrossRef]

46. Cassidy, A.; Minihane, A.-M. The role of metabolism (and the microbiome) in defining the clinical efficacy of dietary flavonoids. *Am. J. Clin. Nutr.* **2017**, *105*, 10–22. [CrossRef]

47. Heřmánková, E.; Zatloukalová, M.; Biler, M.; Sokolová, R.; Bancířová, M.; Tzakos, A.G.; Křen, V.; Kuzma, M.; Trouillas, P.; Vacek, J. Redox properties of individual quercetin moieties. *Free Radic. Biol. Med.* **2019**, *143*, 240–251. [CrossRef]

48. Martel, C.; Cointe, S.; Maurice, P.; Matar, S.; Ghitescu, M.; Théroux, P.; Bonnefoy, A. Requirements for membrane attack complex formation and anaphylatoxins binding to collagen-activated platelets. *PLoS ONE* **2011**, *6*, e18812. [CrossRef]

49. Staniewska, A. Safety of use of micronized diosmin at daily doses up to 2000 mg per day. *Pol. Merkur. Lek.* **2016**, *41*, 188–191.

50. Russo, R.; Chandradhara, D.; De Tommasi, N. Comparative Bioavailability of Two Diosmin Formulations after Oral Administration to Healthy Volunteers. *Molecules* **2018**, *23*, 2174. [CrossRef]

51. Ratnam, D.V.; Ankola, D.D.; Bhardwaj, V.; Sahana, D.K.; Kumar, M.N.V.R. Role of antioxidants in prophylaxis and therapy: A pharmaceutical perspective. *J. Control. Release* **2006**, *113*, 189–207. [CrossRef] [PubMed]

52. Najmanová, I.; Vopršalová, M.; Saso, L.; Mladěnka, P. The pharmacokinetics of flavones. *Crit Rev. Food Sci Nutr* **2019**, 1–17. [CrossRef] [PubMed]

53. Bogucka-Kocka, A.; Woźniak, M.; Feldo, M.; Kockic, J.; Szewczyk, K. Diosmin–isolation techniques, determination in plant material and pharmaceutical formulations, and clinical use. *Nat. Prod. Commun* **2013**, *8*, 545–550. [CrossRef] [PubMed]

54. Lima, T.S.; Gov, L.; Lodoen, M.B. Evasion of Human Neutrophil-Mediated Host Defense during Toxoplasma gondii Infection. *mBio* **2018**, *9*, 02027-17. [CrossRef] [PubMed]

 *molecules*

Communication

Potential of Flavonoid-Inspired Phytomedicines against COVID-19

Wilfred Ngwa [1,2,*], Rajiv Kumar [3,4], Daryl Thompson [5], William Lyerly [5], Roscoe Moore [5], Terry-Elinor Reid [6], Henry Lowe [7,8] and Ngeh Toyang [7,8]

[1] Brigham and Women's Hospital, Harvard Medical School, Boston, MA 02115, USA
[2] Dana-Farber Cancer Institute, Harvard Medical School, Boston, MA 02115, USA
[3] Northeastern University, Boston, MA 02115, USA; rajivtondak@gmail.com
[4] R&D Biomedical Materials, Millipore Sigma, Milwaukee, WI 02115, USA
[5] Global Research and Discovery Group Sciences, Winter Haven, FL 33884, USA;
    d.thompson@globalrdg.com (D.T.); william.lyerly@gmail.com (W.L.); rmoore@phrockwood.com (R.M.)
[6] School of Pharmacy, Concordia University of Wisconsin, Mequon, WI 53097, USA;
    Terry-Elinor.Reid@cuw.edu
[7] Vilotos Pharmaceuticals Inc, Baltimore, MD 21202, USA; lowebiotech@gmail.com (H.L.);
    ngeh.toyang@flavocure.com (N.T.)
[8] Flavocure Biotech Inc, Baltimore, MD 21202, USA
* Correspondence: wngwa@bwh.harvard.edu

Academic Editor: H.P. Vasantha Rupasinghe
Received: 19 May 2020; Accepted: 5 June 2020; Published: 11 June 2020

Abstract: Flavonoids are widely used as phytomedicines. Here, we report on flavonoid phytomedicines with potential for development into prophylactics or therapeutics against coronavirus disease 2019 (COVID-19). These flavonoid-based phytomedicines include: caflanone, Equivir, hesperetin, myricetin, and Linebacker. Our in silico studies show that these flavonoid-based molecules can bind with high affinity to the spike protein, helicase, and protease sites on the ACE2 receptor used by the severe acute respiratory syndrome coronavirus 2 to infect cells and cause COVID-19. Meanwhile, in vitro studies show potential of caflanone to inhibit virus entry factors including, ABL-2, cathepsin L, cytokines (IL-1β, IL-6, IL-8, Mip-1α, TNF-α), and PI4Kiiiβ as well as AXL-2, which facilitates mother-to-fetus transmission of coronavirus. The potential for the use of smart drug delivery technologies like nanoparticle drones loaded with these phytomedicines to overcome bioavailability limitations and improve therapeutic efficacy are discussed.

Keywords: flavonoids and their derivatives; phytomedicine; COVID-19; SARS-COV-2; smart nanoparticles

---

1. Introduction

The 2019 outbreak of the novel coronavirus disease (COVID-19) caused by severe acute respiratory syndrome coronavirus 2 (SARS-CoV-2) catapulted governments and scientists in a race to find a vaccine or drug for the prevention or treatment of this deadly pandemic. As scientists scramble to find effective therapies, doctors are trying cocktails of unproven therapies and desperate populations are trying different phytomedicines [1]. Many clinical trials to test different drugs and phytomedicines for COVID 19 have been launched [2].

Phytomedicine, the use of medicinal plants for prevention and treatment of disease, is of growing importance worldwide [3], but has been largely absent in global health as in the case of COVID-19. A number of studies have surfaced on the use of Chinese traditional medicines including phytomedicines on the treatment of COVID-19 and there are anecdotal reports on phytomedicines discovered for the treatment of COVID-19 in Africa, particularly in Madagascar [4,5]. Phytomedicine of proven quality, safety, and efficacy, contributes to the World Health Organization (WHO)'s global health priority of ensuring that all people have access to quality healthcare. However, the use of phytomedicine is largely driven by anecdotal evidence, and is limited by poor bio-availability. Cross-disciplinary research collaborations could overcome these barriers and help accelerate development of effective evidence-based phytomedicines for COVID-19, that would also be widely accessible in developing countries where over 80% of the population uses phytomedicine. Flavonoids have been reported to possess various disease prevention and treatment potential including against viruses [6,7]. Here we report new findings on a number of flavonoid phytomedicines with potential for development into COVID-19 therapeutics or prophylactics.

2. Results and Discussions

It is now well established that SARS-CoV-2 uses the angiotensin-converting enzyme 2 (ACE2) receptors abundant in the respiratory tract and lungs to infect cells [8–10]. CoV-2 spike glycoprotein binds ACE2 cellular receptors to facilitate fusion and ultimately entry into cells. Chloroquine (CLQ) is being investigated as a prophylactic or therapeutic and is reported to inhibit the activity of CoV-2, thus launching the investigation into possible binding sites to ACE2 [11]. Molecular docking studies were employed to investigate different flavonoid molecules' potential binding efficacy to ACE2 metallopeptidase domain. We are however aware that there are other possible binding sites of these molecules, including the CARS-CoV-2RBD domain of ACE2 [12].

Figure 1 highlights flavonoids investigated by us including hesperetin [7], myricetin [13], Linebacker, and caflanone (FBL-03G) [14]. Binding energy results are highlighted in Figure 2 in comparison to that of CLQ, which is currently in clinical trials as a potential prophylactic and treatment of COVID-19. These preliminary works indicate that flavonoids could bind with high affinity to the spike protein, helicase, and protease sites on the ACE2 receptor causing conformational change to inhibit viral entry of coronaviruses. For comparison, the results indicate that the investigated flavonoids could bind equally or more effectively than CLQ. The binding results highlight potential for considering these flavonoids as prophylactic against SARS-CoV-2.

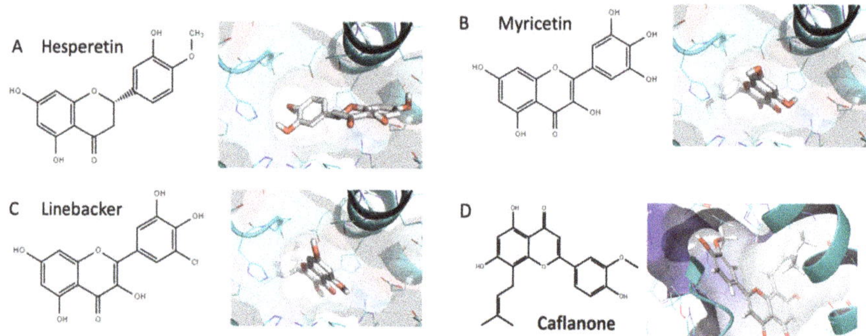

Figure 1. Docking studies for (**A**) Hesperetin; (**B**) Myricetin; (**C**) Linebacker; (**D**) Caflanone (FBL-03G).

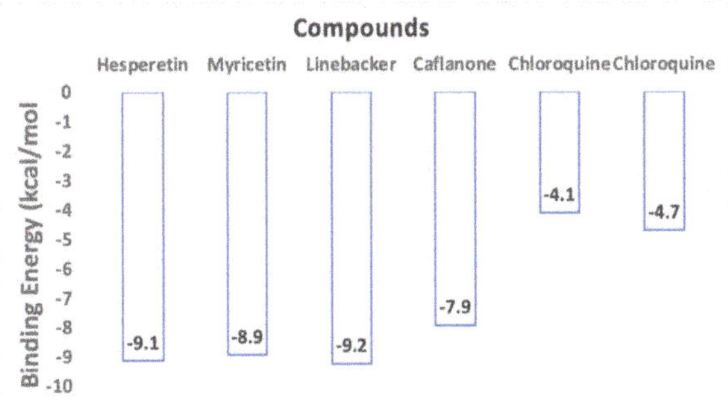

Figure 2. Docking study results show flavonoid molecules can bind very effectively to the ACE2 receptor used by SARS-CoV-2 to infect cells. Two binding energies are shown for chloroquine due to two possible binding poses in different regions of the metallopeptidase domain.

A close look at the interaction of CLQ seen in Figure 3A,B indicates two possible binding poses in different regions of the metallopeptidase domain. Although CLQ seen in Figure 3A exhibits a lower affinity for the site, it is making a key pie-cation interaction with His374 that in turn coordinates with zinc (ZN). The second predicted pose of CLQ inverts the structure allowing for binding deeper into the catalytic site and farther away from zinc. As a result, the chloroquinoline group of CLQ makes pie stacking interactions with the residue of Thr371 and a salt bridge interaction with Glu406 (Figure 3B). The structurally diverse caflanone also interacts within the catalytic site of ACE2 (Figure 3C) and makes multiple favorable interactions with Glu375, Glu402 that coordinates with ZN, Phe274, and Arg273.

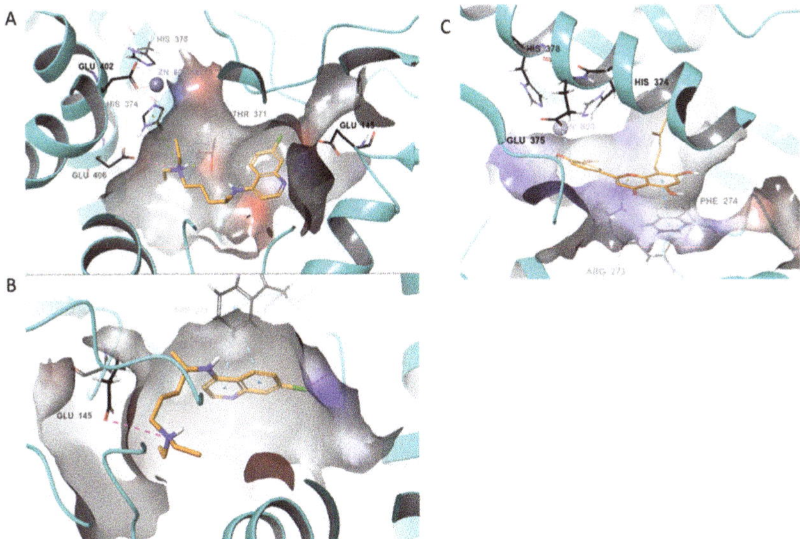

Figure 3. Predicted binding poses of caflanone and chloroquine (CLQ) within the zinc metallopeptidase domain of ACE2 (cyan cartoon and electrostatic potential surface). (**A,B**) Two predicted poses of CLQ (orange), one of which interacts with HIS374, which is coordinated with Zn and the other binds further into the cleft of the catalytic domain, away from Zn. (**C**) Caflanone (orange) is predicted to bind within the catalytic domain similar to CLQ in A.

Moreover, in vitro study results (Table 1) for caflanone show potential to inhibit virus entry factors including, ABL-2, [15] cathepsin L, [16] cytokines (IL-1β, IL-6, IL-8, Mip-1α, TNF-α), [17], and PI4Kiiiβ [18], as well as AXL-2, which facilitates mother-to-fetus transmission of coronavirus [19].

Table 1. Antiviral activity of caflanone against human coronavirus (-OC43) and viral/host factors *.

Bioactivity	EC$_{50}$/IC$_{50}$ (µM)
hCov-OC43 beta virus	0.42
ABL-2	0.27
AXL	<5.0
Cathepsin L	3.28
IL-1β	2.4
IL-6	9.1
IL-8	9.9
Mip-1α	8.9
TNF-α	8.7
CK2a2	0.038
JAK2	1.85
MNK2	0.549
PI4Kiiiβ	0.136

* The hCov-OC43 virus was cultured in RD (Rhabdomyosarcoma) cells while the Hotspot kinase profiling assay was used to assess kinase inhibition. Proinflammatory cytokine inhibition was measured in human PBMC (Peripheral blood mononuclear cells) cells stimulated with LPS (Liposaccharide). ABL2 (Abelson Murine Leukemia Viral Oncogene Homolog 2), AXL (AXL Receptor Tyrosine Kinase), IL-1β (Interleukin 1β), IL-6 (Interleukin 6), IL-8 (Interleukin 8), Mip-1α (macrophage inflammatory protein 1α), TNF-α (Tumor Necrosis Factor-α), CK2a2 (Casein kinase 2), JAK2 (Janus kinase 2), MNK2 (Mitogen-activated protein kinase signal-integrating kinase), PI4Kiiiβ (Phosphatidylinositol 4-kinase III beta).

Similarly, Equivir, a blend of natural bioflavonoids: hesperetin, myricetin, and piperine demonstrated antiviral efficacy. Equivir has a synergistic effect by inhibiting ICAM1 (Intercellular adhesion molecule 1), ATPase, helicase, polymerase and neuraminidase to reduce or prevent viral entry, transcription, replication, and budding. It has potential as a treatment and a preventative prophylactic. The components hesperetin, myricetin, and piperine are all FDA listed as Generally Recognized as Safe (GRAS). Other studies [7,13] have reported other flavonoid compounds with preliminary results showing anti-coronavirus and anti-inflammatory activity. Altogether, these results provide significant impetus for further investigation of these phytomedicines in the fight against COVID-19.

Despite the promising in silico and in vitro results, a major limitation of flavonoids in in vivo and clinical translation studies is the poor bioavailability of flavonoids [20]. Thus, there is often need to enhance the bioavailability of these flavonoids in-vivo. To this end, approaches to increase bioavailability include the use of absorption enhancers, improving metabolic stability, changing the site of absorption from large intestine to small intestine, and the use of drug delivery systems [20].

Smart nanoparticles present as viable drug delivery systems. The SARS-CoV-2 itself is the quintessential smart nanoparticle of 60–140 nm [8,19,21], and has together with other viruses been described as smart and capable nanoparticles that cause human loss, havoc, and devastation [22]. One approach to counter this is to also develop anti-COVID-19 (AC) smart nanoparticles (nanodrones). Figure 4A illustrates such an AC nanodrone with the capacity to target the ACE2 receptor and deliver both hydrophobic and hydrophilic payloads. The flavonoids can be encapsulated in the nanodrones using microfluidic approaches [23]. The flavonoids can also be conjugated as targeting moieties on the AC nanodrones. Our preliminary work with nanodrones for targeting lung lesions (Figure 4B–E) suggests that such a pulmonary (INH) route of delivery may be advantageous with higher amounts of payload reaching the lung [24–26], which is ground zero for COVID-19 [1]. Here the targeting moieties conjugated on the nanoparticle drones could be flavonoid molecules with targeting SARS-CoV-2 infection in the lungs.

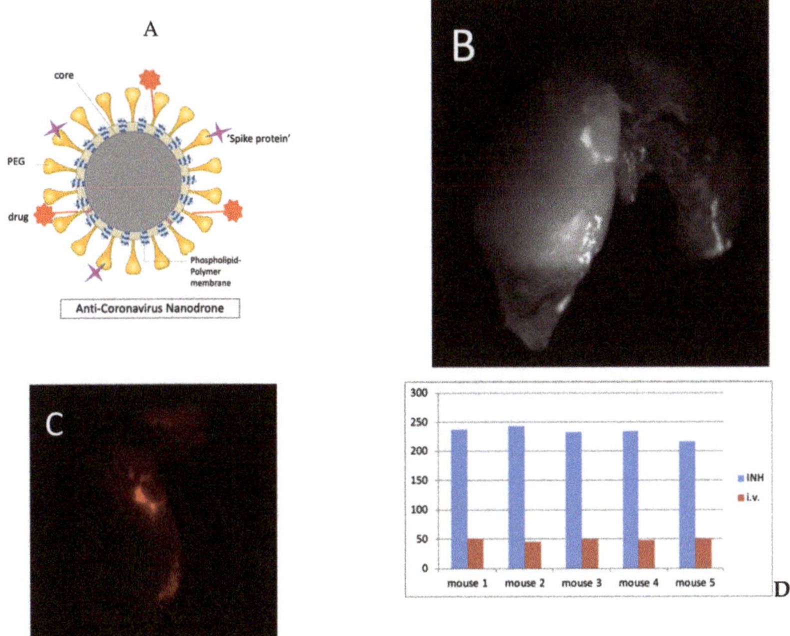

Figure 4. (**A**) A schematic of the anti-COVID-19 (AC) nanodrone designed to target ACE2 receptor allowing for image-guided monitoring of space-time distribution and treatment monitoring; (**B–C**) results of smart nanoparticles (RGD-pGNPs) targeting lung lesions; (**D**) Fluorescence intensity of smart nanoparticles targeting lung lesions in mice administered via INH (pulmonary delivery) versus via i.v. (intravenous delivery).

Beside addressing bioavailability, use of AC nanodrones could allow the ability for image-guided targeting and monitoring including via computed tomography (CT) or magnetic resonance imaging (MRI). While nanodrones [23,26,27] can use gold nanoparticles as core material, this could be readily replaced with iron or gadolinium for MRI. Using the latter would allow them to be directed with magnetic fields for superior targeting to disease sites.

In perspective, previous studies on CLQ have shown that chloroquine can block endocytosis of nanoparticles [28]. The docking study results above suggest that smart nanoparticles with flavonoids might be more effective in binding the ACE2 receptor and could represent a significant improvement or advance in targeting such indications. Smart nanoparticles can also be constructed along the lines of the multi-compartment liposome [29], which have hydrophobic and hydrophilic compartments to carry flavonoids or other drugs for more effective targeting to enhance their effectiveness. In general, advantages of using such smart technology includes better disease targeting, ability to cross membranes and enter cells, longer duration drug action, lower doses, and reduced side effects, which is of vital importance.

3. Conclusions

In summary, our preliminary results highlight significant potential for COVID-19 phytomedicines in the war against the novel coronavirus pandemic. While the activity of other flavonoids like hesperetin and myricetin are known, the results for flavonoids like caflanone and linebacker are presented as new candidates for further investigation. Overall, the findings justify further cross-disciplinary research collaborations investigating such viable flavonoids and leveraging advances in smart nanomaterials for targeted delivery. This could accelerate the development of effective phytomedicines for COVID-19 that could save innumerable lives, with capacity for expedient adaptation to combat likely future pandemics.

4. Materials and Methods

4.1. In-Silico Studies

Molecular Docking is an effective and competent tool for in silico screening, which plays an important and ever-increasing role in rational drug design. Molecular modeling approaches are increasingly being used to accelerate the drug discovery process. Molecular docking is an effective molecular modeling tool that is used to identify key interactions of ligands within the active site of a target protein and assigns a predicted binding affinity. Docking was performed using the Schrödinger software suite (Maestro, version 11.8.012, Schrödinger New York, NY, USA)following well-established approaches described in previous studies [13,30,31]. Compounds were extracted from the PubChem database in the SDF format and were combined in one file. The file was then imported into Maestro and prepared for docking using LigPrep. Ioniser was used to generate an ionized state of all compounds at the target pH 7 $\pm$ 2. This prepared low-energy conformers of the ligand were taken as the input for an induced—fit docking. Prior to docking, the atomic coordinates of the crystal structure of SARS-CoV were retrieved from the Protein Data Bank (PDB 1R4L). The protein was prepared with the Protein Preparation Wizard where bond orders and hydrogens were assigned, missing side chains and loops were filled, heteroatomic states generated and restrained minimization with OPLS_2005 force field conducted. The induced-fit docking protocol was run from the graphical user interface accessible within Maestro 11.8.012 and Maestro 12.1. Receptor sampling and refinement were performed on residues within 5Å of each ligand for each of the ligand–protein complexes. Induced-fit receptor conformations were generated for each of the ligands. Re-docking was performed with the test ligands into their respective structures that are within 30 kcal/mol of their lowest energy structure. Finally, the ligand poses were scored using GlideScore scoring functions.

4.2. In Vitro Studies

4.2.1. Inhibition of Hcov-OC43 Human Coronavirus

For in-vitro studies, test compounds were prepared in 20-mM solutions and sent to Institute for Viral Research (USU) for analysis. Compounds were serially diluted using eight half-log dilutions in test medium (MEM supplemented with 5% FBS and 50 µg/mL gentamicin) so that the starting (high) test concentration was 100 µM. Each dilution was added to 5 wells of a 96-well plate with 80–100% confluent RD cells. Three wells of each dilution were infected with virus, and two wells remained uninfected as toxicity controls. Six wells were infected and untreated as virus controls, and six wells were uninfected and untreated as cell controls. hCoV-OC43 virus was prepared to achieve the lowest possible multiplicity of infection (MOI) that would yield >80% cytopathic effect (CPE) within 5 days. M128533 was tested in parallel as a positive control. Plates were incubated at 37 $\pm$ 2 °C, 5% CO_2.

On day 5 post-infection, once untreated virus control wells reached maximum CPE, plates were stained with neutral red dye for approximately 2 h ($\pm$15 min). Supernatant dye was removed and wells rinsed with PBS, and the incorporated dye was extracted in 50:50 Sorensen citrate buffer/ethanol for >30 min and the optical density was read on a spectrophotometer at 540 nm. Optical densities were converted to percent of cell controls and normalized to the virus control, then the concentration of test compound required to inhibit CPE by 50% (EC50) was calculated by regression analysis. The concentration of compound that would cause 50% cell death in the absence of virus was similarly calculated (CC50). The selective index (SI) is the CC50 divided by EC50.

4.2.2. Kinase Inhibition Assay

The HotSpot kinase profiling assay platform (Reaction Biology Corporation) was used to kinase inhibitory properties of FBL-03G. The detailed protocol for the assay was previously described by Anastassiadis et al., 2011 [32]. In brief, for each reaction, kinase and substrate were mixed in a buffer containing 20 mM HEPES (pH 7.5), 10 mM $MgCl_2$, 1 mM EGTA, 0.02% Brij35, 0.02 mg/mL BSA, 0.1 mM Na_3VO_4, 2 mM DTT, and 1% DMSO. Compounds were then added to each reaction mixture. After a 20-min incubation, ATP (Sigma-Aldrich, Saint Louis, MO, USA) and [g-33P] ATP (PerkinElmer, Waltham, MA, USA) were added at a final total concentration of 10 mM. Reactions were carried out at room temperature for 2 h and spotted onto P81 ion exchange cellulose chromatography paper (Whatman). Filter paper was washed in 0.75% phosphoric acid to remove unincorporated ATP. The percent remaining kinase activity relative to a vehicle-containing (DMSO) kinase reaction was calculated for each kinase/inhibitor pair. Outliers were identified and removed. IC_{50} values were calculated using GraphPad Prism 5 (GraphPad Software, San Diego, CA, USA).

4.2.3. Proinflammatory Cytokine Inhibition

The Human Cytokine/Chemokine Magnetic bead panel from Millipore (Catalog # HCYTOMAG-60K) was used in this study according manufacturer's instructions. Monocyte/macrophage inhibition in human PBMC cells stimulated with LPS was analyzed. Cryopreserved human PBMC cells were seeded in 96 well plates at density of 5×10^4 cells/well in 150 μL RPMI culture media. The plates were incubated for 1 h at 37 °C with 5% CO_2. To the wells, 10 μL diluted compounds, Dexamethasone, and vehicle controls were added to appropriate wells according to the plate map. The plates were incubated at 37 °C with 5% CO_2 for 1 h. To appropriate wells, 40 μL LPS (*E. coli*, 50 pg/mL) was added and plates further incubated at 37 °C, 5% CO_2, for 24 h. The plates were centrifuged at 1000 rpm for 10 min and supernatant harvested. The supernatants were transferred to a clean plate and stored at -80 °C until ready to analyze on the Luminex and cytokine (IL-1β, IL-6, IL-8, MIP-1α and TNFα) levels measured. Levels of cytokine induction were interpolated off a standard curve using a 5-point non-linear regression analysis. The interpolated data was normalized to vehicle control and analyzed to generate IC_{50} values.

4.3. Smart Nanoparticles Studies

Smart nanoparticles discussed in this study were made using: gold (III) chloride trihydrate ($HAuCl_4.3H_2O$), THPC- tetrakis (hydroxymethyl) phosphnium chloride (80% solution in water), sodium hydroxide, dimethyl sulfoxide (DMSO), and HPLC grade water procured from Sigma Aldrich and used without further purifications. Mono-functional mPEG-thiol (Mw: 2000 Da) and hetero bi-functional anime-PEG-thiol (Mw: 3,400 Da) and carboxymethyl-PEG-thiol (Mw: 2000 Da) from Laysan Bio Inc were used for the PEGylation of as-prepared gold nanoparticles. Fluorophore, Alexa Fluor 647 (AF647) carboxylic acid, succinimidyl ester was purchased from Invitrogen (Carlsbad, CA, USA). For purification of nanoparticle, cellulose dialysis membrane (12–14 kDa) and microfuge membrane filters (NANOSEP 100K OMEGA) were procured from Sigma Aldrich (Saint Louis, MO, USA) and Pall Corporation (Westborough, MA, USA), respectively.

THPC capped nanoparticles were synthesized by controlled reduction of auric (III) chloride as reported previously by our group. Briefly, 0.5 mL of freshly prepared 1 M NaOH was added to 45 mL of HPLC grade water followed by addition of 1 mL of THPC solution (prepared by adding 12 μL of 80% THPC in water to 1 mL of HPLC water). The solution was stirred vigorously at room temperature. After 5 min of stirring, 2 mL of 25 mM $HAuCl_4$ was added directly into the stirring mixture. A rapid change of color of the stirring mixture (from yellow to dark brown) indicated the reduction of $HAuCl_4$

and formation of stabilized nanoparticles. The stirring was continued for another 15 min. For grafting the PEG/functionalized PEG on the surface of as-prepared nanoparticles, ligand exchange method was used by adding 2 mL of 3.75 mg/mL mPEG-thiol, 2 mL of carboxymethyl-PEG-thiol, and 4 mL of amine-PEG-thiol to the 45 mL of as-prepared nanoparticles. The reaction mixture was gently stirred overnight at room temperature. The PEGylated nanoparticles were purified using dialysis against distilled water for 24 h using a cellulose membrane with a cut-off size of 12–14 kDa. The dialyzed nanoparticles were sterile filtered using a 0.45-μm syringe filter and lyophilized to obtain dried PEGylated gold nanoparticles (pGNPs). pGNPs were stored at 4 °C for future use. The fluorophore AF647 was conjugated to the amine groups of the PEG on the surface of the pGNPs. Briefly, 2 mg of pGNPs were dispersed in 1 mL of carbonate buffer (0.1 M, pH 8.8) followed by addition of 5 μL of 10 mg/mL AF647 succinimidyl ester in DMSO. The reaction mixture was gently stirred for 2 h at room temperature. The unbound or free AF647 was separated from the AF647 conjugated pGNPs using 100 kDa centrifugal spin filters. The purified AF647 conjugated pGNPs (AF647-pGNPs) were washed three times with HPLC grade water to ensure complete removal of free fluorophore. For active targeting to lung lesions, RGD peptide was conjugated to carboxyl groups of the PEGylated pGNPs via carbodiimide chemistry, and the purified nanoparticles were lyophilized for future use.

All in vivo experiments performed in this study were approved by the Animal Care and Use Committee of the Dana-Farber Cancer Institute (protocol number 17007). Transgenic mouse models bearing single-nodule lung lesions were generated as described in recent work [33]. RGD targeted pGNPs were administered to cohort A mice via INH, while the same concentration of RGD-pGNPs was administered to cohort B mice via i.v. The nanoparticles distribution was measured via fluorescence imaging and ex-vivo electron microcopy methods. The optical fluorescence imaging was done using Maestro GNIR FLEX fluorescence imaging system (CRI) using a 633-nm excitation filter. Mice were dissected and entire lungs were imaged 24 h post nanoparticles administration. The images were acquired with the same acquisition time of 88 ms for all the experiments. A quantitative estimation of the fluorescence intensity was done using the Maestro software and Image J.

Author Contributions: Conceptualization, W.N., R.M., D.T., and H.L.; investigation, D.T., T.-E.R., N.T., W.N., N.T., and R.K.; writing—review and editing, W.N., R.K., D.T., W.L., R.M., H.L., and N.T. All authors have read and agreed to the published version of the manuscript.

Funding: This research received no external funding.

References

1. Ledford, H. How does COVID-19 kill? Uncertainty is hampering doctors' ability to choose treatments. *Nature* **2020**, *580*, 311–312. [CrossRef] [PubMed]

2. Maxmen, A. More than 80 clinical trials launch to test coronavirus treatments. *Nature* **2020**, *578*, 347–348. [CrossRef]

3. Chan, D.M. *WHO Traditional Medicine Strategy: 2014–2023*; World Health Organization: Geneva, Switzerland, 2013.

4. Cui, H.-T.; Li, Y.-T.; Guo, L.-Y.; Liu, X.-G.; Wang, L.-S.; Jia, J.-W.; Liao, J.-B.; Miao, J.; Zhang, Z.-Y.; Wang, L.; et al. Traditional Chinese medicine for treatment of coronavirus disease 2019: A review. *Tradit. Med. Res.* **2020**, *16*, 1708–1717.

5. Tih, F. WHO to Study Madagascar's Drug to Treat COVID-19. Available online: https://www.aa.com.tr/en/africa/who-to-study-madagascars-drug-to-treat-covid-19-/1840971 (accessed on 15 May 2019).

6. Zakaryan, H.; Arabyan, E.; Oo, A.; Zandi, K. Flavonoids: Promising natural compounds against viral infections. *Arch. Virol.* **2017**, *162*, 2539–2551. [CrossRef] [PubMed]

7. Jo, S.; Kim, S.; Shin, D.H.; Kim, M.S. Inhibition of SARS-CoV 3CL protease by flavonoids. *J. Enzyme Inhib. Med. Chem.* **2020**, *35*, 145–151. [CrossRef] [PubMed]

8. Wan, Y.; Shang, J.; Graham, R.; Baric, R.S.; Li, F. Receptor recognition by novel coronavirus from Wuhan: An analysis based on decade-long structural studies of SARS. *J. Virol.* **2020**, *94*, e00127-20. [CrossRef]

9. Yan, R.; Zhang, Y.; Li, Y.; Xia, L.; Guo, Y.; Zhou, Q. Structural basis for the recognition of the SARS-CoV-2 by full-length human ACE2. *Science* **2020**, *367*, 1444–1448. [CrossRef]

10. Moore, J.B.; June, C.H. Cytokine release syndrome in severe COVID-19. *Science* **2020**, *368*, 473–474. [CrossRef]

11. Fantini, J.; Di Scala, C.; Chahinian, H.; Yahi, N. Structural and molecular modelling studies reveal a new mechanism of action of chloroquine and hydroxychloroquine against SARS-CoV-2 infection. *Int. J. Antimicrob. Agents.* **2020**, *55*, 105960. [CrossRef]

12. Lan, J.; Ge, J.; Yu, J.; Shan, S.; Zhou, H.; Fan, S.; Zhang, Q.; Shi, X.; Wang, Q.; Zhang, L.; et al. Structure of the SARS-CoV-2 spike receptor-binding domain bound to the ACE2 receptor. *Nature* **2020**, *581*, 215–220. [CrossRef]

13. Guerrero, L.; Castillo, J.; Quiñones, M.; Garcia-Vallvé, S.; Arola, L.; Pujadas, G.; Muguerza, B. Inhibition of Angiotensin-Converting Enzyme Activity by Flavonoids: Structure-Activity Relationship Studies. *PLoS ONE* **2012**, *7*, e49493. [CrossRef] [PubMed]

14. Moreau, M.; Ibeh, U.; Decosmo, K.; Bih, N.; Yasmin-Karim, S.; Toyang, N.; Lowe, H.; Ngwa, W. Flavonoid Derivative of Cannabis Demonstrates Therapeutic Potential in Preclinical Models of Metastatic Pancreatic Cancer. *Front. Oncol.* **2019**, *9*, 66. [CrossRef] [PubMed]

15. Sisk, J.M.; Frieman, M.B.; Machamer, C.E. Coronavirus S. protein-induced fusion is blocked prior to hemifusion by Abl kinase inhibitors. *J. Gen. Virol.* **2018**, *99*, 619–630. [CrossRef] [PubMed]

16. Simmons, G.; Gosalia, D.N.; Rennekamp, A.J.; Reeves, J.D.; Diamond, S.L.; Bates, P. Inhibitors of cathepsin L prevent severe acute respiratory syndrome coronavirus entry. *Proc. Natl. Acad. Sci. USA* **2005**, *102*, 11876–11881. [CrossRef] [PubMed]

17. Cheung, C.Y.; Poon, L.L.M.; Ng, I.H.Y.; Luk, W.; Sia, S.-F.; Wu, M.H.S.; Chan, K.-H.; Yuen, K.-Y.; Gordon, S.; Guan, Y.; et al. Cytokine Responses in Severe Acute Respiratory Syndrome Coronavirus-Infected Macrophages In Vitro: Possible Relevance to Pathogenesis. *J. Virol.* **2005**, *79*, 7819–7826. [CrossRef]

18. Yang, N.; Ping, M.; Lang, J.; Zhang, Y.; Deng, J.; Ju, X.; Zhang, G.; Jiang, C. Phosphatidylinositol 4-kinase IIIβ is required for severe acute respiratory syndrome coronavirus spike-mediated cell entry. *J. Biol. Chem.* **2012**, *28*, 8457–8467. [CrossRef]

19. Li, M.; Chen, L.; Xiong, C.; Li, X. The ACE2 expression of maternal-fetal interface and fetal organs indicates potential risk of vertical transmission of SARS-COV-2. *BioRxiv* **2020**, *15*, e0230295.

20. Thilakarathna, S.H.; Vasantha Rupasinghe, H.P. Flavonoid bioavailability and attempts for bioavailability enhancement. *Nutrients* **2013**, *5*, 3367–3387. [CrossRef]

21. Yang, X.; Yu, Y.; Xu, J.; Shu, H.; Xia, J.; Liu, H.; Wu, Y.; Zhang, L.; Yu, Z.; Fang, M.; et al. Clinical course and outcomes of critically ill patients with SARS-CoV-2 pneumonia in Wuhan, China: A single-centered, retrospective, observational study. *Lancet Respir. Med.* **2020**, *8*, 475–481. [CrossRef]

22. Kostarelos, K. Nanoscale nights of COVID-19. *Nat. Nanotechnol.* **2020**, *5*, 343–344. [CrossRef]

23. Zhang, H.; Cui, W.; Qu, X.; Wu, H.; Qu, L.; Zhang, X.; Mäkilä, E.; Salonen, J.; Zhu, Y.; Yang, Z.; et al. Photothermal-responsive nanosized hybrid polymersome as versatile therapeutics codelivery nanovehicle for effective tumor suppression. *Proc. Natl. Acad. Sci. USA* **2019**, *116*, 7744–7749. [CrossRef] [PubMed]

24. Mayr, N.A.; Hu, K.S.; Liao, Z.; Viswanathan, A.N.; Wall, T.J.; Amendola, B.E.; Calaguas, M.J.; Palta, J.R.; Yue, N.J.; Rengan, R.; et al. International outreach: What is the responsibility of ASTRO and the major international radiation oncology societies? *Int. J. Radiat. Oncol. Biol. Phys.* **2014**, *89*, 481–484. [CrossRef] [PubMed]

25. Mueller, R.; Yasmin-Karim, S.; Hesser, J.; Ngwa, W. Nanoparticle Drones to Label, Kill and Track Circulating Tumor Cells During Radiotherapy. Available online: https://w3.aapm.org/meetings/2019am/programinfo/programabs.php?sid=7980&aid=44735. (accessed on 18 May 2020).

26. Ngwa, W.; Kumar, R.; Moreau, M.; Dabney, R.; Herman, A. Nanoparticle Drones to Target Lung Cancer with Radiosensitizers and Cannabinoids. *Front. Oncol.* **2017**, *7*, 208. [CrossRef] [PubMed]

27. Kumar, R.; Korideck, H.; Ngwa, W.; Berbeco, R.I.; Makrigiorgos, G.M.; Sridhar, S. Third generation gold nanoplatform optimized for radiation therapy. *Transl. Cancer. Res.* **2013**, *2*, 1–6.

28. Hu, T.Y.; Frieman, M.; Wolfram, J. Insights from nanomedicine into chloroquine efficacy against COVID-19. *Nat. Nanotechnol.* **2020**, *15*, 247–249. [CrossRef]

29. Wholey, W.Y.; Mueller, J.L.; Tan, C.; Brooks, J.F.; Zikherman, J.; Cheng, W. Synthetic Liposomal Mimics of Biological Viruses for the Study of Immune Responses to Infection and Vaccination. *Bioconjug. Chem.* **2020**, *31*, 685–697. [CrossRef]

30. Minuesa, G.; Albanese, S.K.; Xie, W.; Kazansky, Y.; Worroll, D.; Chow, A.; Schurer, A.; Park, S.M.; Rotsides, C.Z.; Taggart, J.; et al. Small-molecule targeting of MUSASHI RNA-binding activity in acute myeloid leukemia. *Nat. Commun.* **2019**, *10*, 2691. [CrossRef]

31. Springer, T.I.; Reid, T.E.; Gies, S.L.; Feix, J.B. Interactions of the effector ExoU from pseudomonas aeruginosa with short-chain phosphatidylinositides provide insights into ExoU targeting to host membranes. *J. Biol. Chem.* **2019**, *294*, 19012–19021. [CrossRef]

32. Anastassiadis, T.; Deacon, S.W.; Devarajan, K.; Ma, H.; Peterson, J.R. Comprehensive assay of kinase catalytic activity reveals features of kinase inhibitor selectivity. *Nat. Biotechnol.* **2011**, *29*, 1039. [CrossRef]

33. Herter-Sprie, G.S.; Korideck, H.; Christensen, C.L.; Herter, J.M.; Rhee, K.; Berbeco, R.I.; Bennett, D.G.; Akbay, E.A.; Kozono, D.; Mak, R.H.; et al. Image-guided radiotherapy platform using single nodule conditional lung cancer mouse models. *Nat. Commun.* **2014**, *5*, 5870. [CrossRef]

Sample Availability: Samples of the compounds caflanone is available from the authors.

Review

Cranberry Polyphenols and Prevention against Urinary Tract Infections: Relevant Considerations

Dolores González de Llano *, M. Victoria Moreno-Arribas and Begoña Bartolomé

Institute of Food Science Research (CIAL), CSIC-UAM, Nicolás Cabrera, 9, Campus de Cantoblanco, 28049 Madrid, Spain; victoria.moreno@csic.es (M.V.M.-A.); b.bartolome@csic.es (B.B.)
* Correspondence: d.g.dellano@csic.es; Tel.: +34-91-0017909

Academic Editor: H.P. Vasantha Rupasinghe
Received: 16 July 2020; Accepted: 29 July 2020; Published: 1 August 2020

Abstract: Cranberry (*Vaccinium macrocarpon*) is a distinctive source of polyphenols as flavonoids and phenolic acids that has been described to display beneficial effects against urinary tract infections (UTIs), the second most common type of infections worldwide. UTIs can lead to significant morbidity, especially in healthy females due to high rates of recurrence and antibiotic resistance. Strategies and therapeutic alternatives to antibiotics for prophylaxis and treatment against UTIs are continuously being sought after. Different to cranberry, which have been widely recommended in traditional medicine for UTIs prophylaxis, probiotics have emerged as a new alternative to the use of antibiotics against these infections and are the subject of new research in this area. Besides uropathogenic *Escherichia coli* (UPEC), the most common bacteria causing uncomplicated UTIs, other etiological agents, such as *Klebsiella pneumoniae* or Gram-positive bacteria of *Enterococcus* and *Staphylococcus* genera, seem to be more widespread than previously appreciated. Considerable current effort is also devoted to the still-unraveled mechanisms that are behind the UTI-protective effects of cranberry, probiotics and their new combined formulations. All these current topics in the understanding of the protective effects of cranberry against UTIs are reviewed in this paper. Further progresses expected in the coming years in these fields are also discussed.

Keywords: cranberry; urinary tract infections; UTIs; uropathogenic *Escherichia coli*; UPEC; flavan-3-ols; A-type proanthocyanidins; phenolic metabolites; antiadhesive activity; probiotics

---

1. Introduction

Urinary tract infections (UTIs) are the second most common type of bacterial infections worldwide, behind otitis media. UTIs incidence peaks in individuals are in their early 20s and after age 85 [1]. The lifetime risk for acquiring a symptomatic UTIs is about 12% in men and 50% in women, with a rate of recurrence after six months of about 40% [2]. Approximately 20–30% of young women will have a recurrent UTIs, named relapses, which can be caused by the same microorganism or by a different microorganism. Microbial host-associated reservoirs at the underlying bladder tissue or gastrointestinal tract could cause reinfection, even after an intensive treatments and subsequent negative urine culture [3]. All of this means that UTIs account for several millions of outpatient hospital visits and millions of emergency room visits with a large cost for the Primary Care and Public Health systems affecting the national economy [4].

According to the anatomical location of the bacteria, UTIs are categorized as pyelonephritis and kidney infection when they affect the upper part of the urinary tract (ureters and kidney parenchyma), and as cystitis and urethritis when they affect the bladder or urethra (lower infection tract). Clinically, uncomplicated UTIs affect healthy individuals without urinary tract anomalies, while UTIs associated with factors that compromise the urinary tract or host defense are referred as complicated UTIs, including urinary obstruction or retention caused by neurological disease, renal failure or transplantation, pregnancy and the presence of foreign bodies such as kidney stones, indwelling catheters or other drainage devices [5]. Among the most common symptoms of UTIs as cystitis, there is a frequent and urgent need to urinate and pain or burning sensation in the urethra during urination. Urine is usually cloudy, sometimes even pink color due to the presence of blood. If infection rises to the kidneys, patients could also be suffer fatigue, fever, nausea and muscle and abdominal pain [6]. Pyelonephritis is usually a more serious problem as it can lead to irreversible kidney damage or sepsis [7].

Uropathogenic *Escherichia coli* (UPEC) is the primary etiological agent for the majority of these infections, causing around 85% of cystitis, but other Gram-negative bacteria, such as *Klebsiella pneumoniae*, and some Gram-positive cocci, such as some staphylococcal and enterococcal species, seem to be also implicated in the etiopathogeny of the remaining infections [8,9]. Polymicrobial infections involving Gram-positive bacteria may be more prevalent than previously appreciated and may also impact on UTI output. As routine care, UTIs were diagnosed and treated based on symptomatology without performing a urine culture but that UTI empiric treatment commonly led to a UTI misdiagnosis [10].

Conventionally, the use of antibiotics to treat this pathology has claimed to be very operative, but they can cause prevalence of resistance among uropathogens and other adverse side effects, such as damage in intestinal microbiota. For these reasons, there is a growing interest in the search of natural therapies for UTI prevention and treatment to face increasing bacterial resistance to antibiotics and high recurrence rates [9,11]. To prevent such infections, different alternatives, such as the use of antiadhesive components, probiotics or vaccines, have been investigated.

Although the consumption of cranberries (*Vaccinium macrocarpon*) has been extensively recommended for UTIs prophylaxis and relief of adverse symptoms, the UTIs preventive activity of cranberry has been debated in the literature and numerous clinical studies have been carried out, including some recent meta-analyses [4,12,13]. Several studies have shown a protective effect of cranberry against UTIs [14–18]; nevertheless, others have not found significant effects [19,20]. This current controversy about conflicting results of the clinical and cost effectiveness of cranberry supplements has been attributed to different manufactured cranberry based products and doses, as well as a lack of systematic protocol for the selection of subjects and clinical assay [13].

The red cranberry is rich in several groups of flavonoids, particularly proanthocyanidins, anthocyanidins, and flavonols, together with phenolic acids and benzoates [21,22]. Among other possible mechanisms behind the protective effects of cranberries against UTIs is the capacity of cranberry polyphenols to act as antiadhesive agents in preventing/inhibiting the adherence of pathogens to uroepithelial cell receptors, which appears to be a major step in the pathogenesis of these infections [23]. Cell culture methodologies have also evidenced anti-adhesive capacity against the UPEC of flavonoids present in cranberry and their metabolites and of urine samples collected after cranberry consumption extracts [24–28].

Probiotic bacteria are considered another promising therapy in UTI prevention and treatment [29,30], and many human intervention studies have evaluated if the consumption of specific strains, such as *Lactobacillus* spp. can prevent or treat UTIs [11,31]. Moreover, the combination of cranberry with some probiotic strains has also been proposed to be effective for the management of recurrent urinary tract infections [32].

In the light of recent scientific publications on the matter, this review tends to recapitulate all these aspects related to protective effects of cranberry against urinary tract infections. After a brief presentation of UTI pathogenesis (Section 2) and UPEC and other uropathogenic bacteria (Section 3), the review covers the main aspects related to the use of cranberries in UTI prophylaxis (Section 4). Particular attention is given to the antiadhesive activity derived from cranberry consumption that has been long considered as a main mechanism involved in the protective effects of cranberry against UTIs (Section 5), and to other potential mechanisms involved in cranberry efficiency against UTIs, including interaction with gut microbiota and renal metabolism (Section 6). Finally, the potential of probiotics against UTIs and their combined action with cranberries are also considered as a promising strategy for future treatments against UTIs (Section 7). Global conclusions from all sections are finally drawn (Section 8).

2. UTIs Pathogenesis

A bacterial ethology associated with the rise in faecal microorganisms to the urinary tract is related to most UTIs, but there are other different risk causes, including transmission by person-to-person direct contact and the faecal-oral route [1]. The higher incidence in women is mainly due to their anatomy: the female urethra and the tract to ascend to the bladder, shorter than the male. Besides, anatomical and physiological changes that occur in the urinary tract during pregnancy increase women's susceptibility to UTIs. The homeostasis of the vaginal microbial ecosystem, formed mainly by lactobacilli, is important to protect vaginal mucosa against the colonization of pathogens and their potential ascending into the urinary system. Therefore, the use of spermicides and antibiotics as well as the menopause period, altering a woman's vaginal microbiota and decreasing the level of estrogen, favour infections. Incomplete cure or recurrent genitourinary infections also lead to alterations in vaginal microbiota from a predominance of lactobacilli to coliform uropathogens [33]. Moreover, the positive correlation between UTI infection history in first-degree female relatives and UTI risk suggests a genetic component for increased susceptibility [34].

A key step in the pathogenesis of infectious diseases is the pathogenic bacteria adherence to host cells. According to Beerepoot et al. [35], UTIs are settled in three main phases: microbial colonization, bacteria adherence to uroepithelium receptors, and invasion of urinary tract cells. The initial step in the UTIs pathogenesis is the colonization of the urinary tract and the subsequent rise of the pathogen to the bladder to cause cystitis. If left untreated, uropathogens may ascend the ureters to the kidney and establish a secondary infection (acute pyelonephritis). The next phase is the pathogens adherence to uroepithelium, which is mediated by adhesins that recognize the cells' surface receptors, allowing the uropathogens to withstand the urine hydrodynamic flow. In this manner, uropathogens, such as *Escherichia coli*, *Klebsiella pneumoniae* and *Staphylococcus saprophyticus*, have the ability to join directly the bladder epithelium causing non-complicated UTIs. However, complicated UTIs are initiated when the bacteria, predominately pathogens such as *Proteus mirabilis*, *Pseudomonas aeruginosa* and *Enterococcus* spp., are attached to a catheter, to a kidney stone or they are retained in the urinary tract by a physical obstruction [8]. Once installed in bladder cells, pathogens internalize, multiply and produce infection. It has been assessed that the uropathogen *E. coli* (UPEC) can access and grow within the bladder cells to form intracellular bacterial communities, which are detected in exfoliated urothelial cells and associated to suffer recurrent UTIs [11]. Moreover, internalized UPEC can persist in quiescence for long periods without causing clinical symptoms [36].

On the other hand, it is worth mentioning that the uropathogens can form biofilms to adhere more easily and increase its resistance. This has an effect on the rest of microorganisms causing greater adherence and multiplication, being able to form thick biofilms that significantly decrease the effectiveness of antimicrobial treatments. When the biofilm resources are limited, mature bacteria are detached and can colonize a new surface to repeat the cycle [37].

3. UPEC and Other Uropathogenic Bacteria

The uninfected urinary tract has long been assumed to be sterile in healthy individuals and it was thought that the major uropathogenic organisms only included Gram-negative bacteria, mainly *E. coli* [38,39]. Currently, it is widely held that Gram-positive bacteria either alone or along with Gram-negative uropathogens in the urine are likely to be innocuous. Many species, including some known to cause UTIs, survive and multiply within the urinary tract without causing chronic infection.

E. coli is a rod Gram-negative bacterium that colonizes the gastrointestinal tract of humans and animals after birth, forming part of the natural bowel microbiota. In addition, it can be an opportunistic pathogen that uses the intracellular environment to survive and protect itself against the action of antibiotics. Several epidemiological, serological, and bacteriological studies revealed that UPEC is the pathogen most frequently associated with UTIs [40], accounting for three-quarters of all UTIs among outpatients [41,42]. These bacteria have evolved a multitude of virulence factors and strategies that facilitate bacterial growth and persistence within the adverse settings of the host urinary tract.

UPEC's ability to bind the host uroepithelial receptors is facilitated by virulence factors, hair-like organelles called fimbriae or adhesins that form part of its bacterial wall. These structures, pili or fimbriae, allow for the attachment of *E. coli* in the sites where they usually do not live as the uroepitelum, and the invasion of host cells and tissues to trigger the infection later. Although the biogenesis of bacterial adhesins is a tightly controlled process, UPEC can alter binding features in reaction to environmental changes, including temperature, osmolarity, pH, and nutrient availability [36]. In recurrent UTIs, an increased adherence of *E. coli* to urogenital epithelial cells was seen compared to healthy controls [43].

Different types of *E. coli* adhesins have been described, such as the fimbriae type 1 and type S, both sensitive to the α-D-Mannose receptor, and the fimbriae type P and Afa, resistant to the α-D-Mannose receptor [44]. The most common are the fimbriae type 1 and type P, which mainly play a role in the pathogenesis of cystitis and pyelonephritis, respectively. The fimbriae type 1 has the ability to interact with the receptor α-D-mannose, present in most of the cells of the uroepitelium. In 99% of uropathogenic *E. coli* strains, genes to encode type 1 fimbriae are present [45] and they recognize uroplakin, a FimH-binding transmembrane protein of urothelial lining cells. Mannoside-containing host proteins are encoded by the bacterial backbone DNA and are mainly composed of FimA proteins along with FimF, FimG, and FimH. It is known that Tamm-Horsfall protein (THP) or uromodulin, a highly mannosylated glycoprotein, strongly interacts with FimH of UPEC decreasing interaction with uroplakin. Also, most uropathogenic UPEC strains harbour genes encoding adhesive type1 fimbria but its expression is strictly regulated under physiological growth conditions, increasing its expression when UPEC settles on and infects bladder epithelial cells or colonizes catheters [46]. As a result, UPEC in these sessile populations enhances bladder cell adherence and invasion potential, while UPEC low-expression planktonic populations are subsequently released to the urine. Type P fimbriae, encoded by the gene papG, has been associated with acute pyelonephritis and is prevalent among the strains that cause invasive and persistent UTIs [47].

Other factors of virulence to highlight are the production of toxins, including α-hemolysin (HlyA), cytotoxic necrotizing factor1 (CNF1), secreted autotransporter toxin (SAT), cytolysin A, plasmid-encoded toxin (PET), vacuolating autotransporter toxin (VAT), Shigella enterotoxin-1 (ShET-1) and arginine succinyltransferase (AST) [48], and iron-chelating factors (siderophores) that enables UPEC to capture iron, both mechanisms impair host immune system and obtain nutrients and growth factors necessary for their survival. The expression of these virulence factors converts an *E. coli* commensal strain into an uropathogen, and, often, the inhibition of a single adhesin may cost enough to a bacterium to lose its virulence [40]. Moreover, virulence factors located on a bacterial surface, including capsule and lipopolysaccharides, may also contribute to UTIs providing antiphagocytosis and antibactericidal complement activity [40]. In recent years, the understanding of virulence factors and the behavior of this pathogen has increased remarkably, and may facilitate the application of more precise approaches in phenotypic, molecular diagnosis and the development of alternatives strategies to antibiotics based on antivirulence prophylaxis. Recently, Tamadonfar et al. [9] reviewed the development of small-molecule inhibitors and vaccines targeting virulence factors, as bacterial adhesins or structural components of adherence like siderophores in the effort to reduce UTIs burden and multidrug resistance. Moreover, *E. coli* lineages are more likely to cause UTIs exhibiting an antibiotic resistance phenotype and seem to be more persistent in the rectal tract and pandemic. In this way, the ST131 group of extraintestinal *E coli* strains exhibiting multidrug resistance to beta-lactams and fluoroquinolones and have been identified globally [41].

Besides the most causative agent of UTIs, *E. coli*, other Gram-negative bacteria, such as *K. pneumoniae*, and/or Gram-positive bacteria, including *S. saprophyticus*, *Enterococcus faecalis*, and group B Streptococcus, are also implied in UTI pathologic outcomes, especially in polymicrobial infections, but they are systematically overlooked [39]. Traditionally, a bacteriological culture of urine is used to isolate and identify UTI pathogens, such as the aerobic fast-growing organisms *E. coli* and *E. faecalis*. Conversely, strains from *Corynebacterium*, *Lactobacillus* and *Ureaplasma* genera were rarely isolated from the urinary tract as these routine culture techniques are not accurate enough to identify anaerobic slow-growing organisms. However, using advanced detection technologies, these bacteria have been identified as members of the urinary microbiota in multiple studies [38].

It should be remarked that UTI etiologic agents vary according to age, sex, and underlying pathology. Thus, in ambulatory patients, *E. coli* predominates, followed by other Gram-negative bacilli, *Klebsiella* spp. and *Proteus* spp., and Gram-positive cocci, such as *S. saprophyticus*, *Enterococcus* spp. and *Streptococcus agalactiae*. However, in the elderly infections, *Enterococcus* spp. are the most common agent (11.6%) compared to the rest of the population, which only represent 5.3% of cases [49].

4. Cranberry in UTIs Prophylaxis

The increase in global antibiotic resistance and recurrence rates has prompted the exploration and assessment of new therapeutic strategies by antiadhesive components. As a natural alternative, the intake of cranberries as fresh or dried fruits, as well as products derived from them (i.e., juices, extracts, etc.) has been extensively recommended. Cranberry consumption has been indicated to be effective in decreasing the occurrence and severity of UTIs in women and to prevent the adherence of pathogenic bacteria in the urinary tract [13,50,51]. Moreover, it could also decrease UTI related symptoms by suppressing inflammatory cascades as an immunologic response to bacteria invasion [15].

Although numerous epidemiological and intervention studies have proved the efficacy of cranberry products in UTI prophylaxis [16,17], others have shown mixed results [19,52,53]. Therefore, the evidence is still insufficient to define a formal health claim. Apart from differences in composition and doses of the cranberry-based products used in the intervention studies [13], differences among studies have been attributed to the different susceptibility of the UPEC strains to cranberry preventive effects [54]. One of the last meta-analyses concerning this topic reported a large interindividual variability in cranberry efficiency against UTIs, also concluding that patients at some risk of these infections were more susceptible to the beneficial effects of cranberry consumption [4]. Recently, Mantzorou and Giaginis [42] critically analyse the current clinical studies that have evaluated the efficacy of supplementing cranberry products against UTIs in different subpopulations; they conclude that it seems to be prophylactic by preventing infections recurrence; however, it exerts low effectiveness in populations at an increased risk of contracting UTIs. At present, cranberry supplementation can safely be suggested as complementary therapy in women with recurrent UTIs but a lack of cost-effectiveness for cranberry supplementation has been highlighted.

For the most recent decades, flavonoids as proanthocyanidins (PACs), mainly A-type (Figure 1), were assumed to be responsible for these preventive effects against UTIs [21,24]. Nevertheless, this finding has been rebutted due to evidence of the limited absorption of proanthocyanidins and their extensive metabolism by the gut microbiota [22,55–57]. In accordance with this, PAC levels in urine after cranberry intake have been found in the low-concentration range (<nM) [58,59]. Moreover, PACs are widely catabolized by the colon microbiota to give bioactive phenolic metabolites that can be further absorbed and also secreted in faeces and urine. Among them, there are single flavonoids, different conjugates (i.e., glucuronides, O-methyl ethers and sulfates) of phenolic acids (i.e., phenylpropionic, phenylacetic, benzoic and cinnamic acids) and other microbial metabolites such as phenyl-γ-valerolactones in urine collected after cranberry intake [50,55–59]. Therefore, these phenolic

metabolites might be the responsible compounds behind the preventive actions of flavonoids and phenolic acids present in cranberry on UTIs.

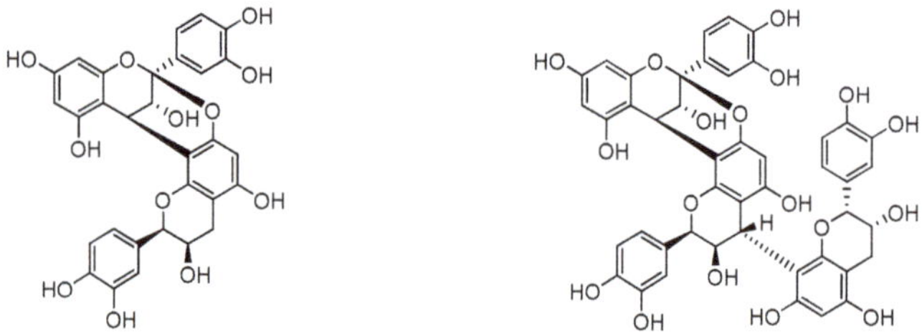

Procyanidin A2   Cinnamtannin B1

Figure 1. Structures of A-type proanthocyanidins found in cranberry.

In addition, complex carbohydrates and sugars, terpenes, as well as organic acids such as quinic, malic, shikimic, and citric are other preponderant cranberry phytochemicals. Among them, D-mannose has also been reported to inhibit the adherence of UPEC to uroepithelial cells in vitro, although new studies will certainly be needed to confirm it [38,60]. Vitamin C (ascorbic acid) and fructose have also been suggested as active compounds against UTIs as they promote changes in urine's physical properties (such as acidification) [11]. Another key characteristic of cranberry juice is the low pH of 2.5 [61]. The unique blend of the organic acids, quinic, malic, shikimic, and citric acid found in cranberries might also lead to antibacterial effects as it has been found in an experimental mouse model of urinary tract infection [62].

5. Antiadherence Activity Derived from Cranberry Consumption

The mechanisms implied in the preventive effects of cranberry consumption against UTIs are not completely established and several leading hypotheses have been proposed. Thus, cranberry polyphenols, in particular, their microbial-derived metabolites, are claimed to operate in the phase of bacterial adherence to the uroepithelial cells, disabling or inhibiting the adherence of UPEC and, therefore, preventing bacterial colonization and the progression of UTIs [23,63]. This potential mechanism is depicted in Figure 2. In fact, numerous *ex vivo* studies have confirmed the antiadhesive activity of urine samples collected from volunteers who consumed cranberry products in comparison to urine samples collected from the placebo group. As example of this, a recent study demonstrates the strong ability of human urine after intake of a cranberry chew compared to a placebo chew to inhibit the ex vivo adherence of both P type and type 1 uropathogenic *E. coli* in a randomized, double-blind, placebo-controlled, crossover design pilot trial [20]. In the same way, Baron et al. [59] reported a significant reduction in the adherence and biofilm formation of a *Candida albicans* strain by urine collected after the intake of a cranberry extract and a lecithin formulation with improved oral bioavailability.

Cell culture methodologies have also applied to assess the antiadhesive activity of cranberry phenolic compounds/fractions/extracts against uropathogens, specifically in bladder epithelial cell lines [23,26,28,64]. It should be noted that some of these in vitro studies have dealt with molecules never appearing or at concentrations far from those found in vivo, and, therefore, the physiological relevance of these studies should be considered with precaution.

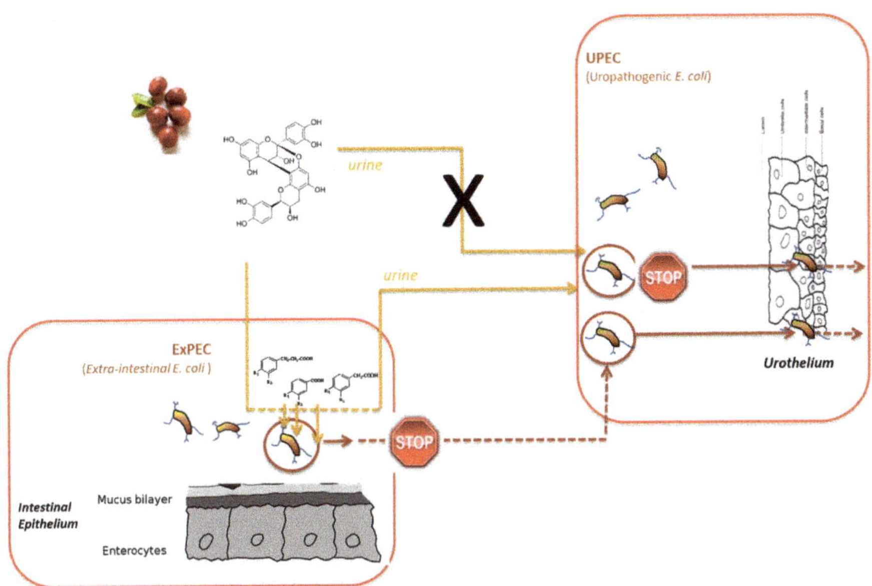

Figure 2. Proposed mechanisms of cranberry polyphenols action against urinary tract infections (UTIs).

Table 1 recompiles comparative data of the antiadhesive activity of different cranberry phenolic compounds and their microbial-derived metabolites against a *E. coli* ATCC® 53503™ into bladder uroepithelial cells (i.e., ATCC® HTB4™ cells) [23,26,28]. As seen from these data, A-type procyanidins (A2 and cinnamtannin B-1) exhibit antiadhesive activity (at concentrations ≥ 250 µM), a feature that is not observed for B-type procyanidins (B2). But, in any case, these concentrations (≥250 µM) are extremely high in comparison to those found in urine (nM range) [62,63], making it unlikely that A-type proanthocyanidins were the compounds responsible for the antiadhesive activity of urine samples collected after cranberry consumption. However, some cranberry-derived phenolic compounds detected in urine after consumption of cranberries and/or cranberry products such as hippuric acid, α-hydroxyhippuric acid, 3,4-dihydroxyphenylacetic acid, and dihydrocaffeic acid 3-*O*-sulfate are able to inhibit the adherence of the uropathogenic bacteria tested at concentrations that may be more physiologically relevant. Additive (and even synergistic) effects among all the cranberry-derived phenolic metabolites present in urine are expected in vivo, which could explain, at least partly, the ex vivo antiadhesive activity of urine samples collected from volunteers who consumed cranberry products [23].

Another interesting finding about the ex vivo capacity to inhibit UPEC adherence after cranberry extract is that of [64] that found, from the data of a cranberry trial with 16 subjects, a positive correlation between urine antiadherence capacity and content in Tamm-Horsfall protein (THP), known as a mechanism of non-specific defense. Authors concluded that the antiadherence effect derived from cranberry consumption not only might be related to the direct inhibition of bacterial adhesins by cranberry-derived metabolites but also to an induction of antiadherence THP in the kidney. Afterwards, they pointed out that cranberry extract stimulates the innate immune defense in the kidney by an increased secretion of THP; on the other hand, the direct inhibition of the bacterial adhesion to host cells has also been proven, which is independent of THP [65]. This is indeed an attractive matter to be further explored.

Table 1. Adherence Inhibition (%) of the of *E. coli* ATCC® 53503™ to ATCC® HTB4™ cells by phenolic compounds (Adapted from González de Llano et al. [23,26,28]).

	Concentration (µM)		
	100	250	500
Flavan-3-ols			
Cinnamtannin B1	1.05	4.11	13.95 *
Procyanidin A2	23.67	30.7	54.5 **
Procyanidin B2	6.79	10.0	−14.7
(−)-Epicatechin	−6.02	−1.21	−5.82
Simple phenols			
1,2-Dihydroxybenzene (catechol/pyrocatechol)	17.0 *	26.0 *	33.2 **
1,3,5-Trihydroxybenzene (phloroglucinol)	−8.53	17.6	−8.15
Benzoic acids			
Benzoic acid	16.5 *	23.3 **	32.2 **
3-Hydroxybenzoic acid	11.1	17.0*	−9.7
3,4-Dihydroxybenzoic acid (protocatechuic acid)	25.5 *	24.0	9.44
4-Hydroxy-3-methoxybenzoic acid (vanillic acid)	18.3 **	24.9 **	29.2 **
3,4,5-Trihydroxybenzoic acid (gallic acid)	−3.72	19.7	40.6**
Phenylacetic acids			
Phenylacetic acid	33.5 *	39.0 **	40.6 **
3-Hydroxyphenylacetic acid	15.0	11.9	19.4
3,4-Dihydroxyphenylacetic acid	18.6	32.5 *	37.0 **
4-Hydroxy-3-methoxyphenylacetic acid	7.11	11.92	12.8
Phenylpropionic acids			
3-Phenylpropionic acid	−11.8	14.7	12.2
3-(3-Hydroxyphenyl)-propionic acid	10.2	18.6	30.5 *
3-(3,4-Dihydroxyphenyl)-propionic acid	6.66	1.19	13.1
3-(3,4-Dihydroxyphenyl)-propionic acid 3-*O*-sulphate sodium salt	6.52	11.22	21.0 *
Dihydroxyphenyl-γ-valerolactones			
5-(3′,4′-Dihydroxyphenyl)-γ-valerolactone	6.79 ± 3.92	9.95 ± 8.28	19.4 ± 10.3 *
5-Phenyl-γ-valerolactone-3′,4′-di-*O*-sulphate	−0.22 ± 0.71	14.7 ± 1.5	30.3 ± 3.6 **
5-(4′-Hydroxyphenyl)-γ-valerolactone-3′-*O*-sulphate	11.9 ± 1.7	10.2 ± 3.9	22.2 ± 5.9 **
5-(3′-Hydroxyphenyl)-γ-valerolactone-4′-*O*-sulphate	10.1 ± 3.1	16.1 ± 6.1*	24.2 ± 3.1 **
Hippuric acids			
Hippuric acids	15.6	14.9	25.5 *
α-Hippuric acid	20.8	23.01	20.0

* Mean significantly different from zero ($p < 0.05$) using a one-sample t-test. ** Mean significantly different from zero ($p < 0.01$) using a one-sample t-test.

6. Other Potential Mechanisms Involved in Cranberry Efficiency against UTIs

Emerging evidence shows that the gut microbiota has a key role in homeostasis, regulating health and disease at distal sites throughout the body [66]. Although microbial profiles and microbial metabolites of the gut and other organs might influence the urinary microbiota, the relationship between these actively metabolizing organisms and urogenital health is yet to be completely elucidated [38,67]. It is now widely accepted that the urinary tract harbours a complex microbial network which is substantially different from the gut populations [68]. Any imbalance in specific bacterial communities is likely to have a profound effect on urologic health owing to their metabolic output and other contributions. Contrary to the urine of an asymptomatic healthy individual, an altered microbiome with specific dominating urotypes was lately reported in subjects with functional disorders of the urinary tract [67]. On the other hand, non-modifiable host factors (e.g., gender and genetic influences associated with UTI) seem to have a role in the UTIs colonization, probably through its influence in intestinal microbiota [69]. Other host factors, such as the response to fibrinogen depositing at the infection site has been found to be critical to establishing catheter-associated UTIs (CAUTI) [9].

Based on this wide influence of gut microbiota in the body, other hypothesis about the mechanisms of action of cranberry flavonoids and phenolic acids against UTIs is that cranberry components (i.e., A-type proanthocyanidins and their metabolites) would interact with gut microbiota, modulating its composition and/or functionality in such a way that it prevents microbial dysbiosis. In accordance with this hypothesis, a pilot study of the human faecal microbiome ($n = 10$) after 2 weeks of consuming dried cranberries (42 g/day) [70] indicates a shift in the Firmicutes:Bacteroidetes ratio, increases in commensal bacteria, and a particular increase in *Akkermansia* in most subjects. Similarity, in a human trial (*n* = 11) evaluating the effect of a cranberry powder (30 g/day) in the context of an animal-based diet, [71] found that cranberries attenuated the impact of the animal-based diet on microbiota composition, bile acids, and SCFA, evidencing their capacity to modulate the gut microbiota.

More specifically in relation to UTIs, and as it is becoming evident that the intestine is a reservoir for uropathogenic bacteria, in vitro studies have indicated that cranberry flavonoids and phenolic acids might interact with extra-intestinal *E. coli* (ExPEC) and decrease its (transient) intestinal colonization, consequently reducing the risk of UTI incidence [72]. This potential mechanism of action of cranberry polyphenols against UTIs is also depicted in Figure 2.

Therefore, the effects of cranberry polyphenols as flavonoids and phenolic acids against UTIs would be influenced by what has been called a "two-way interaction" between polyphenols and gut microbiota [73] in the sense that not only is the gut microbiome involved in the metabolism of cranberry polyphenols, but also the cranberry compounds and their microbial metabolites may impact on gut microbiota and induce preventive effects on intestinal colonization by uropathogens. This two-way interaction between polyphenols and intestinal bacteria has been reviewed focusing on other foods such as tea [74] and wine polyphenols [75].

In other words, the beneficial effects of cranberry against UTIs would be modulated by our intestinal microbiota through its different capacity in metabolizing cranberry flavonoids and phenolic acids into bioactive metabolites. This is the case, for example, in phenyl-γ-valerolactones, one of the most abundant metabolites in the bioactive urine fraction after cranberry intake [56,59] and whose antiadhesive activity has been proven in vitro [28]. Therefore, the inter-individual differences found in the preventive effects of cranberry against UTIs may be attributed, at least partly, to the differences in intestinal microbiota composition among individuals.

7. Combined Action of Cranberry and Probiotics against UTIs

The vaginal microbiota, that has predominance of lactobacilli, plays a dynamic and often critical role in UTI pathology by maintaining a low pH and avoiding uropathogen colonization by competitive exclusion. However, hormonal change due to estrogen deficiency, antimicrobial therapy, contraceptives, incomplete healing and recurrence of UTIs lead to a shift in the local microbiota (dysbiosis), increasing the risk of pathogens triggering pathologies (UTIs, bacterial vaginosis and yeast vaginitis) [69,76]. Therefore, the instillation of lactobacilli in the vaginal cavity is presented as a potential therapy for urinary tract care [29,77]. After their oral administration, selected strains of *Lactobacillus* have been found in vaginal exudates and faecal samples [30,78]; *Lactobacillus* translocation from the intestinal mucosa to distal mucosal surfaces has also been reported [79].

The use of probiotics/lactobacilli against UTIs has been assessed [80]. The inhibition of the adherence of pathogenic bacteria (i.e., UPEC and others) to epithelial vaginal cells by different *Lactobacillus* strains has been proven in many in vitro studies [81–83]. As a recent example, González de Llano et al. [26] reported the inhibitory effects of different strains of *Lactobacillus salivarius*, *Lactobacillus plantarum* and *Lactobacillus acidophilus* on the adherence of various uropathogenics strains from *E. coli*, *Staphylococcus epidermidis* and *E. faecalis* to bladder cells T24. However, the mechanisms behind these inhibitory effects have thus far not been elucidated, with most studies relying on inferred evidence [30,84,85]. The production of biosurfactants, bacteriocins, lactic acid and hydrogen peroxide by *Lactobacillus* spp. seems to inhibit UPEC growth and adversely affects its fimbrial structure and adherence as well as upregulate immunogenic membrane proteins [86,87]. Moreover, the phenolic biotransformation process by the microbiota or probiotic bacteria into low molecular weight cranberry phenolic metabolites can enhance their pharmacological activities by generating additional biologically active metabolites. Recently, the role of *L. rhamnosus* probiotic in the degradation of flavonoids which are present in cranberry pomace fractions, and the effect of the yielded cranberry metabolites on hepatocellular carcinoma HepG2 cells in vitro have been investigated by Rupasinghe et al [88] and they found an inhibitory effect on HepG2 cell proliferation in a dose- and time-dependent manner enhancing their anticancer activity.

Several human intervention studies have evaluated the effectiveness of specific *Lactobacillus* strains' consumption to prevent or treat UTIs [31,35,89]. So far, the results of these clinical studies remain inconclusive as a consequence of the wide diversity of populations, small sample sizes, different probiotic species and dosage forms such as vaginal and oral, and the use of incorrect dosing strategies. Therefore, more randomized clinical trials about the effect of probiotics against UTIs should be conducted to make a stronger recommendation as a probiotic benefit cannot be ruled out [11,35].

New products claiming protection against UTIs and combining cranberry extracts and *Lactobacillus* probiotic preparations have recently been lunched onto the market, being available to the general public without prescription. Many of these products do not have scientific support, although some human studies on the combination of cranberry–probiotics have been reported. In this way, Montorsi et al. [32] have proved, in a pilot study, the efficacy of the combined intake of *Lactobacillus rhamnosus* SGL06, cranberry and vitamin C for the prevention of recurrent UTIs. Polewski et al. [90] have also evaluated the impact of lactobacilli in combination with cranberry A-type proanthocyanidins on reducing the invasiveness of extra-intestinal pathogenic *E. coli* (ExPEC) to intestinal cells. The exploration and incorporation of advanced probiotic formulations and microbiome-targeted treatment strategies within urology is warranted [38].

Lastly, new therapies based on vaccination and small-molecule inhibitors targeting uropathogens virulence factors, are being investigated as future treatments against UTIs [9]. In this lane, Ahumada-Cota et al. [91] have recently outlined the potential treatment and control of recurrent UTIs in adults with autologous bacterial lysates, and its composition shown that different surface components of *E. coli*, mainly outer membrane proteins are potential immunogens. Thus, bacterial lysates could be potential candidates to create a polyvalent protective vaccine against recurrent *E. coli*-associated UTIs. Alternatively, the development of a synthetic urinary microbiota for transplantation might lead to an effective treatment for patients with recurrent UTIs, but still all these new approaches are under development.

It is worthy of note that cranberries, like some traditional medicines, are a well-known functional food, whose efficacy has been assayed in clinical trials. However, the compounds directly responsible for many of cranberries' reported health benefits remain unidentified. They are complex mixtures whose constituents have a synergistic effect by acting at different levels on multiple targets and pathways or reducing toxicity and drug side effects [92]. Nowadays, cranberry characterization and research into single/pure compounds focusing on synergy among pure compounds are essential for understanding their therapeutic effect [92]. In this line, Coleman and Ferreira [93] have recently reported that complex carbohydrates, specifically xyloglucan and pectic oligosaccharides, are significant components of cranberry products, and a new target for focused investigations of possible prebiotic effects that contribute to cranberry bioactivity. However, the use of refined cranberry oligosaccharide constituents *in vivo* may be impractical and ineffective until more is understood about the various mechanisms by which different cranberry components work together to influence overall health.

8. Conclusions

The consumption of cranberry (*Vaccinium macrocarpon*) has been extensively recommended for UTI prophylaxis and the relief of adverse symptoms. *In vivo* evidence has been translated to *in vitro* verification, resulting in the widely recommended use of cranberry juice in traditional medicine for urinary tract pathologies. This review summarizes some relevant considerations about cranberry polyphenols as flavonoids and phenolic acids, and UTIs derived from the last scientific publications. One of our first conclusions is that, besides uropathogenic *Escherichia coli* (UPEC), other etiological agents such as *Klebsiella pneumoniae* or the Gram-positive bacteria of *Enterococcus* and *Staphylococcus* genera seem to be widely involved in UTIs. In particular, complicated UTIs caused by UPEC and *Enterococcus* spp. represent a major health care concern nowadays. The application of advanced techniques for

UTI diagnosis comprising PCR, microarray and next-generation sequencing, together with the study of the individual urological microbiome, may improve diagnosis and treatment of these infections. The other outcome of this review is that the existing clinical trials support substantial evidence about the use of cranberry products as total or partial therapeutic alternatives to antibiotics in UTIs, although it has also been seen that cranberry effectiveness is individual- and/or case-dependent. This variability among individuals and cases has been attributed to different manufactured cranberry-based products and doses, as well as to the lack of systematic protocols for the selection of subjects and clinical assays. However, as more is known about the metabolism of cranberry polyphenols, new factors such as host microbiota, endogenous metabolism, host genetics and the immune system may also play a role in cranberry effectiveness against uropathogens. A question that remains to be fully answered is which cranberry and/or cranberry-derived components are mainly responsible for its protective effects against UTIs. Although recent studies have shed light on the antiadherence activity derived from cranberry consumption, other mechanisms might be jointly implicated that need further investigation. Knowledge of these mechanisms (and the cranberry components involved) might result in a better management of the potential of cranberry products against infections. For instance, nutraceuticals and other forms containing cranberry bioactive compounds and ensuring optimum bioavailability and effectiveness could be designed. Finally, pieces of evidence regarding treatment against UTIs with probiotics and their combinations with cranberry are too limited to draw any conclusions but they seem to indeed be a promising strategy for future treatments. However, any development must be supported by exhaustive research specific to the strain/s and cranberry products, as well as about the mechanisms involved in these potential beneficial effects of probiotics and their combinations with cranberry.

Funding: This research lab is funded by Grants AGL2015-64522-C2-R (Spanish Ministry of Science and Innovation) and ALIBIRD-CM 2020 P2018/BAA-4343 (Comunidad de Madrid).

Conflicts of Interest: The authors declare no conflict of interest.

References

1. Foxman, B. Urinary tract infection syndromes. *Infect. Dis. Clin. N. Am.* **2014**, *28*, 1–13. [CrossRef]
2. Sivick, K.E.; Mobley, H.L.T. Waging war against uropathogenic escherichia coli: Winning back the urinary tract. *Infect. Immun.* **2010**, *78*, 568–585. [CrossRef]
3. Baron, E.J.; Miller, J.M.; Weinstein, M.P.; Richter, S.S.; Gilligan, P.H.; Thomson, R.B.; Bourbeau, P.; Carroll, K.C.; Kehl, S.C.; Dunne, W.M.; et al. A guide to utilization of the microbiology laboratory for diagnosis of infectious diseases: 2013 recommendations by the Infectious Diseases Society of America (IDSA) and the American Society for Microbiology (ASM) a. *Clin. Infect. Dis.* **2013**, *57*, 22–121. [CrossRef]
4. Luís, Â.; Domingues, F.; Pereira, L. Can cranberries contribute to reduce the incidence of urinary tract infections? A systematic review with meta-analysis and trial sequential analysis of clinical trials. *J. Urol.* **2017**, *198*, 614–621. [CrossRef]
5. Spaulding, C.; Hultgren, S. Adhesive pili in UTI pathogenesis and drug development. *Pathogens* **2016**, *5*, 30. [CrossRef]
6. Lane, M.C.; Mobley, H.L.T. Role of P-fimbrial-mediated adherence in pyelonephritis and persistence of uropathogenic Escherichia coli (UPEC) in the mammalian kidney. *Kidney Int.* **2007**, *72*, 19–25. [CrossRef] [PubMed]
7. Scholes, D.; Hooton, T.M.; Roberts, P.L.; Gupta, K.; Stapleton, A.E.; Stamm, W.E. Risk factors associated with acute pyelonephritis in healthy women. *Ann. Intern. Med.* **2005**, *142*, 20. [CrossRef] [PubMed]
8. Flores-Mireles, A.L.; Walker, J.N.; Caparon, M.; Hultgren, S.J. Urinary tract infections: Epidemiology, mechanisms of infection and treatment options. *Nat. Rev. Microbiol.* **2015**, *13*, 269–284. [CrossRef] [PubMed]
9. Tamadonfar, K.O.; Omattage, N.S.; Spaulding, C.N.; Hultgren, S.J. Reaching the end of the line: Urinary tract infections. *Microbiol. Spectr.* **2019**, *7*, 1–16.
10. Nik-Ahd, F.; Lenore Ackerman, A.; Anger, J. Recurrent urinary tract infections in females and the overlap with overactive bladder. *Curr. Urol. Rep.* **2018**, *19*, 94. [CrossRef]

11. Beerepoot, M.; Geerlings, S. Non-antibiotic prophylaxis for urinary tract infections. *Pathogens* **2016**, *5*, 36. [CrossRef] [PubMed]

12. Jepson, R.G.; Williams, G.; Craig, J.C. Cranberries for preventing urinary tract infections. *Cochrane Database Syst. Rev.* **2012**. [CrossRef] [PubMed]

13. Wang, C.-H.; Fang, C.-C.; Chen, N.-C.; Liu, S.S.-H.; Yu, P.-H.; Wu, T.-Y.; Chen, W.-T.; Lee, C.-C.; Chen, S.-C. Cranberry-containing products for prevention of urinary tract infections in susceptible populations. *Arch. Intern. Med.* **2012**, *172*, 988–996. [CrossRef] [PubMed]

14. Blumberg, J.B.; Camesano, T.A.; Cassidy, A.; Kris-etherton, P.; Howell, A.; Manach, C.; Ostertag, L.M.; Sies, H.; Skulas-ray, A.; Vita, J.A. Cranberries and their bioactive constituents in human health. *Adv. Nutr.* **2013**, *4*, 618–632. [CrossRef] [PubMed]

15. Vasileiou, I.; Katsargyris, A.; Theocharis, S.; Giaginis, C. Current clinical status on the preventive effects of cranberry consumption against urinary tract infections. *Nutr. Res.* **2013**, *33*, 595–607. [CrossRef] [PubMed]

16. Occhipinti, A.; Germano, A.; Maffei, M.E. Prevention of urinary tract infection with oximacro, a cranberry extract with a high content of A-type proanthocyanidins: A pre-clinical double-blind controlled study. *Urol. J.* **2016**, *13*, 2640–2649.

17. Maki, K.C.; Kaspar, K.L.; Khoo, C.; Derrig, L.H.; Schild, A.L.; Gupta, K. Consumption of a cranberry juice beverage lowered the number of clinical urinary tract infection episodes in women with a recent history of urinary tract infection. *Am. J. Clin. Nutr.* **2016**, *103*, 1434–1442. [CrossRef]

18. Foxman, B.; Cronenwett, A.E.W.; Spino, C.; Berger, M.B.; Morgan, D.M. Cranberry juice capsules and urinary tract infection after surgery: Results of a randomized trial. *Am. J. Obs. Gynecol.* **2015**, *213*, 194.e1–194.e8. [CrossRef]

19. Stapleton, A.E.; Dziura, J.; Hooton, T.M.; Cox, M.E.; Yarova-Yarovaya, Y.; Chen, S.; Gupta, K. Recurrent urinary tract infection and urinary Escherichia coli in women ingesting cranberry juice daily: A randomized controlled trial. *Mayo Clin. Proc.* **2012**, *87*, 143–150. [CrossRef]

20. Liu, H.; Howell, A.B.; Zhang, D.J.; Khoo, C. A randomized, double-blind, placebo-controlled pilot study to assess bacterial anti-adhesive activity in human urine following consumption of a cranberry supplement. *Food Funct.* **2019**, *10*, 7645–7652. [CrossRef]

21. Pappas, E.; Schaich, K.M. Phytochemicals of cranberries and cranberry products: Characterization, potential health effects, and processing stability. *Crit. Rev. Food Sci. Nutr.* **2009**, *49*, 741–781. [CrossRef] [PubMed]

22. Sánchez-Patán, F.; Bartolomé, B.; Martín-Alvarez, P.J.; Anderson, M.; Howell, A.; Monagas, M. Comprehensive assessment of the quality of commercial cranberry products. Phenolic characterization and in vitro bioactivity. *J. Agric. Food Chem.* **2012**, *60*, 3396–3408. [CrossRef] [PubMed]

23. González de Llano, D.; Liu, H.; Khoo, C.; Moreno-Arribas, M.V.; Bartolomé, B. Some new findings regarding the antiadhesive activity of cranberry phenolic compounds and their microbial-derived metabolites against uropathogenic bacteria. *J. Agric. Food Chem.* **2019**, *67*, 2166–2174. [CrossRef] [PubMed]

24. Howell, A.B. Bioactive compounds in cranberries and their role in prevention of urinary tract infections. *Mol. Nutr. Food Res.* **2007**, *51*, 732–737. [CrossRef] [PubMed]

25. Ermel, G.; Georgeault, S.; Inisan, C.; Besnard, M. Inhibition of adhesion of uropathogenic escherichia coli bacteria to uroepithelial cells by extracts from cranberry. *J. Med. Food* **2012**, *15*, 126–134. [CrossRef]

26. De Llano, D.G.; Esteban-Fernández, A.; Sánchez-Patán, F.; Martín-Álvarez, P.J.; Moreno-Arribas, M.V.; Bartolomé, B. Anti-adhesive activity of cranberry phenolic compounds and their microbial-derived metabolites against uropathogenic escherichia coli in bladder epithelial cell cultures. *Int. J. Mol. Sci.* **2015**, *16*, 12119–12130. [CrossRef]

27. de Llano, D.G.; Arroyo, A.; Cárdenas, N.; Rodríguez, J.M.; Moreno-Arribas, M.V.; Bartolomé, B. Strain-specific inhibition of the adherence of uropathogenic bacteria to bladder cells by probiotic *Lactobacillus* spp. *Pathog. Dis.* **2017**, *75*, 1–8. [CrossRef]

28. Mena, P.; González de Llano, D.; Brindani, N.; Esteban-Fernández, A.; Curti, C.; Moreno-Arribas, M.V.; Del Rio, D.; Bartolomé, B. 5-(3′,4′-Dihydroxyphenyl)-γ-valerolactone and its sulphate conjugates, representative circulating metabolites of flavan-3-ols, exhibit anti-adhesive activity against uropathogenic Escherichia coli in bladder epithelial cells. *J. Funct. Foods* **2017**, *29*, 275–280. [CrossRef]

29. Al-Ghazzewi, F.H.; Tester, R.F. Biotherapeutic agents and vaginal health. *J. Appl. Microbiol.* **2016**, *121*, 18–27. [CrossRef]

30. Reid, G. The development of probiotics for women's health. *Can. J. Microbiol.* **2017**, *63*, 269–277. [CrossRef]

31. Stapleton, A.E.; Au-Yeung, M.; Hooton, T.M.; Fredricks, D.N.; Roberts, P.L.; Czaja, C.A.; Yarova-Yarovaya, Y.; Fiedler, T.; Cox, M.; Stamm, W.E. Randomized, placebo-controlled phase 2 trial of a lactobacillus crispatus probiotic given intravaginally for prevention of recurrent urinary tract infection. *Clin. Infect. Dis.* **2011**, *52*, 1212–1217. [CrossRef] [PubMed]

32. Montorsi, F.; Gandaglia, G.; Salonia, A.; Briganti, A.; Mirone, V. Effectiveness of a combination of cranberries, lactobacillus rhamnosus, and Vitamin C for the management of recurrent urinary tract infections in women: Results of a pilot study. *Eur. Urol.* **2016**, *70*, 912–915. [CrossRef] [PubMed]

33. Barrons, R.; Tassone, D. Use of Lactobacillus probiotics for bacterial genitourinary infections in women: A review. *Clin. Ther.* **2008**, *30*, 453–468. [CrossRef] [PubMed]

34. Scholes, D.; Hawn, T.R.; Roberts, P.L.; Li, S.S.; Stapleton, A.E.; Zhao, L.-P.; Stamm, W.E.; Hooton, T.M. Family history and risk of recurrent cystitis and pyelonephritis in women. *J. Urol.* **2010**, *184*, 564–569. [CrossRef] [PubMed]

35. Beerepoot, M.A.J. Lactobacilli vs. antibiotics to prevent urinary tract infections. *Arch. Intern. Med.* **2012**, *172*, 704. [CrossRef]

36. Mulvey, M.A. Adhesion and entry of uropathogenic Escherichia coli. *Cell. Microbiol.* **2002**, *4*, 257–271. [CrossRef]

37. Jagannathan, V.; Viswanathan, P. Proanthocyanidins—Will they effectively restrain conspicuous bacterial strains devolving on urinary tract infection? *J. Basic Microbiol.* **2018**, *58*, 567–578. [CrossRef]

38. Whiteside, S.A.; Razvi, H.; Dave, S.; Reid, G.; Burton, J.P. The microbiome of the urinary tract—A role beyond infection. *Nat. Rev. Urol.* **2015**, *12*, 81–90. [CrossRef]

39. Kline, K.A.; Lewis, A.L. Gram-positive uropathogens, polymicrobial urinary tract infection, and the emerging microbiota of the urinary tract. *Microbiol. Spectr.* **2016**, *4*, 1–54. [CrossRef]

40. Alam Parvez, S.; Rahman, D. Virulence factors of uropathogenic E. coli. In *Microbiology of Urinary Tract Infections-Microbial Agents and Predisposing Factors*; IntechOpen: Hamilton, NJ, USA, 2019.

41. Ismail, M.D.; Ali, I.; Hatt, S.; Salzman, E.A.; Cronenwett, A.W.; Marrs, C.F.; Rickard, A.H.; Foxman, B. Association of Escherichia coli ST131 lineage with risk of urinary tract infection recurrence among young women. *J. Glob. Antimicrob. Resist.* **2018**, *13*, 81–84. [CrossRef]

42. Mantzorou, M.; Giaginis, C. Cranberry consumption against urinary tract infections: Clinical stateof—the-art and future perspectives. *Curr. Pharm. Biotechnol.* **2019**, *19*, 1049–1063. [CrossRef] [PubMed]

43. Schaeffer, A.J.; Amundsen, S.K.; Jones, J.M. Effect of carbohydrates on adherence of Escherichia coli to human urinary tract epthelial cells. *Infect. Immun.* **1980**, *30*, 531–537. [PubMed]

44. Rossi, R.; Porta, S.; Canovi, B. Overview on cranberry and urinary tract infections in females. *J. Clin. Gastroenterol.* **2010**, *44*, S61–S62. [CrossRef] [PubMed]

45. Vigil, P.D.; Alteri, C.J.; Mobley, H.L.T. Identification of in vivo -induced antigens including an RTX family exoprotein required for uropathogenic Escherichia coli virulence. *Infect. Immun.* **2011**, *79*, 2335–2344. [CrossRef]

46. Stærk, K.; Khandige, S.; Kolmos, H.J.; Møller-Jensen, J.; Andersen, T.E. Uropathogenic Escherichia coli express Type 1 fimbriae only in surface adherent populations under physiological growth conditions. *J. Infect. Dis.* **2016**, *213*, 386–394. [CrossRef]

47. Andreu, A. Patogenia de las infecciones del tracto urinario. *Enferm. Infecc. Microbiol. Clin.* **2005**, *23*, 15–21. [CrossRef]

48. Soltani, S.; Emamie, A.D.; Dastranj, M.; Farahani, A. Role of toxins of uropathogenic Escherichia coli in development of urinary role of toxins of uropathogenic escherichia coli in development of urinary tract infection. *J. Pharm. Res. Int.* **2018**, 1–11. [CrossRef]

49. Matulay, J.T.; Mlynarczyk, C.M.; Cooper, K.L. Urinary tract infections in women: Pathogenesis, diagnosis, and management. *Curr. Bladder Dysfunct. Rep.* **2016**, *11*, 53–60. [CrossRef]

50. Rodriguez-Mateos, A.; Vauzour, D.; Krueger, C.G.; Shanmuganayagam, D.; Reed, J.; Calani, L.; Mena, P.; Del Rio, D.; Crozier, A. Bioavailability, bioactivity and impact on health of dietary flavonoids and related compounds: An update. *Arch. Toxicol.* **2014**, *88*, 1803–1853. [CrossRef]

51. Kimble, L.L.; Mathison, B.D.; Kaspar, K.L.; Khoo, C.; Chew, B.P. Development of a fluorometric microplate antiadhesion assay using uropathogenic escherichia coli and human uroepithelial cells. *J. Nat. Prod.* **2014**, *77*, 1102–1110. [CrossRef]

52. Juthani-Mehta, M.; Van Ness, P.H.; Bianco, L.; Rink, A.; Rubeck, S.; Ginter, S.; Argraves, S.; Charpentier, P.; Acampora, D.; Trentalange, M.; et al. Effect of cranberry capsules on bacteriuria plus pyuria among older women in nursing homes. *JAMA* **2016**, *316*, 1879–1887. [CrossRef] [PubMed]

53. Chughtai, B.; Howell, A.B.; Thomas, D.; Blumberg, J.B. Efficacy of cranberry in preventing recurrent urinary tract infections: Have we learned anything new? *Urology* **2017**, *103*, 2–3. [CrossRef] [PubMed]

54. Rafsanjany, N.; Lechtenberg, M.; Petereit, F.; Hensel, A. Antiadhesion as a functional concept for protection against uropathogenic Escherichia coli: In vitro studies with traditionally used plants with antiadhesive activity against uropathognic Escherichia coli. *J. Ethnopharmacol.* **2013**, *145*, 591–597. [CrossRef] [PubMed]

55. Monagas, M.; Urpi-Sarda, M.; Sánchez-Patán, F.; Llorach, R.; Garrido, I.; Gómez-Cordovés, C.; Andres-Lacueva, C.; Bartolomé, B. Insights into the metabolism and microbial biotransformation of dietary flavan-3-ols and the bioactivity of their metabolites. *Food Funct.* **2010**, *1*, 233–253. [CrossRef] [PubMed]

56. Feliciano, R.P.; Boeres, A.; Massacessi, L.; Istas, G.; Heiss, C.; Ventura, M.R.; Rodriguez-mateos, A. Identi fi cation and quanti fi cation of novel cranberry-derived plasma and urinary (poly) phenols. *Arch. Biochem. Biophys.* **2016**, *599*, 31–41. [CrossRef]

57. Feliciano, R.P.; Mills, C.E.; Istas, G.; Heiss, C.; Rodriguez-Mateos, A. Absorption, metabolism and excretion of cranberry (poly)phenols in humans: A dose response study and assessment of inter-individual variability. *Nutrients* **2017**, *9*, 268. [CrossRef]

58. Peron, G.; Pellizzaro, A.; Brun, P.; Schievano, E.; Mammi, S.; Sut, S.; Castagliuolo, I.; Dall'Acqua, S. Antiadhesive activity and metabolomics analysis of rat urine after cranberry (vaccinium macrocarpon aiton) administration. *J. Agric. Food Chem.* **2017**, *65*, 5657–5667. [CrossRef]

59. Baron, G.; Altomare, A.; Regazzoni, L.; Fumagalli, L.; Artasensi, A.; Borghi, E.; Ottaviano, E.; Del Bo, C.; Riso, P.; Allegrini, P.; et al. Profiling vaccinium macrocarpon components and metabolites in human urine and the urine ex-vivo effect on Candida albicans adhesion and biofilm-formation. *Biochem. Pharm.* **2020**, *173*, 113726. [CrossRef]

60. Domenici, L.; Monti, M.; Bracchi, C.; Giorgini, M.; Colagiovanni, V.; Muzii, L.; Benedetti Panici, P. D-mannose: A promising support for acute urinary tract infections in women. A pilot study. *Eur. Rev. Med. Pharm. Sci.* **2016**, *20*, 2920–2925.

61. Hong, V.; Wrolstad, R.E. Cranberry juice composition. *J. Assoc. Off. Anal. Chem.* **1986**, *69*, 199–207. [CrossRef]

62. Jensen, H.D.; Struve, C.; Christensen, S.B.; Krogfelt, K.A. Cranberry juice and combinations of its organic acids are effective against experimental urinary tract infection. *Front. Microbiol.* **2017**, *8*, 1–6. [CrossRef] [PubMed]

63. Sanchez-Patan, F.; Barroso, E.; Van De Wiele, T.; Jimenez-Giron, A.; Martin-Alvarez, P.J.; Moreno-Arribas, M.V.; Martinez-Cuesta, C.; Pelaez, C.; Requena, T.B. Comparative in vitro fermentations of cranberry and grape seed polyphenols with colonic microbiota. *Food Chem.* **2015**, *183*, 273–282. [CrossRef] [PubMed]

64. Scharf, B.; Sendker, J.; Dobrindt, U.; Hensel, A. Influence of cranberry extract on tamm-horsfall protein in human urine and its antiadhesive activity against uropathogenic Escherichia coli. *Planta Med.* **2019**, *85*, 126–138. [CrossRef] [PubMed]

65. Scharf, B.; Schmidt, T.J.; Rabbani, S.; Stork, C.; Dobrindt, U.; Sendker, J.; Ernst, B.; Hensel, A. Antiadhesive natural products against uropathogenic *E. coli*: What can we learn from cranberry extract ? *J. Ethnopharmacol.* **2020**, *257*, 112889. [CrossRef]

66. Sekirov, I.; Russell, S.L.; Antunes, C.M.; Finlay, B.B. Gut microbiota in health and disease. *Physiol. Rev.* **2010**, *90*, 859–904. [CrossRef]

67. Magistro, G.; Stief, C.G. The urinary tract microbiome: The answer to all our open questions? *Eur. Urol. Focus* **2019**, *5*, 36–38. [CrossRef]

68. Aragón, I.M.; Herrera-Imbroda, B.; Queipo-Ortuño, M.I.; Castillo, E.; Del Moral, J.S.-G.; Gómez-Millán, J.; Yucel, G.; Lara, M.F. The urinary tract microbiome in health and disease. *Eur. Urol. Focus* **2018**, *4*, 128–138. [CrossRef]

69. Stapleton, A.E. The vaginal microbiota and urinary tract infection. *Microbiol. Spectr.* **2016**, *4*. [CrossRef]

70. Bekiares, N.; Krueger, C.G.; Meudt, J.J.; Shanmuganayagam, D.; Reed, J.D. Effect of sweetened dried cranberry consumption on urinary proteome and fecal microbiome in healthy human subjects. *Omics J. Integr. Biol.* **2017**, *21*, 1–9. [CrossRef]

71. Rodríguez-morató, J.; Matthan, N.R.; Liu, J.; De, R.; Chen, C.O. ScienceDirect cranberries attenuate animal-based diet-induced changes in microbiota composition and functionality: A randomized crossover controlled feeding trial. *J. Nutr. Biochem.* **2018**, *62*, 76–86. [CrossRef]

72. Feliciano, R.P.; Meudt, J.J.; Shanmuganayagam, D.; Krueger, C.G.; Reed, J.D. Ratio of "a-type" to "b-type" proanthocyanidin interflavan bonds affects extra-intestinal pathogenic escherichia coli invasion of gut epithelial cells. *J. Agric. Food Chem.* **2014**, *62*, 3919–3925. [CrossRef] [PubMed]

73. van Duynhoven, J.; van der Hooft, J.J.J.; van Dorsten, F.A.; Peters, S.; Foltz, M.; Gomez-Roldan, V.; Vervoort, J.; de Vos, R.C.H.; Jacobs, D.M. Rapid and sustained systemic circulation of conjugated gut microbial catabolites after single-dose black tea extract consumption. *J. Proteome Res.* **2014**, *2*, 2668–2678. [CrossRef] [PubMed]

74. Dueñas, M.; Cueva, C.; Muñoz-gonzález, I.; Jiménez-girón, A.; Sánchez-patán, F.; Santos-buelga, C.; Moreno-arribas, M.V.; Bartolomé, B. Studies on modulation of gut microbiota by wine polyphenols: From isolated cultures to omic approaches. *Antioxidants* **2015**, *4*, 1–21. [CrossRef] [PubMed]

75. Johnson, J.R.; Russo, T.A.; Brown, J.J.; Stapleton, A. papG alleles of Escherichia coli strains causing first-episode or recurrent acute cystitis in adult women. *J. Infect. Dis.* **1998**, *177*, 97–101. [CrossRef]

76. Álvarez-Calatayud, G.; Suárez, E.; Rodríguez, J.M.; Pérez-Moreno, J. La microbiota en la mujer; aplicaciones clínicas de los probióticos. *Nutr. Hosp.* **2015**, *32*, 56–61.

77. Reid, G.; Beuerman, D.; Heinemann, C.; Bruce, A.W. Probiotic Lactobacillus dose required to restore and maintain a normal vaginal flora. *FEMS Immunol. Med. Microbiol.* **2001**, *32*, 37–41. [CrossRef]

78. Rodríguez, J.M. The origin of human milk bacteria: Is there a bacterial entero-mammary pathway during late pregnancy and lactation? *Adv. Nutr.* **2014**, *5*, 779–784. [CrossRef]

79. Kontiokari, T.; Laitinen, J.; Järvi, L.; Pokka, T.; Sundqvist, K.; Uhari, M. Dietary factors protecting women from urinary tract infection. *Am. J. Clin. Nutr.* **2003**, *77*, 600–604. [CrossRef]

80. Osset, J.; Bartolomé, R.M.; García, E.; Andreu, A. Assessment of the capacity of lactobacillus to inhibit the growth of uropathogens and block their adhesion to vaginal epithelial cells. *J. Infect. Dis.* **2001**, *183*, 485–491. [CrossRef]

81. Castro, J.; Henriques, A.; Machado, A.; Henriques, M.; Jefferson, K.K.; Cerca, N.; Whiteside, S.A.; Razvi, H.; Dave, S.; Reid, G.; et al. Effect of two probiotic strains of Lactobacillus on in vitro adherence of listeria monocytogenes, streptococcus agalactiae, and staphylococcus aureus to vaginal epithelial cells. *Arch. Gynecol. Obs.* **2014**, *12*, 479–489.

82. Delley, M.; Bruttin, A.; Richard, M.; Affolter, M.; Rezzonico, E.; Brück, W.M. In vitro activity of commercial probiotic Lactobacillus strains against uropathogenic Escherichia coli. *FEMS Microbiol. Lett.* **2015**, *362*, 1–7. [CrossRef] [PubMed]

83. Bruce, A.W.; Reid, G. Intravaginal instillation of lactobacilli for prevention of recurrent urinary tract infections. *Can. J. Microbiol.* **1988**, *34*, 339–343. [CrossRef] [PubMed]

84. Saxelin, M.; Pessi, T.; Salminen, S. Fecal recovery following oral administration of Lactobacillus strain GG (ATCC 53103) in gelatine capsules to healthy volunteers. *Int. J. Food Microbiol.* **1995**, *25*, 199–203. [CrossRef]

85. Hagan, E.C.; Mobley, H.L.T. Heme acquisition is facilitated by a novel receptor Hma and required by uropathogenic Escherichia coli for kidney infection. *Mol. Microbiol.* **2009**, *71*, 79–91. [CrossRef]

86. Cadieux, P.A.; Burton, J.; Devillard, E.; Reid, G. Lactobacillus by-products inhibit the growth and virulence of uropathogenic Escherichia coli. *J. Physiol. Pharm.* **2009**, *60* (Suppl. 6), 13–18.

87. Rupasinghe, H.P.V.; Parmar, I.; Neir, S.V. Biotransformation of cranberry proanthocyanidins to probiotic metabolites by lactobacillus rhamnosus enhances their anticancer activity in HepG2 cells in vitro. *Oxid. Med. Cell Longev.* **2019**, *2019*, 1–14. [CrossRef]

88. Peter, M.G.; Paulina, M.K.; Waleed, A.; Alison, E.F.-R. Lactobacillus for preventing recurrent urinary tract infections in women: Meta-analysis. *Can. J. Urol.* **2013**, *20*, 6607–6614.

89. Schwenger, E.M.; Tejani, A.M.; Loewen, P.S. Probiotics for preventing urinary tract infections in adults and children. *Cochrane Database Syst. Rev.* **2015**, *12*. [CrossRef]

90. Ahumada-cota, R.E.; Hernandez-chiñas, U.; Mili, F.; Ch, E.; Navarro-ocaña, A.; Mart, D. Effect and analysis of bacterial lysates for the treatment of recurrent urinary tract infections in adults. *Pathogens* **2020**, *9*, 102. [CrossRef]

91. Polewski, M.A.; Krueger, C.G.; Reed, J.D.; Leyer, G. Ability of cranberry proanthocyanidins in combination with a probiotic formulation to inhibit *in vitro* invasion of gut epithelial cells by extra-intestinal pathogenic *E. coli*. *J. Funct. Foods* **2016**, *25*, 123–134. [CrossRef]

92. Yuan, H.; Ma, Q.; Cui, H.; Liu, G.; Zhao, X.; Li, W.; Piao, G. How can synergism of traditional medicines benefit from network pharmacology? *Molecules* **2017**, *22*, 1135. [CrossRef] [PubMed]

93. Coleman, C.M.; Ferreira, D. Oligosaccharides and complex carbohydrates: A new paradigm for cranberry bioactivity. *Molecules* **2020**, *25*, 881. [CrossRef] [PubMed]

 *molecules*

Article

Benefits of Anthocyanin-Rich Black Rice Fraction and Wood Sterols to Control Plasma and Tissue Lipid Concentrations in Wistar Kyoto Rats Fed an Atherogenic Diet

Aneta Kopeć [1], Jerzy Zawistowski [2] and David D. Kitts [2,*

[1] Department of Human Nutrition and Dietetics, Faculty of Food Technology, University of Agriculture in Krakow, Balicka 122, 31-149 Kraków, Poland; aneta.kopec@urk.edu.pl

[2] Faculty of Land and Food Systems, University of British Columbia 209-2205 East Mall, Vancouver, BC V6T 1Z4, Canada; jerzy.zawistowski@ubc.ca

* Correspondence: david.kitts@ubc.ca; Tel.: +604-822-5560

Academic Editors: H. P. Vasantha Rupasinghe and Celestino Santos-Buelga

Received: 8 October 2020; Accepted: 15 November 2020; Published: 17 November 2020

Abstract: Background: This study reports on the relative effects of administrating a cyanidin-3-*O*-glucoside-rich black rice fraction (BRF), a standardized wood sterol mixture (WS), and a combination of both to lower plasma and target tissue lipid concentrations in Wistar Kyoto (WKY) rats fed atherogenic diets. Methods: Male WKY ($n = 40$) rats were randomly divided into five groups, which included a nonatherogenic control diet and atherogenic diets that included a positive control and atherogenic diets supplemented with BRF or WS, respectively, and a combination of both BRF + WS. Plasma and target tissue liver, heart and aorta cholesterol, and triacylglycerides (TAG) content were also measured. Results: Rats fed atherogenic diets exhibited elevated hyperlipidemia compared to counterparts fed nonatherogenic diets ($p < 0.001$); this effect was mitigated by supplementing the atherogenic diets with BRF and WS, respectively ($p < 0.05$). Combining BRF with WS to enrich the supplement lowered cholesterol similar to the WS effect ($p < 0.05$) and lowered TAG characteristic to the BRF effect ($p < 0.05$). Conclusions: Rats fed diets containing BRF or WS effectively mitigate the hypercholesterolemia and elevated TAG induced by feeding an atherogenic diet. The benefit of adding BRF + WS together is relevant to the lipid parameter measured and is target tissue-specific.

Keywords: black rice cyanidin-3-*O*-glucoside; wood sterols; dyslipidemia; CVD

---

1. Introduction

Cardiovascular disease (CVD) represents a major cause of death in Europe, North America, and in some parts of Asia [1–4]. According to the World Health Organization 2016 records, CVD accounted for 31% of the total global mortality [5]. This knowledge has empowered today's consumers to be selective at identifying foods that have both adequate essential nutrient contents to meet daily requirements while also providing distinct health benefits attributed to plant-based bioactives that can protect against CVD. Examples of bioactive food constituents that lower serum cholesterol are plant sterols [6–8] and, also, pigmented anthocyanins that display antioxidant and anti-inflammatory properties [9–12], the combination of which could have a greater role in protecting against CVD. Hypercholesterolemia, along with a disturbed antioxidant status and onset of inflammation, are major underlying causes of CVD.

Phytosterols are natural ingredients present in vegetable oils and, also, can be recovered from wood sources. Phytosterol intake from a traditional Western diet depends on personal habits and geographical location [13]. A typical Western diet contains about 300 mg/day of phytosterols [13–15].

Plant sterols and stanols are structurally similar to cholesterol (Figure 1) [16] and are effective at lowering the plasma total and low-density lipoprotein cholesterol [7,17,18].

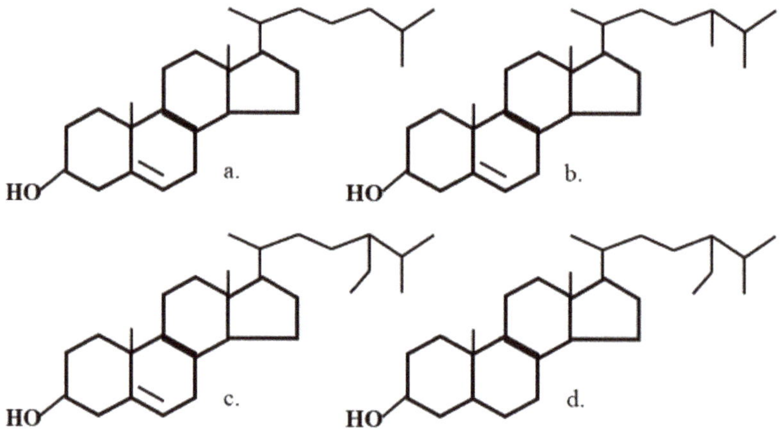

Figure 1. Chemical structures of (**a**) cholesterol, (**b**) campesterol, (**c**) β-sitosterol, and (**d**) β-sitostanol.

Black rice, on the other hand, is an excellent source of anthocyanidins, in addition to dietary fiber, flavonoids, and other polyphenols. Rice oil contains nonatherogenic fatty acids [19,20] and is also a good source of plant sterols, such as oryzanol. These bioactive components are unsaponifiable, nonglyceride components [20], which contribute to cholesterol-lowering effects reported in many animal and human studies [19–24]. The pigment from black rice contains two major anthocyanins—namely, cyanidin-3-*O*-glucoside and peonidin-3-*O*-glucoside—the former being predominant [10,23,25] (Figure 2).

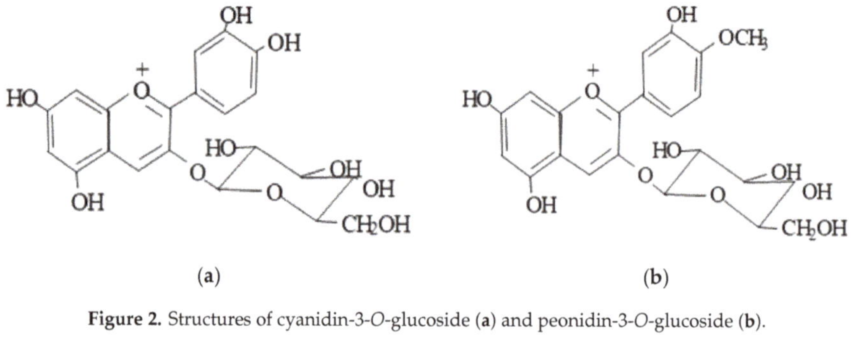

Figure 2. Structures of cyanidin-3-*O*-glucoside (**a**) and peonidin-3-*O*-glucoside (**b**).

Anthocyanins are naturally occurring phenolic compounds that provide color and bioactive properties, such as antioxidants [10] and lowering of serum cholesterol and triacyglycerides in rats, when fed to rats on a daily basis [26].

There are no studies that have examined the effects of consuming a combination of these bioactive agents to mitigate the changes of known CVD risk factors induced by feeding on an atherogenic diet. The objective of the present study was to demonstrate a potential interaction or added effect of combining a standardized cyanidin-3-*O*-glucoside black rice fraction (BRF) and a known wood sterol (WS) mixture to protect against elevated serum lipids and, moreover, target tissue cholesterol deposition in rats fed an atherogenic diet.

2. Results

2.1. Body Weight Gain and Heart and Liver Weights

The body weight gain (Figure 3a) of rats fed experimental atherogenic diets with BRF was significantly higher ($p < 0.05$) compared to other experimental groups, but no differences were observed for the feed efficiency ratio (FER, Figure 3b) between groups. Supplementary Table S1 presents the food intake and tissue organ weights, respectively, of animals fed on atherogenic experimental diets for 11 weeks. We also include data from rats fed a negative cholesterol (NC) diet for comparison of the plasma and tissue lipid concentrations typical of an AIN-76 A control diet. The atherogenic diets supplemented with WS and a combination of WS + BRF lowered the liver weight significantly ($p < 0.05$). The heart weight did not vary significantly between treatments. Feeding atherogenic diets with and without WS and BRF supplements did not affect the hematocrit or hemoglobin concentrations.

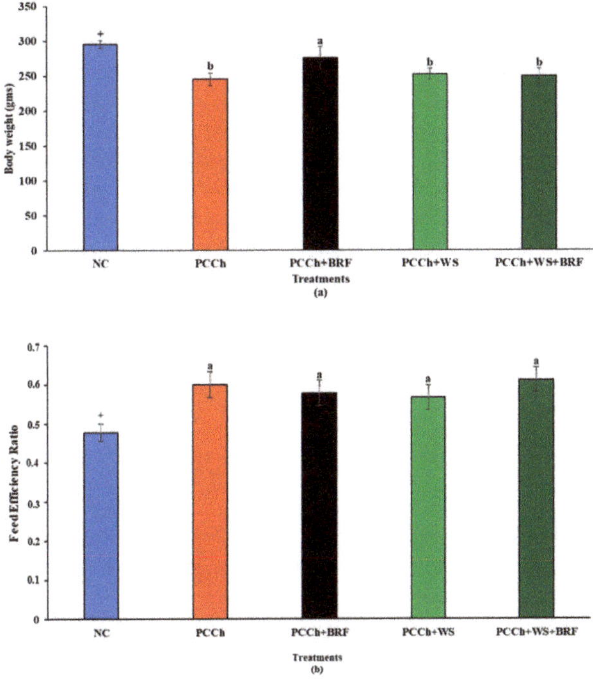

Figure 3. (**a**) Body weight gain of experimental rats. NC- AIN-76A diet. PCCh positive control (0.5% cholesterol, 0.05 cholic acid, 3% butter). BRF-black rice fraction. PCCh +BRF positive control with black rice fraction. PCCh +WS positive control with wood sterols. PCCh +WS+BRF positive control with wood sterols and black rice fraction. Bars represent means SEM/ (n = 8). The + symbol signifies a significant difference ($p < 0.05$) between the negative cholesterol (NC) and positive control group (PCCh). Bars with different superscript letter (a,b) are significantly different ($p \leq 0.05$).Bars with different superscript letter (a,b) are significantly different ($p \leq 0.05$). (**b**). Fed efficiency ratio of rats. NC- AIN-76A diet. PCCh positive control (0.5% cholesterol, 0.05 cholic acid, 3% butter). BRF-black rice fraction. PCCh + BRF positive control with black rice fraction. PCCh + WS positive control with wood sterols. PCCh + WS + BRF positive control with wood sterols and black rice fraction[1]. Bars represent means SEM/ (n = 8). The + symbol signifies a significant difference ($P < 0.05$) between the negative cholesterol (NC) and positive control group (PCCh).

2.2. Blood Constituent Analysis

There was no dietary effect observed that indicated differences in the plasma glucose levels of animals fed the control or experimental diets, respectively. Similarly, the plasma Oxygen Radical Absorbance Capacity (ORAC) and C-Reactive Protein (CRP) were not different between animals fed different diets (data not shown). This result may be explained by the fact that only one-time blood sampling of the blood was made at the time of sacrifice, which was likely insufficient to show meaningful trends in these biochemical parameters.

Animals fed on the experimental atherogenic diets (PCCh) exhibited hyperlipidemia, compared to counterparts fed a NC diet. The total plasma cholesterol concentration (TC) was significantly ($p < 0.05$) lowered in animals fed on experimental PCCh diets that also contained BRF and WS, respectively. The relative degree of lowering was greatest in animals fed on the WS atherogenic diet. Combining BRF + WS in atherogenic diets resulted in changes in the plasma cholesterol that resembled animals fed the WS diet (Table 1). The high-density lipoprotein (HDL)/TC ratio was significantly higher ($p < 0.05$) for all animals fed experimental PCCh diets that included BRF, WS, or the combination of BFR + WS, in comparison to the control PCCh diet (Figure 4). Rats fed the nonatherogenic control diet also exhibited a distinct high-plasma HDL/TC ratio. This observation can be attributed to the fact that the HDL cholesterol concentrations were significantly ($p < 0.05$) higher in WS fed rats on the atherogenic diet (Table 1). Plasma non-HDL cholesterol concentrations were significantly ($p < 0.001$) lower in rats fed atherogenic diets containing BRF, but the greatest reductions were obtained in counterpart animals fed on WS (Table 1). This observation led to the conclusion that the principle component for the cholesterol-lowering effect observed in rats fed on the BRF + WS diet ($p < 0.05$) was indeed attributed to the presence of WS more so than BRF. This trend was reversed for the plasma triacylglyceride (TAG) responses in rats fed the atherogenic diets supplemented with BRF. For example, plasma TAG levels were significantly ($p < 0.01$) lower in rats fed diets with added BRF and lower than those fed WS ($p < 0.05$). Of interest was the observation that the combination of adding BRF + WS produced a significantly lower ($p < 0.05$) plasma TAG level compared to rats fed only WS (Table 1). Hence, rats fed on atherogenic diets that contained both BRF + WS exhibited plasma TAG concentrations that resembled animals fed the BRF supplement. These trends were not observed for plasma phospholipids responses in rats fed the combination of BRF + WS but, rather, resembled moreso the WS group (Table 1).

Table 1. Lipid profile in the plasma of rats fed experimental diets.

Diet Treatment	TC	HDL	Non-HDL	TAG	PL
NC	1.75 ± 0.10 [+]	1.12 ± 0.03 [+]	0.67 ± 0.09 [+]	0.52 ± 0.04	111.8 ± 9.0
PCCh	4.21 ± 0.11 [a]	0.74 ± 0.03 [a]	3.47 ± 0.12 [a]	0.56 ± 0.03 [a]	96.97 ± 4.67 [a]
PCCh + BRF	2.67 ± 0.12 [b]	0.74 ± 0.06 [a]	1.74 ± 0.13 [b]	0.45 ± 0.04 [b]	89.78 ± 2.1 [a]
PCCh + WS	2.07 ± 0.06 [c]	1.08 ± 0.05 [b]	1.05 ± 0.12 [c]	0.74 ± 0.03 [c]	115 ± 4.6 [b]
PCCh + WS + BRF	2.08 ± 0.07 [c]	0.73 ± 0.08 [a]	1.33 ± 0.14 [c]	0.47 ± 0.02 [b]	112.5 ± 5.7 [b]

TC—total cholesterol (mmol/L), HDL—high-density lipoprotein cholesterol (mmol/L), Non-HDL cholesterol (mmol/L), TAG—triacylglycerides (mmol/L), and PL—phospholipids (mg/dL). NC: AIN-76A diet. PCCh: positive control (0.5% cholesterol, 0.05 cholic acid, 3% butter). BRF: black rice fraction. PCCh + BRF: positive control with black rice fraction. PCCh + WS: positive control with wood sterols. PCCh + WS + BRF: positive control with wood sterols and black rice fraction. Data are expressed as means ± SEM ($n = 8$). The + symbol signifies a significant difference ($p < 0.05$) between the negative cholesterol (NC) and positive control group (PCCh). Values in columns with different superscript letters (a, b, and c) are significantly different ($p \leq 0.05$).

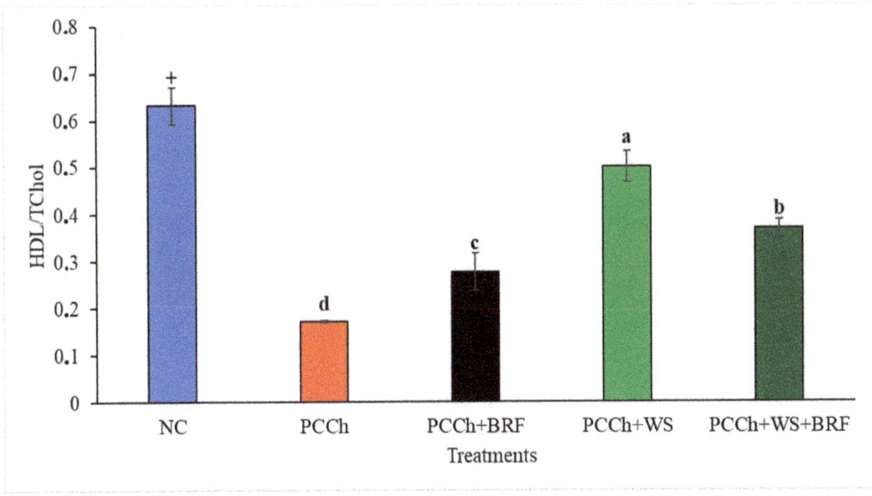

Figure 4. Plasma high-density lipoprotein (HDL) cholesterol/total cholesterol ratio in experimental groups. NC: AIN-76A diet. PCCh: positive control (0.5% cholesterol, 0.05 cholic acid, 3% butter). BRF: black rice fraction. PCCh + BRF: positive control with black rice fraction. PCCh + WS: positive control with wood sterols. PCCh + WS + BRF: positive control with wood sterols and black rice fraction. Bars represent means ± SEM ($n = 8$). The + symbol signifies a significant difference ($p < 0.05$) between the negative cholesterol (NC) and positive control groups (PCCh). Bars with different superscript letters (a, b, c, and d) are significantly different ($p \leq 0.05$).

2.3. Tissue Lipid Concentrations

The crude lipid concentration in liver tissue was significantly ($p < 0.05$) higher in rats fed the atherogenic diet that did not contain added BRF and WS bioactives, respectively (Table 2). The greatest relative reduction in liver cholesterol of rats fed atherogenic diets was observed in rats fed on the WS and BRF + WS diets, respectively ($p < 0.05$) (Table 2). The presence of both BRF + WS to significantly reduce liver ($p < 0.05$) cholesterol in these animals was therefore attributed mostly to the presence of WS in the phytochemical supplement mixture. In addition, the lowest liver triacylglycerol concentrations were observed in rats fed on diets with BRF ($p < 0.05$), which resembled counterparts fed the atherogenic BRF + WS diet.

Table 2. Lipid concentration in selected tissues in experimental rats (mg/g).

Diet Treatment	Liver CL	Liver TC	Liver TAG	Heart TC	Aorta TC
NC	58 ± 2.83 [+]	6.19 ± 0.59 [+]	0.53 ± 0.33 [+]	0.157 ± 0.01 [+]	1.14 ± 0.03 [+]
PCCh	390 ± 20 [a]	15.92 ± 0.56 [a]	9.53 ± 0.64 [a]	0.256 ± 0.03 [a]	5.86 ± 0.83 [a]
PCCh + BRF	280 ± 12 [b]	11.86 ± 1.36 [b]	6.29 ± 0.95 [b]	0.186 ± 0.03 [b]	2.14 ± 0.20 [b]
PCCh + WS	96 ± 15 [c]	8.30 ± 0.58 [c]	10.17 ± 0.50 [a]	0.245 ± 0.01 [a]	0.55 ± 0.10 [d]
PCCh + WS + BRF	120 ± 10 [c]	8.17 ± 0.71 [c]	7.73 ± 0.56 [b]	0.234 ± 0.02 [a]	1.51 ± 0.15 [c]

CL = crude lipid (mg/g ww), TC = total cholesterol (mg/g ww), and TAG = triacylglyceride (mg/g ww). NC-AIN-76A diet. PCCh: positive control (0.5% cholesterol, 0.05 cholic acid, 3% butter). BRF:black rice fraction. PCCh + BRF: positive control with black rice fraction. PCCh + WS: positive control with wood sterols. PCCh + WS + BRF: positive control with wood sterols and black rice fraction. Data are expressed as means ± SEM ($n = 8$). The + symbol signifies a significant difference ($p < 0.05$) between the negative cholesterol (NC) and positive control group (PCCh). Values in columns with different superscript letters (a, b, and c) are significantly different ($p \leq 0.05$).

In heart tissue, cholesterol accumulation in animals fed atherogenic diets was higher ($p < 0.05$) compared to NC diet-fed control animals. The only exception to this was with the atherogenic rats supplemented with BRF, which exhibited characteristically low heart cholesterol accumulation compared to the WS counterparts ($p < 0.05$) and more closely resembled the NC controls. The presence of both BRF and WS in the atherogenic diet was not sufficient to reduce heart cholesterol to the level observed for rats fed the BRF-only diet. Of interest was the fact that measuring the cholesterol content in extracted aortic tissue produced dramatically different results. Rats fed the atherogenic diet produced a five-fold increase in aortic cholesterol content compared to the NC diet-fed animals. The low aortic cholesterol concentration found in NC controls was also observed in animals fed on the BRF atherogenic diet, but this was dramatically lower ($p < 0.05$) compared to counterparts fed on the WS atherogenic diet. Combining both BRF and WS supplements in the atherogenic diet indeed produced a lower cholesterol accumulation ($p < 0.05$) compared to the atherogenic control, but the extent of the cholesterol reduction did not match what was obtained when WS and BRF were fed individually (Table 2).

3. Discussion

The present study shows the relative efficacy of supplementing experimental diets that are formulated to hyperlipidemic, with a standardized anthocyanin-rich BRF and sterol-rich WS, to characteristically mitigate both plasma and tissue lipid increases that are attributable risk factors for CVD. Moreover, in some specific instances, the combination of having both BRF + WS included in the supplemented diet produced an effect that was more or less typical of BRF or WS, added individually, but this was dependent on the plasma lipid measurement and the specific tissue sampled.

In this study, adding 1% BRF to the atherogenic diet produced a greater weight gain in rats, indicating a total acceptance of the diet that included BRF. Other workers reported weight loss in mice and rabbits when fed on BRF-containing diets [9,27]. BRF has been shown to have nutritional qualities outside their antioxidant capacity that include high calcium and protein bioavailabilities [21,27]. Animals fed atherogenic diets that were formulated to contain added cholesterol and lard as the fat source, along with the defined anthocyanin mixture that composed BRF supplementation, showed a lower body weight gain reported by others [28]. In contrast, other studies reported no change in body weight gain in a variety of different experimental animals fed atherogenic diets that included BRF [29,30] or plant sterols [31].

The atherogenic diets fed to rats in this study were composed of butter and canola oil, along with cholesterol and cholic acid. The observed greater liver cholesterol content of control rats corresponded to the noted hypercholesterolemia that occurred in these animals. Supplementing the atherogenic diets with BRF and WS produced lower liver weights, respectively. We attribute the effect associated with BRF to the fact that it is a good source of soluble dietary fiber and, also, contains trace amounts of oryzanol. These bioactive constituents are well-known to form complexes with cholesterol directly in the intestine and, also, bind to bile acids that are required for cholesterol absorption, the net effect being reduced cholesterol absorption that promotes excretion. Oryzanol, a phytosterol also recovered from the black rice outer layer [32], is also known to be effective at reducing cholesterol absorption and increasing the excretion of bile acids [33], thus potentially contributing to lower liver weights of BRF-fed rats.

Similarly, the higher total plasma cholesterol concentrations observed in rats fed on the atherogenic control diet, compared to the NC diet-fed rodents, were reduced when the atherogenic diets were supplemented with BRF and WS, respectively. Similar results have been reported with black rice in C57BL/6J mice fed a high-fat diet containing an anthocyanin-rich extract derived from a similar rice source [9]. Frequent exposure to anthocyanins has been associated with the stimulation of transporters that involve anthocyanin absorption [34]. Thus, daily consumption could potentially decrease CVD risk, associated with conditions of inflammation and atherosclerosis [9,10,24,35,36]. Black rice and outer layer fractions also significantly reduce serum lipids [35] and atherosclerotic plaque formation in hypercholesterolemic rabbits and Apo-E-deficient mice [9,22,27]. These authors reported that the supplementation of experimental diets with the black rice outer layer fraction decreased the level of LDL cholesterol in Apo-E-deficient mice. This observation was not confirmed in other studies with rabbits fed black rice [27]. On the other hand, rats fed diets containing WS alone showed a greater reduction in plasma cholesterol compared to those fed diets containing only BRF. The primary mechanism for the hypocholesterolemic effect attributed to phytosterol intake is related to the inhibition of the micellar cholesterol solubilization of cholesterol, which results in reduced cholesterol absorption from the intestinal lumen and promotes excretion [18,37,38]. The consumption of about 2 g of sterols per day reduces LDL cholesterol by 8–15% and, consequently, reduces the risk of cardiovascular disease in humans [39]. Currently, also, the anti-inflammatory effect of plant sterols is possible; albeit, these data are inconsistent [38,40,41]. The mechanism for WS lowering plasma cholesterol is well-defined to include a competition to form mixed micelles that are required for cholesterol absorption in the small intestine, due to the similarities in structure. The limited solubility of WS in the intestinal chyme also interferes with the solubility of cholesterol, which, in turn, contributes to coprecipitation in the intestinal lumen and low cholesterol bioavailability [18]. WS also competes with cholesterol for various transporters at the apical surface of enterocytes—namely, Niemann-Pick C1-like 1 protein (NPC1L1) and adenosine triphosphate (ATP) cassette protein-binding transporters (ABCG5/G8), which regulate the influx and efflux of cholesterol and sterols between the intestinal lumen and the intestinal mucosa [41]. The greater decrease in plasma cholesterol in animals fed diets supplemented with a combination of WS and BRF paralleled a similar reduction in the total cholesterol contents of the liver, heart, and aorta tissues. For the most part, the effect on cholesterol lowering was most dramatic when WS were present in the atherogenic diet. This was also true when both BRF + WS were combined, with the expected added effect to reduce cholesterol even further not realized but, rather, attributed mainly to the presence of WS in the mixture. However, having BRF present in the mixture may be contributed, to some extent, by other mechanisms due to the presence of both anthocyanins and dietary fiber. In addition to the reported benefits of having the antioxidant activity of anthocyanins to protect against cardiovascular diseases [42,43], these compounds also inhibit rate-limiting enzyme activities for cholesterol synthesis—namely, 3-hydroxy-3-methyl-glutaryl-coenzyme A (HMG-CoA) and cholesteryl ester transfer protein (CEPT) enzymes, which are linked to changes in endogenous cholesterol synthesis [43–45]. Pong et al. [46] showed that mulberry extracts contain anthocyanins, decreased hepatic lipids, and reduce the level of HMG-CoA reductase in hamsters. A similar cascade of lipid specific metabolic reactions were suppressed in rats fed anthocyanin-rich juice [26]. The soluble fiber present in back rice also has cholesterol-lowering properties [47,48], attributed to a reduced dietary cholesterol bioaccessibility for absorption. Moreover, soluble dietary fiber derived from BRF could also be a substrate for bifidobacteria fermentation in the colon, with a subsequent production of short-chain fatty acids (SFCA) that may suppress cholesterol biosynthesis, thus further contributing to low cholesterol levels in plasma and tissue [49,50].

The presence of WS, BRF, or the combination of both in experimental atherogenic diets fed to rats increased the plasma HDL cholesterol. Great interest exists in the strong inverse and independent relationship beween HDL and clinical atherosclerosis [49]. HDL transports cholesterol or cholesterol esters from peripheral tissues to the liver, where cholesterol is metabolized and transformed into bile acids; thus, this route of cholesterol disposal has an important role in protecting against atherosclerosis [27,49]. Researchers have shown that plant sterols slightly increase plasma HDL by elevating a specific subfraction of HDL, HDL3. The mechanism is not completely elucidated and may depend on many related factors, such as the composition of the food matrix and baseline blood lipid profile of the clinical subjects [18]. Anthocyanins may also contribute to increasing the level of HDL cholesterol by the inhibition of cholesteryl ester transfer proteins [51]. On the other hand, plasma non-HDL cholesterol levels were reduced in rats fed on atherogenic diets that contained both BRF and WS. LDL, a well-documented risk factor for cardiovascular disease, is involved in the transportation of cholesterol or cholesterol esters from the liver to peripheral tissues. Increasing the hepatocyte LDL receptor activity due to the presence of dietary WS and BRF would result in an increased influx of cholesterol from the plasma to the hepatocyte, thereby resulting in an increased compartmentalization of cholesterol that would lower both the plasma non-HDL cholesterol and total cholesterol concentrations.

Our results showing BRF to reduce serum TAG concentrations are supported by other studies that demonstrated bioactive constituents present in rice were involved in TAG digestion and metabolism [23]. Our finding that the presence of BRF to lower plasma triacylglyceride levels in rats was more effective than adding WS could be related to the dietary fiber content, which effectively increases the gut transit time and, as consequence, the degree of contact with digestive enzymes such as pancreatic lipase. Studies conducted in rabbits fed a black rice outer layer reported no changes in the plasma triacylglycerols concentrations [27]. Nevertheless, apo E-deficient mice fed on an anthocyanin-rich BRF for 20 weeks exhibited reductions in serum TAG content [23]. Roohinejad et al. [52] also reported a similar finding in rats fed on pregerminated brown rice, and Ausman et al. [20] reported that rice bran oil was effective at decreasing plasma triacylglyceride (TAG) in hypercholesterolemic hamsters. Our study also showed that feeding WS effectively increased plasma TAG. This result agrees with Pritchard et al. [31] in studies conducted with apo E-deficient mice fed on an atherogenic diet containing WS. Vaskonen et al. [37] reported no significant changes in the TAG level in plasma-obese Zucker rats fed WS. Human clinical studies that administrated sterols in an encapsulated form at intakes of 1.8 g/day for four weeks reported a 7.4% decrease in plasma TAG in hypercholesterolemic subjects [47].

Although anthocyanins present in BRF possess antioxidant and anti-inflammatory activities [10,24,43,45], this was not an apparent factor in our experimental results. CRP is an early biomarker for inflammation and a good indicator of atherosclerosis risk [53]. Ma et al. and [54] and Oi et al. [55] showed that dietary fiber decreases the concentration of CRP in humans. Moreover, Devaraj et al. [38] reported that the consumption of orange juice with added WS for eight weeks resulted in lower serum CRP levels in human subjects. A meta-analysis by Rocha and coworkers [8] showed that the CRP level was not affected by WS in diets in various human studies. Unfortunately, our sampling protocal may not have enabled us to detect possible differences in CRP between rats fed different experimental diets. Future studies are required to include more frequent blood sampling to determine if, indeed, the inclusion of BRF and WS in atherogenic diets can promote a positive effect on CRP.

4. Materials and Methods

4.1. Animals and Diets

Wistar Kyoto (WKY) male rats, all 5 weeks of age, were purchased from Charles River, Montreal, PQ, Canada. Animals were acclimatized for 1 week on standard lab chow (Ralston Purina, St. Louis, MO, USA). After acclimatization, WKY ($n = 40$) rats were randomly divided into five groups ($n = 8$/group) and fed an atherogenic diet (with 0.5% cholesterol and 0.05% cholic acid) and atherogenic diets supplemented with 1% black rice extract (BRF) in the presence or absence of 2% wood sterols (WS) for 11 weeks. The BRF was obtained from Guangdong Province, China. The basic chemical composition of BRF is shown in Table 3 and Figure 2. The WS was obtained from Forbes Medi-Tech Inc. (Vancouver, BC, Canada), previously prepared from tall oil soaps via a proprietary extraction and purification protocol under the GMP food standards. The purity of the final product was over 99% and contained a mixture of sterols and stanols (Table 4 and Figure 1).

Table 3. Basic chemical composition of the black rice fraction and selected bioactive contents.

Ingredient	g/100 g d.m.
Protein	17
Fat	15
Ash	8.3
Digestible carbohydrates	50.02
Soluble dietary fiber	3.08
Insoluble dietary fiber	6.60
Anthocyanins (%)	
cyanidin-3-*O*-glucoside	97.86
peonidin-3-*O*-glucoside	2.14
Total anthocyanins (mg/g)	31.3
Oryzanol (% of total fat)	3.83

Table 4. Wood sterols composition (%).

Total Sterol Content	99.6
Sitostanol	16
Sitosterol	73
Campasterol	6
Stigmasterol	1
Other sterols	2

The composition of experimental diets is shown in Table 5. Experimental diets were prepared based on an AIN-76A diet [56]. The basal (g/kg) diet contained 250-g/kg casein (ICN Biochemical, Cleveland, OH, USA), 470-g/kg cornstarch (Neptune Food Service, Richmond, BC, Canada), 50-g/kg alphacel nondigestible fiber (ICN Biochemical, Cleveland, OH, USA), 30-g/kg sucrose (Neptune Food Service, Richmond, BC, Canada), 30-g/kg mineral mix (ICN Biochemical, Cleveland, OH, USA), 30-g/kg vitamin mix (ICN Biochemical, Cleveland, OH, USA), 2.0-g/kg choline chloride (ICN Biochemical, Cleveland, OH, USA), 3.0-g/kg D-L methionine (ICN Biochemical, Cleveland, OH, USA), 30-g/kg monofos (Van Waters and Rogers, Abbotsford, BC, Canada), 0.5-g/kg cholic acid (ICN Biochemical Cleveland, OH, USA), 5.0-g/kg cholesterol (ICN Biochemical, Cleveland, OH, USA), 30-g/kg butter (Dairyworld Foods, Burnaby, BC, Canada), and 70-g/kg canola oil (Neptune Food Service, Richmond, BC, Canada) to provide an adequate supply of essential fatty acids.

Table 5. Experimental diet compositions (g/100 g).

Ingredient	Negative Control (NC)	PCCh	PCCh + BRF	PCCh + WS	PCCh + WS + BRF
Casein	25.5	25.5	25.5	25.5	25.5
Corn starch	47	45.95	45.5	43.5	43.5
Sucrose	3.0	3.0	3.0	3.0	3.0
Cellulose	5.0	5.0	5.0	5.0	5.0
Monophosphate	3.0	3.0	3.0	3.0	3.0
Vitamin mix	3.0	3.0	3.0	3.0	3.0
Choline chloride	0.20	0.20	0.20	0.20	0.20
D-L Methionine	0.30	0.30	0.30	0.30	0.30
Mineral mix	3.0	3.0	3.0	3.0	3.0
Canola Oil	10	7.0	7.0	7.0	7.0
Butter	-	3.0	3.0	3.0	3.0
Wood sterols	0	-	-	2.0	2.0
Cholesterol	0	0.50	0.50	0.50	0.50
Cholic acid	0	0.050	0.05	0.05	0.05
BRF	-	-	1.0	-	1.0

NC: cholesterol and bile acid-free diet; negative control. PCCh: positive control (0.5% cholesterol, 0.05 cholic acid, 3% butter). BRF: black rice fraction. PCCh + BRF: positive control, atherogenic diet with added black rice fraction. PCCh + WS: positive control, atherogenic diet with added wood sterols. PCCh + WS + BRF: positive control, atherogenic diet with added wood sterols and black rice fraction.

Animals were individually housed in stainless-steel cages under 25 °C and lighting (14:10 light: dark cycle) conditions and fed diets for 12 h per day for 11 weeks. Daily feed intakes and the weekly body weight gains of animals were routinely recorded throughout the experimental period. Animal care protocols were in accordance with the principles of the Guide to the Care and Use of Experimental Animals, Vol. 1 of the Council of Animal Care [57], as approved by the University of British Columbia Animal Care Committee (H04-0095: 05/02/11).

4.2. Experimental Design and Plasma Analysis

At 16 weeks of age, nonfasted rats were killed by exsanguination under halothane anesthesia at 13:00 h, and blood was drawn into heparinized tubes prior to centrifugation for recover plasma ($1000 \times g$ for 5 min, 4 °C). Aliquots of plasma were analyzed for hemoglobin (Cayman Chemical, Ann Arbor, MI, USA), total cholesterol, HDL cholesterol, triacylglycerol, phospholipids glucose, and CRP using commercial kits from Sigma Aldrich (Sigma, St. Louis, MO, USA). HDL was recovered by precipitating non-HDL particles with 0.5-mM sodium phosphotungstic acid and 25-mM magnesium choride. After centrifugation, the supernatant containing the HDL particles was analyzed for cholesterol and triacylglyceride using CHOD-PAP and GPO-PAP, respectively (Boehringer, Mannheim, FRG). The antioxidant capacity of the plasma was measured with the method described by Kitts and Hu [58].

4.3. Liver, Heart, and Aorta Tissue Analyses

Liver, heart, and aorta tissues were removed, washed in 4 °C 0.9% NaCl, dried, and weighed. The arotic tissue was removed from the bachicephalic arteries to the bifurcation and the aorta to the iliac branching. Lipids were extracted using the Folch method [59]. Crude lipids content was measured with the Folch method under nitrogen with subsequent drying at 105 °C. The levels of total cholesterol and triacylglycerol were measured using the methods described by Carlson and Goldfarb [60].

4.4. Statistics

All data are expressed as mean ± SEM. A paired t-test was used to determine if a difference existed between animals fed on the noncholesterol (NC diet) diet and the positive control, atherogenic-formulated (PCCh) diet ($p < 0.05$). One-way analysis of variance (ANOVA; SPSS/Win Inc., IBM Co., Armonk N.Y. USA) was used to test for differences between experimental treatments (PCCh-). Where differences did exist, the source of the differences was identified by the Student–Newman–Keuls multiple range test at a $p \leq 0.05$ significance level.

5. Conclusions

In this study, we presented the effects of known bioactive phytochemical components—namely, an anthocyanin-rich BRF, standardized mixture of WS, and a combination of both—to control hyperlipidemia in rats fed on an atherogenic diet. Feeding WS alone produced lower plasma cholesterol concentrations that were related to lower tissue cholesterol notably in liver and aorta tissues. Similar benefits in controlling plasma and tissue TAGs were observed in rats fed BRF. The combination of adding both WS and BRF to the atherogenic diet also produced an improvement in both cholesterols: HDL cholesterol and non-HDL cholesterol, and TAGs, which were attributed more to the presence of WS or BRF, respectively, than the combination of both. These findings support developing novel diets containing value-added BRF and WS components to enable benefits associated with reduced modifiable atherosclerosis risk factors attributed to hyperlipidemia.

Supplementary Materials: The following are available online. Table S1: Body gain, heart, liver, weight, hematocrit, and hemoglobin in rats.

Author Contributions: Conceptualization, D.D.K., J.Z. and A.K.; methodology, A.K., J.Z. and D.D.K.; formal analysis, A.K.; resources, D.D.K. and J.Z.; data curation, A.K., J.Z. and D.D.K.; writing—original draft preparation, A.K., J.Z. and D.D.K.; writing—review and editing, D.D.K., A.K. and J.Z.; project administration, J.Z. and D.D.K.; and funding acquisition, J.Z. and D.D.K. All authors have read and agreed to the published version of the manuscript.

Funding: This research was funded by NSERC (grant number NSERC -RGPIN 38349-12; D.D.K.) and supported by Forbes Medi-Tec Inc. (J.Z.) and the (A.K.) Dekaban Foundation University of British Columbia, Vancouver, Ca.

Acknowledgments: The authors thank the scientists at Guangdong Province, China for the BRF and Forbes Medi-Tec Inc. for the WS mixture and Lina Madilao for analytical assistance to standardize the BRF anthocyanin mixture composition.

Conflicts of Interest: The authors declare no conflict of interest.

References

1. Roger, V.L.; Go, A.S.; Lloyd-Jones, D.M.; Adams, R.J.; Berry, J.D.; Brown, T.M.; Carnethon, M.R.; Dai, S.; de Simone, G.; Ford, E.S.; et al. Heart Disease and Stroke Statistics—2011 Update. A Report from the American Heart Association. *Circulation* **2011**, *123*, e18–e209. [CrossRef] [PubMed]

2. World Health Organization. (WHO). *Noncommunicable Diseases Country Profiles 2018*; World Health Organization: Geneva, Switzerland, 2018.

3. Prabhakaran, D.; Jeemon, P.; Roy, A. Cardiovascular diseases in India: Current epidemiology and future directions. *Circulation* **2016**, *133*, 1605–1620. [CrossRef] [PubMed]

4. Li, Y.; Wang, D.D.; Ley, S.H.; Green-Howard, A.; He, Y.; Lu, Y.; Danaei, G.; Hu, F.B. Potential impact of time trend of life-style factors on cardiovascular disease burden in China. *J. Am. Col. Cardiol.* **2016**, *68*, 818–833. [CrossRef]

5. World Health Organization. (WHO). Global Health Observatory (GHO) Data: Raised Cholesterol. Situation and Trends. 2017. Available online: https://www.who.int/gho/ncd/risk_factors/cholesterol_text/en/ (accessed on 21 June 2020).

6. Gylling, H.; Plat, J.; Turley, S.; Ginsberg, H.N.; Ellegård, L.; Jessup, W.; Jones, P.J.; Lütjohann, D.; Maerz, W.; Masana, L.; et al. Plant sterols and plant stanols in the management of dyslipidaemia and prevention of cardiovascular disease. *Atherosclerosis* **2014**, *232*, 346–360. [CrossRef]

7. Rocha, V.Z.; Ras, R.T.; Gagliardi, A.C.; Mangili, L.C.; Trautwein, E.A.; Santos, R.D. Effects of phytosterols on markers of inflammation: A systematic review and meta-analysis. *Atherosclerosis* **2016**, *248*, 76–83. [CrossRef] [PubMed]

8. Jones, P.J.H.; Shamloo, M.; MacKay, D.S.; Rideout, T.C.; Myrie, S.B.; Plat, J.; Roullet, J.-B.; Baer, D.J.; Calkins, K.L.; Davis, H.R.; et al. Progress and perspectives in plant sterol and plant stanol research. *Nutr. Rev.* **2018**, *76*, 725–746. [CrossRef] [PubMed]

9. Xia, M.; Ling, W.H.; Kitts, D.D.; Zawistowski, J. Supplementation of diets with black rice pigment fraction attenuates atherosclerotic plaque formation in apolipoprotein E deficient mice. *J. Nutr.* **2003**, *133*, 744–751. [CrossRef]

10. Hu, C.; Zawistowski, J.; Ling, W.; Kitts, D.D. Black rice (Oryza *sativia* L. *indica*) pigmented fraction suppresses both reactive oxygen species and nitric oxide in chemical and biological model system. *J. Agric. Food Chem.* **2003**, *51*, 5271–5277. [CrossRef]

11. Yousuf, B.; Gul, K.; Abas Wani, A.; Singh, P. Health benefits of anthocyanins and their encapsulation for potential use in food systems: A review. *Crit. Rev. Food Sci. Nutr.* **2016**, *13*, 2223–2230. [CrossRef]

12. Wallace, T.C.; Slavin, M.; Frankenfeld, C.L. Systematic review of anthocyanins and markers of cardiovascular disease. *Nutrients* **2016**, *32*, 32. [CrossRef]

13. Cabral, C.E.; Klein, M.R.S.T. Phytosterols in the treatment of hypercholesterolemia and prevention of cardiovascular diseases. *Arq. Bras. Cardiol.* **2017**, *109*, 475–482. [CrossRef] [PubMed]

14. Maki, K.C.; Lawless, A.L.; Reeves, M.S.; Kelley, K.M.; Dicklin, M.R.; Jenks, B.H.; Shneyvas, E.; Brooks, J.R. Lipid effects of a dietary supplement softgel capsule containing plant sterols/stanols in primary hypercholesterolemia. *Nutrition* **2013**, *29*, 96–100. [CrossRef] [PubMed]

15. Jaceldo-Siegel, K.; Lütjohann, D.; Sirirat, R.; Mashchak, A.; Fraser, G.E.; Haddad, E. Variations in dietary intake and plasma concentrations of plant sterols across plant-based diets among North American adults. *Mol. Nutr. Food Res.* **2017**, *61*, 1–10. [CrossRef] [PubMed]

16. Vanstone, C.A.; Raeini-Sarjaz, M.; Parson, W.E.; Jones, P.J.H. Unesterified plant sterols and stanols lower LDL-cholesterol concentrations equivalently in hypercholesterolemic persons. *Am. J. Clin. Nutr.* **2002**, *76*, 1272–1278. [CrossRef] [PubMed]

17. Rouyanne, T.; Ras, R.T.; Geleijnse, J.M.; Trautwein, E.A. LDL-cholesterol-lowering effect of plant sterols and stanols across different dose ranges: A meta-analysis of randomized controlled studies. *Br. J. Nutr.* **2014**, *28*, 214–219.

18. Kalliny, S.; Zawistowski, J. Phytosterols and phytosterols. In *Encyclopedia of Food Chemistry*; Aluko, R., Birch, E.J., Larsen, D., Melton, L., Rogers, M., Shahidi, F., Stadler, R., Sun-Waterhouse, D., Varelis, P., Eds.; Elsevier: Amsterdam, The Netherlands, 2019; Volume 3, pp. 289–299.

19. Wilson, T.A.; Idreis, H.M.; Taylor, C.M.; Nicolosi, R.J. Whole fat rice bran reduces the development of aortic atherosclerosis in hypercholesterolemic hamsters compared with wheat bran. *Nutr. Res.* **2002**, *22*, 1319–1332. [CrossRef]

20. Ausman, L.M.; Rong, N.; Nicolosi, R.J. Hypocholesterolemic effect of physically refined rice bran oil: Studies of cholesterol metabolism and early atherosclersis in hyprcholesterolemic hamsters. *J. Nutr. Biochem.* **2005**, *16*, 521–529. [CrossRef]

21. Zawistowski, J.; Kopeć, A.; Kitts, D.D. Effect of a black rice extract (*Oryza sativa* L. *indica*) on cholesterol levels and plasma lipid parameters in Wistar Kyoto rats. *J. Funct. Foods* **2009**, *1*, 50–56. [CrossRef]

22. Ling, W.H.; Cheng, Q.X.; Ma, J.; Wang, T. Red and black rice decrease atherosclerotic plaque formation and increase antioxidant status in rabbits. *J. Nutr.* **2001**, *131*, 1421–1426. [CrossRef]

23. Xia, X.; Ling, W.H.; Ma, J.; Xia, M.; Hou, M.; Wang, Q.; Zhu, H. An anthocyanin-rich extract from black rice enhances atherosclerosis plaque stabilization in apolipoprotein E-deficient mice. *J. Nutr.* **2006**, *136*, 2220–2225. [CrossRef]

24. Novotny, J.A.; Baer, D.J.; Khoo, K.; Gebauer, S.K.; Charron, C.S. Cranberry juice consumption lowers markers of cardiometabolic risk, including blood pressure and circulating c-reactive protein, triglyceride, and glucose concentrations in adults. *J. Nutr.* **2015**, *45*, 1185–1193. [CrossRef] [PubMed]

25. Hou, Z.; Qin, P.; Zhang, Y.; Cui, S.; Ren, G. Identification of anthocyanins isolated from black rice (Oryza sativa L.) and their degradation kinetics. *Food Res. Int.* **2013**, *50*, 691–697. [CrossRef]

26. Graf, D.; Seifert, S.; Jaudszus, A.; Bub, A.; Watzl, B. Anthocyanin-rich juice lowers serum cholesterol, leptin and resistin and improves plasma fatty acid composition in Fischer rats. *PLoS ONE* **2013**, *8*, e66690. [CrossRef] [PubMed]

27. Ling, W.H.; Wang, L.L.; Ma, J. Supplementation of the black rice outer layer fraction to rabbits decreases atherosclerotic plaque formation and increases antioxidant status. *J. Nutr.* **2002**, *132*, 20–26. [CrossRef]

28. Yang, Y.; Andrews, M.C.; Hu, Y.; Wang, D.; Qin, Y.; Zhu, Y.; Ni, H.; Ling, W. Anthocyanin extract from black rice significantly ameliorates platelet hyperactivity and hypertriglyceridemia in dyslipidemic rats induced by high fat diets. *J. Agric. Food. Chem.* **2011**, *59*, 6759–6764. [CrossRef]

29. Guo, H.; Ling, W.; Wang, Q.; Liu, C.; Hu, Y.; Xia, M.; Feng, X.; Xia, X. Effect of anthocyanin-rich extract from black rice (Oryza sativa L. indica) on hyperlipidemia and insulin resistance in fructose-fed rats. *Plant Foods Human Nutr.* **2007**, *62*, 1–6. [CrossRef]

30. Jang, H.H.; Park, M.Y.; Kim, H.W.; Leen, Y.M.; Hwang, K.A.; Park, J.H.; Park, D.S.; Kwon, O. Black rice (Oryza sativa L.) extract attenuates hepatic steatosis in C57BL/6 J mice fed a high-fat diet via fatty acid oxidation. *Nutr. Metab.* **2012**, *9*, 27–38. [CrossRef]

31. Pritchard, P.H.; Li, M.; Zamfir, K.; Lukic, T.; Novak, E.; Moghadasin, M. Comparison of cholesterol-lowering efficacy and anti-atherogenic properties of hydrogenated versus non-hydrogenated (Phytrol^{TM}) tall oil-derived phytrosterols in apo E-deficient mice. *Cardiovasc. Drugs Ther.* **2003**, *17*, 443–449. [CrossRef]

32. Fang, N.; Yu, S.; Bader, T. Characterization of triterpene alkohol and serol ferulates in rice bran using LC-MS/MS. *J. Agric. Food Chem.* **2003**, *51*, 3260–3267. [CrossRef]

33. He, W.S.; Zhu, H.; Chen, Z.Y. Plant Sterols: Chemical and enzymatic structural modifications and effects on their cholesterol-lowering activity. *J. Agric. Food Chem.* **2018**, *66*, 3047–3062. [CrossRef] [PubMed]

34. Redan, B.W.; Albaugh, G.P.; Charron, C.S.; Novotny, J.A.; Ferruzzi, M.G. Adaptation in Caco-2 human intestinal cell differentiation and phenolic transport with chronic exposure to blackberry (*Rubus sp*) extract. *J. Agric. Food Chem.* **2017**, *65*, 2694–2701. [CrossRef] [PubMed]

35. Wu, X.; Pittman, H.E.; Prior, R.L. Fate anhocyanins and antioxidant capacity in contents of the gastrointestinal tract of weanling pigs following black raspberry consumption. *J. Agric. Food Chem.* **2006**, *54*, 583–589. [CrossRef] [PubMed]

36. Elisia, I.; Kitts, D.D. Anthocyanins inhibit peroxyl radical-induced apoptosis in Caco-2 cells. *Mol. Cell Biochem.* **2008**, *312*, 139–145. [CrossRef]

37. Vaskonsen, T.; Mervaala, E.; Sumuvuori, V.; Seppän-Laakso, T.; Karppanen, H. Effect of calcium and plant sterols on serum lipids in obese Zucker rats on a low-fat diet. *Br. J. Nutr.* **2002**, *87*, 239–245. [CrossRef]

38. Devaraj, R.; Autret, B.C.; Jialal, I. Reduced-calorie orange juice beverage with plant sterols lowers C-reactive protein concentrations and improves the lipid profile in human volunteers. *Am. J. Clin. Nutr.* **2006**, *84*, 756–761. [CrossRef]

39. AbuMweis, S.S.; Vanstone, C.A.; Ebine, N.; Kassis, A.; Ausman, L.M.; Jones, P.J.H.; Lichtenstein, A.H. Intake of single morning dose of standard and novel plant sterol preparations for 4 weeks odes not dramatically affected plasma lipid concentratins in humans. *J. Nutr.* **2006**, *136*, 1012–1016. [CrossRef]

40. Fumeron, F.; Bard, J.M.; Lecerf, J.M. Interindividual variability in the cholesterol-lowering effect of supplementation with plant sterols or stanols. *Nutr. Rev.* **2017**, *75*, 134–145. [CrossRef]

41. Calpe-Berdiel, L.; Escola-Gil, J.C.; Blanco-Vaca, F. Phytosterol-mediated inhibition of intestinal cholesterol absorption is independent of ATP-binding cassette transporter A1. *Br. J. Nutr.* **2006**, *95*, 618–622. [CrossRef]

42. He, J.; Giusti, M.M. Anthocyanins: Natural colorants with health-promoting properties. *Annu. Rev. Food Sci. Technol.* **2010**, *1*, 163–168. [CrossRef]

43. Reis, J.F.; Monteiro, V.V.S.M.; de Souza Gomes, R.; do Carmo, M.M.; da Costa, G.V.; Ribera, P.C.; Chagas Monteiro, M. Action mechanism and cardiovascular effect of anthocyanins: A systematic review of animal and human studies. *J. Transl. Med.* **2016**, *14*, 315–331. [CrossRef]

44. Amiot, M.J.; Riva, C.; Vinet, A. Effects of dietary polyphenols on metabolic syndrome features in humans: A systematic review. *Obes. Rev.* **2016**, *17*, 573–586. [CrossRef] [PubMed]

45. Pojer, E.; Mattivi, F.; Johnson, D.; Stockley, C.S. The case for anthocyanin consumption to promote human health: A review. *Comp. Rev. Food Sci. Food Saf.* **2013**, *12*, 483–508. [CrossRef]

46. Pong, C.H.; Liu, L.K.; Chuang, C.M.; Chyau, C.C.; Huang, C.N.; Wang, C.J. Mulberry water extracts possess an anti-obesity effect and ability to inhibit hepatic lipogenesis and promote lipolysis. *J. Agric. Food Chem.* **2011**, *59*, 2663–2671. [CrossRef] [PubMed]

47. Mekki, N.; Dubois, C.; Charbonier, M.; Cara, L.; Senft, M.; Pauli, A.M.; Portugal, H.; Gassin, A.M.; Lafon, H.; Lairon, D. Effects of lowering fat and increasing dietary fiber on fasting and postprandial plasma lipids in hypercholsterolemic subjects consuming a mixed Mediterranean-Western diet. *Am. J. Clin. Nutr.* **1997**, *66*, 1443–1451. [CrossRef]

48. Behall, K.M.; Scholfield, D.J.; Hallfrisch, J. Diets containing barley significantly reduce lipids in mildly hypercholesterolemic men and women. *Am. J. Clin. Nutr.* **2004**, *80*, 1185–1193. [CrossRef]

49. Kontush, A. HDL-mediated mechanisms of protection in cardiovascular disease. *Cardiovasc. Res.* **2014**, *103*, 341–349. [CrossRef]

50. Morgan, A.E.; Mooney, K.M.; Wilkinson, S.J.; Pickles, N.A.; Mc Auley, M.T. Cholesterol metabolism: A review of how ageing disrupts the biological mechanisms responsible for its regulation. *Aging Res. Rev.* **2016**, *27*, 108–124. [CrossRef]

51. Qin, Y.; Xia, M.; Ma, J.; Hao, Y.T.; Liu, J.; Mou, H.Y.; Cao, L.; Ling, W.H. Anthocyanin supplementation improves serum LDL- and HDL-cholesterol concentrations associated with the inhibition of cholesteryl ester transfer protein in dyslipidemic subjects. *Am. J. Clin. Nutr.* **2009**, *90*, 485–492. [CrossRef]

52. Roohinejad, S.; Omidizadeh, A.; Mirhosseini, H.; Saari, N.; Mustafa, S.; Yusof, R.M.; Hussin, A.S.M.; Hamid, A.; Yazid, M.; Manap, A. Effect of pre-germination time of brown rice on serum cholesterol levels of hypercholesterolaemic rats. *J. Sci. Food Agric.* **2010**, *90*, 245–251. [CrossRef]

53. Libby, P.; Ridker, M.P.; Hansson, G.K. Progress and challenges in transating the biology of artherosclerosis. *Nature* **2011**, *473*, 317–325. [CrossRef]

54. Ma, Y.; Griffith, J.A.; Chasan-Taber, L.; Olendzki, B.C.; Jackson, E.; Stanek, E.J.; Li, W.; Pagoto, S.L.; Hafner, A.R.; Ockene, I.S. Association between dietary fiber and serum C-reactive protein. *Am. J. Clin. Nutr.* **2006**, *83*, 754–759. [CrossRef]

55. Oi, L.; van Dam, R.M.; Liu, S.; Franz, M.; Mantzoros, C.; Hu, F.B. Whole-grain, bran, and cereal fiber intakes and markers of systemic inflammation in diabetic women. *Diabetes Care* **2006**, *29*, 207–211.

56. Reeves, P.G.; Nielsen, F.H.; Fahey, G.C. AIN-93 purified diets for laboratory rodents: Final report of the American Institute of Nutrition Ad Hoc Writing Committee on the Reformulation of the AIN-76A Rodent Diet. *J. Nutr.* **1993**, *123*, 1939–1951. [CrossRef] [PubMed]

57. Canadian Council of Animal Care. *Guide to the Care and Use of Experimental Animals*; CCAC: Ottawa, ON, Canada, 1983; Volume 2, p. 180.

58. Kitts, D.D.; Hu, C. Biological and chemical assessment of antioxidant activity of sugar-lysine model Maillard Reaction Products. *Ann. N. Y. Acad. Sci.* **2005**, *1043*, 501–505. [CrossRef] [PubMed]

59. Folch, J.; Lees, M.; Sloane-Stanley, G.H. A simple method for the isolation and purification of total lipids from animal tissues. *J. Biol. Chem.* **1957**, *226*, 497–509.

60. Carlson, S.E.; Goldfarb, S. A sensitive enzymatic method for the determination of free and esterified tissue cholesterol. *Clin. Chim. Acta* **1979**, *79*, 575–584. [CrossRef]

Sample Availability: Samples of the compounds are not available from the authors.

Publisher's Note: MDPI stays neutral with regard to jurisdictional claims in published maps and institutional affiliations.

Article

Effects of Grape Polyphenols on the Life Span and Neuroinflammatory Alterations Related to Neurodegenerative Parkinson Disease-Like Disturbances in Mice

Maria A. Tikhonova [1,2], Nadezhda G. Tikhonova [3,*], Michael V. Tenditnik [2], Marina V. Ovsyukova [2], Anna A. Akopyan [2], Nina I. Dubrovina [2], Tamara G. Amstislavskaya [2] and Elena K. Khlestkina [1,3]

1   Federal Research Center "Institute of Cytology and Genetics", Siberian Branch of the Russian Academy of Sciences, 630090 Novosibirsk, Russia; tikhonovama@physiol.ru (M.A.T.); khlest@bionet.nsc.ru (E.K.K.)
2   Federal State Budgetary Scientific Institution "Scientific Research Institute of Neurosciences and Medicine" (SRINM), 630117 Novosibirsk, Russia; m.v.tenditnik@physiol.ru (M.V.T.); maryov@ngs.ru (M.V.O.); annaaleksanovna@mail.ru (A.A.A.); dubrov@physiol.ru (N.I.D.); amstislavskaya@yandex.ru (T.G.A.)
3   N.I. Vavilov All-Russian Institute of Plant Genetic Resources, 190000 St. Petersburg, Russia
*   Correspondence: n.g.tikhonova@vir.nw.ru

Academic Editor: H.P. Vasantha Rupasinghe
Received: 24 September 2020; Accepted: 9 November 2020; Published: 16 November 2020

Abstract: Functional nutrition is a valuable supplementation to dietary therapy. Functional foods are enriched with biologically active substances. Plant polyphenols attract particular attention due to multiple beneficial properties attributed to their high antioxidant and other biological activities. We assessed the effect of grape polyphenols on the life span of C57BL/6 mice and on behavioral and neuroinflammatory alterations in a transgenic mouse model of Parkinson disease (PD) with overexpression of the A53T-mutant human α-synuclein. C57BL/6 mice were given a dietary supplement containing grape polyphenol concentrate (GPC—1.5 mL/kg/day) with drinking water from the age of 6–8 weeks for life. Transgenic PD mice received GPC beginning at the age of 10 weeks for four months. GPC significantly influenced the cumulative proportion of surviving and substantially augmented the average life span in mice. In the transgenic PD model, the grape polyphenol (GP) diet enhanced memory reconsolidation and diminished memory extinction in a passive avoidance test. Behavioral effects of GP treatment were accompanied by a decrease in α-synuclein accumulation in the frontal cortex and a reduction in the expression of neuroinflammatory markers (IBA1 and CD54) in the frontal cortex and hippocampus. Thus, a GP-rich diet is recommended as promising functional nutrition for aging people and patients with neurodegenerative disorders.

Keywords: flavonoids; cognition; passive avoidance test; memory extinction; mice; microglia; neuroprotection

1. Introduction

There is growing evidence that diets rich in polyphenols decrease the risk of chronic diseases such as obesity, diabetes, heart disease, and cancer [1–5]. Polyphenols are the most abundant group of biologically active molecules and natural antioxidants that can protect human cells from oxidative damage thereby reducing the risk of developing various degenerative diseases associated with oxidative stress. The main sources of polyphenols are fruits, tea, coffee, cacao, and grapes. These compounds are also found in vegetables and seeds of different crops, but in lower concentrations [6,7]. In plants, polyphenols are secondary metabolites participating in plant protection from a wide range of biotic and abiotic stress factors [8,9].

Even with a standard diet, we consume about 1 g of polyphenols daily, which is 10 and 100 times higher than the daily doses of vitamins C and E (as well as carotenoids), respectively [10]. In addition to antioxidant activity, polyphenols depending on their chemical structure, manifest a wide range of biological activities, such as anti-inflammatory, capillary-strengthening, hepatoprotective, diuretic, antiallergic, antimicrobial and antitumor properties [5,11–13].

There is a special interest in reusing winemaking waste, such as grape seed, fruit skin or vine. These waste products are rich in polyphenols, specifically, catechins, quercetin, anthocyanins, proantocyanidins, phenolic acids, and resveratrol [14,15] having most of the activities described above [5,12,13], so they may be used to produce valuable extracts. Furthermore, the photoprotective effect of polyphenolic compounds extracted from red grapes was observed during in vitro human cell culture studies [7]. Red grapes effectively inhibit UV-A-induced synthesis of type III collagen both at the RNA and protein levels. This confirms the potential of grape polyphenols in slowing down the skin photoaging [7]. In addition, polyphenols can overcome the blood–brain barrier and produce a neuroprotective effect on the central nervous system [16,17]. Hence, the application of grape-derived polyphenols as an adjunctive treatment paradigm to prevent neuropathologies including such neurodegenerative disorders as Alzheimer and Parkinson diseases is widely discussed [18]. Parkinson disease (PD) is the second most common neurodegenerative disorder, placing huge economic and social burdens on societies all over the world. The main risk factor for the development of PD is aging. Current therapies for PD are symptomatic and limited they do not cope with the disease onset or progression as they do not address multiple overlapping mechanisms involved in the PD pathogenesis. Moreover, PD patients develop drug tolerance and suffer from serious side effects of the drugs due to uninterrupted long-term treatment. Nutritional supplementation with polyphenols is regarded as a promising prophylactic treatment for neurodegenerative disorders, as it potently and simultaneously targets inflammatory and oxidative pathways [18]. It is noteworthy that a recent study by Ben Youssef et al. [19] evidenced the neuroprotective effect of grape seed and skin extract on a mouse PD model induced by 6-OHDA neurotoxin through reducing apoptosis, oxidative stress, and inflammation. Pathological aggregation and accumulation of α-synuclein in neurons and Lewy bodies appear to play a core role in the pathogenesis of PD [20]. However, the potential impact of grape polyphenols on this mechanism is scantily studied.

In the current study, we assessed the effect of grape polyphenols on the life span of C57BL/6 mice and on behavioral and neuroinflammatory alterations in a transgenic mouse model of PD with overexpression of the A53T-mutant human α-synuclein.

2. Results

2.1. Diet Tolerance and the Effects on Life Expectancy and Body Weight Gain

In mice of the C57Bl/6 strain born and reared at the conventional animal facility, grape polyphenol concentrate (GPC) significantly influenced the cumulative proportion of surviving ($p < 0.01$; Figure 1A) and substantially augmented the average life span ($p < 0.01$; Figure 1B).

The last mouse in the experiment was a GPC-treated male who died at the age of 1031 days. Moreover, in two-year-old mice the general condition of GPC-treated animals appeared to be better than of the controls, including the state of their fur and eyes (Figure 1C). Mice of the B6.Cg-Tg (Prnp-SNCA*A53T)23Mkle/J strain (further mut(PD)), a genetic model of PD, and control wild-type (WT) mice that had been born and reared under SPF conditions also endured the GPC-supplemented diet well. No significant difference was found between the groups in body weight gain after four months of the experiment (Figure 2).

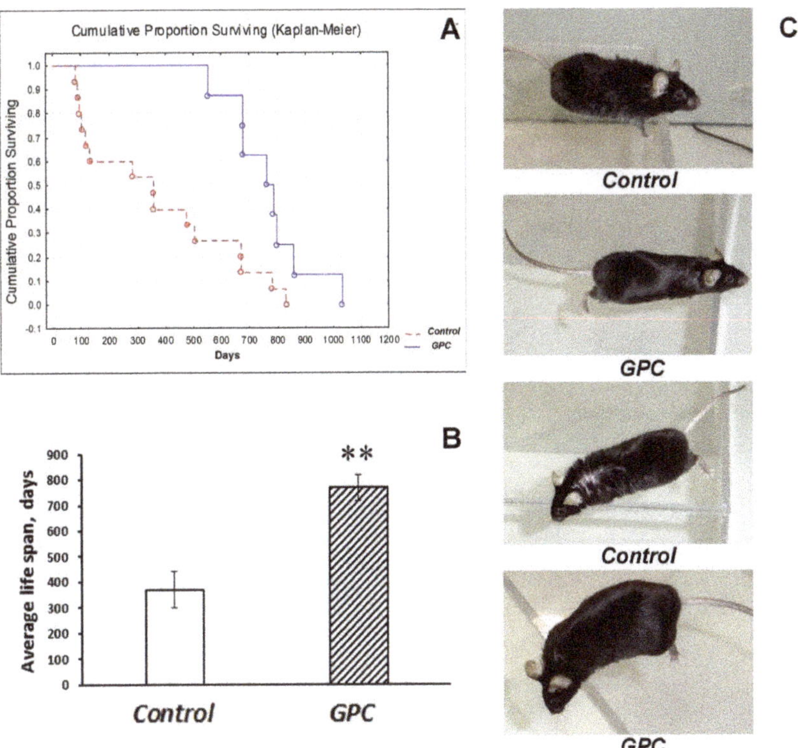

Figure 1. The effect of dietary supplementation with *GPC* on the cumulative proportion of surviving (**A**), average life span (**B**), and general condition (**C**) in mice of the C57Bl/6 strain. The data are presented as a Kaplan–Meier diagram (**A**) or the means ± SEMs (**B**) of the values obtained in an independent group of animals (n = 8–15 per group). Statistically significant differences: ** $p < 0.01$ vs. controls. (**C**) photographs of control and *GPC*-treated mice at the age of two years.

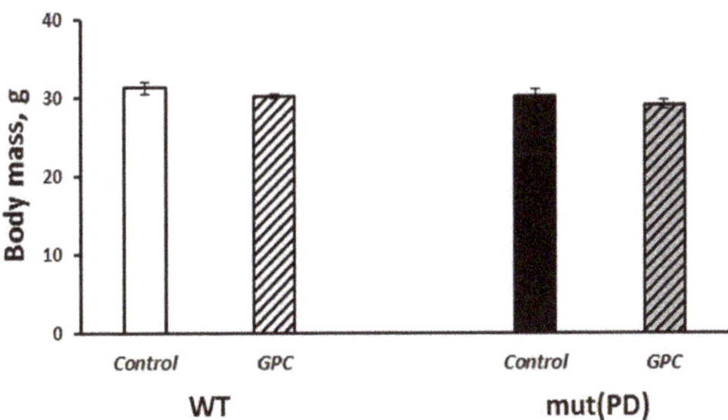

Figure 2. Effects of the overexpression of A53T-mutant α-synuclein and dietary supplementation with *GPC* for four months on body weight gain in mice.The data are expressed as the means ± SEMs of the values obtained in an independent group of animals (n = 5–9 per group).

2.2. Behavioral Effects

2.2.1. The Open Field Test

An open field test was performed to assess general locomotion, vertical locomotor and exploratory activity, anxiety, and emotionality in mice (Table 1).

Table 1. Effects of *GPC* supplementation and overexpression of α-synuclein (genetic Parkinson's disease (PD) model) on the behavior of mice in the open field test.

Parameter	Group				F, p
	WT		**Mut (PD)**		
	Control	*GPC*	*Control*	*GPC*	
Distance travelled, cm	3773 ± 266	3704 ± 297	4964 ± 602	5236 ± 492 [&&]	G: $F(1, 28) = 10.8, p < 0.01$ D: $F(1, 28) < 1$ D × G: $F(1, 28) < 1$
Rearings, n	74.7 ± 6.3	67.0 ± 8.9	76.4 ± 7.3	91.8 ± 12.3	G: $F(1, 28) = 1.8, p > 0.05$ D: $F(1, 28) < 1$ D × G: $F(1, 28) = 1.4, p > 0.05$
Time in the center, s	31.5 ± 4.8	33.2 ± 4.5	36.0 ± 9.8	30.6 ± 6.6	G: $F(1, 28) < 1$ D: $F(1, 28) < 1$ D × G: $F(1, 28) < 1$
Fecal boli, n	2.22 ± 0.74	2.89 ± 0.75	2.8 ± 1.16	1.0 ± 0.58	G: $F(1, 28) < 1$ D: $F(1, 28) < 1$ D × G: $F(1, 28) = 2.4, p > 0.05$

G—genotype factor, D—diet factor. Data are presented as the Mean ± S.E.M. of the values obtained in an independent group of animals (n = 5–9 per group). Statistically significant differences: [&&] $p < 0.01$ vs. a WT + GPC group.

When testing a PD model, two-way ANOVA followed by an LSD post-hoc test revealed a significant effect of the genotype on the distance travelled. Mut(PD) mice had higher horizontal locomotion than WT mice, which is quite in line with previous studies on this PD model [21–23]. The other measured parameters were not significantly affected by the genotype factor or the diet factor or their interaction.

2.2.2. The Passive Avoidance Test

We recorded a significant effect of the repeated measures (learning) factor ($F(1, 28) = 96.4, p < 0.001$) on the step-through latency when evaluating contextual memory retrieval in mice (Figure 3A).

Latency to enter a dark compartment during training (before the foot shock) did not differ significantly among the experimental groups. As evidence of learning and acquisition of the conditioned passive avoidance reaction on testing day, 24 h after receiving the foot shock, mice of all groups demonstrated increased step-through latencies, often ~10-times greater than latencies on the training day. However, memory extinction was influenced not only by the repeated measures (time) factor ($F (11, 297) = 14.6, p < 0.001$)) but also by the interactions between the genotype and diet factors ($F (1, 27) = 6.03, p < 0.05$) and between the repeated measures and diet factors ($F (11, 297) = 3.9, p < 0.001$) (Figure 3B). With exposure to the context in the absence of additional shocks, the fear response gradually diminishes which is called memory extinction [24]. In the WT control, WT + GPC, and mut(PD) control groups, the values of step-through latency remained significantly increased for seven, seven, and two days, respectively, compared to the training day. A significant decrease in step-through latency was observed to start from the 6th, 8th, and 4th day of the extinction phase compared to the test day in WT control, WT + *GPC*, and mut(PD) control group, respectively. Hence, extinction was more pronounced in the mut(PD) control group. At the same time, the values of step-through latency stayed significantly increased for ten days of the extinction phase as compared to the training day in a mut(PD) + *GPC* group. We failed to observe a substantial reduction in step-through latency in mice of the mut(PD) + *GPC* group within ten days of the extinction phase. Thus, the mut (PD) mice demonstrated *GPC* enhanced memory reconsolidation and diminished memory extinction.

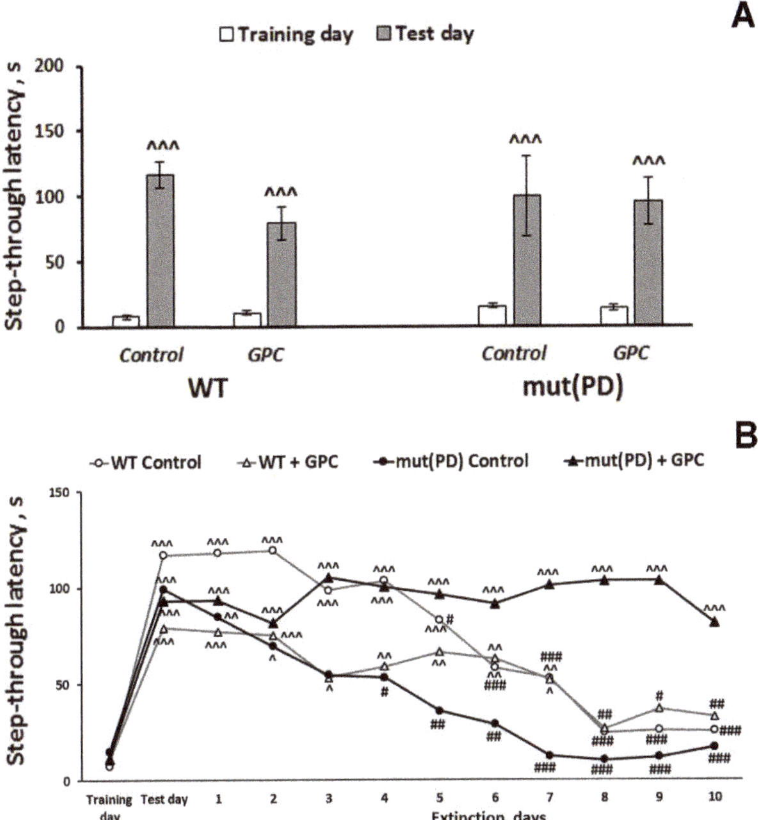

Figure 3. Effects of the overexpression of A53T-mutant α-synuclein and dietary supplementation with *GPC* for four months on memory retrieval (**A**) and memory extinction (**B**) in mice in the passive avoidance test. The data are expressed as the means ± SEMs (**A**) or means (**B**) of the values obtained in an independent group of animals (n = 5–9 per group). Statistically significant differences: ^ $p < 0.05$, ^^ $p < 0.01$, ^^^ $p < 0.001$ compared to values of the same group on the training day; # $p < 0.05$, ## $p < 0.01$, ### $p < 0.001$ compared to values of the same group on the test day.

2.3. Immunohistochemical Analysis

First, we evaluated the accumulation of human α-synuclein in the mouse brain. We detected immunofluorescence against human α-synuclein only in the frontal cortex of seven-month-old transgenic mut(PD) mice (Figure 4B). Both the genotype (F (1, 8) = 35.2, $p < 0.001$) and diet (F (1, 8) = 7.5, $p < 0.05$) factors had a significant effect on α-synuclein accumulation in the 2nd layer of the frontal cortex (Figure 4A).

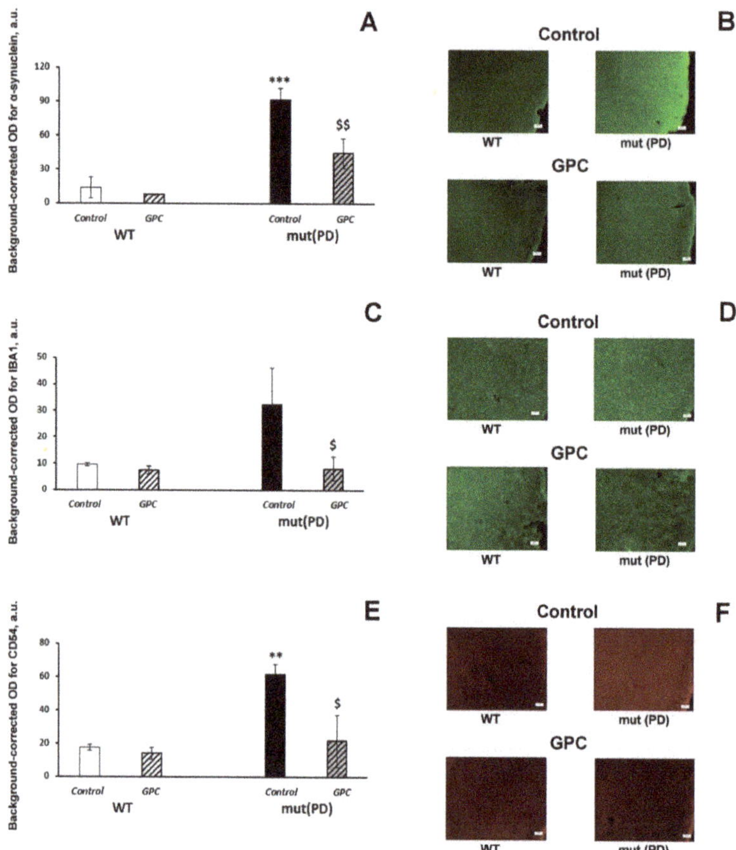

Figure 4. Effects of the overexpression of A53T-mutant α-synuclein and dietary supplementation with *GPC* for four months on α-synuclein accumulation (**A,B**) and expression of the microglial marker IBA1 (**C,D**) or inflammatory marker CD54 (**E,F**) in the frontal cortex of mice. (**A,C,E**): quantitative results. The data are expressed as the means ± SEMs of the values obtained in an independent group of animals ($n = 3$ per group). Statistically significant differences: ** $p < 0.01$, *** $p < 0.001$ vs. WT controls; $^{\$}$ $p < 0.05$, $^{\$\$}$ $p < 0.01$ vs. mut (PD)controls. (**B**) α-synuclein immunoreactivity in the frontal cortex. (zoom: 100×; bar: 100 μm.) (**D**) IBA1 immunoreactivity in the frontal cortex. (zoom: 200×; bar; 50 μm.) (**F**) CD54 immunoreactivity in the frontal cortex. (zoom: \200×; bar: 50 μm).

Dietary supplementation with *GPC* significantly reduced the accumulation of human α-synuclein in the frontal cortex of mut(PD) mice ($p < 0.01$). Indices of neuroinflammation (microglial marker IBA1, Figure 4C; inflammatory marker CD54, Figure 4D) were also increased in the frontal cortex of transgenic mut(PD) mice, while *GPC* treatment significantly suppressed them back to the levels of WT mice. Similar effects on the expression of the inflammatory markers were recorded in the hippocampus (Figure 5; Supplementary Materials: Figure S1).

The diet factor (F $(1, 8) = 23.5$, $p < 0.01$) produced a significant effect on the expression of IBA1 in the CA1 hippocampal area. The parameter was markedly decreased in both WT ($p < 0.05$) and mut(PD) ($p < 0.01$) mice by *GPC* supplementation (Figure 5A). The expression of CD54 was significantly influenced by the diet factor (F(1, 8) $= 5.6$, $p < 0.05$) in the CA3 hippocampal area (Figure 5E) and by the genotype factor (F(1, 8) $= 11.96$, $p < 0.01$), diet factor (F(1, 8) $= 13.1$, $p < 0.01$), and interaction between the factors (F(1, 8) $= 6.8$, $p < 0.05$) in the dentate gyrus of the hippocampus (Figure 5F).

Dietary supplementation with *GPC* significantly reduced the CD54 expression in the CA3 area ($p < 0.05$) and dentate gyrus ($p < 0.01$) in mut(PD) mice.

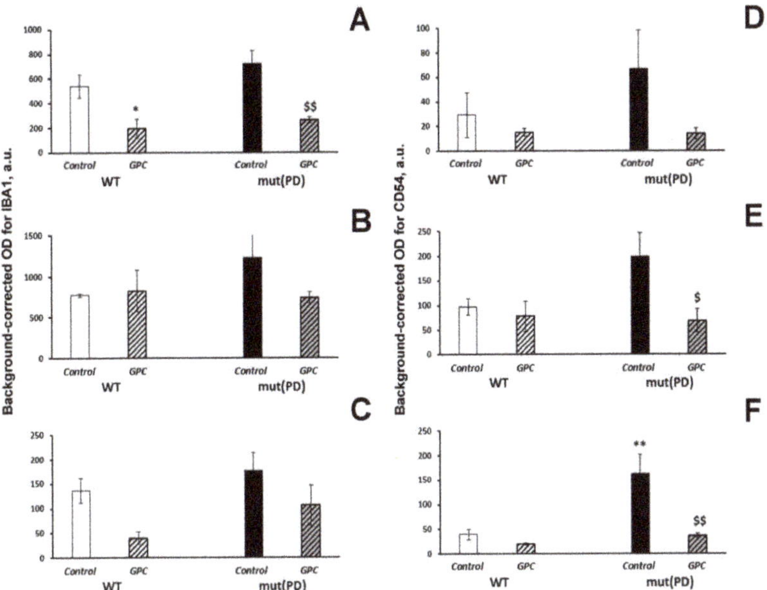

Figure 5. Effects of the overexpression of A53T-mutant α-synuclein and dietary supplementation with *GPC* for four months on the expression of the microglial marker IBA1 (**A–C**) or inflammatory marker CD54 (**D–F**) in the hippocampus (**A,D** in the CA1 area; **B,E** in the CA3 area; **C,F** in the dentate gyrus) in mice. The data are expressed as the means ± SEMs of the values obtained in an independent group of animals ($n = 3$ per group). Statistically significant differences: * $p < 0.05$, ** $p < 0.01$ vs. WT controls; $^\$ p < 0.05$, $^{\$\$} p < 0.01$ vs. mut(PD) controls.

3. Discussion

3.1. Safety of Long-Term Supplementation with Grape Polyphenols and Their Effect on Life Span in Mice

Pallauf et al. [25] reviewed the data on life span extension by flavonoids in worms, flies and mice, describing possible mechanisms that may underpin longer life spans of model organisms treated with flavonoids: energy-restriction-like effects, inhibition of insulin-like-growth-factor signaling, induction of antioxidant activity, hormesis and antimicrobial properties.

In the experiments, with mice, their life span increased when the diets were supplemented with combined extracts of blueberry, green tea and pomegranate powder [26] or green tea polyphenols containing approximately 70% epigallocatechins, epicatechins and gallocatechins [27]. In other studies, however, experimenting with blackcurrant juice (containing anthocyanins, quercetin and quercetin glycosides), epimedium flavonoids (containing 20% icariin) [28], green tea extracts [29,30], blueberry extract [30], triple combination of green tea extract, black tea extract and morin [30] or pomegranate powder [30] failed to induce a longer life span in mice. Overall, several issues should be taken into account when the potential effect of flavonoids on animal or human life span in discussed. When different flavonoids are consumed simultaneously, they may have an additive or even synergistic impact on life expectancy. Moreover, other compounds in extracts may either attenuate or improve the life-extending effect of flavonoids. Besides, synergy was observed between polyphenols, drugs, and hormones [16]. The duration of treatment and the age when the supplementation starts also matter and may lead to different outcomes.

In our study, grape polyphenol concentrate (containing a wide range of anthocyanins, flavan-3-ols, hydroxycinnamic acids, high molecular weight, oligomeric and condensed procyanidins plus 121.2 mg/L, of quercetin, 46.0 mg/L of quercetin-3-O-glycoside,928.4 mg/L of gallic acid and 5.6 mg/L of trans-resveratrol; Supplementary Materials: Table S1) increased life span (Figure 1). A conclusion can be made that *GPC* was safe for mice during almost life-long supplementation starting from an early age. The long-term treatment did not produce any visible harmful effects on the general condition of the animals, nor disturbed their behavioral performance or cognitive function. This observation is quite important, since a preventive long-term intervention at the asymptomatic preclinical and early stages of the disease progression is considered the most promising in the management of neurodegenerative disorders [31].

3.2. Polyphenols Attenuate Neuropathological Changes Associated with Aging and PD-Like Disturbances

Our experiments with a transgenic PD model, demonstrated the grape polyphenols enhanced memory reconsolidation and abated memory extinction in the passive avoidance test (Figure 3). It should be mentioned that the open field test revealed no significant alterations in the *GPC*-treated groups. Hence, the observed effect of *GPC* on cognitive functions was specific and did not depend on general changes in locomotor or exploratory behavior. The beneficial effect of grape polyphenols on cognitive functions agrees well with the previous findings on the restoration of impaired cognition in the mouse models of aging or Alzheimer disease [32–35].

Low bioavailability of flavonoids questioned their direct effects on the central nervous system. However, when their bioactive metabolites are taken into account, including those associated with the activity of the gut microbiota and interaction products, bioavailability appears to be much higher [36]. For example, anthocyanins and their metabolites were found in almost all organs and tissues including the brain of animals fed with anthocyanin-rich feeds. The latter observation attests to active absorption and the ability to overcome the blood–brain barrier [37–39]. Hence, the results obtained on biological activity and impact mechanisms of the intact compounds produced in vitro should be interpreted with a certain caution and need to be confirmed by in vivo findings. Moreover, the dosage is also an essential issue. For example, high doses of resveratrol applied to overcome its low bioavailability caused various side effects [40]. Moreover, some components of the *GPC* mixture, such as resveratrol and epigallocatechin-3-gallate, are regarded as pan-assay-interference (PAINS) compounds [41]. Hence, their effect on an organism might be nonspecific, including cell membrane perturbations, rather than specific protein binding [42], especially when high supraphysiological doses are applied.

Nevertheless, preclinical animal studies revealed that grape polyphenols might affect certain pathogenetic mechanisms involved in the aging-related cognitive decline and neurodegenerative disorders, such as neuroinflammatory response, oxidative stress, protein homeostasis, and apoptotic signaling [17,32,33]. A systematic review of 43 publications performed by de Andrade Teles et al. [43] summarized that the main targets of action for the flavonoid-based PD therapy were the reduction of the cellular oxidative potential and activation of neuronal death mechanisms. Strathearn et al. [44] suggested that anthocyanin- and proanthocyanidin-rich plant extracts could alleviate PD-induced neurodegeneration by enhancing the mitochondrial function.

Although pronounced motor disturbances occur in transgenic mice with overexpression of mutant human α-synuclein at the age of 9–13 months [23], certain behavioral and cognitive alterations appear at early stages of the pathology course including memory deficit [22]. Those non-motor symptoms are associated with the accumulation of α-synuclein and neuroinflammation in the forebrain regions. Indeed, we revealed the deposits of α-synuclein and enhanced expression of inflammatory markers (IBA1 and CD54) in the frontal cortex of transgenic mice. *GPC* supplementation significantly decreased the α-synuclein accumulation and reduced the expression of the neuroinflammatory markers in the frontal cortex (Figure 4). We did not find any d α-synuclein deposition in the hippocampus while the neuroinflammatory response was less pronounced. However, *GPC* treatment produced e similar effects attenuating the inflammatory markers in the hippocampal regions (Figure 5). Many studies reported on the anti-inflammatory effect of polyphenols, including the recent study of a mouse PD model induced by 6-OHDA neurotoxin [20]. At the same time, the effect of polyphenols on α-synuclein aggregation and neurotoxicity was observed in vitro in cellular models [45,46]. Specifically, the major metabolite among anthocyanins cyanidin 3-glucoside inhibited aggregation and fibril formation of α-synuclein [47,48]. The present study confirmed the neuroprotective activity of grape polyphenols against α-synuclein in vivo. Thus, *GPC* supplementation has a potential therapeutic effect in preventing and treating Parkinson disease.

4. Materials and Methods

4.1. Experimental Animal and Design

Experiments were performed using male mice: (1) non-SPF mice of the C57Bl/6 strain born and reared in a conventional animal facility at the Federal State Budgetary Scientific Institution "Scientific Research Institute of Neurosciences and Medicine" (SRINM; former Scientific Research Institute of Physiology and Basic Medicine) (Novosibirsk, Russia); (2) mice of the B6.Cg-Tg(Prnp-SNCA*A53T)23Mkle/J) strain (hereinafter: mut(PD)) and control WT mice acquired from the SPF-vivarium of the Institute of Cytology and Genetics SB RAS (Novosibirsk, Russia). Hemizygous mut (PD) mice were produced by inserting the human A53T missense mutant form of alpha-synuclein cDNA in the mouse genome downstream of a mouse prion Prnp promoter [49].

Animals were housed in groups of 4–5 animals per cage (40 × 25 × 15 cm) under standard conditions (light–dark cycle: 14 h light and 10 h dark (lights off at 15:00); temperature: 18–22 °C; relative humidity: 50–60%). All experimental procedures were carried out in accordance with the guidelines of the NIH Guide for the Care and Use of Laboratory Animals and were approved by the Institutional Animal Care and Use Committee of the SRINM. Every effort was made to minimize the number of animals used and their suffering.

In each experiment, mice of each strain were subdivided into two groups and prescribed one of the following diets. The mice of the control groups received a standard granulated chow for laboratory mice (Ssniff R/M-H V1534-300, Soest, Germany) and pure water (Rosinka, Novosibirsk, Russia) ad libitum. The mice of the *GPC* groups were given the standard chow and the grape polyphenol aqueous solution of the concentrate *Enoant* (produced by RESSFOOD, Company, Yalta, Russia) [50] ad libitum. *Enoant* was diluted with the pure water to a concentration of 0.5–0.8% taking into account a mouse body weight and liquid consumption to provide the daily averaged dosage of 1.5 mL/kg. Fresh solution was prepared every other day. To adjust the dosage, liquid consumption per cage was measured daily and mice were weighed weekly. Concentrations of polyphenolic compounds in *Enoant* are described in Supplementary Materials: Table S1.

The experiment with non-SPF mice of the C57Bl/6 strain born and reared at the conventional animal facility started when mice reached an age of 6–8 weeks and lasted until the death of the last mouse (control, *n* = 15; *GPC*, *n* = 8). Mut(PD) mice, a genetic model of PD, and control WT mice born and reared under SPF conditions were fed with the *GPC* supplement since the age of six weeks (WT control, *n* = 9; WT + GPC, *n* = 9; mut(PD) control, *n* = 8; mut(PD) + *GPC*, *n* = 9). After four months of *GPC* feeding mut(PD) and WT mice were tested for behavior and then sacrificed for further immunohistochemical analysis of their brains.

4.2. Behavioral Tests

Each animal was handled for 5 min/day on three consecutive days, prior to being taken into the experiment. Open field and passive avoidance tests were performed. Observations were made during the dark phase between 15:00 and 22:00 h. For behavioral testing, the animals were placed individually in a clean cage (25 × 40 × 20 cm), and transported to a dim observation room (28 lux of the red light) with sound isolation reinforced by a masking white noise of 70 dB. Performance in the behavioral tests was monitored using a video camera Panasonic WV-CL930 (Panasonic System Networks Suzhou

Co.,Ltd., Suzhou, China) positioned above an apparatus and processed with original EthoVision XT software (Noldus, Wageningen, The Netherlands). The test equipment was cleaned using 20% ethanol and thoroughly dried before each test trial.

4.2.1. The Open Field Test

This test was carried out in an apparatus with a square arena (40 × 40 cm) and plastic walls 37.5 cm high brightly lit from above (1000 lux). A mouse was placed in the center of the arena, and its movements were recorded for 10 min. The following parameters were assessed: general locomotion (distance travelled in cm); vertical locomotor and exploratory activity (number of rearing); anxiety (time spent in the central part of the arena); and emotionality (number of defecations).

4.2.2. The Passive Avoidance Test

Training on the passive avoidance reaction was performed by a standard single-session method in an experimental chamber with dark and light compartments and an automated Gemini Avoidance System apparatus (San Diego Instruments, San Diego, CA, USA)) as described in detail earlier [51]. The Gemini software automatically recorded the latency of the transfer to the dark compartment and the data of testing served as a measure of acquisition of the conditioned passive avoidance reaction. Memory extinction was measured during the next ten days.

4.3. Immunohistochemical Analysis

On the day of euthanasia, mice were culled with CO_2. The animals were perfused transcardially with phosphate-buffered saline (PBS) followed by 4% paraformaldehyde in PBS, then the brains were rapidly excised and postfixed in PBS containing 30% sucrose at 4 °C until further neuromorphological analysis. The analysis was performed on 30-μm-thick cryosections according to a protocol described in detail previously [18]. Coronal slices along the frontal cortex (AP: 2.93 to −2.45 mm) or hippocampus (AP: −2.03 to −2.15 mm) of each mouse brain were made. We applied a rabbit polyclonal antibody (NB110-61645, 1:1000 dilution, Novus Biologicals, Littleton, CO, USA) as the primary antibody to detect human α-synuclein, a goat polyclonal antibody (NB100-1028, 1:200 dilution, Novus Biologicals, Littleton, CO, USA) as the primary antibody to detect the AIF-1/IBA1 microglial marker, or a rat monoclonal antibody (catalog # 16-0542-81, 1:300 dilution, Invitrogen, Carlsbad, CA, USA) as the primary antibody to detect the CD54(ICAM-1) inflammatory marker. A fluorescently labeled (Alexa Fluor 488-conjugated) goat anti-rabbit IgG antibody (ab150077, 1:600 dilution, Abcam, Cambridge, UK), Alexa Fluor 488-conjugated donkey anti-goat IgG antibody (ab150129, 1:200 dilution, Abcam, Cambridge, UK), or Alexa Fluor 594-conjugated goat anti-rat IgG antibody (ab150160, 1:500 dilution, Abcam, Cambridge, UK) served as the secondary antibodies, respectively. Fluorescent images were finally obtained by means of an Axioplan 2 (Carl Zeiss) imaging microscope and then analyzed in Image Pro Plus Software 6.0 (Media Cybernetics, Inc., Rockville, MD, USA). Fluorescence intensity was measured as background-corrected optical density (OD) with subtraction of staining signals of the non-immunoreactive regions in the images converted to grayscale. The area of interest was 7423 μm^2 (IBA1 or CD54) or 30,014 μm^2 (α-synuclein) in the frontal cortex; 19,353 μm^2, 26,100 μm^2, and 50,868 μm^2 in the hippocampal CA1, CA3 areas, and dentate gyrus, respectively.

4.4. Data Analysis

Survival analysis was performed using Gehan's Wilcoxon test and presented as a Kaplan–Meier diagram; average life span for each diet was compared with the Mann–Whitney U-test. The results on the PD model were presented as mean ± SEM and compared using a two-way ANOVA followed by Fisher LSD post-hoc test. The independent variables for the two-way ANOVA were Genotype (WT or mut(PD)) and Diet (control or *GPC*). Repeated Measures ANOVA followed by Fisher LSD post-hoc comparison was applied to analyze the data of the passive avoidance test with Genotype and Diet as between-subject variables and Time (Training, Test, or Extinction days) as a repeated measure.

The level of significance was defined as $p < 0.05$. STATISTICA 10.0 software was used to perform all statistical analyses.

5. Conclusions

GPC proved safe for mice during almost life-long supplementation starting from an early age. The long-term *GPC* feeding did not produce any visible harmful effects on the general condition of the animals, nor disturbed their behavioral performance or cognitive function. Hence, such treatment seems to be applicable as a preventive long-term intervention at the asymptomatic preclinical and early stages of neurodegenerative diseases. *GPC* affected the cognitive function in transgenic PD mice by modulating their memory-specifically, by enhancing memory reconsolidation and abating memory extinction. The behavioral and cognitive effects of *GPC* were accompanied by a reduction in α-synuclein accumulation in the frontal cortex and neuroinflammatory response in the frontal cortex and hippocampus. Our findings confirmed the neuroprotective activity of grape polyphenols against α-synuclein in vivo. Thus, a diet supplemented with grape polyphenol concentrate is suggested as promising functional nutrition for aging people and patients with neurodegenerative disorders, especially Parkinson's disease.

Supplementary Materials: The following are available online at Figure S1: Effects of the overexpression of A53T-mutant α-synuclein and dietary supplementation with *GPC* for four months on the immunoreactivity against the IBA1 microglial marker or CD54 inflammatory marker in the hippocampal CA1 area, CA3 area, and dentate gyrus (DG) in mice, Table S1: Content (mg/L) of active polyphenols in the grape concentrate *Enoant* (*GPC*) and dry wine material (DWM) from Vitis vinifera L. cv. 'Cabernet Sauvignon' according to Zaitsev et al. (2010).

Author Contributions: Conceptualization, M.A.T., T.G.A., E.K.K.; methodology, M.A.T., T.G.A.; validation, M.V.T., N.I.D.; investigation, M.V.O., A.A.A.; writing—original draft preparation, M.A.T., N.G.T., E.K.K.; writing—review and editing, M.A.T., N.G.T., E.K.K.; visualization, M.A.T.; supervision E.K.K.; project administration, E.K.K. All authors have read and agreed to the published version of the manuscript.

Funding: The work was supported by Russian Science Foundation (grant No. 16-14-00086).

Acknowledgments: We thank Konstantin S. Pavlov for technical support in animal treatment and behavioral testing. The studies were implemented using the equipment of the Federal State Budgetary Scientific Institution "Scientific Research Institute of Neurosciences and Medicine" (theme No. AAAA-A16-116021010228-0).

Conflicts of Interest: The authors declare that there is no conflict of interest.

References

1. Li, W.G.; Zhang, X.Y.; Wu, Y.J.; Tian, X. Anti-inflammatory effect and mechanism of proanthocyanidins from grape seeds. *Acta Pharmacol. Sin.* **2001**, *22*, 1117–1120.
2. Ren, W.; Qiao, Z.; Wang, H.; Zhu, L.; Zhang, L. Flavonoids: Promising anticancer agents. *Med. Res. Rev.* **2003**, *23*, 519–534. [CrossRef]
3. Seo, K.-H.; Kim, H.; Chon, J.-W.; Kim, D.-H.; Nah, S.-Y.; Arvik, T.; Yokoyama, W. Flavonoid-rich Chardonnay grape seed flour supplementation ameliorates diet-induced visceral adiposity, insulin resistance, and glucose intolerance via altered adipose tissue gene expression. *J. Funct. Foods* **2015**, *17*, 881–891. [CrossRef]
4. Rasouli, H.; Farzaei, M.H.; Khodarahmi, R. Polyphenols and their benefits: A review. *Int. J. Food Prop.* **2017**, *20* (Suppl. S2), 1–42. [CrossRef]
5. Tartaglione, L.; Gambuti, A.; De Cicco, P.; Ercolano, G.; Ianaro, A.; Taglialatela-Scafati, O.; Moio, L.; Forino, M. NMR-based phytochemical analysis of Vitis vinifera cv Falanghina leaves. Characterization of a previously undescribed biflavonoid with antiproliferative activity. *Fitoterapia* **2018**, *125*, 13–17. [CrossRef]
6. Scalbert, A.; Manach, C.; Morand, C.; Rémésy, C.; Jiménez, L. Dietary Polyphenols and the Prevention of Diseases. *Crit. Rev. Food Sci. Nutr.* **2005**, *45*, 287–306. [CrossRef]
7. Di Francesco, S.; Savio, M.; Bloise, N.; Borroni, G.; Stivala, L.A.; Borroni, R.G. Red grape (*Vitis vinifera* L.) flavonoids down-regulate collagen type III expression after UV-A in primary human dermal blood endothelial cells. *Exp. Dermatol.* **2018**, *27*, 973–980. [CrossRef]
8. Khlestkina, E.K. The adaptive role of flavonoids: Emphasis on cereals. *Cereal Res. Commun.* **2013**, *41*, 185–198. [CrossRef]

9.	Amato, A.; Cavallini, E.; Walker, A.R.; Pezzotti, M.; Bliek, M.; Quattrocchio, F.; Koes, R.; Ruperti, B.; Bertini, E.; Zenoni, S.; et al. The MYB5-driven MBW complex recruits a WRKY factor to enhance the expression of targets involved in vacuolar hyper-acidification and trafficking in grapevine. *Plant J.* **2019**, *99*, 1220–1241. [CrossRef]

10.	Manach, C.; Scalbert, A.; Morand, C.; Rémésy, C.; Jiménez, L. Polyphenols: Food sources and bioavailability. *Am. J. Clin. Nutr.* **2004**, *79*, 727–747. [CrossRef]

11.	Vélez, M.L.; Martínez-Martínez, F.; Del Valle-Ribes, C. The Study of Phenolic Compounds as Natural Antioxidants in Wine. *Crit. Rev. Food Sci. Nutr.* **2003**, *43*, 233–244. [CrossRef]

12.	Jesus, M.S.; Genisheva, Z.; Romaní, A.; Pereira, R.N.C.; Teixeira, J.A.; Domingues, L. Bioactive compounds recovery optimization from vine pruning residues using conventional heating and microwave-assisted extraction methods. *Ind. Crop. Prod.* **2019**, *132*, 99–110. [CrossRef]

13.	Jesus, M.S.; Ballesteros, L.F.; Pereira, R.N.C.; Genisheva, Z.; Carvalho, A.C.; Pereira-Wilson, C.; Teixeira, J.A.; Domingues, L. Ohmic heating polyphenolic extracts from vine pruning residue with enhanced biological activity. *Food Chem.* **2020**, *316*, 126298. [CrossRef]

14.	Shukitt-Hale, B.; Carey, A.; Simon, L.; Mark, D.A.; Joseph, J.A. Effects of Concord grape juice on cognitive and motor deficits in aging. *Nutrition* **2006**, *22*, 295–302. [CrossRef]

15.	Mullen, W.; Marks, S.C.; Crozier, A. Evaluation of Phenolic Compounds in Commercial Fruit Juices and Fruit Drinks. *J. Agric. Food Chem.* **2007**, *55*, 3148–3157. [CrossRef]

16.	Magrone, T.; Magrone, M.; Russo, M.A.; Jirillo, E. Recent Advances on the Anti-Inflammatory and Antioxidant Properties of Red Grape Polyphenols: In Vitro and In Vivo Studies. *Antioxidants* **2019**, *9*, 35. [CrossRef]

17.	Chou, L.-M.; Lin, C.-I.; Chen, Y.-H.; Liao, H.; Lin, S.-H. A diet containing grape powder ameliorates the cognitive decline in aged rats with a long-term high-fructose-high-fat dietary pattern. *J. Nutr. Biochem.* **2016**, *34*, 52–60. [CrossRef]

18.	Herman, F.; Westfall, S.; Brathwaite, J.; Pasinetti, G.M. Suppression of Presymptomatic Oxidative Stress and Inflammation in Neurodegeneration by Grape-Derived Polyphenols. *Front. Pharmacol.* **2018**, *9*, 867. [CrossRef]

19.	Ben Youssef, S.; Brisson, G.; Doucet-Beaupré, H.; Castonguay, A.-M.; Gora, C.; Amri, M.; Lévesque, M. Neuroprotective benefits of grape seed and skin extract in a mouse model of Parkinson's disease. *Nutr. Neurosci.* **2019**, 1–15. [CrossRef]

20.	Rocha, E.M.; De Miranda, B.; Sanders, L.H. Alpha-synuclein: Pathology, mitochondrial dysfunction and neuroinflammation in Parkinson's disease. *Neurobiol. Dis.* **2018**, *109*, 249–257. [CrossRef]

21.	Farrell, K.F.; Krishnamachari, S.; Villanueva, E.; Lou, H.; Alerte, T.N.M.; Peet, E.; Drolet, R.E.; Perez, R.G. Non-motor parkinsonian pathology in aging A53T α-Synuclein mice is associated with progressive synucleinopathy and altered enzymatic function. *J. Neurochem.* **2014**, *128*, 536–546. [CrossRef] [PubMed]

22.	Paumier, K.L.; Rizzo, S.J.S.; Berger, Z.; Chen, Y.; Gonzales, C.; Kaftan, E.; Li, L.; Lotarski, S.; Monaghan, M.; Shen, W.; et al. Behavioral Characterization of A53T Mice Reveals Early and Late Stage Deficits Related to Parkinson's Disease. *PLoS ONE* **2013**, *8*, e70274. [CrossRef] [PubMed]

23.	Unger, E.L.; Eve, D.J.; Perez, X.A.; Reichenbach, D.K.; Xu, Y.; Lee, M.K.; Andrews, A.M. Locomotor hyperactivity and alterations in dopamine neurotransmission are associated with overexpression of A53T mutant human α-synuclein in mice. *Neurobiol. Dis.* **2006**, *21*, 431–443. [CrossRef] [PubMed]

24.	Garelick, M.G.; Storm, D.R. The relationship between memory retrieval and memory extinction. *Proc. Natl. Acad. Sci. USA* **2005**, *102*, 9091–9092. [CrossRef] [PubMed]

25.	Pallauf, K.; Duckstein, N.; Rimbach, G. A literature review of flavonoids and lifespan in model organisms. *Proc. Nutr. Soc.* **2017**, *76*, 145–162. [CrossRef]

26.	Aires, D.J.; Rockwell, G.; Wang, T.; Frontera, J.; Wick, J.; Wang, W.; Tonkovic-Capin, M.; Lu, J.; Lezi, E.; Zhu, H.; et al. Potentiation of dietary restriction-induced lifespan extension by polyphenols. *Biochim. Biophys. Acta.* **2012**, *1822*, 522–526. [CrossRef]

27.	Kitani, K.; Osawa, T.; Yokozawa, T. The effects of tetrahydrocurcumin and green tea polyphenol on the survival of male C57BL/6 mice. *Biogerontology* **2007**, *8*, 567–573. [CrossRef]

28.	Zhang, S.-Q.; Cai, W.-J.; Huang, J.-H.; Wu, B.; Xia, S.-J.; Chen, X.-L.; Zhang, X.-M.; Shen, Z. Icariin, a natural flavonol glycoside, extends healthspan in mice. *Exp. Gerontol.* **2015**, *69*, 226–235. [CrossRef]

29.	Strong, R.; Miller, R.A.; Astle, C.M.; Baur, J.A.; De Cabo, R.; Fernandez, E.; Guo, W.; Javors, M.; Kirkland, J.L.; Nelson, J.F.; et al. Evaluation of Resveratrol, Green Tea Extract, Curcumin, Oxaloacetic Acid, and Medium-Chain Triglyceride Oil on Life Span of Genetically Heterogeneous Mice. *J. Gerontol. Ser. A Biol. Sci. Med Sci.* **2012**, *68*, 6–16. [CrossRef]

30. Spindler, S.R.; Mote, P.L.; Flegal, J.; Teter, B. Influence on Longevity of Blueberry, Cinnamon, Green and Black Tea, Pomegranate, Sesame, Curcumin, Morin, Pycnogenol, Quercetin, and Taxifolin Fed Iso-Calorically to Long-Lived, F1 Hybrid Mice. *Rejuvenation Res.* **2013**, *16*, 143–151. [CrossRef]

31. Frozza, R.L.; Lourenco, M.V.; De Felice, F.G. Challenges for Alzheimer's Disease Therapy: Insights from Novel Mechanisms Beyond Memory Defects. *Front. Neurosci.* **2018**, *12*, 37. [CrossRef] [PubMed]

32. Ma, Y.; Ma, B.; Shang, Y.; Yin, Q.; Wang, D.; Xu, S.; Hong, Y.; Hou, X.; Liu, X. Flavonoid-Rich Ethanol Extract from the Leaves of Diospyros kaki Attenuates D-Galactose-Induced Oxidative Stress and Neuroinflammation-Mediated Brain Aging in Mice. *Oxidative Med. Cell. Longev.* **2018**, *2018*, 1–12. [CrossRef] [PubMed]

33. Souza, L.C.; Antunes, M.S.; Filho, C.B.; Del Fabbro, L.; De Gomes, M.G.; Goes, A.T.R.; Donato, F.; Prigol, M.; Boeira, S.P.; Jesse, C. Flavonoid Chrysin prevents age-related cognitive decline via attenuation of oxidative stress and modulation of BDNF levels in aged mouse brain. *Pharmacol. Biochem. Behav.* **2015**, *134*, 22–30. [CrossRef] [PubMed]

34. Wang, Y.-J.; Thomas, P.; Zhong, J.-H.; Bi, F.-F.; Kosaraju, S.; Pollard, A.; Fenech, M.; Zhou, X.-F. Consumption of Grape Seed Extract Prevents Amyloid-β Deposition and Attenuates Inflammation in Brain of an Alzheimer's Disease Mouse. *Neurotox. Res.* **2009**, *15*, 3–14. [CrossRef] [PubMed]

35. Bensalem, J.; Dudonné, S.; Gaudout, D.; Servant, L.; Calon, F.; Desjardins, Y.; Layé, S.; Lafenetre, P.; Pallet, V. Polyphenol-rich extract from grape and blueberry attenuates cognitive decline and improves neuronal function in aged mice. *J. Nutr. Sci.* **2018**, *7*, e19. [CrossRef] [PubMed]

36. Czank, C.; Cassidy, A.; Zhang, Q.; Morrison, D.J.; Preston, T.; A Kroon, P.; Botting, N.P.; Kay, C.D. Human metabolism and elimination of the anthocyanin, cyanidin-3-glucoside: A 13C-tracer study. *Am. J. Clin. Nutr.* **2013**, *97*, 995–1003. [CrossRef]

37. Celli, G.B.; Ghanem, A.; Brooks, M.S.-L. A theoretical physiologically based pharmacokinetic approach for modeling the fate of anthocyanins in vivo. *Crit. Rev. Food Sci. Nutr.* **2017**, *57*, 3197–3207. [CrossRef]

38. Sandoval-Ramírez, B.A.; Catalán, Ú.; Fernández-Castillejo, S.; Rubió, L.; Macià, A.; Solà, R. Anthocyanin Tissue Bioavailability in Animals: Possible Implications for Human Health. A Systematic Review. *J. Agric. Food Chem.* **2018**, *66*, 11531–11543. [CrossRef]

39. Janle, E.M.; Lila, M.A.; Grannan, M.; Wood, L.; Higgins, A.; Yousef, G.G.; Rogers, R.B.; Kim, H.; Jackson, G.S.; Ho, L.; et al. Pharmacokinetics and Tissue Distribution of 14C-Labeled Grape Polyphenols in the Periphery and the Central Nervous System Following Oral Administration. *J. Med. Food* **2010**, *13*, 926–933. [CrossRef]

40. Kisková, T.; Kubatka, P.; Büsselberg, D.; Kassayova, M. The Plant-Derived Compound Resveratrol in Brain Cancer: A Review. *Biomolecules* **2020**, *10*, 161. [CrossRef]

41. Baell, J.B.; Walters, M.A. Chemistry: Chemical con artists foil drug discovery. *Nat. Cell Biol.* **2014**, *513*, 481–483. [CrossRef] [PubMed]

42. Ingólfsson, H.I.; Thakur, P.; Herold, K.F.; Hobart, E.A.; Ramsey, N.B.; Periole, X.; De Jong, D.H.; Zwama, M.; Yilmaz, D.; Hall, K.; et al. Phytochemicals Perturb Membranes and Promiscuously Alter Protein Function. *ACS Chem. Biol.* **2014**, *9*, 1788–1798. [CrossRef] [PubMed]

43. de Andrade Teles, R.B.; Diniz, T.C.; Pinto, T.C.C.; de Oliveira Júnior, R.G.; E Silva, M.G.; de Lavor, É.M.; Fernandes, A.W.C.; de Oliveira, A.P.; de Almeida Ribeiro, F.P.R.; da Silva, A.A.M.; et al. Flavonoids as Therapeutic Agents in Alzheimer's and Parkinson's Diseases: A Systematic Review of Preclinical Evidences. *Oxidative Med. Cell. Longev.* **2018**, *2018*, 1–21. [CrossRef] [PubMed]

44. Strathearn, K.E.; Yousef, G.G.; Grace, M.H.; Roy, S.L.; Tambe, M.A.; Ferruzzi, M.G.; Wu, Q.-L.; Simon, J.E.; Lila, M.A.; Rochet, J.-C. Neuroprotective effects of anthocyanin- and proanthocyanidin-rich extracts in cellular models of Parkinson's disease. *Brain Res.* **2014**, *1555*, 60–77. [CrossRef]

45. Macedo, D.; Tavares, L.; McDougall, G.J.; Miranda, H.V.; Stewart, D.; Ferreira, R.B.; Tenreiro, S.; Outeiro, T.F.; Santos, C.N. (Poly)phenols protect from α-synuclein toxicity by reducing oxidative stress and promoting autophagy. *Hum. Mol. Genet.* **2014**, *24*, 1717–1732. [CrossRef]

46. Macedo, D.; Jardim, C.; Figueira, I.; Almeida, A.F.; McDougall, G.; Stewart, D.; Yuste, J.E.; Tomás-Barberán, F.A.; Tenreiro, S.; Outeiro, T.F.; et al. (Poly)phenol-digested metabolites modulate alpha-synuclein toxicity by regulating proteostasis. *Sci. Rep.* **2018**, *8*, 6965. [CrossRef]

47. Hornedo-Ortega, R.; Álvarez-Fernández, M.A.; Cerezo, A.B.; Richard, T.; Troncoso, A.; García-Parrilla, M.C. Protocatechuic Acid: Inhibition of Fibril Formation, Destabilization of Preformed Fibrils of Amyloid-β and α-Synuclein, and Neuroprotection. *J. Agric. Food Chem.* **2016**, *64*, 7722–7732. [CrossRef]

48. Pogačnik, L.; Pirc, K.; Palmela, I.; Skrt, M.; Kim, K.S.; Brites, D.; Brito, M.A.; Ulrih, N.P.; Silva, R.F. Potential for brain accessibility and analysis of stability of selected flavonoids in relation to neuroprotection in vitro. *Brain Res.* **2016**, *1651*, 17–26. [CrossRef]

49. The Jackson Laboratory. Available online: https://www.jax.org/ (accessed on 13 November 2020).

50. RESSFOOD. Available online: http://enoant.info/ (accessed on 13 November 2020).

51. Pupyshev, A.B.; Tikhonova, M.A.; Akopyan, A.A.; Tenditnik, M.V.; Dubrovina, N.I.; Korolenko, T.A. Therapeutic activation of autophagy by combined treatment with rapamycin and trehalose in a mouse MPTP-induced model of Parkinson's disease. *Pharmacol. Biochem. Behav.* **2019**, *177*, 1–11. [CrossRef]

Sample Availability: Samples of the compounds are not available from the author.

Publisher's Note: MDPI stays neutral with regard to jurisdictional claims in published maps and institutional affiliations.

 *molecules*

Article

Antiangiogenic Activity of Flavonoids: A Systematic Review and Meta-Analysis

Mai Khater [1,2], Francesca Greco [1] and Helen M. I. Osborn [1,*]

[1] School of Pharmacy, University of Reading, Whiteknights, Reading RG6 6AD, UK; m.a.a.khater@pgr.reading.ac.uk (M.K.); f.greco@reading.ac.uk (F.G.)

[2] Therapeutic Chemistry Department, Pharmaceutical & Drug Industries Research Division, National Research Centre, Cairo 12622, Egypt

* Correspondence: h.m.i.osborn@reading.ac.uk

Academic Editor: H.P. Vasantha Rupasinghe

Received: 18 August 2020; Accepted: 10 October 2020; Published: 14 October 2020

Abstract: An imbalance of angiogenesis contributes to many pathologies such as cancer, arthritis and retinopathy, hence molecules that can modulate angiogenesis are of considerable therapeutic importance. Despite many reports on the promising antiangiogenic properties of naturally occurring flavonoids, no flavonoids have progressed to the clinic for this application. This systematic review and meta-analysis therefore evaluates the antiangiogenic activities of a wide range of flavonoids and is presented in two sections. The first part of the study (Systematic overview) included 402 articles identified by searching articles published before May 2020 using ScienceDirect, PubMed and Web of Science databases. From this initial search, different classes of flavonoids with antiangiogenic activities, related pathologies and use of in vitro and/or in/ex vivo angiogenesis assays were identified. In the second part (Meta-analysis), 25 studies concerning the antiangiogenic evaluation of flavonoids using the in vivo chick chorioallantoic membrane (CAM) assay were included, following a targeted search on articles published prior to June 2020. Meta-analysis of 15 out of the 25 eligible studies showed concentration dependent antiangiogenic activity of six compared subclasses of flavonoids with isoflavones, flavonols and flavones being the most active (64 to 80% reduction of blood vessels at 100 µM). Furthermore, the key structural features required for the antiangiogenic activity of flavonoids were derived from the pooled data in a structure activity relationship (SAR) study. All in all, flavonoids are promising candidates for the development of antiangiogenic agents, however further investigations are needed to determine the key structural features responsible for their activity.

Keywords: flavonoids; angiogenesis; inflammation; cancer; in-vivo angiogenesis; CAM assay; SAR

1. Introduction

Angiogenesis is the process of forming new blood vessels. Physiologically, angiogenesis is pivotal for tissue growth and regeneration [1] which is beneficial for many processes including embryogenesis and wound healing. Regulation of angiogenesis is complex and is maintained by the balance between endogenous stimulators (e.g., vascular endothelial growth factor (VEGF), platelet derived growth factors (PDGFs) and hypoxia-inducible factors (HIFs)), and inhibitors (e.g., angiostatin and endostatin). Other body conditions also contribute to the regulation of angiogenesis under physiological conditions. For example, certain metabolic demands such as the need for more oxygen can induce VEGF secretion and angiogenesis in heart and brain tissues [2,3]. Since angiogenesis affects many organs and tissues in the body, an imbalance in its regulation has been associated with different pathologies [4]. For instance, cancer, rheumatoid arthritis and diabetic retinopathy feature an upregulation of proangiogenic factors [5]. Conversely, if antiangiogenic factors were upregulated, several cardiovascular diseases are more likely to happen [6]. The use of drugs like Bevacizumab

(Avastin®, Genentech) and Aflibercept (Eylea®, Regeneron) for the treatment of cancer and ocular diseases, emphasizes the imperative medicinal applications of antiangiogenic agents [7,8].

Flavonoids are widely distributed in fruits, vegetables and nuts. They are one of the most important chemical classes of natural compounds showing various pharmacological profiles that include anticancer [9–11], anti-inflammatory [12], cardioprotective [13] and neuroprotective activities [14].

The antiangiogenic activity of flavonoids has been extensively studied over the last two decades. Several studies document the ability of flavonoids to inhibit the proliferation and migration of endothelial cells by interfering with key angiogenesis signaling cascades such as the mitogen activated protein kinase (MAPK) and phosphoinositide 3-kinase (PI3K) pathways. Nevertheless, they can inhibit the expression of major proangiogenic factors such as VEGF and matrix metalloproteinases (MMPs) [2,7,15].

Researchers rely on different in vitro and in/ex vivo assays to quantitatively assess the effects of chemical compounds on angiogenesis [16,17]. Each of these assays can probe one or more of the different steps involved in the angiogenesis process such as cell proliferation, migration and tubulogenesis.

Despite considerable research concerning the antiangiogenic activities of flavonoids, to date they have neither progressed to the market nor clinical trials for that purpose. Therefore, the aim of this review is to systematically assess the antiangiogenic activities of flavonoids to provide greater insight into their potential as therapeutic agents. This study is comprised of two parts: Section 1 provides a systematic overview of the classes of flavonoids that have been investigated for their antiangiogenic activities, along with a summary of the different in vitro and/or in/ex vivo angiogenesis assays that have been used; Section 2 is a meta-analysis study of a quantitatively comparative subset of data, based on the in vivo chick chorioallantoic membrane (CAM) assay, to statistically evaluate the antiangiogenic effects of flavonoids.

2. Results

2.1. Section 1: Systematic Overview

2.1.1. Search Results

For Section 1 of the study, 3708 records were initially identified in three electronic databases (1555 from ScienceDirect, 1984 from PubMed and 169 from Web of Science). Search results were then limited to research articles, review articles, short communications and systematic reviews and the remaining 3380 articles were subjected to title and abstract screening. 2556 records were found to be irrelevant of the subject in focus or did not fulfill the inclusion criteria. After the removal of duplicates (422), 402 articles were finally included in the qualitative analysis for Section 1 of this study (Figure 1).

2.1.2. Study Characteristics

The pool of studies included was classified with respect to: (a) flavonoid class (Figure 2), (b) flavonoid name, (c) disease, (d) in vitro test and (e) in/ex vivo tests. Characteristics of the included studies are summarized in Table 1 (flavonols are used as a representative example in Table 1 and Table S1 contains similar data for all classes of flavonoids). A total of 402 research and review articles were considered. All of the included articles reported angiogenesis related in vitro and/or in/ex vivo assays for different classes of flavonoids.

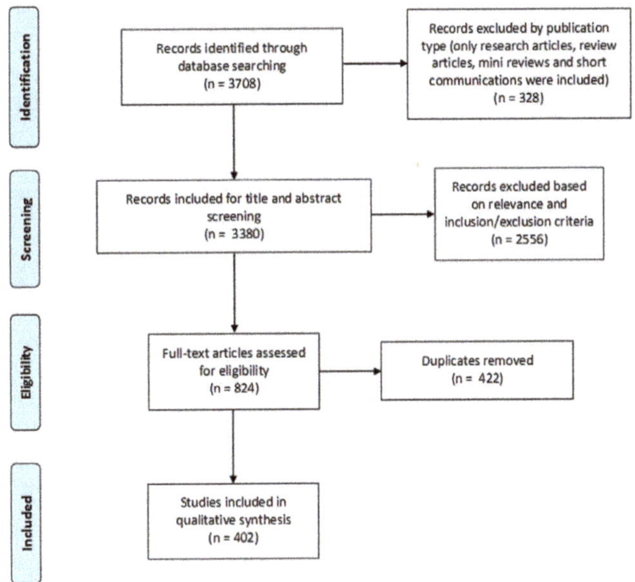

Figure 1. PRISMA flow diagram of study search and selection process of Section 1.

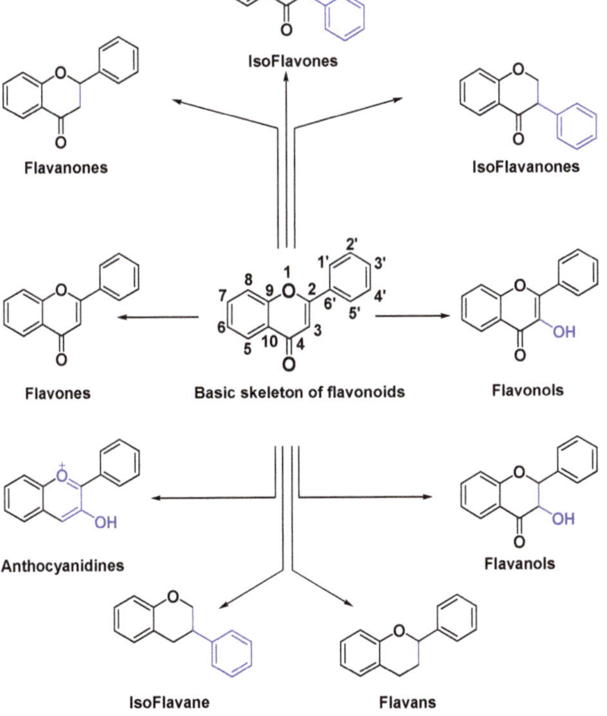

Figure 2. Chemical structures of classes of flavonoids.

Table 1. Characteristics of the studies included in Section 1 for flavonols subclass (see Table S1 for all 9 subclasses).

Flavonol	Disease	In Vitro Tests	In/Ex Vivo Tests	Author, Year
Beturetol	Angiogenesis		CAM	Hisanori Hattori, 2011 [18]
Casticin	Cancer *			Shanaya Ramchandani, 2020 [19]
Denticulatain	Lung Cancer		ZFM	Da Song Yang, 2015 [20]
Dihydrokaempferide	Angiogenesis		CAM	Hisanori Hattori, 2011 [18]
Fisetin	Cancer *			Dharambir Kashyap, 2018 [21]
	Cancer *			Thamaraiselvan Rengarajan, 2016 [22]
	Cancer *			Deeba N.Syed, 2016 [23]
	Cancer *			Lall K. Rahul, 2016 [24]
	Breast Cancer	In		Cheng Fang Tsai, 2018 [25]
	Breast Cancer	WH, In		Xu Sun, 2018 [26]
	Breast Cancer	WH, In	Mets in mice	Jie Li, 2018 [27]
	Cervical Cancer	In		Ruey Hwang Chou, 2013 [28]
	Glioma	In		Chien Min Chen, 2015 [29]
	Hepatic Cancer	In		Xiang Feng Liu, 2017 [30]
	Leukemia	In		Anna Klimaszewska-Wiśniewska, 2019 [31]
	Lung Cancer	WH, In		Saba Tabasum, 2019 [32]
	Lung Cancer	WH, In, Ad		Junjian Wang, 2018 [33]
	Prostate Cancer	WH, In, Ad		Chi Sheng Chien, 2010 [34]
	Renal Cancer	In		Yih Shou Hsieh, 2019 [35]
	Retinopathy		RbCN	A M Joussen, 2000 [36]
Galangin	Hepatic Cancer *			Dengyang Fang, 2019 [37]
	Angiogenesis	TF, Ad		Jong Deog Kim, 2006 [38]
	Glioma	TF, In	CAM, MD in mice	Daliang Chen, 2019 [39]
	Glioma	In		Deqiang Lei, 2018 [40]
	Hepatic Cancer	WH, In, Ad		Shang Tao Chien, 2015 [41]
	Ovarian Cancer	TF	CAM	Haizhi Huang, 2015 [42]
	Renal Cancer	WH, In		Jingyi Cao, 2016 [43]
	Renal Cancer	In		Yun Zhu, 2018 [44]
Gossypin	Gastric Cancer	In		Wang Li, 2019 [45]
Herbacetin	Melanoma	In		Lei Li, 2019 [46]
Hyperoside	Arthritis	WH, In	CIAM in mice	Xiang Nan Jin, 2016 [47]
Icariin	Bone disease *			Xin Zhang, 2014 [48]
	Cancer *			Meixia Chen, 2016 [49]
	Angiogenesis	TF, In	RAR	Byung Hee Chung, 2008 [50]
	Esophageal Cancer	In		Zhen Fang Gu, 2017 [51]
	Ovarian Cancer	WH		Pengzhen Wang, 2019 [52]
	Wound healing		EWM in rats	Wangkheirakpam Ramdas Singh, 2019 [53]

Table 1. *Cont.*

Flavonol	Disease	In Vitro Tests	In/Ex Vivo Tests	Author, Year
Icariside	Cancer *			Meixia Chen, 2016 [49]
	Glioma	WH, In		Kai Quan, 2017 [54]
Isoviolanthin	Hepatic Cancer	WH, In		Shangping Xing, 2018 [55]
Isosakuranetin	Angiogenesis		CAM	Hisanori Hattori, 2011 [18]
Kaempferol	Cancer *			Allen Y. Chen, 2013 [56]
	Cancer *			Dharambir Kashyap, 2017 [57]
	Angiogenesis	WH, TB, In		Hsien Kuo Chin, 2018 [58]
	Angiogenesis	WH, TB	ZFM	Fang Liang, 2015 [59]
	Angiogenesis		CAM	Shigenori Kumazawa, 2013 [60]
	Angiogenesis	TF, Ad		Jong Deog Kim, 2006 [38]
	Diabetes		EWM in rats	Yusuf Özay, 2019 [61]
	Glioma	WH		Vivek Sharma, 2007 [62]
	Glioma	In	Mets in mice	S.C. Shen, 2006 [63]
	Hepatic Cancer	WH, In	Mets in mice	Youyou Qin, 2015 [64]
	Hepatic Cancer	In		Genglong Zhu, 2018 [65]
	Lung Cancer	WH, In		Eunji Jo, 2015 [66]
	Medulloblastoma	Ad		David Labbé, 2009 [67]
	Oral Cancer	In		Chiao Wen Lin, 2013 [68]
	Osteosarcoma	WH, In, Ad		Hui Jye Chen, 2013 [69]
	Ovarian Cancer		CAM	Haitao Luo, 2009 [70]
	Pancreatic Cancer	In		Jungwhoi Lee, 2016 [71]
	Renal Cancer	WH, In	Mets in mice	Tung Wei Hung, 2017 [72]
	Retinal Vascularization	WH, In		Hsiang Wen Chien, 2019 [73]
Kaempferol-3-O-[(6-caffeoyl)-β-glucopyranosyl (1→3) α-rhamnopyranoside]-7-O-α-rhamnopyranoside	Angiogenesis	WH		Marco Clericuzio, 2012 [74]
Kaempferide	Angiogenesis		CAM	Hisanori Hattori, 2011 [18]
Morin	Arthritis	WH, TB	CIAM in rats	Ni Zeng, 2015 [75]
	Arthritis	WH, TB	CIAM in rats	Mengfan Yue, 2018 [76]
	Leukemia	Ad		Nagaja Capitani, 2019 [77]
	Melanoma	WH		Hua Wen Li, 2016 [78]
Myricetin	Melanoma *			Nam Joo Kang, 2011 [79]
	Angiogenesis	TF, Ad		Jong Deog Kim, 2006 [38]
	Breast Cancer	In	CAM, MD in mice, RAR	Zhiqing Zhou, 2019 [80]
	Breast Cancer	WH, In, Ad	Mets in mice	Yingqian Ci, 2018 [81]
	Glioma	WH, In		Wen Ta Chiu, 2010 [82]
	Hepatic Cancer	In		Noriko Yamada, 2020 [83]
	Hepatic Cancer	WH, In		Hongxin Ma, 2019 [84]
	Lung Cancer	WH, In, Ad		Yuan Wei Shih, 2009 [85]
	Medullobalstoma	In, Ad		David Labbé, 2009 [67]
	Ovarian Cancer	TF	CAM	Haizhi Huang, 2015 [42]

Table 1. *Cont.*

Flavonol	Disease	In Vitro Tests	In/Ex Vivo Tests	Author, Year
	Breast Cancer *			Maryam Ezzati, 2020 [86]
	Cancer *			Si-min Tang, 2020 [87]
	Cancer *			Dharambir Kashyap, 2016 [88]
	Colorectal Cancer *			Saber G. Darband, 2018 [89]
	Angiogenesis	WH, In		Nu Ry Song, 2014 [90]
	Angiogenesis	WH, TB	ZFM	Chen Lin, 2012 [91]
	Angiogenesis	TF, Ad		Jong Deog Kim, 2006 [38]
	Bladder Cancer	WH, In		Yu Hsiang Lee, 2019 [92]
	Breast Cancer	WH		Divyashree Ravishankar, 2015 [93]
	Breast Cancer		MD in mice	Xin Zhao, 2016 [94]
	Breast Cancer		CAM	Soo Jin Oh, 2010 [95]
	Breast Cancer	WH, In		Asha Srinivasan, 2016 [96]
	Breast Cancer	WH, In		Cheng Wei Lin, 2008 [97]
	Breast Cancer	In		Amilcar Rivera Rivera, 2016 [98]
	Cancer	TF	ZFM	Daxian Zhao, 2014 [99]
	Cancer	TF, In	CAM	Wen Fu Tan, 2003 [100]
	Cancer		MD in mice	Xiangpei Zhao, 2012 [101]
	Cancer	WH, In		Lung Ta Lee, 2004 [102]
	Cancer	WH		Dong Eun Lee, 2013 [103]
	Colorectal Cancer	WH, In	Mets in mice	Ji Ye Kee, 2016 [104]
	Glioma	WH		Hong Chao Pan, 2015 [105]
	Glioma	WH, In		Wen Ta Chiu, 2010 [82]
	Glioma	WH, In		Yue Liu, 2017 [106]
	Glioma	In		Jonathan Michaud-Levesque, 2012 [107]
	Glioma	WH		Alessandra Bispo da Silva, 2020 [108]
Quercetin	Glioma	WH, TB, In		Yue Liu, 2017 [109]
	Hepatic Cancer	In		Noriko Yamada, 2020 [83]
	Hepatic Cancer	WH, In		Jun Lu, 2018 [110]
	Lung Cancer	WH		Anna Klimaszewska-Wiśniewska, 2017 [111]
	Lung Cancer	In		Tzu Chin Wu, 2018 [112]
	Lung Cancer	In		Yo Chuen Lin, 2013 [113]
	Medulloblastoma	In, Ad		David Labbé, 2009 [67]
	Melanoma	In		Mun Kyung Hwang, 2009 [114]
	Melanoma	In		Hui Hui Cao, 2015 [115]
	Melanoma	WH, In	Mets in mice	Hui Hui Cao, 2014 [116]
	Oral Cancer	In		Junfang Zhao, 2019 [117]
	Osteoblasts	In		Tae Wook Nam, 2008 [118]
	Osteosarcoma	WH, In, Ad		Shenglong Li, 2019 [119]
	Osteosarcoma	WH, In	Mets in mice	Haifeng Lan, 2017 [120]
	Osteosarcoma	WH, Ad		Kersten Berndt, 2013 [121]
	Pancreatic Cancer	WH, In		Ying Tang Huang, 2005 [122]
	Pancreatic Cancer	WH, In		Yu Dinglai 2017 [123]
	Prostate Cancer	WH, In		Firdous Ahmad Bhat, 2014 [124]
	Prostate Cancer	TF, In	MD in mice	Feiya Yang, 2016 [125]
	Retinoblastoma	In		Wei Song, 2017 [126]

Table 1. *Cont.*

Flavonol	Disease	In Vitro Tests	In/Ex Vivo Tests	Author, Year
Quercetin-3-*O*-[(6-caffeoyl)-β-glucopyranosyl(1→3)α-rhamnopyranoside]-7-*O*-α-rhamnopyranoside	Angiogenesis	WH		Marco Clericuzio, 2012 [74]
Rutin	Angiogenesis		CAM	César Muñoz Camero, 2018 [127]
	Angiogenesis		CAM	Shigenori Kumazawa, 2013 [60]
	Cancer	WH, In, Ad		Mohamed ben Sghaier, 2016 [128]
	Glioma	WH		Alessandra Bispo da Silva, 2020 [108]
	Neuroblastoma	WH, In		Hongyan Chen, 2013 [129]

* Review article; TB, Tube Formation; WH, Wound Healing; In, Invasion; Ad, Adhesion; Mets, Metastasis; CAM, Chick Chorioallantoic Membrane; MPA, Matrigel Plug Assay; RAR, Rat Aortic Ring; EWM, Excision Wound Model; SF, Skin Flap; RRN, Rat Retinal Neovascularization; MAR, Mice Aortic Ring; MD, Microvessel Density; MRN, Mice Retinal Neovascularization; MCN, Mice Corneal Neovascularization; RbCN, Rabbit Corneal Neovascularization; ZFM, Zebra Fish Model; RCN, Rat Corneal Neovascularization; CIAM, Collagen Induced Arthritis Model; DASM, Dorsal Air Sac Model; IWM, Incision Wound Model.

2.1.3. Data Analysis

The majority of articles (332, 82%) focused on the implications of angiogenesis on cancer growth and metastasis. 7% of the articles studied antiangiogenic effects of flavonoids on other diseases such as diabetes, bone and eye diseases, whilst 11% focused on the antiangiogenic activity of flavonoids without application to a specific pathology (Figure 3a). A profiling of the studies retrieved with respect to chemical class of flavonoids is shown in Figure 3b.

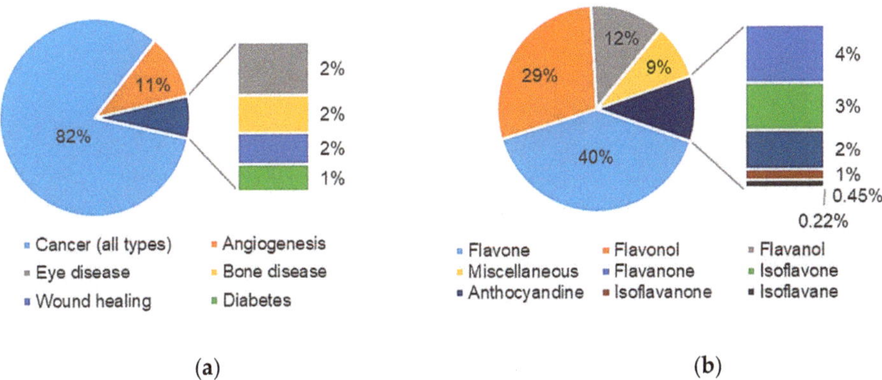

Figure 3. Profiling of papers retrieved in Section 1 with respect to: (**a**) pathology type; (**b**) chemical class of flavonoid.

Figure 4 summarizes the types of in vitro and in vivo assays that were utilized in the studies. From a pool of 342 research articles included in this study, 152 articles (44%) reported a combination of in vitro and in/ex vivo assays in their studies. The percentage of research articles that depended only on in/ex vivo tests to evaluate antiangiogenic activity of flavonoids were comparatively low compared to those conducting only in vitro assays (3% vs. 53%, respectively).

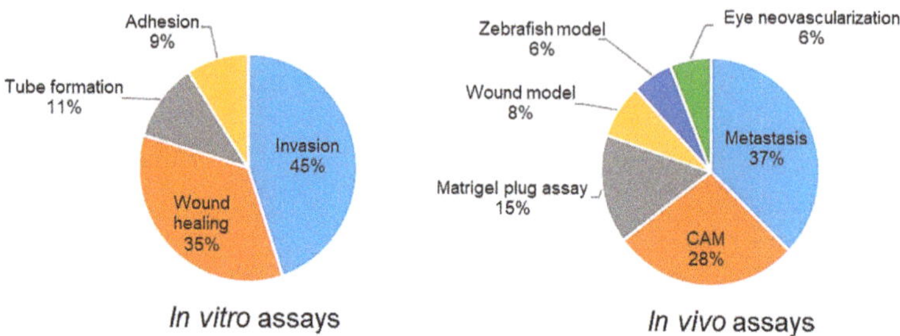

Figure 4. Types of assays used for in vitro and in vivo antiangiogenic evaluation of flavonoids.

2.2. Section 2: Meta-Analysis

2.2.1. Search Results

The second subset search, which is the basis of the meta-analysis forming Section 2 of this study, followed the same general methodology as detailed in the initial overview. 960 records were identified from four electronic databases (381 from ScienceDirect, 496 from PubMed, 65 from Web of Science and 18 from Google Scholar). 25 research articles were finally included in the quantitative analysis after the sequential steps of screening and sifting, as shown in Figure 5.

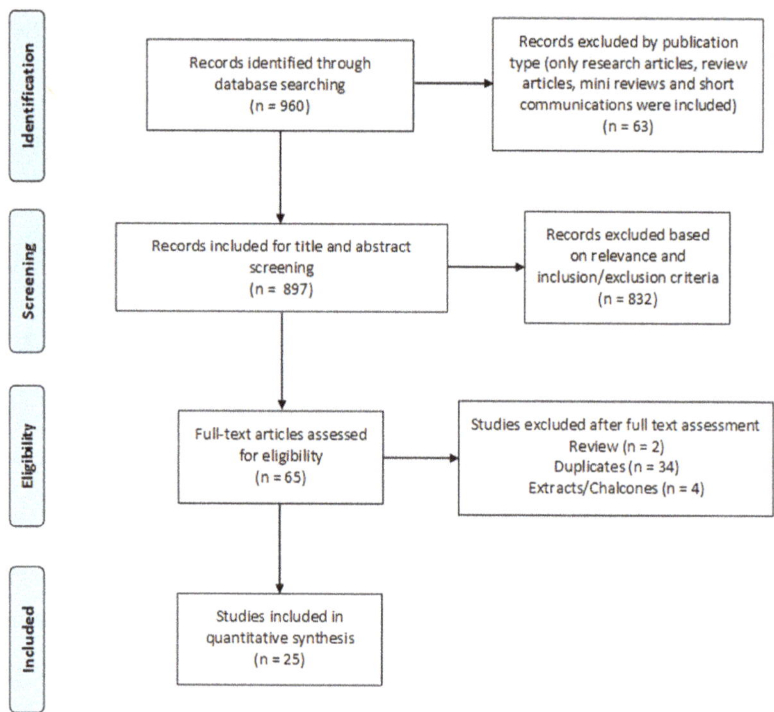

Figure 5. PRISMA flow diagram of study search and selection process of Section 2.

2.2.2. Study Characteristics

The main study characteristics of the research articles included in Section 2 for the meta-analysis are summarized in Table 2 by study name.

Table 2. Characteristics of the studies included in Section 2.

Author, Year	Flavonoid	Angiogenesis Promoter	Cell Line	Concentration	Time, Duration of Treatment	Results Representation	n
Soo Jin Oh, 2010 [95]	Quercetin	NA	TAMR-MCF-7	3, 10, 30 µM	NA	Number of branches	5 to 7
Chiu-Mei Lin, 2006 [130]	Wogonin	LPS (1µg/mL)	NA	$10^{-5}, 10^{-6}, 10^{-7}, 10^{-8}$ M	10th day, 48 h	Percentage of vascular counts (%)	3
Ling-Zhi Liu, 2011 [131]	Acacetin	NA	OVCAR-3	10 µM	9th day, 4 days	Relative angiogenesis	5
Kai Zhao, 2018 [132]	Wogonin, LW-215	NA	NA	Wogonin: 80 ng/CAM, LW-215: 2, 4, 8 ng/CAM	10th day, 48 h	The number of new vessels (% of control)	3
Haizhi Huang, 2015 [42]	Galangin, myricetin	NA	OVCAR-3	G: 40 µM, M: 20 µM	9th day, 5 days	Blood vessels (%)	6
Olga Viegas, 2019 [133]	Cyanidin, C-3-*O*-glucoside, delphinidin, D-3-*O*-glucoside	NA	NA	20, 40, 80, 100, 200 µM	11th day, 48 h	% of control	5
Wen-fu Tan, 2003 [100]	Quercetin	NA	NA	25, 50, 100 nmol/10 µL/CAM	10th day, 48 h	Microscopic pictures	10
Rajesh Gacche, 2010 [134]	Flavone, 3/5/6/7/-Hydroxy flavone	NA	NA	10, 50, 100 µM	10th day, 48 h	Antiangiogenic activity (%) of selected flavonoids	8
R.N. Gacche, 2011 [135]	3, 6-DHF, 3, 7-DHF, 5, 7-DHF, apigenin, genistein, kaempferol, luteolin, fisetin, rutin, quercetin	NA	NA	10, 50, 100 µM in 0.05% DMSO/20 µL/CAM	10th day, 48 h	Antiangiogenic activity (%) of selected flavonoids	8
R.N. Gacche, 2015 [136]	4′-Methoxy flavone, 3-Hydroxy-7-methoxy flavone, Formononetin, Biochanin-A, Diosmin, Hesperitin, Hesperidin, 2′-Hydroxy flavanone, 4′-Hydroxy flavanone, 7-Hydroxy flavanone, Myricetin, Taxifolin, Silibinin, Silymarin, Naringenin, Naringin, Catechin	NA	NA	10, 50, 100 µM in 0.05% DMSO/20 µL/CAM	10th day, 48 h	Antiangiogenic activity (%) of selected flavonoids	8
Yan Chen, 2010 [137]	LYG-202	NA	NA	2.4, 12, 60 ng/CAM	10th day, 48 h	Percentage of vascular counts (% of control)	10

Table 2. *Cont.*

Author, Year	Flavonoid	Angiogenesis Promoter	Cell Line	Concentration	Time, Duration of Treatment	Results Representation	n
Hisanori Hattori, 2011 [18]	Beturetol, isosakuranetin	NA	NA	300 ng/CAM	5th day, 7 days	Inhibition % of angiogenesis at 300 ng/CAM.	10
Yujie Huang, 2019 [138]	Wogonoside	NA	MDA-MB-231, MDA-MB-468	50, 100, 200 ng/CAM	10th day, 48 h	Number of new vessels (% cells)	3
Yan Chen, 2009 [139]	Wogonoside	LPS (1μg/mL)	NA	1.5, 15, 150 ng/CAM	10th day, 48 h	Number of vessels (% of LPS)	10
Xiaobo Li, 2017 [140]	Luteolin	Gas6 (300 ng/mL)	NA	10, 20 μM	6th day, 48 h	Relative vascular density (% of control)	3
Siva Prasad Panda, 2019 [141]	TMF	NA	EAT	10, 17, 25 μg/mL	5th day, 11 days	Microscopic pictures	5
Yujie Huang, 2016 [142]	Wogonoside	NA	MCF-7	50, 100, 200 ng/CAM	10th day, 48 h	Number of new vessels (% MCF-7)	3
Tariq A. Bhat, 2013 [143]	Acacetin	NA	NA	50 μM	6th day, (every 48 h for 8 days)	% capillary formation	5 independent areas on CAMs for each treatment
Jing Fang, 2007 [144]	Apigenin	NA	OVCAR-3, PC-3	OVCAR-3: 7.5, 15 μM, PC-3: 10, 20 μM	9th day, 4 days	Quantification of blood vessels on the CAM	8
Jianchu Chen, 2015 [145]	Nobiletin	NA	A2780	20 μM	9th day, 5 days	Blood vessel count	10
Poyil Pratheeshkumar, 2012 [146]	Luteolin	NA	NA	20, 40 μM	8th day, 48 h	Relative vascular density	3
Chiu-Mei Lin, 2006 [147]	Wogonin	IL-6 (10 ng/mL)	NA	$10^{-5}, 10^{-6}, 10^{-7}, 10^{-8}$ M	10th day, 48 h	Percentage of vascular count (%)	3
Dongqing Zhu, 2016 [148]	Baicalin, baicalein	NA	NA	0.5, 2, 10, 50 μg/mL and 0.2, 1, 5 mg/mL	7.5th day, 48 h	Number of new blood vessels	30
Haitao Luo, 2009 [70]	Kaempferol	NA	OVCAR-3	20 μM	9th day, 5 days	Blood vessel count	5
Laure Favot, 2003 [149]	Delphinidin	NA	NA	2, 10, 25, 50 μg	8th day, 48 h	Microscopic pictures	5

n = number of CAMs used in each experiment; NA, Not available; DHF, Dihydroxyflavone; TMF, Trimethoxyflavonoid; TMAR, Tamoxifen breast cancer resistant cell line; MCF-7, Breast cancer cell line; LPS, Lipopolysaccharide; OVCAR-3, Ovarian cancer cell line; MDA-MB-231, MDA-MB-468, Triple negative breast cancer cell lines; Gas6, Growth arrest specific 6; EAT, Mouse breast carcinoma (Ehrlich-Lettre Ascites); PC-3, Prostate cancer cell line; A2780, ovarian cancer cell line; IL-6, Interleukin 6.

2.2.3. Meta-Analysis (Antiangiogenic Effect of Flavonoids on CAMs)

25 studies reporting the CAM assay for the in vivo evaluation of flavonoids were eligible for the meta-analysis. The number of blood vessels relative to the control was used as the outcome measure, the lower the ratio the higher the antiangiogenic activity. The studies were grouped into 3 sub-sets based on the controls used. In the first set (12 studies), the normal vasculature of the CAM was used as a control without any interventions that would induce angiogenesis. The second and third sets, however, tested the antiangiogenic activity of flavonoids on CAMs with abnormal angiogenesis using either proangiogenic factors for set 2 (4 studies) or cancer cell lines for set 3 (9 studies). 10 studies [18,95,100,132,137,139–141,145,149] out of the 25 eligible studies were not included in any of the conducted meta-analyses as they failed to report the required data outcomes or did not fit under any particular subgroup.

Set 1: Antiangiogenic effect of flavonoids under normal conditions

To ensure consistency in our comparison, for the meta-analysis of set 1, the concentrations were grouped into three ranges i.e., low (10–20 µM), medium (40–50 µM) and high (100 µM). Flavonoids were sub grouped based on their chemical class as shown in Figure 6. Pooled results indicate that all subclasses, except for anthocyanidines, demonstrate concentration dependent antiangiogenic activity expressed as a reduction in the number of blood vessels in a CAM. For the flavonols subgroup, for instance, the overall means ratios (summary estimates of antiangiogenic activity of a subgroup of flavonoids relative to control) were 0.74 (95%CI: 0.69, 0.79; p-value < 0.00001), 0.50 (95%CI: 0.46, 0.56; p-value < 0.00001) and 0.26 (95%CI: 0.19, 0.35; p-value < 0.00001) for the low, mid and high concentrations, respectively. On the other hand, the anthocyanidines subgroup exhibited only a minor overall reduction of 18% at the highest concentration and a slightly proangiogenic effect (overall means ratio: 1.07; 95%CI: 0.86, 1.33; p-value: 0.53) at 20 µM.

In addition to the forest plot analysis that gives a general idea about the overall in vivo antiangiogenic activity of flavonoids and identifies trends of activity among the different subclasses, some structure activity relationship (SAR) conclusions were drawn from the pooled results (Figure 7).

First, there was no correlation between the number of hydroxyl groups and antiangiogenic activity. However, the position of the hydroxyl groups appeared to be of importance as most of the highly active flavonoids had hydroxyl groups at positions 3, 5 and 7 and/or 4′ (e.g., as demonstrated for 3-OH flavone, acacetin, biochanin A, apigenin, silibinin and kaempferol). The 7-OH group can be considered to be of greatest importance for activity since 7-OH flavone showed higher activity in the low and medium concentrations compared to the 5-OH analogue. Absence of the 3-OH group caused up to a 14% decrease in activity at the 50 and 100 µM concentrations, as demonstrated, for example, for 3-OH flavone vs. flavone, kaempferol vs. apigenin and 3,7-diOH flavone vs. 7-OH flavone. This was also true for quercetin vs. luteolin but with only a trivial drop of activity of 1 to 2%. However, this was not the case for 3,6-diOH flavone vs. 6-OH flavone where removal of the 3-OH group slightly increased the activity by 1 to 5% at the mid and high concentrations.

Secondly, unsaturation of the C2 and C3 bond is a common feature of most of the highly active flavonoids and is important for activity. 7-OH flavone and 7-OH flavanone are two good examples that exemplify this, as demonstrated by a reduction of the number of vessels: 27%, 32% and 52% for 7-OH flavone and 10%, 22% and 39% for 7-OH flavanone at 10 µM, 50 µM, and 100 µM, respectively.

Third, there are examples of where the presence of a methoxy group at position 4′ increases activity (e.g., biochanin A, diosmin and formononetin). However, the presence of a methoxy group at C7 caused a decrease in the activity when compared to the unsubstituted analogue (ie for the 3-OH flavone vs. 3-OH-7-OCH$_3$ flavone, reduction of number of vessels: 35% and 20% at 10 µM, 64% and 42% at 50 µM, 79% and 69% at 100 µM, respectively). On the other hand, conversion of the 7-OH group in 3,7-diOH flavone to a 7-OCH$_3$ group in 3-OH-7-OCH$_3$ flavone slightly improved the activity (reduction of number of vessels) from 18% to 20% at 10 µM and from 63% to 69% at 100 µM, respectively. Finally, glycosylation at positions 3 or 7 showed neither a pronounced nor a consistent effect on the antiangiogenic activity of flavonoids. While a decrease in activity was observed with quercetin vs. rutin, hesperitin vs. hesperidin and cyanidin vs. cyanidin-3-*O*-glucoside, an increase was observed in the cases of naringin vs. naringenin and delphinidin vs. delphinidin-3-*O*-glucoside.

Study or Subgroup	log[Means Ratio]	SE	Weight	Means Ratio IV, Random, 95% CI
1.1.1 Isoflavones				
Gacche 2011 (Genistein)	-0.42	0.01	33.1%	0.66 [0.64, 0.67]
Gacche 2015 (Biochanin-A)	-0.33	0.01	33.1%	0.72 [0.70, 0.73]
Gacche 2015 (Formononetin)	-0.29	0.004	33.7%	0.75 [0.74, 0.75]
Subtotal (95% CI)			100.0%	0.71 [0.66, 0.76]
Heterogeneity: Tau² = 0.00; Chi² = 149.97, df = 2 (P < 0.00001); I² = 99%				
Test for overall effect: Z = 8.85 (P < 0.00001)				
1.1.2 Flavones				
Gacche 2010 (5-Hydroxy flavone)	-0.22	0.02	9.9%	0.80 [0.77, 0.83]
Gacche 2010 (6-Hydroxy flavone)	-0.21	0.01	10.5%	0.81 [0.79, 0.83]
Gacche 2010 (7-Hydroxy flavone)	-0.32	0.01	10.5%	0.73 [0.71, 0.74]
Gacche 2010 (Flavone)	-0.4	0.01	10.5%	0.67 [0.66, 0.68]
Gacche 2011 (5,7-Dihydroxy flavone)	-0.36	0.01	10.5%	0.70 [0.68, 0.71]
Gacche 2011 (Apigenin)	-0.4	0.01	10.5%	0.67 [0.66, 0.68]
Gacche 2011 (Luteolin)	-0.36	0.01	10.5%	0.70 [0.68, 0.71]
Gacche 2015 (4'-Methoxy flavone)	-0.26	0.004	10.7%	0.77 [0.77, 0.78]
Gacche 2015 (Diosmin)	-0.27	0.01	10.5%	0.76 [0.75, 0.78]
Pratheeshkumar 2012 (Luteolin)	-0.3	0.06	6.0%	0.74 [0.66, 0.83]
Zhu 2016 (Baicalein)	-0.13	1.4	0.0%	0.88 [0.06, 13.65]
Zhu 2016 (Baicalin)	0.22	1.6	0.0%	1.25 [0.05, 28.67]
Subtotal (95% CI)			100.0%	0.73 [0.70, 0.77]
Heterogeneity: Tau² = 0.00; Chi² = 480.27, df = 11 (P < 0.00001); I² = 98%				
Test for overall effect: Z = 13.94 (P < 0.00001)				
1.1.3 Flavonols				
Gacche 2010 (3-Hydroxy flavone)	-0.43	0.02	10.8%	0.65 [0.63, 0.68]
Gacche 2011 (3,6-Dihydroxy flavone)	-0.13	0.01	11.1%	0.88 [0.86, 0.90]
Gacche 2011 (3,7-Dihydroxy flavone)	-0.2	0.01	11.1%	0.82 [0.80, 0.83]
Gacche 2011 (Fisetin)	-0.36	0.01	11.1%	0.70 [0.68, 0.71]
Gacche 2011 (Kaempferol)	-0.39	0.01	11.1%	0.68 [0.66, 0.69]
Gacche 2011 (Quercetin)	-0.39	0.01	11.1%	0.68 [0.66, 0.69]
Gacche 2011 (Rutin)	-0.34	0.01	11.1%	0.71 [0.70, 0.73]
Gacche 2015 (3-Hydroxy-7-methoxy flavone)	-0.22	0.004	11.2%	0.80 [0.80, 0.81]
Gacche 2015 (Myricetin)	-0.27	0.01	11.1%	0.76 [0.75, 0.78]
Subtotal (95% CI)			100.0%	0.74 [0.69, 0.79]
Heterogeneity: Tau² = 0.01; Chi² = 878.21, df = 8 (P < 0.00001); I² = 99%				
Test for overall effect: Z = 9.71 (P < 0.00001)				
1.1.4 Flavanols				
Gacche 2015 (Silibinin)	-0.45	0.01	50.0%	0.64 [0.63, 0.65]
Gacche 2015 (Taxifolin)	-0.15	0.01	50.0%	0.86 [0.84, 0.88]
Subtotal (95% CI)			100.0%	0.74 [0.55, 0.99]
Heterogeneity: Tau² = 0.04; Chi² = 450.00, df = 1 (P < 0.00001); I² = 100%				
Test for overall effect: Z = 2.00 (P = 0.05)				
1.1.5 Flavanones				
Gacche 2015 (2'-Hydroxy flavanone)	-0.16	0.02	13.6%	0.85 [0.82, 0.89]
Gacche 2015 (4'-Hydroxy flavanone)	-0.12	0.01	14.4%	0.89 [0.87, 0.90]
Gacche 2015 (7-Hydroxy flavanone)	-0.11	0.01	14.4%	0.90 [0.88, 0.91]
Gacche 2015 (Hesperidin)	-0.12	0.01	14.4%	0.89 [0.87, 0.90]
Gacche 2015 (Hesperitin)	-0.14	0.01	14.4%	0.87 [0.85, 0.89]
Gacche 2015 (Naringenin)	-0.24	0.01	14.4%	0.79 [0.77, 0.80]
Gacche 2015 (Naringin)	-0.29	0.01	14.4%	0.75 [0.73, 0.76]
Subtotal (95% CI)			100.0%	0.84 [0.80, 0.89]
Heterogeneity: Tau² = 0.01; Chi² = 288.24, df = 6 (P < 0.00001); I² = 98%				
Test for overall effect: Z = 6.02 (P < 0.00001)				
1.1.6 Anthocyanidines				
Viegas 2019 (Cyanidin)	-0.11	0.1	24.8%	0.90 [0.74, 1.09]
Viegas 2019 (Cyanidin-3-O-glucoside)	0.26	0.08	26.8%	1.30 [1.11, 1.52]
Viegas 2019 (Delphinidin)	0.26	0.13	21.7%	1.30 [1.01, 1.67]
Viegas 2019 (Delphinidin-3-O-glucoside)	-0.11	0.08	26.8%	0.90 [0.77, 1.05]
Subtotal (95% CI)			100.0%	1.07 [0.86, 1.33]
Heterogeneity: Tau² = 0.04; Chi² = 16.02, df = 3 (P = 0.001); I² = 81%				
Test for overall effect: Z = 0.63 (P = 0.53)				

Test for subgroup differences: Chi² = 30.99, df = 5 (P < 0.00001), I² = 83.9%

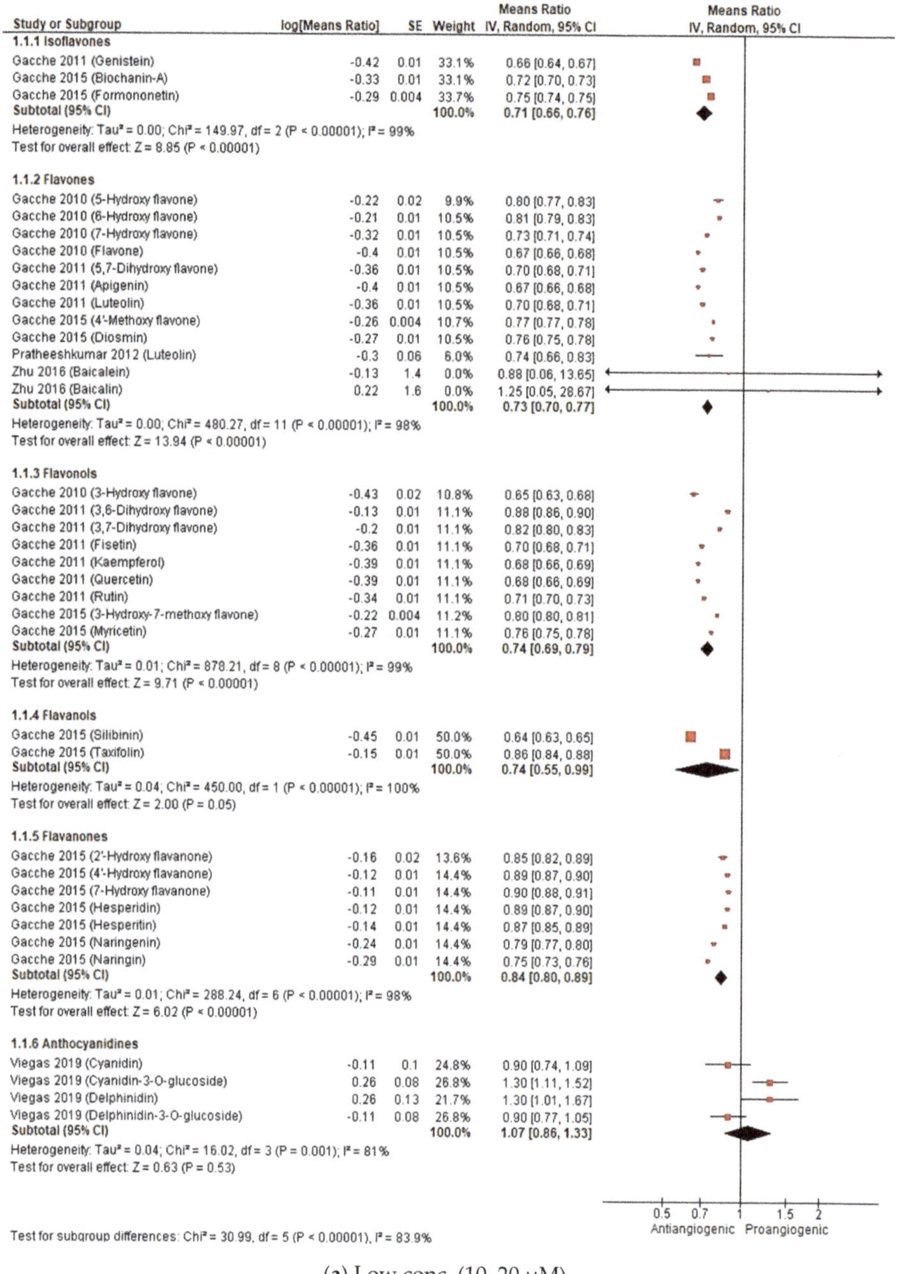

(**a**) Low conc. (10–20 μM)

Figure 6. *Cont.*

Study or Subgroup	log[Means Ratio]	SE	Weight	Means Ratio IV, Random, 95% CI
2.1.1 Isoflavones				
Gacche 2011 (Genistein)	-0.76	0.01	40.5%	0.47 [0.46, 0.48]
Gacche 2015 (Biochanin-A)	-0.82	0.01	40.5%	0.44 [0.43, 0.45]
Gacche 2015 (Formononetin)	-0.67	0.05	19.1%	0.51 [0.46, 0.56]
Subtotal (95% CI)			100.0%	0.46 [0.44, 0.49]
Heterogeneity: Tau² = 0.00; Chi² = 23.65, df = 2 (P < 0.00001); I² = 92%				
Test for overall effect: Z = 26.04 (P < 0.00001)				
2.1.2 Flavonols				
Gacche 2010 (3-Hydroxy flavone)	-1.03	0.02	11.0%	0.36 [0.34, 0.37]
Gacche 2011 (3,6-Dihydroxy flavone)	-0.39	0.01	11.1%	0.68 [0.66, 0.69]
Gacche 2011 (3,7-Dihydroxy flavone)	-0.62	0.01	11.1%	0.54 [0.53, 0.55]
Gacche 2011 (Fisetin)	-0.6	0.01	11.1%	0.55 [0.54, 0.56]
Gacche 2011 (Kaempferol)	-0.73	0.01	11.1%	0.48 [0.47, 0.49]
Gacche 2011 (Quercetin)	-0.69	0.01	11.1%	0.50 [0.49, 0.51]
Gacche 2011 (Rutin)	-0.62	0.01	11.1%	0.54 [0.53, 0.55]
Gacche 2015 (3-Hydroxy-7-methoxy flavone)	-0.54	0.003	11.2%	0.58 [0.58, 0.59]
Gacche 2015 (Myricetin)	-0.94	0.01	11.1%	0.39 [0.38, 0.40]
Subtotal (95% CI)			100.0%	0.50 [0.46, 0.56]
Heterogeneity: Tau² = 0.02; Chi² = 2701.66, df = 8 (P < 0.00001); I² = 100%				
Test for overall effect: Z = 13.23 (P < 0.00001)				
2.1.3 Flavanols				
Gacche 2015 (Silibinin)	-0.97	0.01	50.0%	0.38 [0.37, 0.39]
Gacche 2015 (Taxifolin)	-0.3	0.01	50.0%	0.74 [0.73, 0.76]
Subtotal (95% CI)			100.0%	0.53 [0.27, 1.02]
Heterogeneity: Tau² = 0.22; Chi² = 2244.50, df = 1 (P < 0.00001); I² = 100%				
Test for overall effect: Z = 1.90 (P = 0.06)				
2.1.4 Flavones				
Bhat 2013 (Acacetin)	-1.24	0.01	9.1%	0.29 [0.28, 0.30]
Gacche 2010 (5-Hydroxy flavone)	-0.03	0.01	9.1%	0.97 [0.95, 0.99]
Gacche 2010 (6-Hydroxy flavone)	-0.4	0.01	9.1%	0.67 [0.66, 0.68]
Gacche 2010 (7-Hydroxy flavone)	-0.39	0.02	9.1%	0.68 [0.65, 0.70]
Gacche 2010 (Flavone)	-0.8	0.02	9.1%	0.45 [0.43, 0.47]
Gacche 2011 (5,7-Dihydroxy flavone)	-0.6	0.01	9.1%	0.55 [0.54, 0.56]
Gacche 2011 (Apigenin)	-0.67	0.01	9.1%	0.51 [0.50, 0.52]
Gacche 2011 (Luteolin)	-0.67	0.01	9.1%	0.51 [0.50, 0.52]
Gacche 2015 (4'-Methoxy flavone)	-0.46	0.01	9.1%	0.63 [0.62, 0.64]
Gacche 2015 (Diosmin)	-0.71	0.01	9.1%	0.49 [0.48, 0.50]
Pratheeshkumar 2012 (Luteolin)	-0.65	0.07	8.7%	0.52 [0.46, 0.60]
Zhu 2016 (Baicalein)	-0.11	1.4	0.5%	0.90 [0.06, 13.93]
Subtotal (95% CI)			100.0%	0.55 [0.45, 0.67]
Heterogeneity: Tau² = 0.11; Chi² = 8370.34, df = 11 (P < 0.00001); I² = 100%				
Test for overall effect: Z = 5.98 (P < 0.00001)				
2.1.5 Flavanones				
Gacche 2015 (2'-Hydroxy flavanone)	-0.27	0.02	14.0%	0.76 [0.73, 0.79]
Gacche 2015 (4'-Hydroxy flavanone)	-0.31	0.01	14.3%	0.73 [0.72, 0.75]
Gacche 2015 (7-Hydroxy flavanone)	-0.25	0.01	14.3%	0.78 [0.76, 0.79]
Gacche 2015 (Hesperidin)	-0.2	0.004	14.4%	0.82 [0.81, 0.83]
Gacche 2015 (Hesperitin)	-0.22	0.01	14.3%	0.80 [0.79, 0.82]
Gacche 2015 (Naringenin)	-0.42	0.01	14.3%	0.66 [0.64, 0.67]
Gacche 2015 (Naringin)	-0.48	0.01	14.3%	0.62 [0.61, 0.63]
Subtotal (95% CI)			100.0%	0.74 [0.68, 0.80]
Heterogeneity: Tau² = 0.01; Chi² = 1007.20, df = 6 (P < 0.00001); I² = 99%				
Test for overall effect: Z = 7.07 (P < 0.00001)				
2.1.6 Anthocyanidines				
Viegas 2019 (Cyanidin)	0	0.1	10.4%	1.00 [0.82, 1.22]
Viegas 2019 (Cyanidin-3-O-glucoside)	0	0.05	41.7%	1.00 [0.91, 1.10]
Viegas 2019 (Delphinidin)	0.14	0.13	6.2%	1.15 [0.89, 1.48]
Viegas 2019 (Delphinidin-3-O-glucoside)	-0.02	0.05	41.7%	0.98 [0.89, 1.08]
Subtotal (95% CI)			100.0%	1.00 [0.94, 1.07]
Heterogeneity: Tau² = 0.00; Chi² = 1.32, df = 3 (P = 0.72); I² = 0%				
Test for overall effect: Z = 0.01 (P = 0.99)				

Test for subgroup differences: Chi² = 343.73, df = 5 (P < 0.00001), I² = 98.5%

(**b**) Mid conc. (40–50 μM)

Figure 6. *Cont.*

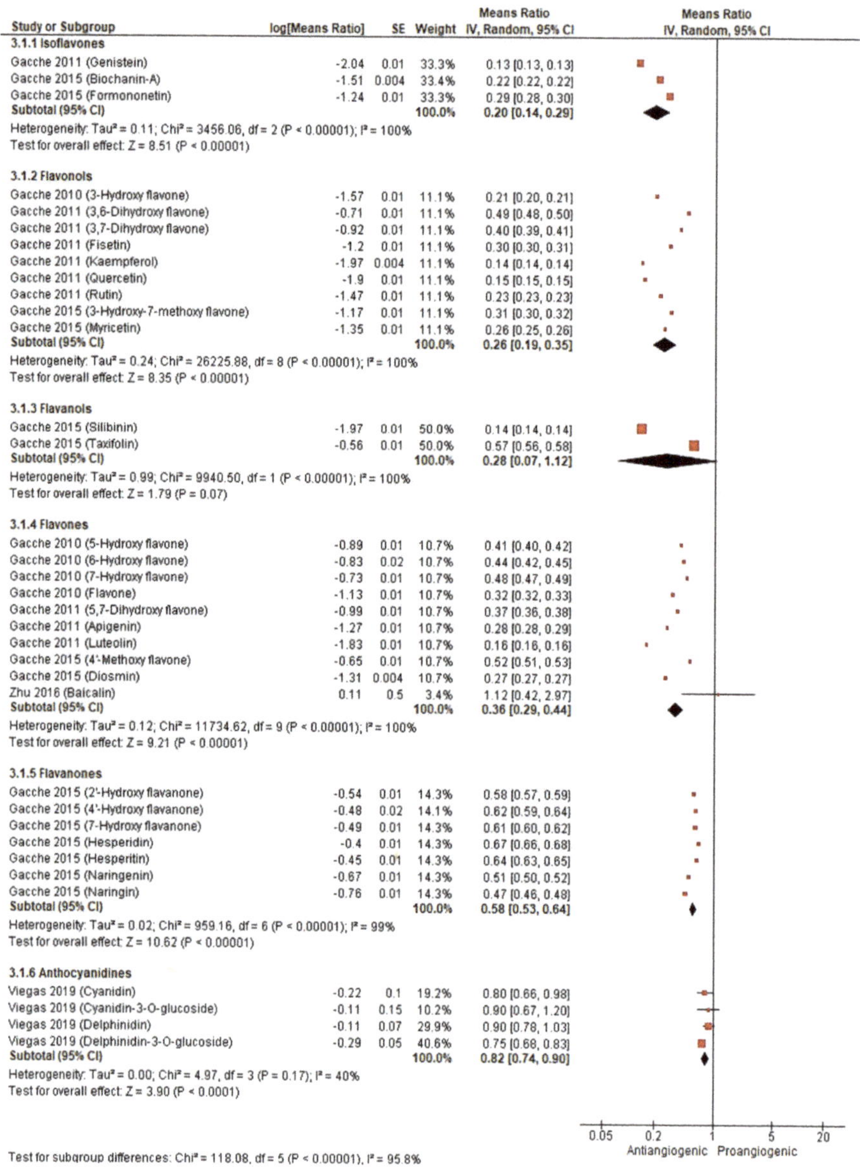

(**c**) High conc. (100 µM)

Figure 6. Forest plots of means ratio and 95% confidence interval (CI) of number of blood vessels relative to control at 3 concentration ranges as calculated by inverse variance (IV) method: (**a**) low (10–20 µM); (**b**) medium (40–50 µM); (**c**) high (100 µM).

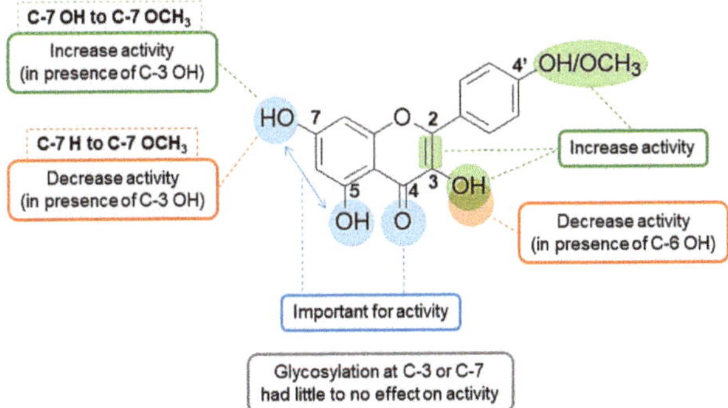

Figure 7. Summary of antiangiogenic SAR of flavonoids.

Set 2: Antiangiogenic effect of flavonoids under inflammatory conditions

Lin et al. evaluated the antiangiogenic activity of the flavone wogonin on LPS (the main component of gram negative bacterial membrane) and IL-6 induced angiogenesis in two reports [130,147]. The documented reduction in the number of CAM blood vessels by wogonin was shown to be dose dependent in both cases but more prominent in the case of IL-6 induced angiogenesis (75% as opposed to 38% in the case of LPS induced angiogenesis at 100 µM) (Figure 8). The authors also probed the possible mechanisms of wogonin's inhibition of this inflammation-induced angiogenesis through different in vitro techniques such as western blotting and polymerase chain reaction (PCR) in which both LPS and IL-6 resulted in an upregulation of the IL-6/IL-6R pathway [130,147]. Although wogonin attenuated the IL-6/IL-6R pathway and levels of VEGF in both cases, it exhibited different expression of downstream vascular endothelial growth factor receptors (VEGFRs). Only VEGFR2 expression was downregulated with wogonin LPS-induced angiogenesis inhibition as opposed to VEGFR1 downregulation with IL-6 induced angiogenesis inhibition. This data needs further investigation in order to understand why these two similar mechanisms lead to the downregulation of two different downstream receptors (VGFR2 and VEGFR1) and to address the impact of this on the antiangiogenic potency. Inhibition of LPS-induced angiogenesis was also reported for wogonoside, which is the 7-glucuronic acid of wogonin, by Chen et al. [139] 150 ng/CAM of wogonoside reduced neo-vascularization of CAMs by 43%. Additionally, wogonoside downregulated mammalian toll-like receptor (TLR4), extracellular signal-regulated kinase (ERK1/2) and p38MAPK in a western blotting assay [139].

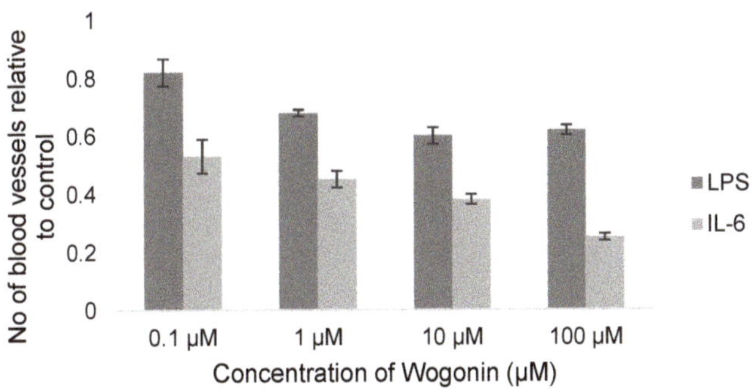

Figure 8. Reported antiangiogenic effect of wogonin on LPS and IL-6 induced angiogenesis ± SEM.

Set 3: Antiangiogenic effect of flavonoids under tumor conditions.

Since angiogenesis plays a vital role in tumor growth and metastasis, several studies have focused on the antiangiogenic evaluation of promising cytotoxic agents. Figure 9 shows the estimated antiangiogenic effect of the 4 flavonoids apigenin, myricetin, acacetin and keampferol on the ovarian cancer cell line (OVCAR-3) at 10–20 μM. The reduction in the number of CAM blood vessels ranged from 30 to 60% with an overall summary outcome of 0.35 (95%CI: 0.27, 0.45; *p*-value < 0.00001). HIFα and VEGF were significantly downregulated, as evidenced by immunoblotting analysis of CAM OVCAR-3 tissues that were treated with apigenin or acacetin [131,144]. The antiangiogenic activity of the flavone wogonoside was evaluated on the estrogen receptor positive (MCF-7) and two triple negative breast (MDA-MB-231 and MDA-MB-468) cancer cell lines by Huang et al. [138,142]. At 50 ng/CAM, wogonoside's effect on the 3 cell lines was not prominent (Figure 10). However, a 55% reduction of the number of blood vessels was observed at 100 ng/CAM for the MDA-MB-468 cell line. A two-fold increase in the concentration of wogonoside to 200 ng/CAM did not, however, result in an increased antiangiogenic effect on the same cell line. On the other hand, reduction of the neo-vascularization for the MDA-MB-231 cell line increased from 32% to 77% upon increasing the concentration from 100 to 200 ng/CAM. Huang et al. demonstrated the ability of wogonoside to target the Hedgehog signaling pathway, which is upregulated in triple negative breast cancer, in MDA-MB-231 and MDA-MB-468 cell lines [138]. Expression of the Hedgehog downstream transmembrane protein smoothened (SMO) and glioma-associated oncogene homolog protein (Gli), is significantly increased in triple negative breast cancer [150] leading to an elevation in VEGF levels [151]. According to Huang and his colleagues, wogonoside promoted SMO degradation and inhibited Gli1 activity as well as expression of VEGF [138].

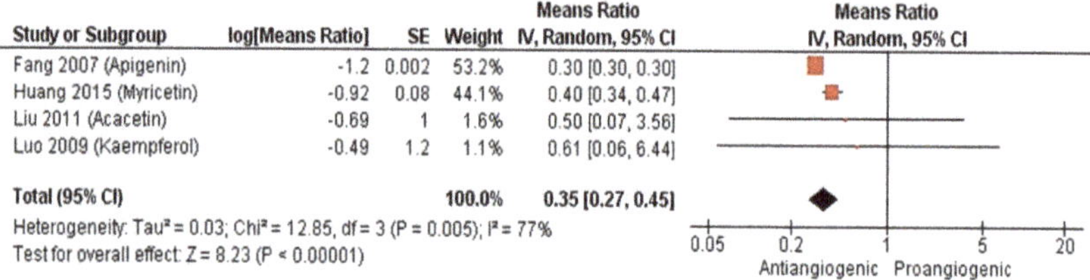

Figure 9. Forest plot of means ratio and 95% confidence interval (CI) of number of blood vessels relative to control of flavonoids on OVCAR-3 cell lines.

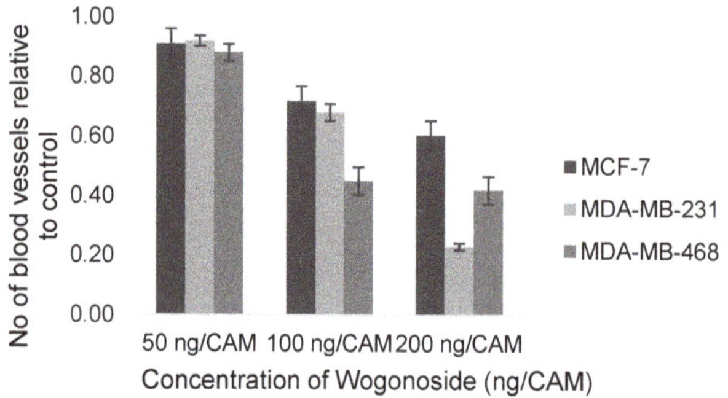

Figure 10. Reported antiangiogenic effect of wogonoside on breast cancer cell lines; MCF-7, MDA-MB-231 and MDA-MB-468 ± SEM.

2.2.4. Sensitivity Analysis

The high heterogeneity ($I^2 > 80\%$) observed for all subgroups in the generated forest plots, except for the anthocyanidines subgroup at the mid and high concentrations analyses ($I^2 = 0\%$ and 40%, respectively), was expected given that each class included different flavonoid molecules. In that context, a sensitivity analysis was conducted by a leave-one-out strategy to assess the robustness of the results and determine the contribution of each flavonoid to heterogeneity. Overall, the results showed good robustness and the overall summary estimates did not show significant changes upon the systematic removal of individual studies (Tables S2–S4). This was the case in all subgroups with the exception of the flavanol subgroup which showed some difference in the overall summary at all concentrations. At the 40–50 µM range for instance, the overall pooled means ratio changed from 0.53 (95%CI: 0.27, 1.02, $I^2 = 100\%$) to 0.74 (95%CI: 0.73, 0.76, NA) and 0.38 (95%CI: 0.37, 0.39, NA) upon removal of the Gacche 2015 (Silibinin) and Gacche 2015 (Taxifolin) flavonoids, respectively (Table S3). This indicates that data provided on the flavanols subgroup is not sufficient to draw meaningful conclusions. Likewise, heterogeneity (I^2) of the subgroups totals did not show significant change, with very few exceptions, upon implementation of the leave-one-out strategy (Tables S2–S4). This might be due to the fact that most of the flavonoids in a single subgroup belong to the same study, consequently, there are no differences in their experimental designs. In that case heterogeneity is believed to be either of clinical or statistical origin.

3. Discussion

Flavonoids have been reported to modulate several angiogenic factors and cascades in either a proangiogenic or an antiangiogenic manner which is postulated to be dose dependent [2,148]. A good illustration of this dual effect is demonstrated by the flavone baicalin; low doses were reported to stimulate angiogenesis [152] whilst high doses showed an inhibitory effect [153]. Due to the emerging importance of the use of angiogenesis modulators in the treatment of various pathological conditions including cancer, diabetes, bone, eye, cardiovascular and neurological disorders, the identification of flavonoids altering angiogenesis has gained new significance [2,154]. To the best of our knowledge, no systematic reviews have been conducted to quantitatively assess the antiangiogenic effects of flavonoids, despite the potential of such a study to have a positive impact on the treatment of serious health issues like cancer and rheumatoid arthritis. Given the breadth of the literature related to the antiangiogenic effects of flavonoids, a systematic search of the literature was initially conducted in this research program to identify (a) the extent to which angiogenesis modulation effects had been proposed for flavonoids and (b) the most widely used in vitro and in/ex vivo assays to determine the antiangiogenic activities of flavonoids.

Various study designs have been used in the literature to report on the antiangiogenic activity of chemical compounds. There are a number of comprehensive reviews in the literature comparing the different available angiogenesis assay models [16,17,155,156]. Although in vitro studies are less expensive and quicker to perform than in vivo studies, the results do not always convert into the same effect, in vivo. In vitro assays usually focus on monitoring the individual steps of angiogenesis such as migration or proliferation of endothelial cells rather than the collective formation of new tube-like structures [16]. In vivo assays offer the considerable advantage of mimicking more closely the body's physiological conditions which is particularly important in angiogenic studies due to the complex nature of the process. While in vivo angiogenesis assays can be more informative, they present some cost, time and experimental design limitations. Inflammation resulting from the trauma that is caused by some assays, for instance, can stimulate several proangiogenic factors which compromise the sensitivity and specificity of the results [17]. Hence, it is recommended that a combination of in vitro and in vivo assays is used to provide consistent and complementary results. In relation to this, 44% of the research articles included in the conducted preliminary search reported a combination of in vitro and in/ex vivo assays.

Herein, a meta-analysis study was carried out in order to quantitatively evaluate the antiangiogenic effects of flavonoids. Only articles implementing the CAM assay in their study design were included. This is because the CAM assay is currently the most widely used in vivo angiogenic assay and, as such, it allows a comparison across different flavonoid types and offers many advantages over other angiogenic assays [157–159]. For instance, it is fairly simple, inexpensive, suitable for large scale screening and also offers the important advantage of expressing almost all of the known angiogenic factors [17,156]. Set-up of the assay is briefly as follows: fertilized chicken eggs are incubated at 37 °C for 3 days, a small hole is made in the egg shell to remove some of the albumin in order to facilitate detachment of the CAM from the shell. Compounds under investigation are added to approximately 5 to 10 day old chicks on specific carriers, such as matrigel or sterile filter/plastic discs, through a small window cut in the egg shell. After 48 to 72 h, existing blood vessels or tubules can be visualized and evaluated by light or electron microscopy [17,156]. Nevertheless, the CAM test comes with certain limitations such as sensitivity to oxygen tension and difficulty of visualization of newly formed vessels due to the presence of pre-existing ones [157].

Meta-analysis of results of the antiangiogenic evaluation of flavonoids via the in vivo CAM assay showed increasing activities with increasing concentrations. The evaluated flavonoids also demonstrated antiangiogenic activities of varying potencies. In light of this, results were inspected to gain some insights on the SAR of antiangiogenic activity of flavonoids (Figure 7). Although SARs of chemical compounds change based on the sought pharmacological activity, there are some common structural features of flavonoids that are recognized as important for activity [160]. Combination of the C2=C3 double bond and a 4-C=O is favorable for the antiviral/bacterial [161], anticancer [162,163], cardioprotective [164], anti-inflammatory [165], and antioxidant [164] activities of flavonoids. This conjugation maintains the planarity of the molecule and helps with the electron delocalization between rings A and C which is important for interaction with several targets [160]. Similarly, the 5, 7 di-OH is important for many of the biological activities of flavonoids [164,166–168]. This can be explained by the fact that flavonoids exert different pharmacological activities that have mutual and/or overlapping mechanisms. For example, the antioxidant activity of flavonoids contributes to their anti-inflammatory activity and both contribute to their anticancer activity. Moreover, several targets in the body have structurally similar binding sites and this is a phenomenon that is partially responsible for drug promiscuity or polypharmacology (binding of a drug to multiple targets). This was, in fact, observed for binding of the flavonoid quercetin with phosphatidylinositol 4,5-bisphosphate 3-kinase (PI3KCG) and the serine/threonine proto oncogene, PIM1 kinase [169].

With respect to the antiangiogenic activity of flavonoids, limited SAR studies have been reported. Lam et al. tested the antiangiogenic activity of a number of polymethoxylated flavonoids in vitro and in vivo [170]. The authors concluded that methylation of C5, C6, C7 and/or C4′ OH groups increased the activity which is in agreement with Ravishankar et al. [93] who reported the in vitro antiangiogenic activity of a number of quercetin and luteolin derivatives. Our results also suggest that the presence of a 4′-OCH$_3$ increases the antiangiogenic activity. Despite this, there were some discrepancies between the aforementioned SAR conclusions. In this SAR analysis we showed that the presence of a 3-OH group enhanced the antiangiogenic activity, which is in contrast to the report from Ravishankar et al. that noted that the same 3-OH caused a drop in the activity yet methylation of that OH increased the activity [93]. A study by Lam et al. reported that glycosylation at C7 dramatically decreased the activity [170] while our study showed such modification to cause a minor or no decrease in the activity and even a slight increase in some cases. These inconsistencies are likely to be a result of the different experimental methodologies and flavonoid concentrations used in each study. Additionally, the different evaluated flavonoids might exert their antiangiogenic activities by binding to different targets that require different structural features. This highlights the need for larger scale studies to more fully probe the antiangiogenic SAR of flavonoids taking in consideration the employed mechanisms of action.

Since the relation between inflammation and angiogenesis is well established and many flavonoids possess anti-inflammatory activities, several studies assessed the antiangiogenic effects of flavonoids on inflammation-induced angiogenesis. Inflammatory cells like T-lymphocytes and macrophages secrete cytokines that can control the survival, proliferation, activation and migration of endothelial cells [171,172]. Endothelial cells can additionally produce several cytokines and chemokines themselves [173]. Flavonoids such as baicalin, quercetin and kaempferol caused a reduction in both inflammatory and angiogenic markers in cultured macrophages and human umbilical vein endothelial cells (HUVECs) [174,175].

Bacterial infections also trigger angiogenesis through inflammatory pathways. In that context, binding of LPS to the TLR4 receptor located on the surface of endothelial cells leads to upregulation of ERK1/2 and p38MAPK pathways and increases production of pro-inflammatory cytokines like IL-6 [176,177]. Pro-inflammatory cytokines like IL-6 and tumor necrosis factor α (TNFα) can interact with VEGF expression and promote angiogenesis [178,179]. The flavone, wogonin, and its glucoside, wogonoside, showed promising antiangiogenic activity against LPS induced angiogenesis [130]. Wogonin also inhibited IL-6 induced angiogenesis in a concentration dependent manner where it was reported to downregulate VEGFR1 not VEGFR2 genetic expression [147]. While VEGR2 is the main receptor for VEGF and is downregulated by many flavonoids [91,180], VEGFR1's role in angiogenesis is still not fully understood and needs further investigation.

As mentioned earlier, cancer is one of the most serious pathologies related to angiogenesis. When cells grow malignantly beyond a certain size, they need more vascularization to receive oxygen and nutrients i.e., tumors depend on angiogenesis to grow above a certain limit, and to metastasize [181]. The tumor vasculature is characterized by an imbalance between pro and anti-angiogenic factors where several angiogenic stimulators like VEGF and HIF are overexpressed. The HIFs are major regulators of angiogenesis and orchestrate many of the steps involved [182]. Under physiological conditions, HIFs are released in response to low oxygen levels in the blood (hypoxia) and stimulate angiogenesis at various levels from endothelial cell proliferation to activating the transcription of angiogenic genes like VEGF and platelet derived growth factor (PDGF). During malignancy, HIF dependent angiogenesis is activated either in response to the predominant hypoxic environment or by the genetic transformations caused by cancer. Flavonoids can downregulate HIFα and VEGF in different cancer cell lines such as OVCAR-3, A2780, MCF-7 and PC-3 [42,70,95,131,144,145]. Many studies have also reported the ability of the flavonoids 3-hydroxy flavone, hesperidin, apigenin, fisetin and many others to reduce tumor size, capillary density and metastasis of different cancers, such as osteosarcoma, melanoma, lung and breast cancers, in xenograft mice [26,183–187].

Although this meta-analysis demonstrated the overall promising in vivo antiangiogenic activity of flavonoids whether in normal, inflammatory or tumor conditions, there were some limitations to the study. First, the standard forms and guidelines used in a systematic analysis are only applicable for clinical or animal trials. Consequently, the quality of the retrieved studies and publication bias were not taken into account here, as this would be methodologically inappropriate. As such, large scale animal studies and meta-analyses evaluating the antiangiogenic activity of flavonoids are much needed in the future to provide more definitive conclusions about the role of flavonoids in angiogenesis.

Second, despite subgrouping flavonoids based on their chemical class and using the random effects model, heterogeneity remained high in this study. There are three types of heterogeneity as defined by the Cochrane handbook for systematic reviews, (i) clinical: differences in participants, interventions or outcomes, (ii) methodological: differences in study design, risk of bias and (iii) statistical: variation in intervention effects or results [188]. Looking deeper into the generated forest plots we concluded the cause of heterogeneity to be clinical and/or statistical. This is mainly because most of the flavonoids in a single subgroup are from the same study hence methodological heterogeneity was excluded. This was further supported by the fact that no single flavonoid was found to solely contribute to the heterogeneity when applying the leave-one-out strategy in the sensitivity analysis. In that case, heterogeneity is mainly due to the different flavonoids used in the study (variation in interventions) in addition to other factors like variable outcomes (number of blood vessels). This clinical heterogeneity can lead to a statistical heterogeneity manifested as a variation among the effects or results (ratio of means of number of blood vessels).

4. Materials and Methods

This review and meta-analysis were conducted according to Preferred Reporting Items for systematic reviews and Meta Analyses (PRISMA) guidelines [189].

4.1. Search Strategy

For Section 1, a literature search was conducted using ScienceDirect, PubMed and Web of Science databases between 3 April 2020 and 23 April 2020 with no time limits. The first set of keywords, (flavonoid, flavone, flavonol, flavanol, anthocyanidin, polyphenol) was combined systematically using the Boolean operator AND with the second set, (angiogenesis, antiangiogenic, proangiogenic, "cell migration", "wound healing") in all databases (Table S5).

With regards to the detailed meta-analysis for Section 2, the literature search was carried out using ScienceDirect, PubMed, Web of Science and Google Scholar databases between 8 June 2020 and 10 June 2020 with no time limits. The first set of keywords, (flavonoid, flavone, flavonol, flavanol, anthocyanidin, polyphenol) was combined systematically using the Boolean operator AND with the second set, (angiogenesis, "chick chorioallantoic membrane", "in vivo angiogenesis") in all databases (Table S6).

4.2. Inclusion and Exclusion Criteria

Studies were included in the Section 1 overview search if they met the following eligibility criteria: (i) natural or synthetic flavonoids (ii) in vitro, in vivo and/or ex vivo angiogenesis assays (iii) focus on cancer, diabetes, bone regeneration or eye diseases. For the meta-analysis Section 2, the inclusion criteria were: (i) natural or synthetic flavonoids (ii) in vivo CAM angiogenesis assays. Articles not written in English and/or focusing on chalcones, plant extracts/total flavonoids content, combination of compounds, nanoformulations, prodrugs, neurological disorders or cardiovascular diseases were excluded from both searches. This systematic review and meta-analysis followed PRISMA guidelines (Table S7).

4.3. Data Extraction

Initially, articles' titles and abstracts were screened based on relevance and inclusion/exclusion criteria. Full texts were checked in some cases when abstracts failed to provide a detailed description. Eligible articles were retrieved and data extracted into a specially designed form. The first set of extracted data for Section 1 included title, publication type, year of publication, flavonoid, disease of focus and conducted in vitro and/or in/ex vivo angiogenesis assays. The second set of data were extracted for the meta-analysis Section 2 study and included title, year of publication, flavonoid, angiogenesis promotor, cancer cell line, concentration, time and duration of flavonoid treatment, results representation and number of CAMs used for each test concentration (n).

4.4. Data Analysis

Means of the number of blood vessels in a CAM relative to control were used as the outcome measure. Concentrations were reported in μM in all analyses except for analysis of wogonoside's antiangiogenic effect on breast cancer cell lines in which ng/CAM was used. Values are represented as means ratio ± standard error of means (SEM). For studies reporting standard deviation (SD), the SEM was calculated by dividing SD by square root of the corresponding study sample size. Pool effect size was expressed as means ratio and 95% CI and was calculated using the inverse variance (IV) method. The random effects model was used because it accounts for between study variability. Heterogeneity was assessed using Higgins' I^2 measure where $I^2 \geq 50\%$ indicates substantial heterogeneity [190]. Sensitivity analysis was applied to evaluate the effect of each flavonoid on summary effect size and on heterogeneity. It is based on the sequential removal of one study at a time. Statistical analysis was performed using Review Manager Version 5.1 (The Nordic Cochrane Centre, The Cochrane Collaboration, Copenhagen, Denmark) and Microsoft Excel 2016.

5. Conclusions

Despite the promising antiangiogenic activity of flavonoids presented in many literature studies, no flavonoids have reached clinical trials for this application. This systematic review and meta-analysis therefore aimed to provide further insight into this area by evaluating the in vivo antiangiogenic activity of flavonoids as determined by the widely reported, clinically relevant CAM assay. A comprehensive overview of the antiangiogenic activities of flavonoids with regards to the class of flavonoids, pathology and assays used was presented. Results have shown that the biggest fraction of studies focused on the flavone subclass, cancer related angiogenesis, and in vitro assays. Furthermore, an overall evaluation of the in vivo antiangiogenic activity of flavonoids was offered focusing on SAR and mechanistic considerations. Isoflavones, flavonols and flavones were found to be the most active classes of flavonoids where antiangiogenic activity was dose dependent. Several structural features were considered, from which it was concluded that the position of the hydroxyl substituents and the degree of unsaturation are key for high activity. Even though there were some limitations such as the miscellany of the studied flavonoids and the high heterogeneity, this study provided substantial information that will underpin further investigations by addressing current gaps in the literature regarding the antiangiogenic activity of flavonoids, and highlighting their future prospective as potentially clinically active antiangiogenic agents.

Supplementary Materials: The following are available online, Table S1: Study characteristics of Section 1, Tables S2–S4: Sensitivity analysis, Tables S5 and S6: Database search results, Table S7: PRISMA checklist.

Author Contributions: M.K., F.G. and H.M.I.O. designed the study. The literature search, documentation, data extraction and analysis were carried out by M.K. and supervised by F.G. and H.M.I.O. M.K. wrote the first draft of the manuscript. M.K., F.G. and H.M.I.O. edited and revised the manuscript and approved the final version. All authors have read and agreed to the published version of the manuscript.

Funding: This research was funded by the Newton-Mosharafa Fund, through a scholarship to MK.

Acknowledgments: We are grateful to the Newton-Mosharafa Fund for a scholarship that has funded MK's PhD studies.

Conflicts of Interest: The authors declare no conflict of interest.

References

1. Carmeliet, P.; Jain, R.K. Molecular Mechanisms and Clinical Applications of Angiogenesis. *Nature* **2011**, *473*, 289–307. [CrossRef]
2. Diniz, C.; Suliburska, J.; Ferreira, I.M. New Insights into the Antiangiogenic and Proangiogenic Properties of Dietary Polyphenols. *Mol. Nutr. Food Res.* **2017**, *61*, 1600912. [CrossRef]
3. Yancopoulos, G.D.; Davis, S.; Gale, N.W.; Rudge, J.S.; Wiegand, S.J.; Holash, J. Vascular-Specific Growth Factors and Blood Vessel Formation. *Nature* **2000**, *407*, 242–248. [CrossRef]
4. Shibuya, M. VEGF-VEGFR System as a Target for Suppressing Inflammation and Other Diseases. *Endocr. Metab. Immune Disord. Targets* **2015**, *15*, 135–144. [CrossRef]
5. Gacche, R.N.; Meshram, R.J. Angiogenic Factors as Potential Drug Target: Efficacy and Limitations of Anti-Angiogenic Therapy. *Biochim. Biophys. Acta Rev. Cancer* **2014**, *1846*, 161–179. [CrossRef]
6. Majewska, I.; Gendaszewska-Darmach, E. Proangiogenic Activity of Plant Extracts in Accelerating Wound Healing a New Face of Old Phytomedicines. *Acta Biochim. Polonica* **2011**, *58*, 449–460. [CrossRef]
7. Mirossay, L.; Varinská, L.; Mojžiš, J. Antiangiogenic Effect of Flavonoids and Chalcones: An Update. *Int. J. Mol. Sci.* **2018**, *19*, 27. [CrossRef]
8. Sulaiman, R.S.; Basavarajappa, H.D.; Corson, T.W. Natural Product Inhibitors of Ocular Angiogenesis. *Exp. Eye Res.* **2014**, *129*, 161–171. [CrossRef]
9. Raffa, D.; Maggio, B.; Raimondi, M.V.; Plescia, F.; Daidone, G. Recent Discoveries of Anticancer Flavonoids. *Eur. J. Med. Chem.* **2017**, *142*, 213–228. [CrossRef]

10. Ravishankar, D.; Rajora, A.K.; Greco, F.; Osborn, H.M.I. Flavonoids as Prospective Compounds for Anti-Cancer Therapy. *Int. J. Biochem. Cell Biol.* **2013**, *45*, 2821–2831. [CrossRef]

11. Abotaleb, M.; Samuel, S.; Varghese, E.; Varghese, S.; Kubatka, P.; Liskova, A.; Büsselberg, D. Flavonoids in Cancer and Apoptosis. *Cancers* **2018**, *11*, 28. [CrossRef]

12. Peluso, I.; Raguzzini, A.; Serafini, M. Effect of Flavonoids on Circulating Levels of TNF-α and IL-6 in Humans: A Systematic Review and Meta-Analysis. *Mol. Nutr. Food Res.* **2013**, *57*, 784–801. [CrossRef]

13. Wang, X.; Ouyang, Y.Y.; Liu, J.; Zhao, G. Flavonoid Intake and Risk of CVD: A Systematic Review and Meta-Analysis of Prospective Cohort Studies. *Br. J. Nutr.* **2014**, *111*, 1–11. [CrossRef]

14. Beking, K.; Vieira, A. Flavonoid Intake and Disability-Adjusted Life Years Due to Alzheimers and Related Dementias: A Population-Based Study Involving Twenty-Three Developed Countries. *Public Health Nutr.* **2010**, *13*, 1403–1409. [CrossRef]

15. Mojzis, J.; Varinska, L.; Mojzisova, G.; Kostova, I.; Mirossay, L. Antiangiogenic Effects of Flavonoids and Chalcones. *Pharmacol. Res.* **2008**, *57*, 259–265. [CrossRef]

16. Jain, R.K.; Schlenger, K.; Höckel, M.; Yuan, F. Quantitative Angiogenesis Assays: Progress and Problems. *Nat. Med.* **1997**, *3*, 1203–1208. [CrossRef]

17. Norrby, K. In Vivo Models of Angiogenesis. *J. Cell. Mol. Med.* **2006**, *10*, 588–612. [CrossRef]

18. Hattori, H.; Okuda, K.; Murase, T.; Shigetsura, Y.; Narise, K.; Semenza, G.L.; Nagasawa, H. Isolation, Identification, and Biological Evaluation of HIF-1-Modulating Compounds from Brazilian Green Propolis. *Bioorg. Med. Chem.* **2011**, *19*, 5392–5401. [CrossRef]

19. Ramchandani, S.; Naz, I.; Lee, J.H.; Khan, M.R.; Ahn, K.S. An Overview of the Potential Antineoplastic Effects of Casticin. *Molecules* **2020**, *25*, 1287. [CrossRef]

20. Yang, D.S.; Li, Z.L.; Peng, W.B.; Yang, Y.P.; Wang, X.; Liu, K.C.; Li, X.L.; Xiao, W.L. Three New Prenylated Flavonoids from Macaranga Denticulata and Their Anticancer Effects. *Fitoterapia* **2015**, *103*, 165–170. [CrossRef]

21. Kashyap, D.; Sharma, A.; Sak, K.; Tuli, H.S.; Buttar, H.S.; Bishayee, A. Fisetin: A Bioactive Phytochemical with Potential for Cancer Prevention and Pharmacotherapy. *Life Sci.* **2018**, *194*, 75–87. [CrossRef]

22. Rengarajan, T.; Yaacob, N.S. The Flavonoid Fisetin as an Anticancer Agent Targeting the Growth Signaling Pathways. *Eur. J. Pharmacol.* **2016**, *789*, 8–16. [CrossRef]

23. Syed, D.N.; Adhami, V.M.; Khan, N.; Khan, M.I.; Mukhtar, H. Exploring the Molecular Targets of Dietary Flavonoid Fisetin in Cancer. *Semin. Cancer Biol.* **2016**, *40*, 130–140. [CrossRef]

24. Lall, R.K.; Adhami, V.M.; Mukhtar, H. Dietary Flavonoid Fisetin for Cancer Prevention and Treatment. *Mol. Nutr. Food Res.* **2016**, *60*, 1396–1405. [CrossRef]

25. Tsai, C.F.; Chen, J.H.; Chang, C.N.; Lu, D.Y.; Chang, P.C.; Wang, S.L.; Yeh, W.L. Fisetin Inhibits Cell Migration via Inducing HO-1 and Reducing MMPs Expression in Breast Cancer Cell Lines. *Food Chem. Toxicol.* **2018**, *120*, 528–535. [CrossRef]

26. Sun, X.; Ma, X.; Li, Q.; Yang, Y.; Xu, X.; Sun, J.; Yu, M.; Cao, K.; Yang, L.; Yang, G.; et al. Anti-cancer Effects of Fisetin on Mammary Carcinoma Cells via Regulation of the PI3K/Akt/MTOR Pathway: In Vitro and in Vivo Studies. *Int. J. Mol. Med.* **2018**, *42*, 811–820. [CrossRef]

27. Li, J.; Gong, X.; Jiang, R.; Lin, D.; Zhou, T.; Zhang, A.; Li, H.; Zhang, X.; Wan, J.; Kuang, G.; et al. Fisetin Inhibited Growth and Metastasis of Triple-Negative Breast Cancer by Reversing Epithelial-to-Mesenchymal Transition via PTEN/Akt/GSK3β Signal Pathway. *Front. Pharmacol.* **2018**, *9*, 772. [CrossRef]

28. Chou, R.H.; Hsieh, S.C.; Yu, Y.L.; Huang, M.H.; Huang, Y.C.; Hsieh, Y.H. Fisetin Inhibits Migration and Invasion of Human Cervical Cancer Cells by Down-Regulating Urokinase Plasminogen Activator Expression through Suppressing the P38 MAPK-Dependent NF-KB Signaling Pathway. *PLoS ONE* **2013**, *8*, e71983. [CrossRef]

29. Chen, C.M.; Hsieh, Y.H.; Hwang, J.M.; Jan, H.J.; Hsieh, S.C.; Lin, S.H.; Lai, C.Y. Fisetin Suppresses ADAM9 Expression and Inhibits Invasion of Glioma Cancer Cells through Increased Phosphorylation of ERK1/2. *Tumor Biol.* **2015**, *36*, 3407–3415. [CrossRef]

30. Liu, X.F.; Long, H.J.; Miao, X.Y.; Liu, G.L.; Yao, H.L. Fisetin Inhibits Liver Cancer Growth in a Mouse Model: Relation to Dopamine Receptor. *Oncol. Rep.* **2017**, *38*, 53–62. [CrossRef]

31. Klimaszewska-Wiśniewska, A.; Grzanka, D.; Czajkowska, P.; Hałas-Wiśniewska, M.; Durślewicz, J.; Antosik, P.; Grzanka, A.; Gagat, M. Cellular and Molecular Alterations Induced by Low-dose Fisetin in Human Chronic Myeloid Leukemia Cells. *Int. J. Oncol.* **2019**, *55*, 1261–1274. [CrossRef]

32. Tabasum, S.; Singh, R.P. Fisetin Suppresses Migration, Invasion and Stem-Cell-like Phenotype of Human Non-Small Cell Lung Carcinoma Cells via Attenuation of Epithelial to Mesenchymal Transition. *Chem. Biol. Interact.* **2019**, *303*, 14–21. [CrossRef]

33. Wang, J.; Huang, S. Fisetin Inhibits the Growth and Migration in the A549 Human Lung Cancer Cell Line via the ERK1/2 Pathway. *Exp. Ther. Med.* **2018**, *15*, 2667–2673. [CrossRef]

34. Chien, C.S.; Shen, K.H.; Huang, J.S.; Ko, S.C.; Shih, Y.W. Antimetastatic Potential of Fisetin Involves Inactivation of the PI3K/Akt and JNK Signaling Pathways with Downregulation of MMP-2/9 Expressions in Prostate Cancer PC-3 Cells. *Mol. Cell. Biochem.* **2010**, *333*, 169–180. [CrossRef]

35. Hsieh, M.-H.; Tsai, J.-P.; Yang, S.-F.; Chiou, H.-L.; Lin, C.-L.; Hsieh, Y.-H.; Chang, H.-R. Fisetin Suppresses the Proliferation and Metastasis of Renal Cell Carcinoma through Upregulation of MEK/ERK-Targeting CTSS and ADAM9. *Cells* **2019**, *8*, 948. [CrossRef]

36. Joussen, A.M. Treatment of Corneal Neovascularization with Dietary Isoflavonoids and Flavonoids. *Exp. Eye Res.* **2000**, *71*, 483–487. [CrossRef]

37. Fang, D.; Xiong, Z.; Xu, J.; Yin, J.; Luo, R. Chemopreventive Mechanisms of Galangin against Hepatocellular Carcinoma: A Review. *Biomed. Pharmacother.* **2019**, *109*, 2054–2061. [CrossRef]

38. Kim, J.D.; Liu, L.; Guo, W.; Meydani, M. Chemical Structure of Flavonols in Relation to Modulation of Angiogenesis and Immune-Endothelial Cell Adhesion. *J. Nutr. Biochem.* **2006**, *17*, 165–176. [CrossRef]

39. Chen, D.; Li, D.; Xu, X.; Qiu, S.; Luo, S.; Qiu, E.; Rong, Z.; Zhang, J.; Zheng, D. Galangin Inhibits Epithelial-Mesenchymal Transition and Angiogenesis by Downregulating CD44 in Glioma. *J. Cancer* **2019**, *10*, 4499–4508. [CrossRef]

40. Lei, D.; Zhang, F.; Yao, D.; Xiong, N.; Jiang, X.; Zhao, H. Galangin Increases ERK1/2 Phosphorylation to Decrease ADAM9 Expression and Prevents Invasion in A172 Glioma Cells. *Mol. Med. Rep.* **2018**, *17*, 667–673. [CrossRef]

41. Chien, S.-T.; Shi, M.-D.; Lee, Y.-C.; Te, C.-C.; Shih, Y.-W. Galangin, a Novel Dietary Flavonoid, Attenuates Metastatic Feature via PKC/ERK Signaling Pathway in TPA-Treated Liver Cancer HepG2 Cells. *Cancer Cell Int.* **2015**, *15*, 15. [CrossRef]

42. Huang, H.; Chen, A.Y.; Rojanasakul, Y.; Ye, X.; Rankin, G.O.; Chen, Y.C. Dietary Compounds Galangin and Myricetin Suppress Ovarian Cancer Cell Angiogenesis. *J. Funct. Foods* **2015**, *15*, 464–475. [CrossRef]

43. Cao, J.; Wang, H.; Chen, F.; Fang, J.; Xu, A.; Xi, W.; Zhang, S.; Wu, G.; Wang, Z. Galangin Inhibits Cell Invasion by Suppressing the Epithelial-Mesenchymal Transition and Inducing Apoptosis in Renal Cell Carcinoma. *Mol. Med. Rep.* **2016**, *13*, 4238–4244. [CrossRef]

44. Zhu, Y.; Rao, Q.; Zhang, X.; Zhou, X. Galangin Induced Antitumor Effects in Human Kidney Tumor Cells Mediated via Mitochondrial Mediated Apoptosis, Inhibition of Cell Migration and Invasion and Targeting PI3K/AKT/MTOR Signalling Pathway. *J. BUON* **2018**, *23*, 795–799.

45. Wang, L.; Wang, X.; Chen, H.; Zu, X.; Ma, F.; Liu, K.; Bode, A.M.; Dong, Z.; Kim, D.J. Gossypin Inhibits Gastric Cancer Growth by Direct Targeting AURKA and RSK2. *Phyther. Res.* **2019**, *33*, 640–650. [CrossRef]

46. Li, L.; Fan, P.; Chou, H.; Li, J.; Wang, K.; Li, H. Herbacetin Suppressed MMP9 Mediated Angiogenesis of Malignant Melanoma through Blocking EGFR-ERK/AKT Signaling Pathway. *Biochimie* **2019**, *162*, 198–207. [CrossRef]

47. Jin, X.N.; Yan, E.Z.; Wang, H.M.; Sui, H.J.; Liu, Z.; Gao, W.; Jin, Y. Hyperoside Exerts Anti-Inflammatory and Anti-Arthritic Effects in LPS-Stimulated Human Fibroblast-like Synoviocytes in Vitro and in Mice with Collagen-Induced Arthritis. *Acta Pharmacol. Sin.* **2016**, *37*, 674–686. [CrossRef]

48. Zhang, X.; Liu, T.; Huang, Y.; Wismeijer, D.; Liu, Y. Icariin: Does It Have an Osteoinductive Potential for Bone Tissue Engineering? *Phyther. Res.* **2014**, *28*, 498–509. [CrossRef]

49. Chen, M.; Wu, J.; Luo, Q.; Mo, S.; Lyu, Y.; Wei, Y.; Dong, J. The Anticancer Properties of Herba Epimedii and Its Main Bioactive Componentsicariin and Icariside II. *Nutrients* **2016**, *8*, 563. [CrossRef]

50. Chung, B.H.; Kim, J.D.; Kim, C.K.; Kim, J.W.; Won, M.H.; Lee, H.S.; Dong, M.S.; Ha, K.S.; Kwon, Y.G.; Kim, Y.M. Icariin Stimulates Angiogenesis by Activating the MEK/ERK- and PI3K/Akt/ENOS-Dependent Signal Pathways in Human Endothelial Cells. *Biochem. Biophys. Res. Commun.* **2008**, *376*, 404–408. [CrossRef]

51. Gu, Z.F.; Zhang, Z.T.; Wang, J.Y.; Xu, B.B. Icariin Exerts Inhibitory Effects on the Growth and Metastasis of KYSE70 Human Esophageal Carcinoma Cells via PI3K/AKT and STAT3 Pathways. *Environ. Toxicol. Pharmacol.* **2017**, *54*, 7–13. [CrossRef]

52. Wang, P.; Zhang, J.; Xiong, X.; Yuan, W.; Qin, S.; Cao, W.; Dai, L.; Xie, F.; Li, A.; Liu, Z. Icariin Suppresses Cell Cycle Transition and Cell Migration in Ovarian Cancer Cells. *Oncol. Rep.* **2019**, *41*, 2321–2328. [CrossRef]

53. Singh, W.R.; Devi, H.S.; Kumawat, S.; Sadam, A.; Appukuttan, A.V.; Patel, M.R.; Lingaraju, M.C.; Singh, T.U.; Kumar, D. Angiogenic and MMPs Modulatory Effects of Icariin Improved Cutaneous Wound Healing in Rats. *Eur. J. Pharmacol.* **2019**, *858*, 172466. [CrossRef]

54. Quan, K.; Zhang, X.; Fan, K.; Liu, P.; Yue, Q.; Li, B.; Wu, J.; Liu, B.; Xu, Y.; Hua, W.; et al. Icariside II Induces Cell Cycle Arrest and Apoptosis in Human Glioblastoma Cells through Suppressing Akt Activation and Potentiating FOXO3A Activity. *Am. J. Transl. Res.* **2017**, *9*, 2508–2519.

55. Xing, S.; Yu, W.; Zhang, X.; Luo, Y.; Lei, Z.; Huang, D.; Lin, J.; Huang, Y.; Huang, S.; Nong, F.; et al. Isoviolanthin Extracted from Dendrobium Officinale Reverses TGF-B1-Mediated Epithelial–Mesenchymal Transition in Hepatocellular Carcinoma Cells via Deactivating the TGF-β/Smad and PI3K/Akt/MTOR Signaling Pathways. *Int. J. Mol. Sci.* **2018**, *19*, 1556. [CrossRef]

56. Chen, A.Y.; Chen, Y.C. A Review of the Dietary Flavonoid, Kaempferol on Human Health and Cancer Chemoprevention. *Food Chem.* **2013**, *138*, 2099–2107. [CrossRef]

57. Kashyap, D.; Sharma, A.; Tuli, H.S.; Sak, K.; Punia, S.; Mukherjee, T.K. Kaempferol—A Dietary Anticancer Molecule with Multiple Mechanisms of Action: Recent Trends and Advancements. *J. Funct. Foods* **2017**, *30*, 203–219. [CrossRef]

58. Chin, H.K.; Horng, C.T.; Liu, Y.S.; Lu, C.C.; Su, C.Y.; Chen, P.S.; Chiu, H.Y.; Tsai, F.J.; Shieh, P.C.; Yang, J.S. Kaempferol Inhibits Angiogenic Ability by Targeting VEGF Receptor-2 and Downregulating the PI3K/AKT, MEK and ERK Pathways in VEGF-Stimulated Human Umbilical Vein Endothelial Cells. *Oncol. Rep.* **2018**, *39*, 2351–2357. [CrossRef]

59. Liang, F.; Han, Y.; Gao, H.; Xin, S.; Chen, S.; Wang, N.; Qin, W.; Zhong, H.; Lin, S.; Yao, X.; et al. Kaempferol Identified by Zebrafish Assay and Fine Fractionations Strategy from *Dysosma versipellis* Inhibits Angiogenesis through VEGF and FGF Pathways. *Sci. Rep.* **2015**, *5*, 14468. [CrossRef]

60. Kumazawa, S.; Kubota, S.; Yamamoto, H.; Okamura, N.; Sugiyama, Y.; Kobayashi, H.; Nakanishi, M.; Ohta, T. Antiangiogenic Activity of Flavonoids from Melia Azedarach. *Nat. Prod. Commun.* **2013**, *8*, 1719–1720. [CrossRef]

61. Özay, Y.; Güzel, S.; Yumrutaş, Ö.; Pehlivanoğlu, B.; Erdoğdu, İ.H.; Yildirim, Z.; Türk, B.A.; Darcan, S. Wound Healing Effect of Kaempferol in Diabetic and Nondiabetic Rats. *J. Surg. Res.* **2019**, *233*, 284–296. [CrossRef]

62. Sharma, V.; Joseph, C.; Ghosh, S.; Agarwal, A.; Mishra, M.K.; Sen, E. Kaempferol Induces Apoptosis in Glioblastoma Cells through Oxidative Stress. *Mol. Cancer Ther.* **2007**, *6*, 2544–2553. [CrossRef]

63. Shen, S.C.; Lin, C.W.; Lee, H.M.; Chien, L.L.; Chen, Y.C. Lipopolysaccharide plus 12-o-Tetradecanoylphorbol 13-Acetate Induction of Migration and Invasion of Glioma Cells in Vitro and in Vivo: Differential Inhibitory Effects of Flavonoids. *Neuroscience* **2006**, *140*, 477–489. [CrossRef]

64. Qin, Y.; Cui, W.; Yang, X.; Tong, B. Kaempferol Inhibits the Growth and Metastasis of Cholangiocarcinoma in Vitro and in Vivo. *Acta Biochim. Biophys. Sin. (Shanghai).* **2015**, *48*, 238–245. [CrossRef]

65. Zhu, G.; Liu, X.; Li, H.; Yan, Y.; Hong, X.; Lin, Z. Kaempferol Inhibits Proliferation, Migration, and Invasion of Liver Cancer HepG2 Cells by down-Regulation of MicroRNA-21. *Int. J. Immunopathol. Pharmacol.* **2018**, *32*, 2058738418814341. [CrossRef]

66. Jo, E.; Park, S.J.; Choi, Y.S.; Jeon, W.K.; Kim, B.C. Kaempferol Suppresses Transforming Growth Factor-B1-Induced Epithelial-to-Mesenchymal Transition and Migration of A549 Lung Cancer Cells by Inhibiting Akt1-Mediated Phosphorylation of Smad3 at Threonine-179. *Neoplasia* **2015**, *17*, 525–537. [CrossRef]

67. Labbé, D.; Provençal, M.; Lamy, S.; Boivin, D.; Gingras, D.; Béliveau, R. The Flavonols Quercetin, Kaempferol, and Myricetin Inhibit Hepatocyte Growth. *J. Nutr. Biochem. Mol. Genet. Mech.* **2009**, *139*, 646–652. [CrossRef]

68. Lin, C.W.; Chen, P.N.; Chen, M.K.; Yang, W.E.; Tang, C.H.; Yang, S.F.; Hsieh, Y.S. Kaempferol Reduces Matrix Metalloproteinase-2 Expression by down-Regulating ERK1/2 and the Activator Protein-1 Signaling Pathways in Oral Cancer Cells. *PLoS ONE* **2013**, *8*, e80883. [CrossRef]

69. Chen, H.J.; Lin, C.M.; Lee, C.Y.; Shih, N.C.; Peng, S.F.; Tsuzuki, M.; Amagaya, S.; Huang, W.W.; Yang, J.S. Kaempferol Suppresses Cell Metastasis via Inhibition of the ERK-P38-JNK and AP-1 Signaling Pathways in U-2 OS Human Osteosarcoma Cells. *Oncol. Rep.* **2013**, *30*, 925–932. [CrossRef]

70. Luo, H.; Rankin, G.O.; Liu, L.; Daddysman, M.K.; Jiang, B.H.; Chen, Y.C. Kaempferol Inhibits Angiogenesis and VEGF Expression through Both HIF Dependent and Independent Pathways in Human Ovarian Cancer Cells. *Nutr. Cancer* **2009**, *61*, 554–563. [CrossRef]

71. Lee, J.; Kim, J.H. Kaempferol Inhibits Pancreatic Cancer Cell Growth and Migration through the Blockade of EGFR-Related Pathway in Vitro. *PLoS ONE* **2016**, *11*, e0155264. [CrossRef]

72. Hung, T.W.; Chen, P.N.; Wu, H.C.; Wu, S.W.; Tsai, P.Y.; Hsieh, Y.S.; Chang, H.R. Kaempferol Inhibits the Invasion and Migration of Renal Cancer Cells through the Downregulation of AKT and FAK Pathways. *Int. J. Med. Sci.* **2017**, *14*, 984–993. [CrossRef]

73. Chien, H.W.; Wang, K.; Chang, Y.Y.; Hsieh, Y.H.; Yu, N.Y.; Yang, S.F.; Lin, H.W. Kaempferol Suppresses Cell Migration through the Activation of the ERK Signaling Pathways in ARPE-19 Cells. *Environ. Toxicol.* **2019**, *34*, 312–318. [CrossRef]

74. Clericuzio, M.; Tinello, S.; Burlando, B.; Ranzato, E.; Martinotti, S.; Cornara, L.; La Rocca, A. Flavonoid Oligoglycosides from *Ophioglossum vulgatum* L. Having Wound Healing Properties. *Planta Med.* **2012**, *78*, 1639–1644. [CrossRef]

75. Zeng, N.; Tong, B.; Zhang, X.; Dou, Y.; Wu, X.; Xia, Y.; Dai, Y.; Wei, Z. Antiarthritis Effect of Morin Is Associated with Inhibition of Synovial Angiogensis. *Drug Dev. Res.* **2015**, *76*, 463–473. [CrossRef]

76. Yue, M.; Zeng, N.; Xia, Y.; Wei, Z.; Dai, Y. Morin Exerts Anti-Arthritic Effects by Attenuating Synovial Angiogenesis via Activation of Peroxisome Proliferator Activated Receptor-γ. *Mol. Nutr. Food Res.* **2018**, *62*, 1800202. [CrossRef]

77. Capitani, N.; Lori, G.; Paoli, P.; Patrussi, L.; Troilo, A.; Baldari, C.T.; Raugei, G.; D'Elios, M.M. LMW-PTP Targeting Potentiates the Effects of Drugs Used in Chronic Lymphocytic Leukemia Therapy. *Cancer Cell Int.* **2019**, *19*, 67. [CrossRef]

78. Li, H.W.; Zou, T.-B.; Jia, Q.; Xia, E.Q.; Cao, W.J.; Liu, W.; He, T.P.; Wang, Q. Anticancer Effects of Morin-7-Sulphate Sodium, a Flavonoid Derivative, in Mouse Melanoma Cells. *Biomed. Pharmacother.* **2016**, *84*, 909–916. [CrossRef]

79. Kang, N.J.; Jung, S.K.; Lee, K.W.; Lee, H.J. Myricetin Is a Potent Chemopreventive Phytochemical in Skin Carcinogenesis. *Ann. N. Y. Acad. Sci.* **2011**, *1229*, 124–132. [CrossRef]

80. Zhou, Z.; Mao, W.; Li, Y.; Qi, C.; He, Y. Myricetin Inhibits Breast Tumor Growth and Angiogenesis by Regulating VEGF/VEGFR2 and P38MAPK Signaling Pathways. *Anat. Rec.* **2019**, *302*, 2186–2192. [CrossRef]

81. Ci, Y.; Zhang, Y.; Liu, Y.; Lu, S.; Cao, J.; Li, H.; Zhang, J.; Huang, Z.; Zhu, X.; Gao, J.; et al. Myricetin Suppresses Breast Cancer Metastasis through Down-Regulating the Activity of Matrix Metalloproteinase (MMP)-2/9. *Phyther. Res.* **2018**, *32*, 1373–1381. [CrossRef]

82. Chiu, W.T.; Shen, S.C.; Chow, J.M.; Lin, C.W.; Shia, L.T.; Chen, Y.C. Contribution of Reactive Oxygen Species to Migration/Invasion of Human Glioblastoma Cells U87 via ERK-Dependent COX-2/PGE2 Activation. *Neurobiol. Dis.* **2010**, *37*, 118–129. [CrossRef]

83. Yamada, N.; Matsushima-Nishiwaki, R.; Kozawa, O. Quercetin Suppresses the Migration of Hepatocellular Carcinoma Cells Stimulated by Hepatocyte Growth Factor or Transforming Growth Factor-α: Attenuation of AKT Signaling Pathway. *Arch. Biochem. Biophys.* **2020**, *682*, 108296. [CrossRef]

84. Ma, H.; Zhu, L.; Ren, J.; Rao, B.; Sha, M.; Kuang, Y.; Shen, W.; Xu, Z. Myricetin Inhibits Migration and Invasion of Hepatocellular Carcinoma MHCC97H Cell Line by Inhibiting the EMT Process. *Oncol. Lett.* **2019**, *18*, 6614–6620. [CrossRef]

85. Shih, Y.W.; Wu, P.F.; Lee, Y.C.; Shi, M.-D.; Chiang, T.A. Myricetin Suppresses Invasion and Migration of Human Lung Adenocarcinoma A549 Cells: Possible Mediation by Blocking the Erk Signaling Pathway. *J. Agric. Food Chem.* **2009**, *57*, 3490–3499. [CrossRef]

86. Ezzati, M.; Yousefi, B.; Velaei, K.; Safa, A. A Review on Anti-Cancer Properties of Quercetin in Breast Cancer. *Life Sci.* **2020**, *248*, 117463. [CrossRef]

87. Tang, S.; Deng, X.; Zhou, J.; Li, Q.; Ge, X.; Miao, L. Pharmacological Basis and New Insights of Quercetin Action in Respect to Its Anti-Cancer Effects. *Biomed. Pharmacother.* **2020**, *121*, 109604. [CrossRef]

88. Kashyap, D.; Mittal, S.; Sak, K.; Singhal, P.; Tuli, H.S. Molecular Mechanisms of Action of Quercetin in Cancer: Recent Advances. *Tumor Biol.* **2016**, *37*, 12927–12939. [CrossRef]

89. Darband, S.G.; Kaviani, M.; Yousefi, B.; Sadighparvar, S.; Pakdel, F.G.; Attari, J.A.; Mohebbi, I.; Naderi, S.; Majidinia, M. Quercetin: A Functional Dietary Flavonoid with Potential Chemo-Preventive Properties in Colorectal Cancer. *J. Cell. Physiol.* **2018**, *233*, 6544–6560. [CrossRef]

90. Song, N.R.; Chung, M.Y.; Kang, N.J.; Seo, S.G.; Jang, T.S.; Lee, H.J.; Lee, K.W. Quercetin Suppresses Invasion and Migration of H-Ras-Transformed MCF10A Human Epithelial Cells by Inhibiting Phosphatidylinositol 3-Kinase. *Food Chem.* **2014**, *142*, 66–71. [CrossRef]

91. Lin, C.; Wu, M.; Dong, J. Quercetin-4′-o-β-d-Glucopyranoside (QODG) Inhibits Angiogenesis by Suppressing VEGFR2-Mediated Signaling in Zebrafish and Endothelial Cells. *PLoS ONE* **2012**, *7*, e31708. [CrossRef]

92. Lee, Y.H.; Tuyet, P.T. Synthesis and Biological Evaluation of Quercetin–Zinc (II) Complex for Anti-Cancer and Anti-Metastasis of Human Bladder Cancer Cells. *Vitr. Cell. Dev. Biol. Anim.* **2019**, *55*, 395–404. [CrossRef]

93. Ravishankar, D.; Watson, K.A.; Boateng, S.Y.; Green, R.J.; Greco, F.; Osborn, H.M.I. Exploring Quercetin and Luteolin Derivatives as Antiangiogenic Agents. *Eur. J. Med. Chem.* **2015**, *97*, 259–274. [CrossRef]

94. Zhao, X.; Wang, Q.; Yang, S.; Chen, C.; Li, X.; Liu, J.; Zou, Z.; Cai, D. Quercetin Inhibits Angiogenesis by Targeting Calcineurin in the Xenograft Model of Human Breast Cancer. *Eur. J. Pharmacol.* **2016**, *781*, 60–68. [CrossRef]

95. Oh, S.J.; Kim, O.; Lee, J.S.; Kim, J.A.; Kim, M.R.; Choi, H.S.; Shim, J.H.; Kang, K.W.; Kim, Y.C. Inhibition of Angiogenesis by Quercetin in Tamoxifen-Resistant Breast Cancer Cells. *Food Chem. Toxicol.* **2010**, *48*, 3227–3234. [CrossRef]

96. Srinivasan, A.; Thangavel, C.; Liu, Y.; Shoyele, S.; Den, R.B.; Selvakumar, P.; Lakshmikuttyamma, A. Quercetin Regulates β-Catenin Signaling and Reduces the Migration of Triple Negative Breast Cancer. *Mol. Carcinog.* **2016**, *55*, 743–756. [CrossRef]

97. Lin, C.W.; Hou, W.C.; Shen, S.C.; Juan, S.H.; Ko, C.H.; Wang, L.M.; Chen, Y.C. Quercetin Inhibition of Tumor Invasion via Suppressing PKCδ/ERK/AP-1-Dependent Matrix Metalloproteinase-9 Activation in Breast Carcinoma Cells. *Carcinogenesis* **2008**, *29*, 1807–1815. [CrossRef]

98. Rivera Rivera, A.; Castillo-Pichardo, L.; Gerena, Y.; Dharmawardhane, S. Anti-Breast Cancer Potential of Quercetin via the Akt/AMPK/Mammalian Target of Rapamycin (MTOR) Signaling Cascade. *PLoS ONE* **2016**, *11*, e0157251. [CrossRef]

99. Zhao, D.; Qin, C.; Fan, X.; Li, Y.; Gu, B. Inhibitory Effects of Quercetin on Angiogenesis in Larval Zebra Fi Sh and Human Umbilical Vein Endothelial Cells. *Eur. J. Pharmacol.* **2014**, *723*, 360–367. [CrossRef]

100. Tan, W.F.; Lin, L.P.; Li, M.H.; Zhang, Y.X.; Tong, Y.G.; Xiao, D.; Ding, J. Quercetin, a Dietary-Derived Flavonoid, Possesses Antiangiogenic Potential. *Eur. J. Pharmacol.* **2003**, *459*, 255–262. [CrossRef]

101. Zhao, X.; Shu, G.; Chen, L.; Mi, X.; Mei, Z.; Deng, X. A Flavonoid Component from *Docynia delavayi* (Franch.) Schneid Represses Transplanted H22 Hepatoma Growth and Exhibits Low Toxic Effect on Tumor-Bearing Mice. *Food Chem. Toxicol.* **2012**, *50*, 3166–3168. [CrossRef]

102. Lee, L.T.; Huang, Y.T.; Hwang, J.J.; Lee, A.Y.L.; Ke, F.C.; Huang, C.J.; Kandaswami, C.; Lee, P.P.H.; Lee, M.T. Transinactivation of the Epidermal Growth Factor Receptor Tyrosine Kinase and Focal Adhesion Kinase Phosphorylation by Dietary Flavonoids: Effect on Invasive Potential of Human Carcinoma Cells. *Biochem. Pharmacol.* **2004**, *67*, 2103–2114. [CrossRef]

103. Lee, D.E.; Chung, M.Y.; Lim, T.G.; Huh, W.B.; Lee, H.J.; Lee, K.W. Quercetin Suppresses Intracellular Ros Formation, MMP Activation, and Cell Motility in Human Fibrosarcoma Cells. *J. Food Sci.* **2013**, *78*. [CrossRef]

104. Kee, J.Y.; Han, Y.H.; Kim, D.S.; Mun, J.G.; Park, J.; Jeong, M.Y.; Um, J.Y.; Hong, S.H. Inhibitory Effect of Quercetin on Colorectal Lung Metastasis through Inducing Apoptosis, and Suppression of Metastatic Ability. *Phytomedicine* **2016**, *23*, 1680–1690. [CrossRef]

105. Pan, H.C.; Jiang, Q.; Yu, Y.; Mei, J.P.; Cui, Y.K.; Zhao, W.J. Quercetin Promotes Cell Apoptosis and Inhibits the Expression of MMP-9 and Fibronectin via the AKT and ERK Signalling Pathways in Human Glioma Cells. *Neurochem. Int.* **2015**, *80*, 60–71. [CrossRef]

106. Liu, Y.; Tang, Z.G.; Lin, Y.; Qu, X.G.; Lv, W.; Wang, G.-B.; Li, C.L. Effects of Quercetin on Proliferation and Migration of Human Glioblastoma U251 Cells. *Biomed. Pharmacother.* **2017**, *92*, 33–38. [CrossRef]

107. Michaud-Levesque, J.; Bousquet-Gagnon, N.; Béliveau, R. Quercetin Abrogates IL-6/STAT3 Signaling and Inhibits Glioblastoma Cell Line Growth and Migration. *Exp. Cell Res.* **2012**, *318*, 925–935. [CrossRef]

108. da Silva, A.B.; Cerqueira Coelho, P.L.; das Neves Oliveira, M.; Oliveira, J.L.; Oliveira Amparo, J.A.; da Silva, K.C.; Soares, J.R.P.; Pitanga, B.P.S.; dos Santos Souza, C.; de Faria Lopes, G.P.; et al. The Flavonoid Rutin and Its Aglycone Quercetin Modulate the Microglia Inflammatory Profile Improving Antiglioma Activity. *Brain. Behav. Immun.* **2020**, *85*, 170–185. [CrossRef]

109. Liu, Y.; Tang, Z.G.; Yang, J.Q.; Zhou, Y.; Meng, L.H.; Wang, H.; Li, C.L. Low Concentration of Quercetin Antagonizes the Invasion and Angiogenesis of Human Glioblastoma U251 Cells. *OncoTargets. Ther.* **2017**, *10*, 4023–4028. [CrossRef]

110. Lu, J.; Wang, Z.; Li, S.; Xin, Q.; Yuan, M.; Li, H.; Song, X.; Gao, H.; Pervaiz, N.; Sun, X.; et al. Quercetin Inhibits the Migration and Invasion of HCCLM3 Cells by Suppressing the Expression of P-Akt1, Matrix Metalloproteinase (MMP) MMP-2, and MMP-9. *Med. Sci. Monit.* **2018**, *24*, 2583–2589. [CrossRef]

111. Klimaszewska-Wiśniewska, A.; Hałas-Wiśniewska, M.; Izdebska, M.; Gagat, M.; Grzanka, A.; Grzanka, D. Antiproliferative and Antimetastatic Action of Quercetin on A549 Non-Small Cell Lung Cancer Cells through Its Effect on the Cytoskeleton. *Acta Histochem.* **2017**, *119*, 99–112. [CrossRef]

112. Wu, T.C.; Chan, S.T.; Chang, C.N.; Yu, P.S.; Chuang, C.H.; Yeh, S.L. Quercetin and Chrysin Inhibit Nickel-Induced Invasion and Migration by Downregulation of TLR4/NF-KB Signaling in A549 cells. *Chem. Biol. Interact.* **2018**, *292*, 101–109. [CrossRef]

113. Lin, Y.C.; Tsai, P.H.; Lin, C.Y.; Cheng, C.H.; Lin, T.H.; Lee, K.P.H.; Huang, K.Y.; Chen, S.H.; Hwang, J.J.; Kandaswami, C.C.; et al. Impact of Flavonoids on Matrix Metalloproteinase Secretion and Invadopodia Formation in Highly Invasive A431-III Cancer Cells. *PLoS ONE* **2013**, *8*, e71903. [CrossRef]

114. Hwang, M.K.; Song, N.R.; Kang, N.J.; Lee, K.W.; Lee, H.J. Activation of Phosphatidylinositol 3-Kinase Is Required for Tumor Necrosis Factor-α-Induced Upregulation of Matrix Metalloproteinase-9: Its Direct Inhibition by Quercetin. *Int. J. Biochem. Cell Biol.* **2009**, *41*, 1592–1600. [CrossRef]

115. Cao, H.H.; Cheng, C.Y.; Su, T.; Fu, X.Q.; Guo, H.; Li, T.; Tse, A.K.W.; Kwan, H.Y.; Yu, H.; Yu, Z.L. Quercetin Inhibits HGF/c-Met Signaling and HGFstimulated Melanoma Cell Migration and Invasion. *Mol. Cancer* **2015**, *14*, 103. [CrossRef]

116. Cao, H.H.; Tse, A.K.W.; Kwan, H.Y.; Yu, H.; Cheng, C.Y.; Su, T.; Fong, W.F.; Yu, Z.L. Quercetin Exerts Anti-Melanoma Activities and Inhibits STAT3 Signaling. *Biochem. Pharmacol.* **2014**, *87*, 424–434. [CrossRef]

117. Zhao, J.; Fang, Z.; Zha, Z.; Sun, Q.; Wang, H.; Sun, M.; Qiao, B. Quercetin Inhibits Cell Viability, Migration and Invasion by Regulating MiR-16/HOXA10 Axis in Oral Cancer. *Eur. J. Pharmacol.* **2019**, *847*, 11–18. [CrossRef]

118. Nam, T.W.; Il Yoo, C.; Kim, H.T.; Kwon, C.H.; Park, J.Y.; Kim, Y.K. The Flavonoid Quercetin Induces Apoptosis and Inhibits Migration through a MAPK-Dependent Mechanism in Osteoblasts. *J. Bone Miner. Metab.* **2008**, *26*, 551–560. [CrossRef]

119. Li, S.; Pei, Y.; Wang, W.; Liu, F.; Zheng, K.; Zhang, X. Quercetin Suppresses the Proliferation and Metastasis of Metastatic Osteosarcoma Cells by Inhibiting Parathyroid Hormone Receptor 1. *Biomed. Pharmacother.* **2019**, *114*, 108839. [CrossRef]

120. Lan, H.; Hong, W.; Fan, P.; Qian, D.; Zhu, J.; Bai, B. Quercetin Inhibits Cell Migration and Invasion in Human Osteosarcoma Cells. *Cell. Physiol. Biochem.* **2017**, *43*, 553–567. [CrossRef]

121. Berndt, K.; Campanile, C.; Muff, R.; Strehler, E.; Born, W.; Fuchs, B. Evaluation of Quercetin as a Potential Drug in Osteosarcoma Treatment. *Anticancer Res.* **2013**, *33*, 1297–1306.

122. Huang, Y.T.; Lee, L.T.; Lee, P.P.H.; Lin, Y.S.; Lee, M.T. Targeting of Focal Adhesion Kinase by Flavonoids and Small-Interfering RNAs Reduces Tumor Cell Migration Ability. *Anticancer Res.* **2005**, *25*, 2017–2025.

123. Yu, D.; Ye, T.; Xiang, Y.; Shi, Z.; Zhang, J.; Lou, B.; Zhang, F.; Chen, B.; Zhou, M. Quercetin Inhibits Epithelial–Mesenchymal Transition, Decreases Invasiveness and Metastasis, and Reverses IL-6 Induced Epithelial–Mesenchymal Transition, Expression of MMP by Inhibiting STAT3 Signaling in Pancreatic Cancer Cells. *OncoTargets. Ther.* **2017**, *10*, 4719–4729. [CrossRef]

124. Bhat, F.A.; Sharmila, G.; Balakrishnan, S.; Arunkumar, R.; Elumalai, P.; Suganya, S.; Raja Singh, P.; Srinivasan, N.; Arunakaran, J. Quercetin Reverses EGF-Induced Epithelial to Mesenchymal Transition and Invasiveness in Prostate Cancer (PC-3) Cell Line via EGFR/PI3K/Akt Pathway. *J. Nutr. Biochem.* **2014**, *25*, 1132–1139. [CrossRef]

125. Yang, F.; Jiang, X.; Song, L.; Wang, H.; Mei, Z.; Xu, Z.; Xing, N. Quercetin Inhibits Angiogenesis through Thrombospondin-1 Upregulation to Antagonize Human Prostate Cancer PC-3 Cell Growth in Vitro and in Vivo. *Oncol. Rep.* **2016**, *35*, 1602–1610. [CrossRef]

126. Song, W.; Zhao, X.; Xu, J.; Zhang, H. Quercetin Inhibits Angiogenesis—Mediated Human Retinoblastoma Growth by Targeting Vascular Endothelial Growth Factor Receptor. *Oncol. Lett.* **2017**, *14*, 3343–3348. [CrossRef]

127. Camero, C.M.; Germanò, M.P.; Rapisarda, A.; D'Angelo, V.; Amira, S.; Benchikh, F.; Braca, A.; De Leo, M. Anti-Angiogenic Activity of Iridoids from Galium Tunetanum. *Braz. J. Pharmacogn.* **2018**, *28*, 374–377. [CrossRef]

128. Ben Sghaier, M.; Pagano, A.; Mousslim, M.; Ammari, Y.; Kovacic, H.; Luis, J. Rutin Inhibits Proliferation, Attenuates Superoxide Production and Decreases Adhesion and Migration of Human Cancerous Cells. *Biomed. Pharmacother.* **2016**, *84*, 1972–1978. [CrossRef]

129. Chen, H.; Miao, Q.; Geng, M.; Liu, J.; Hu, Y.; Tian, L.; Pan, J.; Yang, Y. Anti-Tumor Effect of Rutin on Human Neuroblastoma Cell Lines through Inducing G2/M Cell Cycle Arrest and Promoting Apoptosis. *Sci. World J.* **2013**, *2013*. [CrossRef]

130. Lin, C.M.; Chang, H.; Chen, Y.H.; Li, S.Y.; Wu, I.H.; Chiu, J.H. Protective Role of Wogonin against Lipopolysaccharide-Induced Angiogenesis via VEGFR-2, Not VEGFR-1. *Int. Immunopharmacol.* **2006**, *6*, 1690–1698. [CrossRef]

131. Liu, L.Z.; Jing, Y.; Jiang, L.L.; Jiang, X.E.; Jiang, Y.; Rojanasakul, Y.; Jiang, B.H. Acacetin Inhibits VEGF Expression, Tumor Angiogenesis and Growth through AKT/HIF-1α Pathway. *Biochem. Biophys. Res. Commun.* **2011**, *413*, 299–305. [CrossRef]

132. Zhao, K.; Yuan, Y.; Lin, B.; Miao, Z.; Li, Z.; Guo, Q.; Lu, N. LW-215, a Newly Synthesized Flavonoid, Exhibits Potent Anti-Angiogenic Activity in Vitro and in Vivo. *Gene* **2018**, *642*, 533–541. [CrossRef]

133. Viegas, O.; Faria, M.A.; Sousa, J.B.; Vojtek, M.; Gonçalves-Monteiro, S.; Suliburska, J.; Diniz, C.; Ferreira, I.M.P.L.V.O. Delphinidin-3-O-Glucoside Inhibits Angiogenesis via VEGFR2 Downregulation and Migration through Actin Disruption. *J. Funct. Foods* **2019**, *54*, 393–402. [CrossRef]

134. Rajesh, G.; Harshala, S.; Dhananjay, G.; Jadhav, A.; Vikram, G. Effect of Hydroxyl Substitution of Flavone on Angiogenesis and Free Radical Scavenging Activities: A Structure-Activity Relationship Studies Using Computational Tools. *Eur. J. Pharm. Sci.* **2010**, *39*, 37–44. [CrossRef]

135. Gacche, R.N.; Shegokar, H.D.; Gond, D.S.; Yang, Z.; Jadhav, A.D. Evaluation of Selected Flavonoids as Antiangiogenic, Anticancer, and Radical Scavenging Agents: An Experimental and in Silico Analysis. *Cell Biochem. Biophys.* **2011**, *61*, 651–663. [CrossRef]

136. Gacche, R.N.; Meshram, R.J.; Shegokar, H.D.; Gond, D.S.; Kamble, S.S.; Dhabadge, V.N.; Utage, B.G.; Patil, K.K.; More, R.A. Flavonoids as a Scaffold for Development of Novel Anti-Angiogenic Agents: An Experimental and Computational Enquiry. *Arch. Biochem. Biophys.* **2015**, *577–578*, 35–48. [CrossRef]

137. Chen, Y.; Lu, N.; Ling, Y.; Wang, L.; You, Q.; Li, Z.; Guo, Q. LYG-202, a Newly Synthesized Flavonoid, Exhibits Potent Anti-Angiogenic Activity in Vitro and in Vivo. *J. Pharmacol. Sci.* **2010**, *112*, 37–45. [CrossRef]

138. Huang, Y.; Fang, J.; Lu, W.; Wang, Z.; Wang, Q.; Hou, Y.; Jiang, X.; Reizes, O.; Lathia, J.; Nussinov, R.; et al. A Systems Pharmacology Approach Uncovers Wogonoside as an Angiogenesis Inhibitor of Triple-Negative Breast Cancer by Targeting Hedgehog Signaling. *Cell Chem. Biol.* **2019**, *26*, 1143–1158.e6. [CrossRef]

139. Chen, Y.; Lu, N.; Ling, Y.; Gao, Y.; Wang, L.; Sun, Y.; Qi, Q.; Feng, F.; Liu, W.; Liu, W.; et al. Wogonoside Inhibits Lipopolysaccharide-Induced Angiogenesis in Vitro and in Vivo via Toll-like Receptor 4 Signal Transduction. *Toxicology* **2009**, *259*, 10–17. [CrossRef]

140. Li, X.; Chen, M.; Lei, X.; Huang, M.; Ye, W.; Zhang, R.; Zhang, D. Luteolin Inhibits Angiogenesis by Blocking Gas6/Axl Signaling Pathway. *Int. J. Oncol.* **2017**, *51*, 677–685. [CrossRef]

141. Panda, S.P.; Panigrahy, U.P.; Prasanth, D.; Gorla, U.S.; Guntupalli, C.; Panda, D.P.; Jena, B.R. A Trimethoxy Flavonoid Isolated from Stem Extract of *Tabebuia chrysantha* Suppresses Angiogenesis in Angiosarcoma. *J. Pharm. Pharmacol.* **2020**, jphp.13272. [CrossRef]

142. Huang, Y.; Zhao, K.; Hu, Y.; Zhou, Y.; Luo, X.; Li, X.; Wei, L.; Li, Z.; You, Q.; Guo, Q.; et al. Wogonoside Inhibits Angiogenesis in Breast Cancer via Suppressing Wnt/β-Catenin Pathway. *Mol. Carcinog.* **2016**, *55*, 1598–1612. [CrossRef]

143. Bhat, T.A.; Nambiar, D.; Tailor, D.; Pal, A.; Agarwal, R.; Singh, R.P. Acacetin Inhibits in Vitro and in Vivo Angiogenesis and Downregulates Stat Signaling and VEGF Expression. *Cancer Prev. Res.* **2013**, *6*, 1128–1139. [CrossRef]

144. Fang, J.; Zhou, Q.; Liu, L.Z.; Xia, C.; Hu, X.; Shi, X.; Jiang, B.H. Apigenin Inhibits Tumor Angiogenesis through Decreasing HIF-1α and VEGF Expression. *Carcinogenesis* **2007**, *28*, 858–864. [CrossRef]

145. Chen, J.; Chen, A.Y.; Huang, H.; Ye, X.; Rollyson, W.D.; Perry, H.E.; Brown, K.C.; Rojanasakul, Y.; Rankin, G.O.; Dasgupta, P.; et al. The Flavonoid Nobiletin Inhibits Tumor Growth and Angiogenesis of Ovarian Cancers via the Akt Pathway. *Int. J. Oncol.* **2015**, *46*, 2629–2638. [CrossRef]

146. Pratheeshkumar, P.; Son, Y.O.; Budhraja, A.; Wang, X.; Ding, S.; Wang, L.; Hitron, A.; Lee, J.C.; Kim, D.; Divya, S.P.; et al. Luteolin Inhibits Human Prostate Tumor Growth by Suppressing Vascular Endothelial Growth Factor Receptor 2-Mediated Angiogenesis. *PLoS ONE* **2012**, *7*, e52279. [CrossRef]

147. Lin, C.M.; Chang, H.; Chen, Y.H.; Wu, I.H.; Chiu, J.H. Wogonin Inhibits IL-6-Induced Angiogenesis via down-Regulation of VEGF and VEGFR-1, Not VEGFR-2. *Planta Med.* **2006**, *72*, 1305–1310. [CrossRef]

148. Zhu, D.; Wang, S.; Lawless, J.; He, J.; Zheng, Z. Dose Dependent Dual Effect of Baicalin and Herb Huang Qin Extract on Angiogenesis. *PLoS ONE* **2016**, *11*, e0167125. [CrossRef]

149. Favot, L.; Martin, S.; Keravis, T.; Andriantsitohaina, R.; Lugnier, C. Involvement of Cyclin-Dependent Pathway in the Inhibitory Effect of Delphinidin on Angiogenesis. *Cardiovasc. Res.* **2003**, *59*, 479–487. [CrossRef]

150. Hui, M.; Cazet, A.; Nair, R.; Watkins, D.N.; O'Toole, S.A.; Swarbrick, A. The Hedgehog Signalling Pathway in Breast Development, Carcinogenesis and Cancer Therapy. *Breast Cancer Res.* **2013**, *15*, 1–14. [CrossRef]

151. Cao, X.; Geradts, J.; Dewhirst, M.W.; Lo, H.W. Upregulation of VEGF-A and CD24 Gene Expression by the TGLI1 Transcription Factor Contributes to the Aggressive Behavior of Breast Cancer Cells. *Oncogene* **2012**, *31*, 104–115. [CrossRef]

152. Zhang, K.; Lu, J.; Mori, T.; Smith-Powell, L.; Synold, T.W.; Chen, S.; Wen, W. Baicalin Increases VEGF Expression and Angiogenesis by Activating the ERR/PGC-1 Pathway. *Cardiovasc. Res.* **2011**, *89*, 426–435. [CrossRef]

153. Liu, J.-J.; Huang, T.-S.; Cheng, W.-F.; Lu, F.-J. Baicalein and Baicalin Are Potent Inhibitors of Angiogenesis: Inhibition of Endothelial Cell Proliferation, Migration and Differentiation. *Int. J. Cancer* **2003**, *106*, 559–565. [CrossRef]

154. Morbidelli, L. Polyphenol-Based Nutraceuticals for the Control of Angiogenesis: Analysis of the Critical Issues for Human Use. *Pharmacol. Res.* **2016**, *111*, 384–393. [CrossRef]

155. Cockerill, G.W.; Gamble, J.R.; Vadas, M.A. Angiogenesis: Models and Modulators. *Int. Rev. Cytol.* **1995**, *159*, 113–160. [CrossRef]

156. Ribatti, D.; Vacca, A. Models for Studying Angiogenesis in Vivo. *Int. J. Biol. Markers* **1999**, *14*, 207–213. [CrossRef]

157. Staton, C.A.; Stribbling, S.M.; Tazzyman, S.; Hughes, R.; Brown, N.J.; Lewis, C.E. Current Methods for Assaying Angiogenesis in Vitro and in Vivo. *Int. J. Experimantal Biol.* **2004**, *85*, 233–248. [CrossRef]

158. Nguyen, M.; Shing, Y.; Folkman, J. Quantitation of Angiogenesis and Antiangiogenesis in the Chick Embryo Chorioallantoic Membrane. *Microvasc. Res.* **1994**, *47*, 31–40. [CrossRef]

159. Ribatti, D.; Vacca, A.; Roncali, L.; Dammacco, F. The Chick Embryo Chorioallantoic Membrane as a Model for in Vivo Research on Angiogenesis. *Int. J. Dev. Biol.* **1996**, *40*, 1189–1197. [CrossRef]

160. Wang, T.; Li, Q.; Bi, K. Bioactive Flavonoids in Medicinal Plants: Structure, Activity and Biological Fate. *Asian J. Pharm. Sci.* **2018**, *13*, 12–23. [CrossRef]

161. Nguyen, T.T.H.; Moon, Y.H.; Ryu, Y.B.; Kim, Y.M.; Nam, S.H.; Kim, M.S.; Kimura, A.; Kim, D. The Influence of Flavonoid Compounds on the in Vitro Inhibition Study of a Human Fibroblast Collagenase Catalytic Domain Expressed in *E. coli*. *Enzyme Microb. Technol.* **2013**, *52*, 26–31. [CrossRef]

162. Krych, J.; Gebicka, L. Catalase Is Inhibited by Flavonoids. *Int. J. Biol. Macromol.* **2013**, *58*, 148–153. [CrossRef]

163. Huang, Z.; Fang, F.; Wang, J.; Wong, C.W. Structural Activity Relationship of Flavonoids with Estrogen-Related Receptor Gamma. *FEBS Lett.* **2010**, *584*, 22–26. [CrossRef]

164. Atrahimovich, D.; Vaya, J.; Khatib, S. The Effects and Mechanism of Flavonoid-RePON1 Interactions. Structure-Activity Relationship Study. *Bioorganic Med. Chem.* **2013**, *21*, 3348–3355. [CrossRef]

165. Çelik, H.; Koşar, M. Inhibitory Effects of Dietary Flavonoids on Purified Hepatic NADH-Cytochrome B5 Reductase: Structure-Activity Relationships. *Chem. Biol. Interact.* **2012**, *197*, 103–109. [CrossRef]

166. Sithisarn, P.; Michaelis, M.; Schubert-Zsilavecz, M.; Cinatl, J. Differential Antiviral and Anti-Inflammatory Mechanisms of the Flavonoids Biochanin A and Baicalein in H5N1 Influenza A Virus-Infected Cells. *Antiviral Res.* **2013**, *97*, 41–48. [CrossRef]

167. Kothandan, G.; Gadhe, C.G.; Madhavan, T.; Choi, C.H.; Cho, S.J. Docking and 3D-QSAR (Quantitative Structure Activity Relationship) Studies of Flavones, the Potent Inhibitors of p-Glycoprotein Targeting the Nucleotide Binding Domain. *Eur. J. Med. Chem.* **2011**, *46*, 4078–4088. [CrossRef]

168. Singh, M.; Kaur, M.; Silakari, O. Flavones: An Important Scaffold for Medicinal Chemistry. *Eur. J. Med. Chem.* **2014**, *84*, 206–239. [CrossRef]

169. Haupt, V.J.; Daminelli, S.; Schroeder, M. Drug Promiscuity in PDB: Protein Binding Site Similarity Is Key. *PLoS ONE* **2013**, *8*, e65894. [CrossRef]

170. Lam, I.K.; Alex, D.; Wang, Y.H.; Liu, P.; Liu, A.L.; Du, G.H.; Yuen Lee, S.M. In Vitro and in Vivo Structure and Activity Relationship Analysis of Polymethoxylated Flavonoids: Identifying Sinensetin as a Novel Antiangiogenesis Agent. *Mol. Nutr. Food Res.* **2012**, *56*, 945–956. [CrossRef]

171. Lingen, M.W. Historical Perspective Role of Leukocytes and Endothelial Cells in the Development of Angiogenesis in Inflammation and Wound Healing. *Arch. Pathol. Lab. Med.* **2001**, *125*, 67–71. [CrossRef]

172. Naldini, A.; Carraro, F. Role of Inflammatory Mediators in Angiogenesis. *Curr. Drug Targets Inflamm. Allergy* **2005**, *4*, 3–8. [CrossRef]

173. Kofler, S.; Nickel, T.; Weis, M. Role of Cytokines in Cardiovascular Diseases: A Focus on Endothelial Responses to Inflammation. *Clin. Sci.* **2005**, *108*, 205–213. [CrossRef]

174. Gong, G.; Wang, H.; Kong, X.; Duan, R.; Dong, T.T.X.; Tsim, K.W.K. Flavonoids Are Identified from the Extract of Scutellariae Radix to Suppress Inflammatory-Induced Angiogenic Responses in Cultured RAW 264.7 Macrophages. *Sci. Rep.* **2018**, *8*, 1–11. [CrossRef]

175. Wu, W.-B.; Hung, D.K.; Chang, F.W.; Ong, E.T.; Chen, B.H. Anti-Inflammatory and Anti-Angiogenic Effects of Flavonoids Isolated from *Lycium Barbarum* Linnaeus on Human Umbilical Vein Endothelial Cells. *Food Funct.* **2012**, *3*, 1068–1081. [CrossRef]

176. Zhang, F.X.; Kirschning, C.J.; Mancinelli, R.; Xu, X.P.; Jin, Y.; Faure, E.; Mantovani, A.; Rothe, M.; Muzio, M.; Arditi, M. Bacterial Lipopolysaccharide Activates Nuclear Factor-KB through Interleukin-1 Signaling Mediators in Cultured Human Dermal Endothelial Cells and Mononuclear Phagocytes. *J. Biol. Chem.* **1999**, *274*, 7611–7614. [CrossRef]

177. Wu, S.H.; Liao, P.Y.; Dong, L.; Chen, Z.Q. Signal Pathway Involved in Inhibition by Lipoxin A4 of Production of Interleukins Induced in Endothelial Cells by Lipopolysaccharide. *Inflamm. Res.* **2008**, *57*, 430–437. [CrossRef]

178. Bachelot, T.; Ray-Coquard, I.; Menetrier-Caux, C.; Rastkha, M.; Duc, A.; Blay, J.Y. Prognostic Value of Serum Levels of Interleukin 6 and of Serum and Plasma Levels of Vascular Endothelial Growth Factor in Hormone-Refractory Metastatic Breast Cancer Patients. *Br. J. Cancer* **2003**, *88*, 1721–1726. [CrossRef]

179. Molostvov, G.; Morris, A.; Rose, P.; Basu, S.; Muller, G. The Effects of Selective Cytokine Inhibitory Drugs (CC-10004 and CC-1088) on VEGF and IL-6 Expression and Apoptosis in Myeloma and Endothelial Cell Co-Cultures. *Br. J. Haematol.* **2004**, *124*, 366–375. [CrossRef]

180. Fang, J.; Xia, C.; Cao, Z.; Zheng, J.Z.; Reed, E.; Jiang, B.-H. Apigenin Inhibits VEGF and HIF-1 Expression via PI3K/AKT/P70S6K1 and HDM2/P53 Pathways. *FASEB J.* **2005**, *19*, 342–353. [CrossRef]

181. Carmeliet, P.; Jain, R.K. Angiogenesis in Cancer and Other Diseases. *Nature* **2000**, *407*, 249–257. [CrossRef]

182. Zimna, A.; Kurpisz, M. Hypoxia-Inducible Factor-1 in Physiological and Pathophysiological Angiogenesis: Applications and Therapies. *Biomed. Res. Int.* **2015**, *2015*. [CrossRef]

183. Lu, K.H.; Chen, P.N.; Hsieh, Y.H.; Lin, C.Y.; Cheng, F.Y.; Chiu, P.C.; Chu, S.C.; Hsieh, Y.S. 3-Hydroxyflavone Inhibits Human Osteosarcoma U2OS and 143B Cells Metastasis by Affecting EMT and Repressing u-PA/MMP-2 via FAK-Src to MEK/ERK and RhoA/MLC2 Pathways and Reduces 143B Tumor Growth in Vivo. *Food Chem. Toxicol.* **2016**, *97*, 177–186. [CrossRef]

184. Liu, L.Z.; Fang, J.; Zhou, Q.; Hu, X.; Shi, X.; Jiang, B.H. Apigenin Inhibits Expression of Vascular Endothelial Growth Factor and Angiogenesis in Human Lung Cancer Cells: Implication of Chemoprevention of Lung Cancer. *Mol. Pharmacol.* **2005**, *68*, 635–643. [CrossRef]

185. Byun, E.B.; Kim, H.M.; Song, H.Y.; Kim, W.S. Hesperidin Structurally Modified by Gamma Irradiation Induces Apoptosis in Murine Melanoma B16BL6 Cells and Inhibits Both Subcutaneous Tumor Growth and Metastasis in C57BL/6 Mice. *Food Chem. Toxicol.* **2019**, *127*, 19–30. [CrossRef]

186. Deep, G.; Kumar, R.; Jain, A.K.; Agarwal, C.; Agarwal, R. Silibinin Inhibits Fibronectin Induced Motility, Invasiveness and Survival in Human Prostate Carcinoma PC3 Cells via Targeting Integrin Signaling. *Mutat. Res. Fundam. Mol. Mech. Mutagen.* **2014**, *768*, 35–46. [CrossRef]

187. Zhao, J.; Xu, J.; Lv, J. Identification of Profilin 1 as the Primary Target for the Anti-Cancer Activities of Furowanin A in Colorectal Cancer. *Pharmacol. Rep.* **2019**, *71*, 940–949. [CrossRef]

188. Deeks, J.J.; Higgins, J.P.; Altman, D.G. Analysing Data and Undertaking Meta-analyses. In *Cochrane Handbook for Systematic Reviews of Interventions*; John and Wiley and Sons: Hoboken, NJ, USA, 2019; pp. 241–284. [CrossRef]

189. Moher, D.; Shamseer, L.; Clarke, M.; Ghersi, D.; Liberati, A.; Petticrew, M.; Shekelle, P.; Stewart, L.A.; Estarli, M.; Barrera, E.S.A.; et al. Preferred Reporting Items for Systematic Review and Meta-Analysis Protocols (PRISMA-P) 2015 Statement. *Rev. Esp. Nutr. Hum. Diet.* **2016**, *20*, 148–160. [CrossRef]
190. Huedo-Medina, T.B.; Sánchez-Meca, J.; Marín-Martínez, F.; Botella, J. Assessing Heterogeneity in Meta-Analysis: Q Statistic or I 2 Index? *Psychol. Methods* **2006**, *11*, 193–206. [CrossRef]

Sample Availability: Samples of the compounds are not available.

Publisher's Note: MDPI stays neutral with regard to jurisdictional claims in published maps and institutional affiliations.

Article

Isorhamnetin Has Potential for the Treatment of *Escherichia coli*-Induced Sepsis

Anil Kumar Chauhan, Jieun Kim, Yeongjoon Lee, Pavithra K. Balasubramanian and Yangmee Kim *

Department of Bioscience and Biotechnology, Research Institute for Bioactive-Metabolome Network, Konkuk University, Seoul 05029, Korea; chauhananil48@konkuk.ac.kr (A.K.C.); za3524@konkuk.ac.kr (J.K.); lyj7956@konkuk.ac.kr (Y.L.); Pavithra@konkuk.ac.kr (P.K.B.)
* Correspondence: ymkim@konkuk.ac.kr; Tel.: +822-450-3421; Fax: +822-447-5987

Academic Editors: H.P. Vasantha Rupasinghe and Vincenzo De Feo
Received: 9 September 2019; Accepted: 31 October 2019; Published: 4 November 2019

Abstract: Isorhamnetin is a flavonoid that is abundant in the fruit of *Hippophae rhamnoides* L. It is widely studied for its ability to modulate inflammatory responses. In this study, we evaluated the potential of isorhamnetin to prevent gram-negative sepsis. We investigated its efficacy using an *Escherichia coli*-induced sepsis model. Our study reveals that isorhamnetin treatment significantly enhances survival and reduces proinflammatory cytokine levels in the serum and lung tissue of *E. coli*-infected mice. Further, isorhamnetin treatment also significantly reduces the levels of aspartate aminotransferase, alanine amino transferase and blood urea nitrogen, suggesting that it can improve liver and kidney function in infected mice. Docking studies reveal that isorhamnetin binds deep in the hydrophobic binding pocket of MD-2 via extensive hydrophobic interactions and hydrogen bonding with Tyr102, preventing TLR4/MD-2 dimerization. Notably, binding and secreted alkaline phosphatase reporter gene assays show that isorhamnetin can interact directly with the TLR4/MD-2 complex, thus inhibiting the TLR4 cascade, which eventually causes systemic inflammation, resulting in death due to cytokine storms. We therefore presume that isorhamnetin could be a suitable therapeutic candidate to treat bacterial sepsis.

Keywords: isorhamnetin; flavonoid; bacterial sepsis; toll-like receptor 4; inflammation

---

1. Introduction

Inflammation is an innate immune response against microbial infection. However, if it persists for a long time, it can cause several fatal diseases, including sepsis. Sepsis is a clinical condition defined as a systemic inflammation in the body in response to microbial infection, which eventually leads to multiple organ failure. Sepsis is reported to be the leading cause of mortality across the globe, and thus the World Health Organization (WHO) devoted the year 2017 to prioritize this disease and its treatment strategies [1–3]. The organs and systems that are commonly affected by sepsis include the lung, abdomen, blood and kidneys, with associated incidences of 64%, 20%, 15% and 14%, respectively [3]. Pathophysiological studies also reveal that the activation of toll-like receptors (TLRs) by microbes or microbial peptides represents the principle mechanism underlying the development of sepsis [3,4].

TLRs are transmembrane receptors present in major immune cells, and are composed of an extracellular leucine-rich repeat domain, a transmembrane region and an intracellular toll-interleukin-1 receptor (TIR) domain [5,6].

In total, 10 and 12 TLRs are identifiable in human beings and mice, respectively, and each of them recognizes specific pathogen-associated molecular patterns [7]. However, among these TLRs, TLR4 is responsible for the recognition of bacterial lipopolysaccharide (LPS) and the initiation of an immune response against LPS and/or gram-negative bacteria. Notably, TLR4 requires MD-2 to recognize LPS, and a TLR4–MD-2–LPS complex is essential to activate the TLR4 pathway in macrophages. Once TLR4 recognizes LPS, it undergoes an oligomerization process to recruit downstream adaptor molecules via interactions with the TIR domain. Five adaptor proteins that contain the TIR domain are identifiable, including MyD88 (myeloid differentiation primary response gene 88), TIRAP (TIR domain-containing adaptor protein, also known as Mal, MyD88-adapter-like), TRIF (TIR domain-containing adaptor inducing IFN-β), TRAM (TRIF-related adaptor molecule), and SARM (sterile α and HEAT-Armadillo motifs-containing protein) [8]. These adaptor proteins play crucial roles in the TLR4 pathway, which eventually results in the translocation of NF-κB from the cytoplasm to the nucleus to trigger the production of pro-inflammatory cytokines [7].

Flavonoids are naturally-occurring polyphenols that are found in a variety of edible plants. Chemically, they exist in glycosylated (conjugated with sugar) or aglycone forms (free form) [9]. The role of flavonoids in plants is crucial, as they confer infection resistance to the plant [10], in addition to protecting them from harmful ultraviolet rays [10,11]. Moreover, much research was carried out to explore the role of flavonoids in animal systems, and among the various activities shown by all types of flavonoids, their antioxidant potential is the most notable [12]. They are also reported to have other health-promoting effects, such as anti-inflammatory activity. For example, the cytoprotective efficacy of plant flavanol quercetins is widely explored [13,14]. Isorhamnetin (Figure 1), another plant flavanol, is a 3′-O-methylated metabolite of quercetin, and is found predominantly in the fruit of *Hippophae rhamnoides* L.; it exerts various biological effects, including anti-inflammatory [15], anticancer [16] and antioxidant activities [17]. Further, we previously reported that this flavonoid has potent anti-tuberculosis activity against *Mycobacterium tuberculosis* H37Rv and multi-drug- and extensively drug-resistant clinical isolates [18]. Importantly, the absorption and metabolic stability of methylated flavonoids are believed to be higher than those of unmethylated flavanols [16]. Thus, this property makes isorhamnetin more potent than quercetin.

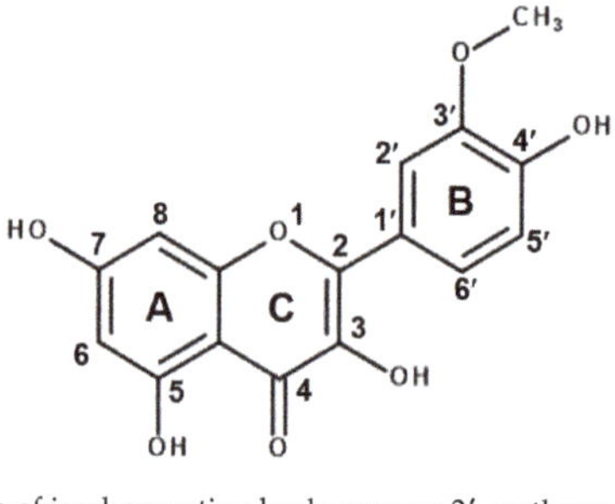

Figure 1. Chemical structure of isorhamnetin also known as 3′-methoxyquercetin (molecular weight: 316.26). The image was drawn using Chemdraw software.

In the present study, we explored the potential of isorhamnetin to protect against *Escherichia coli*-induced sepsis by establishing a murine model. Furthermore, based upon binding affinity and a molecular docking examination, we show for the first time, to our knowledge, that isorhamnetin can bind TLR4/MD-2 directly, thus preventing the activation of the TLR4 cascade, which is responsible for sepsis progression. The outcomes of our study might provide further insights into the natural flavonoid isorhamnetin and its role in sepsis prevention and treatment.

2. Results

2.1. Isorhamnetin Treatment Protects Mice from E. coli-Induced Sepsis

It was previously reported that isorhamnetin can reduce inflammation by downregulating nitric oxide (NO) and cytokine production [19]. Therefore, we sought to examine the ability of isorhamnetin to protect mice from *E. coli*-induced sepsis. To establish a bacterial sepsis model, we used a virulent

E. coli K1 strain to infect isorhamnetin-pretreated mice for 96 h for survival assays and 18 h to examine inflammation. As depicted in Figure 2A, mice in the isorhamnetin-treated group survived until the end of the experiment, but the *E. coli*-treated mice died within 24 h of infection. Further, we assessed the bacterial burden in the visceral organs of the mice (lung, liver, and kidneys) and found that isorhamnetin treatment significantly reduced the bacterial population in all organs compared to that in the *E. coli* only group (Figure 2B). Although isorhamnetin has no antibacterial activity in vitro, it is reported to enhance macrophage functions, such as phagocytosis, by increasing superoxide generation after the engulfment of bacteria [20]. Thus, we speculate that the reduction in bacterial loads in the organs is a result of the immunomodulatory potential of isorhamnetin.

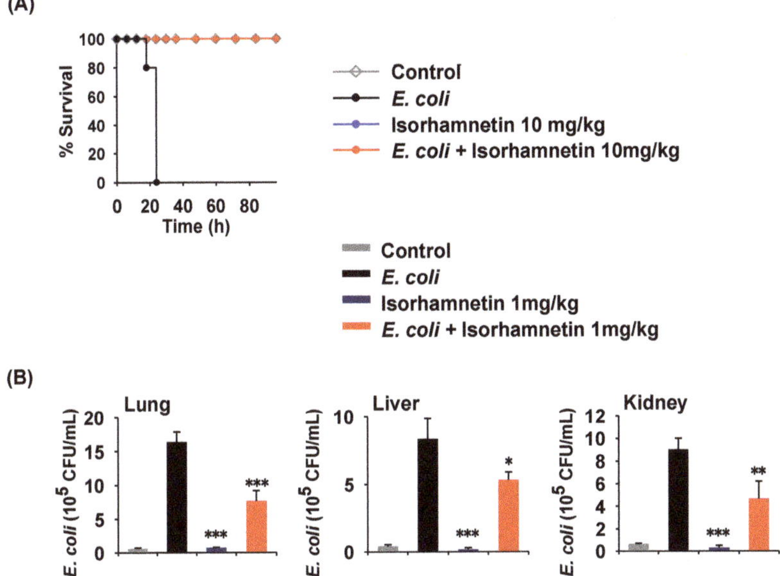

Figure 2. In vivo examination of the effects of isorhamnetin on *Escherichia coli*-induced sepsis. (**A**) Potential of this compound to promote the survival of mice in response to *E. coli*-induced sepsis. (**B**) Evaluation of the bacterial load in the visceral organs of mice. Data are presented as the means ± standard error of the mean (SEM). * $p < 0.05$; ** $p < 0.01$; *** $p < 0.001$ compared to the *E. coli* group.

2.2. Isorhamnetin Treatment Protects Mice from E. coli-Induced Inflammation

The effect of isorhamnetin on *E. coli*-induced myeloperoxidase (MPO) and proinflammatory cytokines (TNF-α and IL-6) was then examined. As depicted in Figure 3A,B, MPO activity is high in untreated mice infected with *E. coli*, whereas treatment with isorhamnetin decreased MPO activity in the kidney, liver and lung. However, the best results were obtained in the lung tissue (Figure 3B). Moreover, the secretion of cytokines is also highly increased in the sera and lung lysates after *E. coli* infection, which was drastically reduced in the isorhamnetin-treated group, suggesting that isorhamnetin can suppress the inflammatory response (Figure 3C,D).

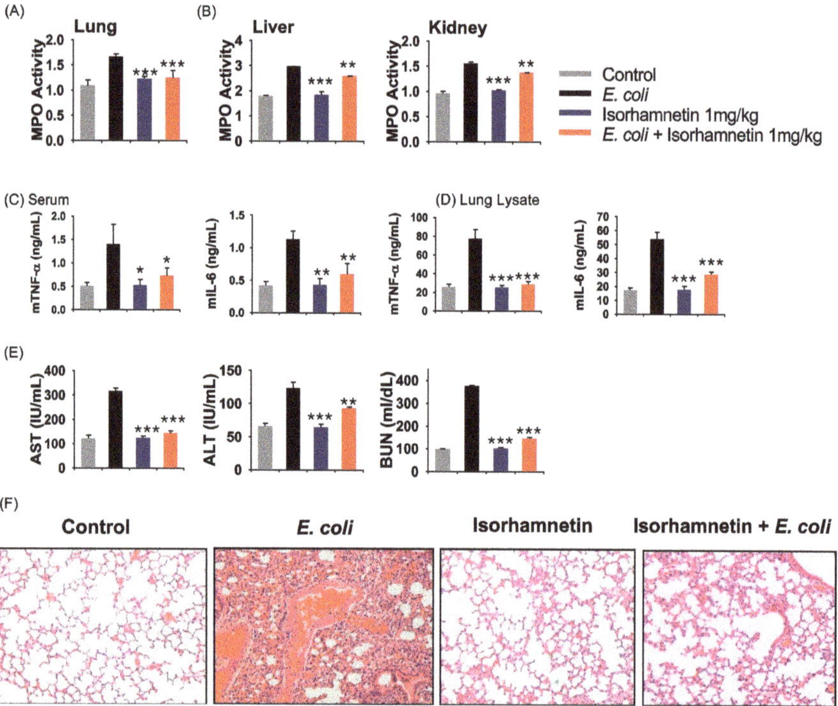

Figure 3. In vivo anti-inflammatory and toxicity evaluation of isorhamnetin. (**A,B**) Examination of myeloperoxidase (MPO) activity in lung, liver and kidney. (**C,D**) Production of cytokines (TNF-α, IL-6) in the sera and lung lysates. (**E**) Aspartate aminotransferase (AST), alanine amino transferase (ALT) and blood urea nitrogen (BUN) levels in the serum. (F) Hematoxylin and eosin staining of the lung tissue. Data are presented as the means ± SEM. * $p < 0.05$; ** $p < 0.01$; and *** $p < 0.001$ compared to the *E. coli* group.

Additionally, aspartate aminotransferase (AST), alanine amino transferase (ALT) and blood urea nitrogen (BUN) assays were carried out to evaluate liver and kidney function, and the results of these assays reveal that untreated mice infected with *E. coli* develop severe liver and kidney dysfunction (Figure 3E). However, isorhamnetin treatment protects the mouse liver and kidneys from *E. coli*-induced inflammation, as indicated by the reduced levels of AST, ALT and BUN (Figure 3E). Notably, we also examined the effect of isorhamnetin on uninfected mice to assess toxicity; here, we observed that this treatment does not alter the levels of AST, ALT and BUN, compared to those in the control group (Figure 3E), clearly indicating its safety. Moreover, we examined the histopathology of the lung tissues to assess neutrophil infiltration; the results indicate that isorhamnetin treatment could reduce the infiltration of neutrophils in the lung, which is noticeable after *E. coli* treatment (Figure 3F).

2.3. Interactions between Isorhamnetin and MD-2 Represent the Principal Mechanism Underlying the Anti-Inflammatory Activity

As we observed that isorhamnetin could successfully promote the survival of *E. coli*-infected mice and ameliorate inflammation, we next determined its probable mechanism of action. Isorhamnetin was previously reported to downregulate the TLR4 pathway [21], which has a crucial role in the development of inflammation. We therefore measured the binding affinity of isorhamnetin to MD-2 by surface plasmon resonance (SPR), as it is reported that MD-2 is responsible for binding hydrophobic LPS and other small molecules [22], which eventually induces the dimerization of TLR4/MD-2.

The binding affinity of isorhamnetin for MD-2 is very strong, specifically 5.41×10^{-6} M (association rate: 2.296×10^{2} M^{-1} s^{-1}; dissociation rate: 1.242×10^{-3} s^{-1}; Figure 4A).

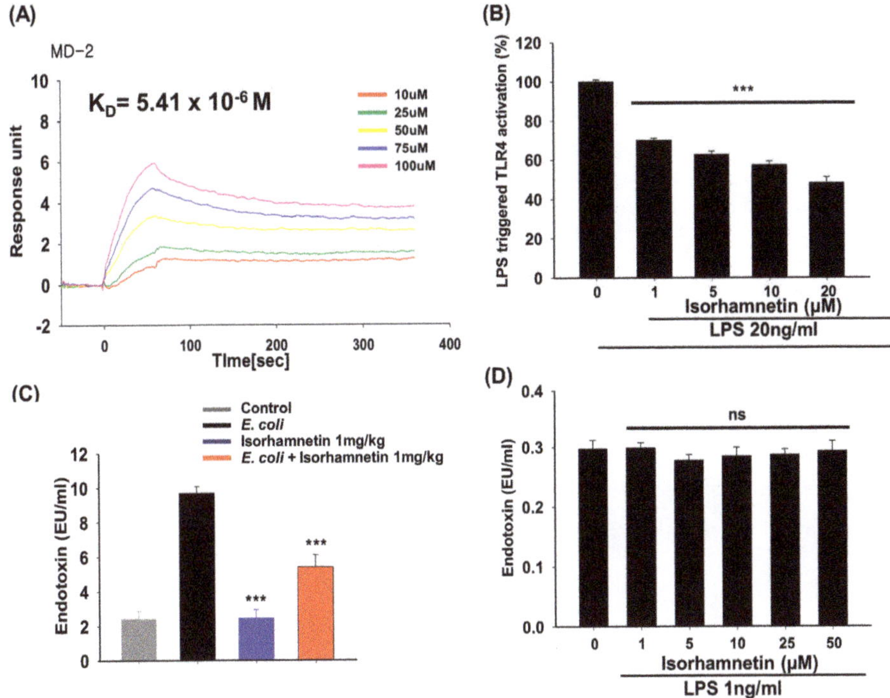

Figure 4. Effect of isorhamnetin on binding to MD-2, LPS-induced SEAP secretion and LPS neutralization (**A**) Surface plasmon resonance (SPR) analysis to determine the binding interaction. (**B**) Secretary alkaline phosphate (SEAP) assay in HEK-Blue™ hTLR4 cells (toll-like receptors (TLRs)). (**C**) Limulus amebocyte lysate (LAL) assay in serum of mice. (**D**) LAL assay to detect LPS neutralization in vitro.

Furthermore, we performed a SEAP (secreted embryonic alkaline phosphatase) assay using HEK-Blue™-hTLR4 cells to confirm the selectivity of isorhamnetin for the TLR4 receptor. SEAP is a protein secreted after the translocation of NF-κB from the cytosol to the nucleus; the amount of SEAP secreted into the medium correlates with the activation of the TLR4 pathway. Results of this experiment reveal that isorhamnetin could reduce SEAP activity in a concentration-dependent manner, with a 52% inhibition rate, indicating its specificity for the TLR4 receptor (Figure 4B).

We also carried out a Limulus amebocyte lysate (LAL) assay to examine the level of LPS in the mice serum and the in vitro LPS neutralizing efficacy of isorhamnetin. As, depicted in Figure 4C the level of LPS in the *E. coli* treatment group is significantly higher than those of the isorhamnetin treatment group ($p < 0.001$), suggesting that isorhamnetin treatment causes the reduction of the *E. coli* count, thus the level of LPS is decreased. However, the in vitro assessment of LPS neutralization reveals that isorhamnetin has no effect on the neutralization of LPS, as the level of LPS is similar in both the isorhamnetin treatment group as well as the LPS-only-treated group (Figure 4D), implying that isorhamnetin does not neutralize LPS, and instead it may give a competition to LPS in binding with MD-2.

2.4. Molecular Docking

To understand the interaction between isorhamnetin and MD-2 at the molecular level, a binding model of isorhamnetin to MD-2 was determined by a docking simulation. A model of the binding between isorhamnetin and MD-2 shows that isorhamnetin was inserted into the hydrophobic binding pocket of MD-2, which is a part of the LPS-binding site (Figure 5a). Isorhamnetin also exhibits extensive hydrophobic interactions with MD-2. Further, the 3′-OCH3 of isorhamnetin bound to the deep hydrophobic pocket of MD-2, forming hydrophobic interactions with Leu74 and Phe147. The B ring is also shown to form two pi–pi interactions with Phe104 and Phe147, whereas Tyr102 forms pi–pi interactions with the A ring. Furthermore, pi interactions were observed between the C ring of isorhamnetin and Leu61 and Ile117, with one between the B ring and I63, L71, L74 and V113, as shown in the two-dimensional plots of Figure 5b,c. Therefore, isorhamnetin displays extensive hydrophobic interactions with the hydrophobic pockets of MD-2, resulting in a tight binding affinity. Furthermore, 7–OH of isorhamnetin forms a hydrogen bond with Tyr102. It was reported that xanthohumol and curcumin also exhibit the same interactions with Tyr102 of MD-2 [23,24]. Docking studies and mutation studies of R90A and Y102A show that Arg90 and Tyr102 are the crucial residues required for the recognition process during inhibitor binding to the MD-2 protein [25–27]. These results confirm that Tyr102 might be the crucial residue required for binding between isorhamnetin and MD-2.

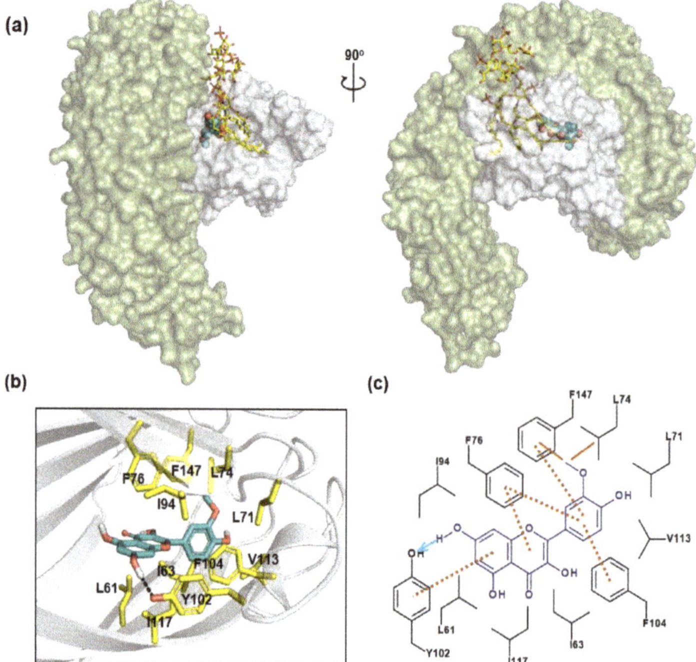

Figure 5. Binding interactions between isorhamnetin and MD-2. (**a**) Isorhamnetin (blue and red sphere) overlapped with LPS (yellow stick) in the hydrophobic binding cavity of MD-2 (gray) together with TLR4 (green), and its 90-degree rotated complex structure is shown. (**b**) Overview of the MD-2–isorhamnetin complex. Isorhamnetin is shown as sky blue sticks, and important residues in MD-2 (gray ribbon cartoon) are shown as yellow sticks. (**c**) Two-dimensional illustration of the MD-2–isorhamnetin docking pose showing hydrophobic interactions and hydrogen bonding interactions.

3. Discussion

Isorhamnetin, a natural flavonoid and metabolite of quercetin, has gained much scientific attention due to its broad range of biological activities. Specifically, it is reported to exert anti-inflammatory [19], anti-gastric cancer [28] and antioxidant [17,29] effects. In a recent study carried out by Yang et al., it is found that isorhamnetin can protect against LPS-induced acute lung injury by inhibiting cyclooxygenase-2 (COX-2) expression [30]. However, to the best of our knowledge, there have been no studies reporting the efficacy of isorhamnetin in protecting mice from bacterial sepsis. Therefore, we explored this aspect in the present study.

Natural products isolated from plants and herbs are categorized in complementary and alternative medicine, and are now replacing synthetic drugs. A report published by the National Health Interview Survey (NHIS) reveals that approximately 38% of adults in America use complementary and alternative medicine [31], which shows the importance of natural product research. Among these natural products, flavonoids have gained increasing attention due to their wide range of biological activities [9,11]. Flavonoids are reported to exert anticancer effects by inducing apoptosis in cancer cells and interfering with signal transduction pathways associated with metastasis [32]. Quercetin, a well-studied flavonoid found abundantly in apples, berries and *Brassica* vegetables, is reported to stabilize mast cells and protect the gastrointestinal tract; moreover, its potent anti-inflammatory and antioxidant activity was also reported [33]. However, the toxicity and efficacy of these compounds have always been an obstacle for their development as alternative medicines. Therefore, we examined the toxicity of isorhamnetin using a mouse model, in which isorhamnetin was administered alone without bacteria, and evaluated the AST, ALT and BUN levels. We found that this compound does not alter these biochemical parameters, suggesting that it is safe to use as a drug. In contrast, in our previous study, rhamnetin, another derivative of quercetin, was found to exhibit anti-inflammatory activity by binding to c-Jun NH2-terminal kinase 1 and p38 MAP kinase [34]; however, it exhibits cytotoxicity towards HEK293 cells. In another study, we report that tamarixetin, another derivative of quercetin, exhibits anti-septic activity similar to that of isorhamnetin [35].

The exaggerated production of cytokines is a major cause of systemic inflammation, often referred to as septic shock. Septic shock or sepsis is the most fatal condition, and here the mortality rate is predominantly high with very few treatment options [3]. As we observe that isorhamnetin protects macrophages from LPS-induced pro-inflammatory cytokine secretion, we next sought to examine the efficacy of isorhamnetin against *E. coli*-induced sepsis using a mouse model. We observed a significantly high survival rate in mice treated with isorhamnetin compared with that in the *E. coli* only group. Moreover, the bacterial load in the visceral organs (lung, liver and kidney) is greatly decreased in the isorhamnetin-treated group, suggesting that this compound protects mice from *E. coli* infection. Further, the secretion of IL-6 and TNF-α is reduced significantly in the sera and lung lysates from the isorhamnetin-treated mice, whereas this is predominantly high in the *E. coli* only group, suggesting that isorhamnetin suppresses *E. coli*-induced inflammation in mice.

To induce systemic inflammation or sepsis, bacteria or their byproducts target TLR signaling, and more specifically, the TLR4 pathway. Natural compounds, and precisely, flavonoids, represent principal candidates that target the TLR4 pathway. For example, epigallocatechin gallate, a polyphenolic flavonoid found in green tea, has been identified as a potent antagonist of the TLR4 pathway [36]. Therefore, flavonoids have always attracted research attention as valuable resources for the development of TLR4 antagonists due to their lack of toxicity and regular consumption in food. Curcumin, for example, has been explored for its various biological effects, including anti-inflammatory activity [37]. Further, cinnamic acid and its derivatives are known for their anti-inflammatory properties [25]. Similarly, cordycepin, a traditional oriental medicine from the *Cordyceps* species, is found to suppress LPS-induced inflammatory signaling by inactivating the MAPK and NF-κB pathways and inhibiting TLR4-mediated signaling [38].

In addition, curcumin, sulforaphane, 6-shagaol, glycyrrhizin, isoliquiritigenin, caffeic acid phenethyl ester, cinnamaldehyde, paclitaxel, morphine, naloxone, chitohexose and xanthohumol are natural small molecular inhibitors that target TLR4 activation and signaling [36]. Among these, xanthohumol and curcumin were further studied for their TLR4 antagonistic activity; it is reported that they interact with MD-2 with high affinity, thus interfering with the binding between LPS with MD-2, which eventually antagonizes TLR4 [23,39].

In this study, we examined the binding efficacy of isorhamnetin to MD-2 in order to regulate the TLR4/MD-2 interaction, which eventually leads to the induction of TLR4 pathway. We carried out the SEAP assay using HEK-Blue™-hTLR4 cells, as the translocation of NF-κB from the cytosol to the nucleus tends to release the SEAP protein in cell supernatant, and is directly associated with the activation of the TLR4 pathway. From this experiment, we observed that isorhamnetin treatment reduces the activation of TLR4 pathway as the secretion of SEAP protein is drastically reduced even at the 1 μM concentration, suggesting that isorhamnetin may inhibit TLR4-inflammatory signaling pathway.

Isorhamnetin, a flavonoid, has potent anti-inflammatory potential [21]; therefore, we sought to examine the target through which it executes its anti-septic activity. To examine this mechanism, we determined the binding affinity of isorhamnetin for MD-2, as the latter is responsible for TLR4/MD-2 dimerization, thereby activating the TLR4 pathway upon exposure to LPS [40]. The SPR method that we applied to examine binding reveals that isorhamnetin has micromolar binding affinity for the MD-2 protein. We further confirmed our results through docking simulations, where we observed that the hydrophobic region of MD-2 is responsible for the interaction with isorhamnetin, resulting in an antagonization of the TLR4 pathway. Especially, the O-methyl group of isorhamnetin is found to exhibit hydrophobic interactions with Leu74 and Phe147 of MD-2, and 7–OH is suggested to form a hydrogen bond with Tyr102 of this protein. Therefore, we confirmed that isorhamnetin binds directly to MD-2 in vitro and in silico.

Additionally, we carried out the LAL assay to examine whether isorhamnetin neutralizes the LPS, because that might also be another way to block the TLR4 pathway. However, our results of this in vitro LAL assay demonstrate that there is no neutralizing effect of isorhamnetin against LPS. On the other hand, there is significant difference in the amount of LPS present in the serum of the *E. coli*-infected group and the isorhamnetin-treated group, suggesting that isorhamnetin treatment exerts the reduction of the bacterial load in blood, and thus decreases the level of LPS. Taken together both in vitro and in vivo, LAL assay results indicate that isorhamnetin does not target LPS neutralization, instead it might give a competition to the LPS while binding with MD-2 in order to inhibit the TLR4/MD-2-LPS interaction which eventually downregulates the activation of the TLR4 pathway. Moreover, it is reported that treatment of isorhamnetin significantly decreases the expression of TLR4 and MyD88 in the LPS-treated cell [21]. Similarly, curcumin is reported to non-covalently interact with MD-2, and this interaction leads to competitive binding for LPS, resulting in the inhibition of TLR4/MD-2 dimerization, which is essential to initiate the TLR4 cascade [39,41]. Therefore, it can be suggested that isorhamnetin binds to MD-2, resulting in the inhibition of the TLR4/MD-2 dimerization, which may eventually downregulate the activation of TLR4 pathway-induced inflammation and sepsis.

In conclusion, this is the first study, to our knowledge, to report that isorhamnetin, a natural flavonoid, protects against *E. coli*-induced sepsis in a mouse model. Moreover, our study, for the first time, reports that from binding affinity, molecular docking and SEAP assay experiments, isorhamnetin directly interacts with MD-2, implying that this compound could be a potent therapeutic candidate for the treatment of TLR4-mediated inflammation and sepsis.

4. Materials and Methods

4.1. Chemicals and Biological Reagents

All chemicals, unless otherwise stated, were of the highest quality (this is a general statement for all chemicals used; we can not specify the exact purity) and were used as supplied. Isorhamnetin (95.0% purity) was purchased from Sigma-Aldrich (St. Louis, MO, USA). We further purified it to 99% purity using high performance liquid chromatography in Korea Basic Science Institute (Ochang,

South Korea). Compounds were dissolved in dimethyl sulfoxide (DMSO) to produce a 10-mg/mL stock solution.

4.2. Animals

Female BALB/c (six weeks old) mice were used for the *E. coli*-induced sepsis model. All mice used in this study were purchased from Orient (Daejeon, Korea) and were housed under specific pathogen-free conditions in a temperature- and humidity-controlled environment for one week prior to the experiments. All procedures were reviewed and approved by the Institutional Animal Care and Use Committee (IACUC) of Konkuk University, South Korea (IACUC number: KU18163-1).

4.3. SEAP Assay

HEK-Blue™-hTLR4 cells (toll-like receptors (TLRs)), obtained via the co-transfection of human TLR4, MD-2 and CD14 co-receptor genes and an inducible SEAP reporter gene into HEK293 cells, were purchased from Invitrogen (San Diego, CA, USA). HEK293 cells that stably expressed TLR4 contained a secreted alkaline phosphatase reporter gene (SEAP) located downstream of the *NF-κB* promoter. HEK-Blue™ hTLR4 cells were seeded in 96-well plates at 2.5×10^4 per well, using HEK-Blue detection media (Invitrogen; San Diego, CA, USA) and these were treated with isorhamnetin. After 1 h, LPS at 20 ng/mL was added for stimulation. After 16 h, SEAP production was determined based on the absorbance of the supernatant, measured at 630 nm, as described previously.

4.4. Measurement of Binding Affinity by SPR

The binding affinity between isorhamnetin and MD-2 was determined by SPR using a Biacore T100 (GE Healthcare, Sweden). Briefly, MD-2 protein was immobilized onto a CM5 chip using a standard EDS/NHS amine coupling method with sodium acetate (pH 5.0, acidic). MD-2 protein at 30 μg/mL was injected onto the surface of the CM5 chip at a resonance value of 2300. To analyze the kinetics, we used a running buffer composed of phosphate-buffered saline (PBS) (pH 8.0, slightly alkali), 0.05% Tween 20 and 1% DMSO. The chip was regenerated with 5 mM sodium hydroxide (NaOH) after the analysis of every sample. Equilibrium dissociation constant (K_D) values for the binding affinity were measured using a 1:1 binding assay with BIA Evaluation 2.0 software (GE Healthcare, CA, USA).

4.5. Molecular Docking

To analyze the TLR4/MD-2/isorhamnetin interaction, molecular docking was carried out using AutoDock Vina implemented by Yasara software (http://www.yasara.org) [42]. The crystal structure of the human TLR4/MD-2/*E. coli* LPS Ra complex was obtained from the protein data bank (PDB ID: 3FXI). The structure of isorhamnetin was acquired from the PubChem compound database (PCID:5281654), and the structure was subjected to energy optimization before docking. Water molecules and hetero atoms were removed from the protein, and polar hydrogen atoms were added. One hundred docking cycles with 500,000 evaluation steps were performed. Based on the cluster information, binding energy, molecular interactions such as hydrogen bonds and hydrophobic interactions, the best docked pose was selected.

4.6. Survival test for the sepsis mouse model

Survival tests were carried out as described previously [43]. Briefly, 20 BALB/c mice were used in each of the four groups. For the control group, mice were administered an intraperitoneal (i.p.) injection of PBS. For the second group, mice were administered an i.p. injection of 0.2 mL of *E. coli* (1×10^7 CFU/mouse). For the third group, mice were administered an i.p. injection of isorhamnetin (1 mg/kg) with 0.2 mL of *E. coli* (1×10^7 CFU/mouse). For the fourth group, mice were administered an i.p. injection of isorhamnetin (1 mg/kg) without *E. coli*. Survival in each group was observed for up to four days (0, 6, 12, 18, 24, 30, 36, 48, 60, 72, 84 and 96 h).

4.7. Cytokine Levels in the Serum and Lung Lysates in the Sepsis Mouse Model

Isorhamnetin (1 mg/kg) was injected 60 min before the *E. coli* (1×10^7 CFU/mouse) injection, and the mice were housed for 12 h. After 12 h, the mice were sacrificed, and blood and lung tissues were collected to examine inflammatory cytokines (TNF-α and IL-6) in the serum or lung lysates, using respective ELISA kits (R&D system), according to the manufacturer's instructions.

4.8. Detection of AST, ALT, and BUN in the mouse serum

The detection of aspartate aminotransferase (AST), alanine amino transferase (ALT) and blood urea nitrogen (BUN) in the serum was performed using a standard kit available from Asan Pharmaceutical, (Gyeonggi-do, South Korea) as per the manufacturer's instructions. The levels of AST, ALT and BUN were determined relative to a standard provided by the kit after subtracting the background levels. PBS was used as the negative control and a standard solution was used as the positive control to calculate the rates.

4.9. Determination of E. coli Counts in Organ Tissues

At the time of sacrifice, the lungs, liver and kidneys were removed aseptically and placed separately in 1 mL of sterile PBS. The tissues were then homogenized on ice using a tissue homogenizer under a vented hood. The lung, liver and kidney homogenates were diluted with PBS to 1:1000. After plating 10 μL of each diluted sample onto Luria Bertani (LB) agar, the plates were incubated at 37 °C for 24 h. We then counted the numbers of *E. coli* colonies, which were used to assess the relative abundances of *E. coli*.

4.10. LAL Assay

To determine the endotoxin concentration in mice serum we have carried out the LAL assay. The assay was carried out using Pierce LAL Chromogenic Endotoxin Quantification Kit (Thermo Scientific, Rockford, IL, USA) according to the manufacturer's instruction. Similarly, to examine the LPS neutralizing efficacy of isorhamnetin in vitro, we incubated the different concentrations of isorhamnetin (0, 1, 5, 10, 25 and 50 μM) with 1ng/mL of LPS at 37 °C in a 96-well plate which was then analyzed by a Pierce LAL Chromogenic Endotoxin Quantification Kit (Thermo Scientific, Rockford, IL, USA) according to the manufacturer's instruction.

4.11. Histopathological Examination

For histopathological examination, all samples were fixed in 10% formalin buffer and dehydrated with graded alcohol. The tissues were then embedded in paraffin blocks, and pathological sections were sliced along the longitudinal axis. From each sample, 5-mm thick sections were obtained, and staining with hematoxylin and eosin was performed to evaluate lung tissue morphology.

4.12. Statistical Analysis

All statistical analyses were carried out using GraphPad Prism software. Dunnett's multiple comparisons test (Prism 7.0, GraphPad Software Inc., La Jolla, CA, USA) was used for the comparisons of multiple groups. Values were considered statistically significant at $p < 0.05$. The error bars represent the mean ± standard error of the mean (SEM) (* $p < 0.05$, ** $p < 0.01$, and *** $p < 0.001$ compared to cells treated with agonist).

Author Contributions: Conceptualization, methodology, supervision, funding acquisition. Y.K.; investigation, data curation, A.K.C., J.K., Y.L., and P.K.B.; resources, Y.K.; writing—original draft preparation, Y.K., A.K.C., and P.K.B.; writing—review and editing, Y.K. and A.K.C.; visualization, Y.L.

Funding: This research was funded by the National Research Foundation of South Korea, grant number 2019R1H1A2079889.

Conflicts of Interest: The authors declare no conflict of interest.

References

1. Tiru, B.; DiNino, E.K.; Orenstein, A.; Mailloux, P.T.; Pesaturo, A.; Gupta, A.; McGee, W.T. The economic and humanistic burden of severe sepsis. *PharmacoEconomics* **2015**, *33*, 925–937. [CrossRef] [PubMed]
2. Fleischmann, C.; Scherag, A.; Adhikari, N.K.; Hartog, C.S.; Tsaganos, T.; Schlattmann, P.; Angus, D.C.; Reinhart, K.; International Forum of Acute Care Trialists. Assessment of global incidence and mortality of hospital-treated sepsis. Current estimates and limitations. *Am. J. Respir. Crit. Care Med.* **2016**, *193*, 259–272. [CrossRef] [PubMed]
3. Hotchkiss, R.S.; Moldawer, L.L.; Opal, S.M.; Reinhart, K.; Turnbull, I.R.; Vincent, J.L. Sepsis and septic shock. *Nat. Rev. Dis. Primers* **2016**, *2*, 16045. [CrossRef] [PubMed]
4. Boyd, J.H.; Russell, J.A.; Fjell, C.D. The meta-genome of sepsis: Host genetics, pathogens and the acute immune response. *J. Innate. Immun.* **2014**, *6*, 272–283. [CrossRef] [PubMed]
5. Ohto, U. Conservation and divergence of ligand recognition and signal transduction mechanisms in toll-like receptors. *Chem. Pharm. Bull.* **2017**, *65*, 697–705. [CrossRef]
6. Bell, J.K.; Mullen, G.E.; Leifer, C.A.; Mazzoni, A.; Davies, D.R.; Segal, D.M. Leucine-rich repeats and pathogen recognition in toll-like receptors. *Trends Immunol.* **2003**, *24*, 528–533. [CrossRef]
7. Lu, Y.C.; Yeh, W.C.; Ohashi, P.S. Lps/tlr4 signal transduction pathway. *Cytokine* **2008**, *42*, 145–151. [CrossRef]
8. O'Neill, L.A.; Bowie, A.G. The family of five: Tir-domain-containing adaptors in toll-like receptor signalling. *Nat. Reviews. Immunol.* **2007**, *7*, 353–364. [CrossRef]
9. Pérez-Cano, F.J.; Massot-Cladera, M.; Rodríguez-Lagunas, M.J.; Castell, M. Flavonoids affect host-microbiota crosstalk through TLR modulation. *Antioxidants* **2014**, *3*, 649–670. [CrossRef]
10. Corradini, E.; Foglia, P.; Giansanti, P.; Gubbiotti, R.; Samperi, R.; Lagana, A. Flavonoids: Chemical properties and analytical methodologies of identification and quantitation in foods and plants. *Nat. Prod. Res.* **2011**, *25*, 469–495. [CrossRef]
11. Monici, M.; Baglioni, P.; Mulinacci, N.; Baldi, A.; Vincieri, F.F. A research model on flavonoids as photoprotectors: Studies on the photochemistry of kaempferol and pelargonidin. *Acta Hortic.* **1994**, *381*, 340–347. [CrossRef]
12. Nijveldt, R.J.; van Nood, E.; van Hoorn, D.E.C.; Boelens, P.G.; van Norren, K.; van Leeuwen, P.A.M. Flavonoids: A review of probable mechanisms of action and potential applications. *Am. J. Clin. Nutr.* **2001**, *74*, 418–425. [CrossRef] [PubMed]
13. Lee, H.N.; Shin, S.A.; Choo, G.S.; Kim, H.J.; Park, Y.S.; Kim, B.S.; Kim, S.K.; Cho, S.D.; Nam, J.S.; Choi, C.S.; et al. Antiinflammatory effect of quercetin and galangin in lpsstimulated raw264.7 macrophages and dncbinduced atopic dermatitis animal models. *Int. J. Mol. Med.* **2018**, *41*, 888–898. [PubMed]
14. Cho, Y.H.; Kim, N.H.; Khan, I.; Yu, J.M.; Jung, H.G.; Kim, H.H.; Jang, J.Y.; Kim, H.J.; Kim, D.I.; Kwak, J.H.; et al. Anti-inflammatory potential of quercetin-3-o-beta-d-("2"-galloyl)-glucopyranoside and quercetin isolated from diospyros kaki calyx via suppression of map signaling molecules in lps-induced raw 264.7 macrophages. *J. Food Sci.* **2016**, *81*, C2447–C2456. [CrossRef]
15. Qi, F.; Sun, J.H.; Yan, J.Q.; Li, C.M.; Lv, X.C. Anti-inflammatory effects of isorhamnetin on lps-stimulated human gingival fibroblasts by activating nrf2 signaling pathway. *Microb. Pathog* **2018**, *120*, 37–41. [CrossRef]
16. Jaramillo, S.; Lopez, S.; Varela, L.M.; Rodriguez-Arcos, R.; Jimenez, A.; Abia, R.; Guillen, R.; Muriana, F.J.G. The flavonol isorhamnetin exhibits cytotoxic effects on human colon cancer cells. *J. Agr. Food Chem.* **2010**, *58*, 10869–10875. [CrossRef]
17. Choi, Y.H. The cytoprotective effect of isorhamnetin against oxidative stress is mediated by the upregulation of the nrf2-dependent ho-1 expression in c2c12 myoblasts through scavenging reactive oxygen species and erk inactivation. *Gen. Physiol. Biophys* **2016**, *35*, 145–154. [CrossRef]
18. Jnawali, H.N.; Jeon, D.; Jeong, M.C.; Lee, E.; Jin, B.; Ryoo, S.; Yoo, J.; Jung, I.D.; Lee, S.J.; Park, Y.M.; et al. Antituberculosis activity of a naturally occurring flavonoid, isorhamnetin. *J. Nat. Prod.* **2016**, *79*, 961–969. [CrossRef]
19. Antunes-Ricardo, M.; Gutierrez-Uribe, J.A.; Martinez-Vitela, C.; Serna-Saldivar, S.O. Topical anti-inflammatory effects of isorhamnetin glycosides isolated from opuntia ficus-indica. *Biomed. Res. Int.* **2015**, *2015*, 847320. [CrossRef]
20. Akbay, P.; Basaran, A.A.; Undeger, U.; Basaran, N. In vitro immunomodulatory activity of flavonoid glycosides from *Urtica dioica* L. *Phytother. Res.* **2003**, *17*, 34–37. [CrossRef]

21. Kim, S.Y.; Jin, C.-Y.; Kim, C.H.; Yoo, Y.H.; Choi, S.H.; Kim, G.-Y.; Yoon, H.M.; Park, H.T.; Choi, Y.H. Isorhamnetin alleviates lipopolysaccharide-induced inflammatory responses in bv2 microglia by inactivating nf-κb, blocking the tlr4 pathway and reducing ros generation. *Int. J. Mol. Med.* **2019**, *43*, 682–692. [CrossRef] [PubMed]

22. Wang, Y.; Shan, X.O.; Chen, G.Z.; Jiang, L.L.; Wang, Z.; Fang, Q.L.; Liu, X.; Wang, J.Y.; Zhang, Y.L.; Wu, W.C.; et al. Md-2 as the target of a novel small molecule, l6h21, in the attenuation of lps-induced inflammatory response and sepsis. *Br. J. Pharmacol.* **2015**, *172*, 4391–4405. [CrossRef] [PubMed]

23. Fu, W.T.; Chen, L.F.; Wang, Z.; Zhao, C.W.; Chen, G.Z.; Liu, X.; Dai, Y.R.; Cai, Y.P.; Li, C.L.; Zhou, J.M.; et al. Determination of the binding mode for anti-inflammatory natural product xanthohumol with myeloid differentiation protein 2. *Drug Des. Dev.* **2016**, *10*, 455–463.

24. Wang, Z.; Chen, G.Z.; Chen, L.F.; Liu, X.; Fu, W.T.; Zhang, Y.L.; Li, C.L.; Liang, G.; Cai, Y.P. Insights into the binding mode of curcumin to md-2: Studies from molecular docking, molecular dynamics simulations and experimental assessments. *Mol. Biosyst* **2015**, *11*, 1933–1938. [CrossRef]

25. Chen, G.; Zhang, Y.; Liu, X.; Fang, Q.; Wang, Z.; Fu, L.; Liu, Z.; Wang, Y.; Zhao, Y.; Li, X.; et al. Discovery of a new inhibitor of myeloid differentiation 2 from cinnamamide derivatives with anti-inflammatory activity in sepsis and acute lung injury. *J. Med. Chem.* **2016**, *59*, 2436–2451. [CrossRef]

26. Wang, Y.; Shan, X.; Dai, Y.; Jiang, L.; Chen, G.; Zhang, Y.; Wang, Z.; Dong, L.; Wu, J.; Guo, G.; et al. Curcumin analog l48h37 prevents lipopolysaccharide-induced tlr4 signaling pathway activation and sepsis via targeting md2. *J. Pharmacol. Exp. Ther.* **2015**, *353*, 539–550. [CrossRef]

27. Zhang, Y.; Wu, J.; Ying, S.; Chen, G.; Wu, B.; Xu, T.; Liu, Z.; Liu, X.; Huang, L.; Shan, X.; et al. Discovery of new md2 inhibitor from chalcone derivatives with anti-inflammatory effects in lps-induced acute lung injury. *Sci. Rep.* **2016**, *6*, 25130. [CrossRef]

28. Ramachandran, L.; Manu, K.A.; Shanmugam, M.K.; Li, F.; Siveen, K.S.; Vali, S.; Kapoor, S.; Abbasi, T.; Surana, R.; Smoot, D.T.; et al. Isorhamnetin inhibits proliferation and invasion and induces apoptosis through the modulation of peroxisome proliferator-activated receptor gamma activation pathway in gastric cancer. *J. Biol. Chem.* **2012**, *287*, 38028–38040. [CrossRef]

29. Pengfei, L.; Tiansheng, D.; Xianglin, H.; Jianguo, W. Antioxidant properties of isolated isorhamnetin from the sea buckthorn marc. *Plant. Foods Hum. Nutr* **2009**, *64*, 141–145. [CrossRef]

30. Yang, B.; Li, X.P.; Ni, Y.F.; Du, H.Y.; Wang, R.; Li, M.J.; Wang, W.C.; Li, M.M.; Wang, X.H.; Li, L.; et al. Protective effect of isorhamnetin on lipopolysaccharide-induced acute lung injury in mice. *Inflammation* **2016**, *39*, 129–137. [CrossRef]

31. Schneider, M. Complementary and alternative medicines during cancer treatments in the united states. *Oncologie* **2007**, *9*, 64–65. [CrossRef]

32. Karak, P. Biological activities of flavonoids: An overview. *Int. J. Pharm Sci Res.* **2019**, *10*, 1567–1574.

33. Li, Y.; Yao, J.Y.; Han, C.Y.; Yang, J.X.; Chaudhry, M.T.; Wang, S.N.; Liu, H.N.; Yin, Y.L. Quercetin, inflammation and immunity. *Nutrients* **2016**, *8*, 167. [CrossRef] [PubMed]

34. Jnawali, H.N.; Lee, E.; Jeong, K.W.; Shin, A.; Heo, Y.S.; Kim, Y. Anti-inflammatory activity of rhamnetin and a model of its binding to c-jun nh2-terminal kinase 1 and p38 mapk. *J. Nat. Prod.* **2014**, *77*, 258–263. [CrossRef] [PubMed]

35. Park, H.J.; Lee, E.J.; Cho, J.; Gharbi, A.; Han, H.D.; Kang, T.H.; Kim, Y.; Lee, Y.; Park, W.S.; Jung, I.D.; et al. Tamarixetin exhibits anti-inflammatory activity and prevents bacterial sepsis by increasing il-10 production. *J. Nat. Prod.* **2018**, *81*, 1435–1443. [CrossRef] [PubMed]

36. Peri, F.; Calabrese, V. Toll-like receptor 4 (tlr4) modulation by synthetic and natural compounds: An update. *J. Med. Chem.* **2014**, *57*, 3612–3622. [CrossRef]

37. Jayaprakasha, G.K.; Jagan, L.; Rao, M.; Sakariah, K.K. Chemistry and biological activities of c-longa. *Trends Food Sci. Tech.* **2005**, *16*, 533–548. [CrossRef]

38. Choi, Y.H.; Kim, G.Y.; Lee, H.H. Anti-inflammatory effects of cordycepin in lipopolysaccharide-stimulated raw 264.7 macrophages through toll-like receptor 4-mediated suppression of mitogen-activated protein kinases and nf-kappa b signaling pathways. *Drug Des. Dev.* **2014**, *8*, 1941–1953. [CrossRef] [PubMed]

39. Gradisar, H.; Keber, M.M.; Pristovsek, P.; Jerala, R. Md-2 as the target of curcumin in the inhibition of response to lps. *J. Leukoc. Biol.* **2007**, *82*, 968–974. [CrossRef]

40. Kang, J.Y.; Lee, J.O. Structural biology of the toll-like receptor family. *Annu. Rev. Biochem.* **2011**, *80*, 917–941. [CrossRef]

41. Molteni, M.; Bosi, A.; Rossetti, C. Natural products with toll-like receptor 4 antagonist activity. *Int. J. Inflamm* **2018**, 2859135. [CrossRef] [PubMed]

42. Kim, J.; Durai, P.; Jeon, D.; Jung, I.D.; Lee, S.J.; Park, Y.M.; Kim, Y. Phloretin as a potent natural TLR2/1 inhibitor suppresses tlr2-induced inflammation. *Nutrients* **2018**, *10*, 868. [CrossRef] [PubMed]

43. Kim, J.; Jacob, B.; Jang, M.; Kwak, C.; Lee, Y.; Son, K.; Lee, S.; Jung, I.D.; Jeong, M.S.; Kwon, S.H.; et al. Development of a novel short 12-meric papiliocin-derived peptide that is effective against gram-negative sepsis. *Sci. Rep.* **2019**, *9*. [CrossRef] [PubMed]

Sample Availability: Isorhamnetin can be purchased from Sigma-Aldrich (St. Louis, MO, USA).

 *molecules*

Article

Optimization of Total Phenolic and Flavonoid Contents of Defatted Pitaya (*Hylocereus polyrhizus*) Seed Extract and Its Antioxidant Properties

Siti Atikah Zulkifli [1], Siti Salwa Abd Gani [1,2,*], Uswatun Hasanah Zaidan [3] and Mohd Izuan Effendi Halmi [4]

[1] Halal Products Research Institute, Universiti Putra Malaysia, Putra Infoport, 43400 Serdang, Selangor, Malaysia; sitiatikahzulkifli@gmail.com
[2] Department of Agriculture Technology, Faculty of Agriculture, Universiti Putra Malaysia, 43400 Serdang, Selangor, Malaysia
[3] Department of Biochemistry, Faculty of Biotechnology and Biomolecular Sciences, Universiti Putra Malaysia, 43400 Serdang, Selangor, Malaysia; uswatun@upm.edu.my
[4] Department of Land Management, Faculty of Agriculture, Universiti Putra Malaysia, 43400 Serdang, Selangor, Malaysia; m_izuaneffendi@upm.edu.my
* Correspondence: ssalwaag@upm.edu.my; Tel.: +60-3-97694945

Academic Editor: H.P. Vasantha Rupasinghe
Received: 28 November 2019; Accepted: 6 January 2020; Published: 12 February 2020

Abstract: The present study was conducted to optimize extraction process for defatted pitaya seed extract (DPSE) adopting response surface methodology (RSM). A five-level central composite design was used to optimize total phenolic content (TPC), total flavonoid content (TFC), ferric reducing antioxidant power (FRAP), and 2,2′-azino-bis (3-ethylbenzothizoline-6-sulfonic acid (ABTS) activities. The independent variables included extraction time (30–60 min), extraction temperature (40–80 °C) and ethanol concentration (60%–80%). Results showed that the quadratic polynomial equations for all models were significant at ($p < 0.05$), with non-significant lack of fit at $p > 0.05$ and R^2 of more than 0.90. The optimized extraction parameters were established as follows: extraction time of 45 min, extraction temperature of 70 °C and ethanol concentration of 80%. Under these conditions, the recovery of TPC, TFC, and antioxidant activity based on FRAP and ABTS were 128.58 ± 1.61 mg gallic acid equivalent (GAE)/g sample, 9.805 ± 0.69 mg quercetin equivalent (QE)/g sample, 1.23 ± 0.03 mM Fe2+/g sample, and 91.62% ± 0.15, respectively. Ultra-high-performance liquid chromatography-quadrupole time-of-flight mass spectrometry (UPLC-QTOF/MS) analysis identified seven chemical compounds with flavonoids constituting major composition of the DPSE.

Keywords: defatted pitaya seed; extraction; phenolic content; flavonoid content; antioxidant activity; response surface methodology

---

1. Introduction

Pitaya, *Hylocereus polyrhizus*, or commonly known as dragon fruit in English or 'Buah Naga' in Malay, is oblong in shape with scaly structures on its outer peels. It belongs to the family, Cactacae from the genus *Hylocereous*. Pitaya originates from Central and Northern South America and presently has become a recent fruit crop of interest being cultivated in Malaysia, Thailand, Vietnam, Australia, Taiwan, and other regions around the world. Other than appealing appearance of the fruit, pitaya has recently gained much attraction due to its high dietary contents [1,2]. The flesh of the fruit is tasty and sugary with numerous tiny and grainy black seeds. Pitaya fruit seeds contain oil like most other fruit seeds such as seeds of grape [3], linseeds [4] and berry including blackberry, blueberry and red raspberry seeds) [5]. Seed oil from pitaya fruit has been successfully extracted and reported to be potentially suitable in food, health and cosmetic applications. However, biological activities of polyphenols extracted from defatted pitaya seed (DPS) as a byproduct after extraction, still remains unexplored.

Polyphenols or phenolic compounds are plants' secondary metabolites produced to shield plants from ultraviolet rays or damaging insect pests or disease pathogens. Present in plant tissues, they play important roles in plants' nutrients assimilation, protein synthesis, enzyme activities, photosynthesis, cell signaling, and protection against adverse environmental conditions [6]. Flavonoids are one of the important classes of natural compounds which encompass the largest group of plant polyphenols. The compounds are made up of fifteen carbon atoms having common basic structures of two aromatic rings bound together by three carbon atoms. This group of compounds is classified as low molecular weight compared to other groups of polyphenols that are widely distributed in plants. There are several classes of flavonoids, such as flavonols, flavones, flavanones, flavanols, isoflavones, and flavanonols.

Flavones and flavonols are the most abundantly present in most plants [7]. A number of studies have reported medicinal features of this group of polyphenols such as anticancer [8], anti-inflammatory, antimicrobial and antioxidant activities [9]. The main mechanism for these activities comprises of the reduction of oxidative stress by scavenging free radicals [10]. It had been documented that the mechanism was either by transferring electron to complete the radicals compound or by breaking them down to make them safer and become more stable [11]. A number of extensive research works had been conducted to extract phenolic and flavonoid compounds from natural sources for their potentially high therapeutic properties.

Extraction process is the primary stage to source crude extract of bioactive compounds from plant materials. Various plant materials require distinct extraction conditions and procedures to yield optimum retrieval of phenolic compounds as every single plant has distinctive characteristics in terms of phenolic constituents [12]. Several factors have been proven to significantly affect extraction yield such as extraction methods, particle size, solvent types, solvent concentrations, solvent-to-solid ratios, extraction temperatures, extraction times, and pH levels [13–15]. One-variable-at-a-time approach is a traditional method of analyzing extraction optimization conditions by changing only one factor at a time while keeping others set at constant values. However, this approach is time-consuming and required a large number of experiments and materials. Moreover, interactive effect between the factors studied cannotbe determined. Thus, experimental design has been used to overcome these problems.

Experimental design is a systematic approach to apply statistical methods in experimental processes for various area of academic research and industry. The most common statistical techniques used to optimize the process is response surface methodology (RSM). RSM is an integrated statistical and mathematical technique to determine the relationship between the independent variables and dependent variable based on experimental design. That is, RSM has become an important tool that is employed for modeling by identifying the influence of several factors on the process. This method was first developed by Box and Wilson [16] to optimize the chemical process and now has been extensively adopted in various field including electronics, biotechnology, aerospace, automotive, life sciences, agricultural settings, and process industries [17]. The advantages offered by the RSM including reduced number of experimental trials, calculating the complex interaction between the independent variables, analysis, and optimization as well as the enhancement of existing design. Hence, optimization of an extraction process using RSM is important to examine suitable conditions to isolate bioactive compounds especially from different food matrices [18].

Thus, the present study was undertaken to optimize extraction conditions for higher total phenolic, flavonoids recovery, as well as determine in-vitro antioxidant activities of DPS extract using response surface methodology (RSM). Polyphenolic compounds from defatted seed extract of pitaya were identified by ultra-high-performance liquid chromatography-quadrupole time-of-flight mass spectrometry (UPLC-QTOF/MS).

2. Results and Discussion

2.1. Fitting Model

Preference for the ideal extraction process which influenced better yield of total phenolic content (TPC), total flavonoid content (TFC), and antioxidant activity (ferric reducing antioxidant power (FRAP) and ABTS activity) of defatted pitaya seed extract (DPSE) was accomplished through response surface methodology. In the present study, the highest order polynomials were selected to choose the models where additional terms were significant and the models were not aliased in accordance to the sequential model sum of square. As suggested by the software, a quadratic polynomial model was chosen and well-fitted for all three independent parameters and responses [19]. The final predictive equations generated by the software were expressed in terms of coded factors as shown in Equations (1)–(4) where the empirical correlation between time of extraction (A) temperatures (B) and ethanol concentrations (C) were established:

$$Y(TPC) = + 109.53 + 2.66A + 17.60B + 9.23C - 13.83AB - 0.36AC - 3.96BC - 9.04A^2 - 7.93B^2 + 4.22C^2 \quad (1)$$

$$Y(TFC) = + 7.62 + 0.22A + 0.71B + 1.88C - 0.45AB - 0.97AC - 0.28BC - 0.24A^2 + 0.11B^2 - 0.18C^2 \quad (2)$$

$$Y(FRAP) = + 1.23 + 0.1152A + 0.2729B - 0.0245C + 0.0375AB - 0.1525AC$$
$$- 0.0600BC - 0.0580A^2 - 0.0050B^2 - 0.0615C^2 \quad (3)$$

$$Y(ABTS) = + 91.85 + 3.82A + 3.69B + 0.8471C - 4.00AB - 2.37AC - 2.29BC - 2.25A^2 - 1.77B^2 + 0.2186C^2 \quad (4)$$

The significance of the model was examined through analysis of variance (ANOVA) where a large F-value and small p-value of each term in the models implied a more significant impact on the respective response variables [20]. The ANOVA for second order polynomial model of TPC, TFC, and FRAP had shown that the model was significant ($p < 0.05$) with a p-value of < 0.0001 for TPC, TFC, FRAP, and ABTS activities (Table 1). Meanwhile, the coefficient of determination (R^2) was employed to evaluate the quality of the fit quadratic model which were 0.9418, 0.9817, 0.9742, and 0.9840 for phenolic and flavonoid content, frap value, and ABTS activity respectively. This suggests that only 5.82%, 1.88%, 2.58%, and 1.60% of the total variations for TPC, TFC, FRAP, and ABTS activities respectively could not be explained by the model. The fitness of the model was identified through lack-of-fit test ($p > 0.05$), which indicated suitability of models to accurately predict the variations [21]. The coefficient variation (CV) is a measure of relative variability from the mean value, claimed that the model was reliable. In general, a CV of less than 10 percent revealed smaller variation in the mean value and indicated a better precision and reproducibility [22]. The CV values obtained for TPC, TFC, FRAP, and ABTS were 7.62, 4.62, 5.57, and 1.21 respectively, demonstrating the high reliability and accuracy of the model.

2.2. Analysis of Response Surface

Response surface methodology was used to estimate the recovery of TPC, TFC, FRAP, and ABTS activities based on varied values of tested factors, while the contours of the plots were used to establish their correlative interactions between the variables involved. According to [23], the best way to figure out the influence of the independent variables on the dependent variables are through drawing surface response plots of the model. The significance or not the mutual interaction between the factors can be determined by several shapes developed from the contour plots. A circular contour plot indicates the interactions between the corresponding parameters are negligible, while an elliptical contour or saddle nature implies the interactions between the corresponding parameters are significant [24]. The three-dimensional response surface and two-dimensional contours designed by the fitted model are demonstrated in Figures 1 and 2. Each diagram shows the effect of two variables on yield of TPC, TFC, FRAP, and ABTS activities while holding other variables at their zero level.

2.3. Optimum Extraction Condition Based on TPC

The recovery yield of total phenolic content, TPC from defatted pitaya seed extract ranged from 52 to 144 mg GAE/g sample, which were higher than from kiwi fruit seeds [25] and defatted marigold *Tagetes erecta* L. residues [26]. In the present study, the mean recorded value was 100.83 mg GAE/g sample of the total extracts. The maximum yield of phenolic content was recorded for Experiment No. 18 whereas the lowest yield of flavonoid content discovered from Experiment No. 7. The ANOVA of the regression coefficient indicated that the two linear parameters, temperature (B) and ethanol concentration (C) were significant at ($p < 0.0001$ and $p < 0.01$) respectively. The quadratic (A^2 and B^2) and interactive effects between extraction time and temperatures (AB) were also significant ($p < 0.05$) on yield of total phenolic content.

Figure 1a shows the interactions between extraction time and temperature on total phenolic contents at concentration of solvent fixed at 70%. It was observed that extraction efficiency was simultaneously enhanced with increase in extraction temperature and time. The recovery of phenolic contents increased rapidly with increase in extraction temperature during a shorter extraction time, while there was a slight increase in phenolic content with increase in extraction time at a higher extraction temperature. Under normal conditions, the temperature had constructive effect on extraction of phenolic compounds from plant sources [27–29]. The phenomenon could be clarified that higher temperature stimulates higher solubility of phenolic compounds in the extraction solvent. Similar trend on effect of higher temperature enhanced extraction efficiency of TPC was reported by [30,31] on an Indian medicinal plant and annatto seeds respectively. Higher temperature increases diffusion of extracted molecules, reduces its viscosity as well as improves mass transfer [32]. High temperatures from solvent have been reported to increase permeability of cell walls by breaking down interaction between phenolic compounds and macromolecules (proteins, polysaccharides) and thus facilitates recovery of phenolic yield in extract.

The interaction between extraction time and ethanol concentration (AC) revealed a significant ($p < 0.05$) positive effect on TPC as shown in Figure 1c. Yield of TPC gradually increased with increases in extraction time and ethanol concentration from 30–90 min and 60–80 °C respectively. However, prolonged extraction time did not significantly improve recovery of TPC. This occurrence could be explained by Fick's second law of diffusion which stated that a final equilibrium is accomplished between the concentration of the solute in the solid matrix (plant matrix) and in bulk solution (solvent) after certain time [15]. Hence, a longer time does not necessarily extract more phenolic compounds. Extending extraction times might contribute to increase risk of phenolic oxidation unless reducing agents are added to the solvent system [33].

The effects of temperature and ethanol concentration at constant extraction time of 60 min increased recovery of TPC as shown in Figure 1e. The effect was probably due to the change in solvent polarity with the addition of certain amount of water to the solvent. Ethanol facilitated increase in TPC recovery by disrupting the bonding between the solutes and plant matrices, while water could enhance the swelling of cell material [34]. Previous findings reported that binary solvent system demonstrated higher yield of phenolic compounds and flavonoids as compared to mono-solvent system containing pure solvent or pure water [12]. Similar effects of this variables have also been reported for phenolic extraction from *Phaleria macrocarpa* (Scheff) Boerl fruits [22].

2.4. Optimum Extraction Condition Based on TFC

The recovery yield of total flavonoid content, TFC from defatted pitaya seed extract ranged from 2.69 to 10.67 mg QE/g sample, which were higher than grape byproducts [35]. The mean recorded value was 7.40 mg QE/g sample of total pitaya seeds extracts. Maximum yield of phenolic content was recorded from Experiment No. 15, whereas the lowest yield of flavonoid content was recovered from Experiment No. 7. The ANOVA of the regression coefficient indicated that all three linear parameters, (A, B, C), interaction parameters (AB, AC, BC) were significant at ($p < 0.05$). Meanwhile, only one quadratic effect A^2 was significant at ($p < 0.05$) on yield of total flavonoid content. A 3D-surface plot of interaction between time (A) and temperature (B) at the fixed ethanol concentration of 70% is as shown in Figure 1b. The figure shows that TFC slightly increased with increase in extraction temperature. TFC increased until reaching an optimum temperature of 80 °C, similar with TPC. The effects of temperature on flavonoids extraction had previously been reported by various other researchers. For example, the highest flavonoid recovery was reported from *Flos populi* using solid-liquid extraction at temperature of 94.66 °C [36]. The achievement was supported by [30] and [37] who documented that elevated temperature enhanced better extraction yield. The effect was attributed to the fact that higher temperature disrupted the structure of plant matrix by weakening the phenolic matrix bonds which then increased the solubility of flavonoids [38]. Nevertheless, results of the present study demonstrated that yield of TFC decreased in long extraction time with increasing ethanol concentration as demonstrated in Figure 1d. Extraction time appeared to be one of the primary factors influencing an extraction process. It is therefore important in attempt at reducing energy cost in extraction procedure and in inhibiting the decomposition of active compounds. Extraction time can possibly be as short as few minutes or extended for up to 24 h [39]. This has been reported to be dependent on the extraction process phase, either rapid phase or slow phase [39]. Rapid phase is explained by the fact that solutes are present on surface sites of plant materials, and a slow phase corresponds to the molecular diffusion of the solute from internal sites through pores [29,40]. The present study achieved contrary to that recorded by [41] in extracting flavonoid from red and brown rice bran who observed an increase in flavonoid content with increase in time. These distinctions in time of extraction could be correlated to the nature of the sample (seed, leaf, rhizome, or bark), particle size, solvent type, and extraction approaches [29,42].

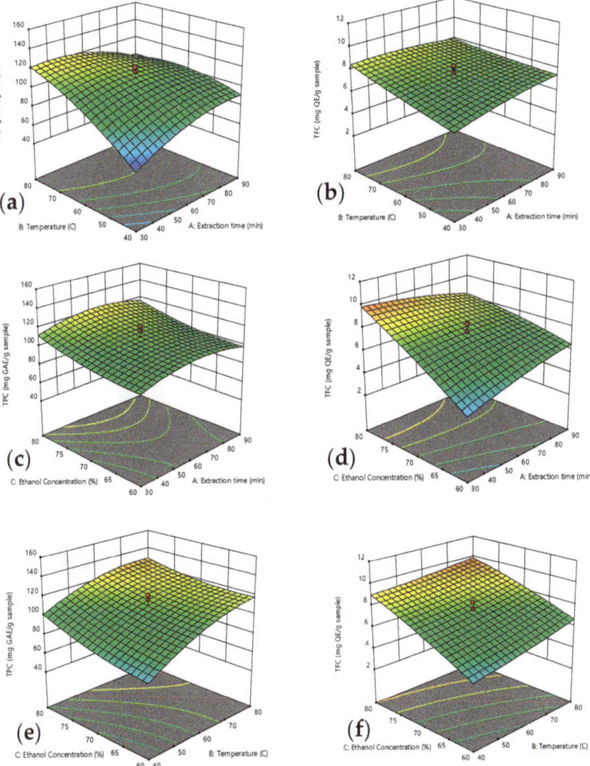

Figure 1. Response surface for: (**a**,**b**) effect of extraction time and temperature; (**c**,**d**) effect of extraction time and ethanol concentration, and (**e**,**f**) effect of temperature and ethanol concentration, on total phenolic content (TPC) (**a**,**c**,**e**) and TFC (**b**,**d**,**f**).

Table 1. Analysis of Variance (ANOVA) of regression equation for optimization of TPC, TFC, FRAP, and ABTS.

Source	TPC				TFC				FRAP				ABTS	
	dF[a]	SS[b]	F-Value	p-Value	dF[a]	SS[b]	F-Value	p-Value	dF[a]	SS[b]	F-Value	p-Value	dF[a]	SS[b]
Model	9	9533.98	17.97	<0.0001	9	67.3	63.87	<0.0001	9	1.53	41.97	<0.0001	9	722.71
A	1	96.90	1.64	0.2288	1	0.6602	5.64	0.039	1	0.1811	44.82	<0.0001	1	199.29
B	1	4231.34	71.76	<0.0001	1	6.95	59.39	<0.0001	1	1.02	251.6	<0.0001	1	185.83
C	1	1163.51	19.73	0.0013	1	48.33	412.77	<0.0001	1	0.0082	2.03	01849	1	9.8
AB	1	1530.98	25.96	0.0005	1	1.66	14.15	0.0037	1	0.0112	2.78	0.1262	1	128.24
AC	1	1.00	0.0170	0.8989	1	7.57	64.63	<0.0001	1	0.1861	46.04	<0.0001	1	44.79
BC	1	125.37	2.13	0.1755	1	0.6161	5.26	0.0447	1	0.0288	7.13	0.0235	1	42
A^2	1	1178.75	19.99	0.0012	1	0.8306	7.09	0.0238	1	0.0485	11.99	0.0061	1	72.79
B^2	1	906.72	15.38	0.0029	1	0.1742	1.49	0.2505	1	0.0004	0.0874	0.7736	1	45.34
C^2	1	256.80	4.36	0.0635	1	0.4853	4.15	0.0691	1	0.0545	13.5	0.0043	1	0.6885
Residual	10	589.64			10	1.17			10	0.0404			10	11.73
Lack of fit	5	323.64	1.22	0.4174	5	0.4582	0.6431	0.6801	5	0.0129	0.4677	0.788	5	9.07
R^2		0.9418				0.9829				0.9742				0.9840
R^2_{Adj}		0.8893				0.9675				0.9510				0.9696
CV		7.62				4.62				5.57				1.21

dF[a]: degree of freedom; SS[b]: sum of square, TPC: total phenolic content; TFC: total flavonoid content; FRAP: ferric reducing antioxidant power (antioxidant activity); ABTS: 2,2′-azino-bis (3-ethylbenzothizoline-6-sulfonic acid (anti-oxidant activity).

In the present study, the interaction effects between temperature (B) and ethanol concentration (C) show that TFC was found to be higher at higher values of variables as shown in Figure 1f. The maximum TFC recorded was when 80% ethanol concentration was used, compared to use of 70% ethanol with increasing extraction time. A general principle in solvent extraction is based on the law of similarity and intermiscibility "like dissolves like", which means that solvents extract phytochemicals with a polarity value near to the polarity of the solvent [43]. The results suggest that the solvent polarity and the solubility of flavonoids compounds in pitaya seed extract were similar.

2.5. Optimum Extraction Condition on Antioxidant Activity

2.5.1. Optimum Extraction Condition Based on FRAP Activity

The reducing power activity of DPSE ranged from 0.53 to 1.75 mM Fe^{2+}/g sample. The mean recorded value was 1.14 mg mM Fe^{2+}/g sample of total DPSE. The highest value was recorded from Experiment No. 20 whereas the lowest value was recorded from Experiment No. 7. The ANOVA on regression coefficient revealed that the two linear parameters, time (A) and temperature (B), interactive effect between time and ethanol concentration (AC) were significant at ($p < 0.0001$). Meanwhile, interaction parameter, (AB) and quadratic parameters (A^2, B^2) were significantly influenced at ($p < 0.05$). Similar effects of temperature have been previously recorded by other researchers on phenolic content of grape cane extracts which subsequently resulted in varied antioxidant activities [44]. The effects of mutual interaction between extraction parameters on the FRAP value of phenolic extract can be seen in Figure 2.

As shown in Figure 2a, the FRAP values increased as extraction time was prolonged up to 90 min. This patterns of responses were probably due to the fact that increasing extraction time provided longer contact of solids with the solvent and this enhanced the diffusion of phenolic compounds linked to the antioxidant activities [45]. Higher extraction time and medium ethanol concentration led to increase in FRAP values. However, extraction time was did not significantly influence total phenolic contents in defatted pitaya seed extracts which were in agreement with previous other reports by [46–48].

Figure 2c illustrates the interactive effect between extraction time and ethanol concentration. When ethanol concentrations were at 60%–80% and the extraction time at 30–90 min, the recorded value of FRAP increased initially and later decreased as the ethanol concentration was increased. This occurrence can be correlated to the changes in polarity of the compound that was responsible for reducing power activity. Therefore, the maximum value for reducing power could be obtained when the lowest ethanol composition was used. It has been reported that the polarity of the solvent used in extraction directly affects not only the quantity of total phenolic compounds, but also the composition and potency of the phenolic compounds as antioxidants [49–51]. Consequently, differences in extracted phenolic compounds result in varied antioxidant activity of the extracts.

The enhancement of FRAP activity significantly influenced by temperature. The FRAP value recorded in the present study gradually increased as extraction temperature was increased over time as shown in Figure 2e. The results implied that an increase in temperature led to an increase in antioxidant activity. This occurrence was probably due to the fact that at low temperatures, the rate of mass transfer was also low and additional time was needed for the phenolic compounds responsible for FRAP activity to dissolve in the solvent. Similar interactive effect showing similar response curve was also recorded by [52] on FRAP activity on tomatoes.

2.5.2. Optimum Extraction Condition Based on ABTS Activity

The ABTS activity of the DPSE ranged from 69.79%–94.34%. The mean value recorded was 89.26% which was higher than ABTS activity reported for methanolic extract of *Adiantum caudatum* leaves [53]. The highest value was recorded for Experiment No. 20 whereas the lowest value was recorded from Experiment No. 1. The ANOVA of the regression coefficient revealed that the two linear parameters, time (A) and temperature (B), interactive effect between time and temperature (AB) and quadratic parameters (A^2, B^2) were significant at ($p < 0.0001$). Meanwhile, ethanol concentration (C), interaction parameter, (AC) and (BC) were significantly influenced at ($p < 0.05$). The effect between independent variables can be seen in Figure 2.

Figure 2b shows increases of ABTS activity of DPSE with increases in extraction time and temperature from 30–60 min and 40–60 °C respectively. Prolonged extraction time for up to 90 min and rise in temperature to 80 °C slightly decreased ABTS activity. The data revealed that the interactive effect between extraction time (A) and temperature (B) on ABTS activity was significant and in good agreement with the results shown in Table 1. Similar interaction was also recorded by [54] on ABTS activity of aqueous extract of *Schizophyllum commune*. Increase in ethanol concentration from 60–70% caused an increase in ABTS activity of DPSE (Figure 2d). This occurrence was probably due to the polarity of the solvent used coincided with the solubility of the phenolic compounds responsible for ABTS activity [22]. In the present study, the highest ABTS activity was recorded when DPS was extracted using 60% ethanol concentration. Similar trend was also obtained by [55], where 60% of ethanol concentration yielded the highest ABTS radical scavenging activity of *Dendropanax morbifera* (*D. morbifera*) Levillis leaves. Figure 2f shows the relationship between temperature and ethanol concentration on ABTS activity at a constant extraction time of 60 min. The data obtained revealed that ABTS activity was significantly influenced by increment in extraction temperature from 40 to 60 °C. This circumstance can be explained by the fact that solubility of solute and extraction efficiency were enhanced by increasing extraction temperature [56].

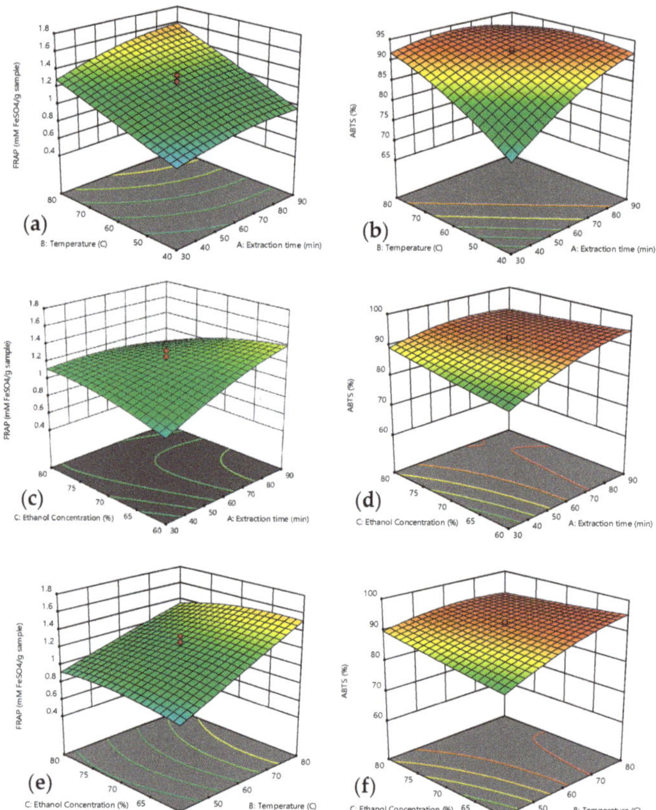

Figure 2. Response surface for: (**a**,**b**): effect of extraction time and temperature; (**c**,**d**): effect of extraction time and ethanol concentration; and (**e**,**f**): effect of temperature and ethanol concentration, on FRAP (**a**,**c**,**e**) and ABTS (**b**,**d**,**f**) activity.

2.6. Optimization of Extracting Conditions for TPC, TFC, FRAP, and ABTS Activity

The numerical optimization for highest recovery of TPC, TFC, FRAP, and ABTS activity of DPSE were determined. The simultaneous optimization using desirability function approach proposed that the optimal extraction conditions for DPSE were at 45 min extraction time, 70 °C extraction temperature and 80% ethanol concentration at desirability of 93.8%. Therefore, extraction of DPS was performed based on the suggested extraction conditions and the data obtained were statistically compared with the predicted values given by the Design Expert 11.0 Software. Table 2 shows results from verification experiment which were in close agreement with predicted values at 93.8% confidence level.

Table 2. Predicted and Experimental Values for Responses of TPC, TFC, FRAP, and ABTS.

	TPC (mg GAE/g Sample)	TFC (mg QE/g Sample)	FRAP (mM Fe^{2+}/g Sample)	ABTS (%)
Predicted	129.75	9.995	1.24	92.87
Experimental	128.58 ± 1.61	9.805 ± 0.69	1.23 ± 0.03	91.62 ± 0.15

TPC: total phenolic content; TFC: total flavonoid content; FRAP: ferric reducing antioxidant power (antioxidant activity); ABTS: 2,2′-azino-bis(3-ethylbenzothizoline-6-sulfonic acid (antioxidant activity).

2.7. Identification of Phytochemical Compound in DPSE

Separation of chemical constituents from DPSE in negative ionization mode was analyzed by ultra-high-performance liquid chromatography coupled with quadrupole time-of-flight mass spectrometry (UPLC-QTOF/MS) method. A summary of identified potential compounds are shown in Table 3. The identification of the detected compounds was made by comparing retention times MS data (neutral and observed mass) and theoretical fragmentation with data reported in literature. The isolated compounds were then categorized as tentative or confirmed. For the tentative category, they were assigned to a compound identified in Waters library with acquisition mass accuracy less than 5 ppm with at least one fragment ion [57]. The confirmed category assigned to a compound identified from Water Library and compared with available standard sample of the same compound. Based on the requirement set for UPLC-QTOF/MS method, seven different compounds were detected with 6 of them tentatively categorized and only rutin was categorized as confirmed and validated with the standard. Within these analyzed compounds, five were classified under flavonoid group while two were classified as phenolic acid group.

Table 3. Compounds identified in defatted Pitaya, *Hylocereus polyrhizus* seed extract using UPLC-QTOF/MS.

No.	A	Component Name	B	C	D	Identification Status and Category
1	FL	Rutin	610.15338	609.146	8.38	Identified, confirmed
2	FL	Kaempferol-3-O-rutinoside	594.15847	593.1514	6.67	Identified, tentative
3	FL	Kaempferol-3-O-β-ᴅ-glucopyranoside	448.10056	447.093	9.32	Identified, tentative
4	FL	Apigenin-7-O-α-ʟ-rhamnose (1→4)-6″-O-acetyl-β-ᴅ-glucoside	620.17412	619.1664	7.98	Identified, tentative
5	FL	Isorhamnetin-3-O-(2G-α-ʟ-rhamnosyl)-rutinoside	770.22694	769.2208	7.74	Identified, tentative
6	PA	Sinapic acid	224.06847	223.0609	5.42	Identified, tentative
7	PA	E-p-Coumatic acid	164.04734	163.0396	11.69	Identified, tentative

A: compound class; FL: flavonoid; PA: phenolic acid, B: natural mass (Da); C: observed *m/x*; D: retention time (Min.). UPLC-QTOF/MS: ultra-high-performance liquid chromatography coupled with quadrupole time-of-flight mass spectrometry.

Rutin is a common flavonoid found in seed extracts such as in the *Euryale ferox* seed shells [58], baobab seeds extract, grape seeds [59], and buckwheat seeds [60]. It was previously reported to possess health benefits and considered as being a good antioxidant agent [58]. MS/MS Spectra of Rutin obtained at low and high collision energy is shown in Figure 3. Different in collision energy is a technique used by mass spectrometry to induce fragmentation of selected ions in the gas phase which can then be analyzed by mass spectrometry [61]. Low collision energy is referring to collisions where the precursor ions have kinetic energies less than 1 kiloelectron volt (1 keV) or in the range of a few eV to a hundred eV. High collision energy referring to the collision where the precursor ion is accelerated to kinetic energies in the kilovolt range normally from 1 keV to 20 keV. Low collision energy led to rearrangement of the ion structure, meanwhile, high collision energy can produce more fragmentation that are not formed in low collision energy. Hence, more structural information can be obtained and easier for result interpretation [62].

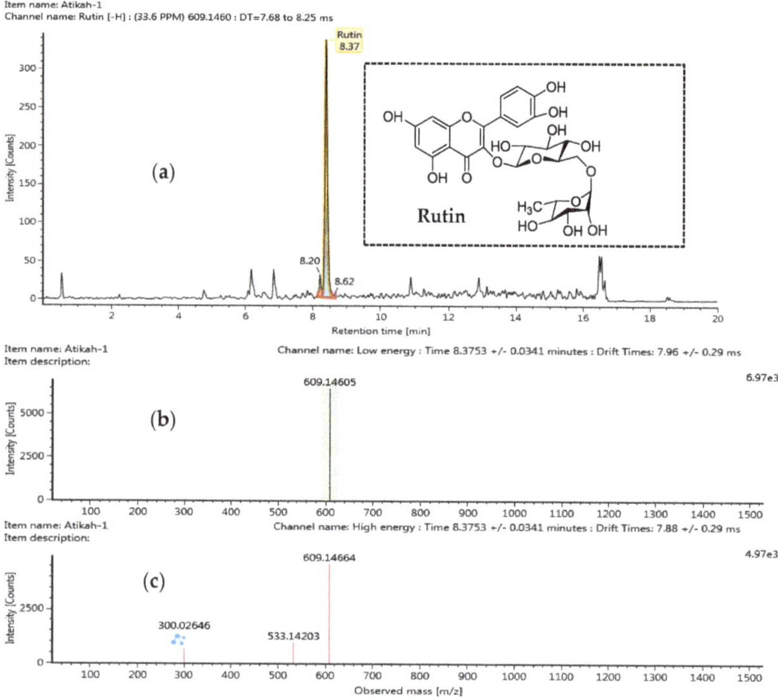

Figure 3. UPLC-QTOF/MS chromatograms of Rutin from defatted pitaya seed extract (DPSE): (**a**) mass spectra (MS/MS) of Rutin obtained at low collision energy; (**b**) mass spectra (MS/MS) of Rutin obtained at high collision energy (**c**).

For other flavonoids 2, 3, 4, and 5 occurred as O-glycoside with sugar bound at certain position. The identified flavonoids in DPSE included the following:

- kaempferol-3-O-rutinoside,
- kaempferol-3-O-β-D-glucopyranoside,
- apigenin-7-O-α-L-rhamnose(1->4)-6″-O-acetyl-β-D glucoside, and
- isohamnetin-3-O-(2G-α-L-rhamnosyl)-rutinoside.

Most of the flavonoids contained either kaempferol, apigenin or isohamnetin core moieties. These illustrate that sugar moieties mostly attached to the basic skeleton flavonoid of DPSE and the total number of sugar moieties and their attachment to the structure of flavonoid effect their antioxidant activity [63]. In addition to the above, phenolic acid metabolites such as sinapic acid and E-p-coumatic acid have been detected in DPSE. These metabolites have been reported found in extract of plant family of *Cactaceae* and reported to have anti-inflammatory activities [64].

3. Materials and Methods

3.1. Materials

Sodium carbonate (Na$_2$CO$_3$), Folin–Ciocaltue's reagent, gallic acid, quercetin, aluminium chloride (AlCl$_3$), iron (II) sulfate 7-hydrate (FeSO$_4$·7H$_2$O), iron (III) chloride 6-hydrate (FeCl$_3$·6H$_2$O), acetic acid, 2,2,6-Tri(2-pyridyl)-*s*-triazine (TPTZ), and hydrochloric acid (HCl) were purchased from Sigma-Aldrich. Only analytical grade of chemicals was used throughout the experiments. Fruits of *Hylocereus polyrhizus* were obtained from a local farm located in Sepang, Malaysia. Seeds were separated from the pulp and washed under running tap water to remove the mucilage and dried in the laboratory. Dried seeds were ground into smaller particles and kept in air-tight bottles. Subsequently, the seeds were defatted by maceration method using *n*-hexane as the solvent. The defatted seeds were left overnight in a fume hood to let residual *n*-hexane evaporates before being used in subsequent polyphenols extraction procedures.

3.2. Extraction of Defatted Pitaya Seed (DPS)

An amount of 3 g of defatted seeds was mixed with extracting solvent at different levels of independent variables including ethanol concentration (60%–80% *v/v*), extraction time (30–90 min) and extraction temperature (40–80 °C) in separate conical flasks and placed in a thermostatic water bath set at a constant shaking of 90 rpm. Parafilm and aluminum foil were used to cover the conical flasks to prevent loss of solvent due to evaporation during the extraction process. The mixture was subsequently centrifuged at 10,000 rpm for 5 min to separate the insoluble materials. The supernatant was filtered using Whatman No. 1 filter papers and vacuum-dried in a rotary evaporator at 60 °C until the solvent was completely removed. All the samples were stored at −4 °C until further analyses.

3.3. Experimental Design

A five-level with three independent variables central composite design (CCD) was employed to determine the optimal extraction condition of defatted pitaya seed (DPS). The experiment was designed using Design-Expert software version 11.1.0.1 (Stat-Ease, MN, USA). Three independent variables (extraction times, temperatures and ethanol concentrations) were selected and their coded values and levels of each variable are presented in Table 4. A total of 20 experimental runs were conducted to determine TPC, TFC and in vitro antioxidant activity based on FRAP and ABTS activity (Table 5). Least square regression was used by CCD to fit the experimental data to a second-degree polynomial model. The model was explained by the following equation:

$$Y = A_0 + \sum_{i=1}^{k} A_i X_i + \sum_{i=1}^{k} A_{ii} X_i^2 + \sum_{i=1}^{k-1} \sum_{j=i+1}^{k} A_{ij} X_i X_j$$

where Y is the predicted response, A_0, A_i, A_{ii}, A_{ij} are the constant, linear coefficient, quadratic coefficient, and interaction coefficient respectively. X_i and X_j are independent variables. k is the number of variables.

Table 4. Independent variables and their levels in central composite design.

Independent Variables	Levels				
	$-\alpha$	−1	0	1	$-\alpha$
Extraction time (A) (min)	−9.54	30	60	90	110.45
Temperature (B) (°C)	26.36	40	60	80	93.64
Ethanol concentration (C) (%)	53.18	60	70	80	86.82

3.4. Determination of Total Phenolics Content (TPC)

Total phenolic content was determined using method described by [65] with several modifications. Firstly, an amount of 100 µL (1 mg/mL) of sample extract was transferred into test tubes. Then, 50 µL of Folin solution previously diluted with 7.9 mL distilled water was added. After 4 min, 1.5 mL of 7.5 *w/v%* sodium carbonate solution was added to the sample tubes and incubated for 2 h in a dark room at room temperature. The absorbance values of the samples were recorded at 765 nm using a UV-VIS microplate reader. All the analyses were performed in triplicates to obtain the main value of the absorbance. Gallic acid at different concentrations was prepared to establish a standard calibration curve. TPCs in the sample extract were calculated by referring to the standard calibration curve and expressed as mg gallic acid equivalent (mg GAE/g) of extract sample.

3.5. Determination of Total Flavonoid Content (TFC)

Flavonoids content in the sample extract was assayed using spectrophotometric method as described by [66] with slight modifications. An amount of 100 µL (1 mg/mL) of sample and 100 µL of 2% $AlCl_3$ were mixed and incubated at room temperature for 15 min. Increase in absorbance was determined using UV-Vis microplate reader at λmax = 415 nm. The same procedure was repeated for the standard solution of quercetin to obtain the standard calibration curve. TFC was calculated based on quercetin calibration curve and expressed as mg quercetin equivalent (mg QE/g) of extract.

3.6. Ferric Reducing Antioxidant Power (FRAP)

FRAP assay was conducted following method of [22] with minor modifications. Firstly, acetate buffer (330 mM, pH 3.6), a 10 mM TPTZ solution in 40 mM of HCl and 20 mM $FeCl_3 \cdot 6H_2O$ solution were mixed in ratio of 10:1:1 (*v/v*) to prepare for FRAP reagent. The working FRAP reagent was freshly prepared prior to use. An amount of 60 µl of the sample extract was mixed with 1.8 ml FRAP reagent and the increase in absorbance was measured in comparison to a blank at 593 nm after 4 min. A standard curve was constructed using Fe_2SO_4 solution ranging from (0.1–1.0 mM) and the results were expressed as mM Fe2+/g dry weight.

3.7. ABTS Radical Scavenging Activity

The ABTS radical scavenging activity of the sample was carried out according to the method [67] with minor modification. Firstly, ABTS radical cation was prepared by reacting 7 mM ABTS with 2.45 mM potassium persulfate in ethanol and kept the mixture in the dark room temperature for 12–16 h before use. Then, the solution was diluted with ethanol to obtain an absorbance of 0.70 ($\pm$0.02) at 734 nm. For the evaluation of antioxidant activity, 2 mL of ABTS radical solution were mixed with 20 µL of the sample. The absorbance changes at 734 nm were measured using UV-Vis microplate reader after 30 min of the initial mixing. Antioxidant activity as the percent inhibition of absorbance at 734 nm was calculated using the following equation:

$$\text{ABTS} \cdot + \text{ scavenging activity} = \% \text{ inhibition: } ((AB - AA))/AB \times 100 \tag{5}$$

where, AB is the absorbance of ABTS radical with ethanol; AA is absorbance of ABTS radical + sample.

Table 5. Experimental design with observed responses of total phenolic content (TPC), total flavonoid content (TFC), and ferric reducing antioxidant power (FRAP) from defatted pitaya seed extract.

Run	Extraction Time, A	Temperature, B	Ethanol Conc., C	TPC	TFC	FRAP	ABTS
	(Min.)	(°C)	(%)	(mg GAE/g)	(mg QE/g)	(mM Fe^{2+}/g)	(%)
1	9.54622	60	70	75.83	6.38	0.85	78.89
2	60	60	70	103.75	7.34	1.34	91.22
3	60	26.3641	70	55.83	6.65	0.77	80.87
4	90	80	80	107	8.54	1.34	89.13
5	110.454	60	70	89.63	7.44	1.25	91.9
6	30	80	60	127.75	5.96	1.21	91.97
7	30	40	60	52	2.69	0.53	69.79
8	90	40	80	102.42	8.2	0.75	92.13
9	60	60	70	105.75	7.64	1.12	91.08
10	60	60	70	102.75	8.29	1.22	92.35
11	60	60	53.1821	96.5	3.88	1.08	91.52
12	60	60	70	108.38	7.2	1.19	91.35
13	60	93.6359	70	115.92	9.15	1.63	92.6
14	60	60	70	116.25	7.68	1.22	92.33
15	30	80	80	126.75	10.67	1.28	92.37
16	90	40	60	89	6.27	1.05	92.03
17	60	60	70	120.75	7.55	1.27	92.82
18	60	60	86.8179	144	10.26	1	93.22
19	30	40	80	77.5	9.35	0.97	83.21
20	90	80	60	98.75	6.88	1.75	-

3.8. Identification of Phytochemical Compounds in Defatted Pitaya Seed Extract (DPSE)

Phytochemical compounds in DPSE were identified using ultra-high-performance liquid chromatography (UPLC-MS). The analysis was completed with Waters Acquity ultra-performance LC system (Waters, Milford, MA, USA). Chromatographic separation was done using a column (ACQUITY UPLC HSS T3, 100 mm × 2.1 mm × 1.8 μm, Waters, Manchester, UK. The UPLC systems was connected to Vion IMS QTof detector (Waters, Milford, MA, USA). The mobile phase used were 0.1% formic acid (A) and acetonitrile (B). The mobile phase composition consisted of the following multistep linear gradient: 0 min, 1% B and 99% A; 0.5 min, 1% B and 99% A; 16.00 min, 35% B and 65% A; 18.00 min, 100% B and 0% A; and 20.00 min, 1% B and 99% A. The injection volume of the sample was 1 μL. The flow rate was set at 0.6 mL/min. The data were obtained in the range of *m/z* 50–1500 at 0.1 s/scan in high-definition mass spectrometry elevated energy (HDMSE) with collision energies (CE) at a fixed 4 eV and at ramped from 10 to 40 eV were required for low energy and high energy scan respectively.

4. Conclusions

Central composite design (CCD) response surface methodology was employed to evaluate the optimized extraction process for the recovery of TPC, TFC. Antioxidant activity was based on FRAP and ABTS from DPSE by analyzing the interaction effects between the independent variables (extraction time, extraction temperature and ethanol concentration). The results revealed that, extraction temperature significantly influenced the extraction process of DPS. The extraction time at 45 min with extraction temperature at 70 °C and 80% of solvent concentration which resulted in 128.58 ± 1.61 mg GAE/g sample, 9.805 ± 0.69 mg QE/g sample, 1.23 ± 0.03 mM Fe^{2+}/g sample, and 91.62% ± 0.15 were found to be the optimized condition for experimental run. Results from the validation experiments were in good agreement with the predicted values. Results from UPLC-QTOF/MS revealed that there were seven phytochemicals identified in DPSE with flavonoid found to be the major compound. The optimized extraction method was helpful in designing the experiment, separation of the chemical compounds which then contributed to further work.

Author Contributions: S.A.Z. conceived the study, designed and performed the experiments, and drafted the manuscript; S.S.A.G. participated in designing the experiments, drafted the manuscript, and provided comment to the paper; U.H.Z. and M.I.E.H. designed and supervised the workflow of the experiment. All authors have read and agreed to the published version of the manuscript.

Funding: This research was funded by the Geran Putra Berimpak of Universiti Putra Malaysia (Grant No. 9550700).

Acknowledgments: The authors gratefully acknowledge the final assistance grant by Universiti Putra Malaysia under Graduate Research Fellowship (GRF).

Conflicts of Interest: The authors declare no conflict of interest.

References

1. Rebecca, O.P.S.; Boyce, A.N.; Chandran, S. Pigment identification and antioxidant properties of red dragon fruit (*Hylocereus polyrhizus*). *Afr. J. Biotechnol.* **2010**, *9*, 1450–1454.

2. Tenore, G.C.; Novellino, E.; Basile, A. Nutraceutical potential and antioxidant benefits of red pitaya (*Hylocereus polyrhizus*) extracts. *J. Funct. Foods* **2012**, *4*, 129–136. [CrossRef]

3. Baydar, N.G.; Özkan, G.; Çetin, E.S. Characterization of grape seed and pomace oil extracts. *Grasas y Aceites* **2007**, *58*, 29–33.

4. Lemcke-Norojärvi, M.; Kamal-Eldin, A.; Appelqvist, L.-Å.; Dimberg, L.H.; Öhrvall, M.; Vessby, B. Corn and Sesame Oils Increase Serum γ-Tocopherol Concentrations in Healthy Swedish Women. *J. Nutr.* **2001**, *131*, 1195–1201. [CrossRef]

5. Van Hoed, V.; De Clercq, N.; Echim, C.; Andjelkovic, M.; Leber, E.; Dewettinck, K.; VerhÉ, R. Berry seeds: A source of specialty oils with high content of bioactives and nutritional value. *J. Food Lipids* **2009**, *16*, 33–49. [CrossRef]

6. Voigt, J. Phenolic Compounds in Food and Their Effects on Health. I. Analysis, Occurrence, and Chemistry (ACS Symposium Series 506). Herausgegeben von Chi-Tang Ho, Chang Y. Lee und Mou-Tuan Huang. 338 Seiten, zahlr. Abb. und Tab. American Chemical Society, Washington. *Food/Nahrung* **1993**, *37*, 185.

7. Hollman, P.C.H.; Katan, M.B. Dietary flavonoids: Intake, health effects and bioavailability. *Food Chem. Toxicol.* **1999**, *37*, 937–942. [CrossRef]

8. Yao, L.H.; Jiang, Y.M.; Shi, J.; Tomás-Barberán, F.A.; Datta, N.; Singanusong, R.; Chen, S.S. Flavonoids in food and their health benefits. *Plant Foods Hum. Nutr.* **2004**, *65*, 79–85. [CrossRef]

9. Evans, C.; Paganga, G.; Miller, J.N. Antioxidant properties of phenolic compounds. *Trends Plant Sci.* **1997**, *2*, 152–159. [CrossRef]

10. Zengin, G.; Uysal, S.; Ceylan, R.; Aktumsek, A. Phenolic constituent, antioxidative and tyrosinase inhibitory activity of *Ornithogalum narbonense* L. from Turkey: A phytochemical study. *Ind. Crop. Prod.* **2015**, *70*, 1–6. [CrossRef]

11. Clifford, M.N. Chlorogenic acids and other cinnamates—Nature, occurrence, dietary burden, absorption and metabolism. *J. Sci. Food Agric.* **2000**, *80*, 1033–1043. [CrossRef]

12. Chirinos, R.; Rogez, H.; Campos, D.; Pedreschi, R.; Larondelle, Y. Optimization of extraction conditions of antioxidant phenolic compounds from mashua (*Tropaeolum tuberosum* Ruíz & Pavón) tubers. *Sep. Purif. Technol.* **2007**, *55*, 217–225.

13. Pinelo, M.; Rubilar, M.; Jerez, M.; Sineiro, J.; Núñez, M.J. Effect of solvent, temperature, and solvent-to-solid ratio on the total phenolic content and antiradical activity of extracts from different components of grape pomace. *J. Agric. Food Chem.* **2005**, *53*, 2111–2117. [CrossRef]

14. Banik, R.M.; Pandey, D.K. Optimizing conditions for oleanolic acid extraction from *Lantana camara* roots using response surface methodology. *Ind. Crop. Prod.* **2008**, *27*, 241–248. [CrossRef]

15. Silva, E.M.; Rogez, H.; Larondelle, Y. Optimization of extraction of phenolics from *Inga edulis* leaves using response surface methodology. *Sep. Purif. Technol.* **2007**, *55*, 381–387. [CrossRef]

16. Box, G.E.P.; Wilson, K.B. On the Experimental Attainment of Optimum Conditions. *J. R. Stat. Soc. Ser. B* **1951**, *13*, 1–38. [CrossRef]

17. Anderson-Cook, C.M.; Borror, C.M.; Montgomery, D.C. Response surface design evaluation and comparison. *J. Stat. Plan. Inference* **2009**, *139*, 629–641. [CrossRef]

18. Andreotti, C.; Ravaglia, D.; Ragaini, A.; Costa, G. Phenolic compounds in peach (*Prunus persica*) cultivars at harvest and during fruit maturation. *Ann. Appl. Biol.* **2008**, *153*, 11–23. [CrossRef]

19. Azahar, N.F.; Gani, S.S.A.; Mohd Mokhtar, N.F. Optimization of phenolics and flavonoids extraction conditions of *Curcuma Zedoaria* leaves using response surface methodology. *Chem. Cent. J.* **2017**, *11*, 96. [CrossRef]

20. Yuan, Y.; Gao, Y.; Mao, L.; Zhao, J. Optimisation of conditions for the preparation of β-carotene nanoemulsions using response surface methodology. *Food Chem.* **2008**, *107*, 1300–1306. [CrossRef]

21. Quanhong, L.; Caili, F. Application of response surface methodology for extraction optimization of germinant pumpkin seeds protein. *Food Chem.* **2005**, *92*, 701–706. [CrossRef]

22. Mohamed Mahzir, K.A.; Abd Gani, S.S.; Hasanah Zaidan, U.; Halmi, M.I.E. Development of *Phaleria macrocarpa* (Scheff.) Boerl Fruits Using Response Surface Methodology Focused on Phenolics, Flavonoids and Antioxidant Properties. *Molecules* **2018**, *23*, 724. [CrossRef] [PubMed]

23. Qu, X.J.; Fu, Y.J.; Luo, M.; Zhao, C.J.; Zu, Y.G.; Li, C.Y.; Wang, W.; Li, J.; Wei, Z.F. Acidic pH based microwave-assisted aqueous extraction of seed oil from yellow horn (*Xanthoceras sorbifolia* Bunge.). *Ind. Crop. Prod.* **2013**, *43*, 420–426. [CrossRef]

24. Lotfi, D.; Anissa, M.; Hafsa, B.; Djamila, B.; Ihdene, Z.; Yelda Bakos, P.; Boudjema, H. Optimization of Hydrolysis Degradation of Neurotoxic Pesticide Methylparathion Using a Response Surface Methodology (RSM). *IOSR J. Appl. Chem. Ver. I* **2015**, *8*, 43–52.

25. Deng, J.; Liu, Q.; Zhang, C.; Cao, W.; Fan, D.; Yang, H. Extraction optimization of polyphenols from waste kiwi fruit seeds (*Actinidia chinensis* Planch.) and evaluation of its antioxidant and anti-inflammatory properties. *Molecules* **2016**, *21*, 832. [CrossRef]

26. Gong, Y.; Hou, Z.; Gao, Y.; Xue, Y.; Liu, X.; Liu, G. Optimization of extraction parameters of bioactive components from defatted marigold (*Tagetes erecta* L.) residue using response surface methodology. *Food Bioprod. Process.* **2012**, *90*, 9–16. [CrossRef]

27. Bucić-Kojić, A.; Planinić, M.; Tomas, S.; Bilić, M.; Velić, D. Study of solid-liquid extraction kinetics of total polyphenols from grape seeds. *J. Food Eng.* **2007**, *81*, 236–242. [CrossRef]

28. Harbourne, N.; Marete, E.; Jacquier, J.C.; O'Riordan, D. Effect of drying methods on the phenolic constituents of meadowsweet (*Filipendula ulmaria*) and willow (*Salix alba*). *LWT-Food Sci. Technol.* **2009**, *42*, 1468–1473. [CrossRef]

29. Spigno, G.; Tramelli, L.; De Faveri, D.M. Effects of extraction time, temperature and solvent on concentration and antioxidant activity of grape marc phenolics. *J. Food Eng.* **2007**, *8*, 200–208. [CrossRef]

30. Lim, Y.Y.; Murtijaya, J. Antioxidant properties of Phyllanthus amarus extracts as affected by different drying methods. *LWT-Food Sci. Technol.* **2007**, *40*, 1664–1669. [CrossRef]

31. Yolmeh, M.; Habibi Najafi, M.B.; Farhoosh, R. Optimisation of ultrasound-assisted extraction of natural pigment from annatto seeds by response surface methodology (RSM). *Food Chem.* **2014**, *155*, 319–324. [CrossRef]

32. Prasad, K.N.; Yang, B.; Zhao, M.; Ruenroengklin, N.; Jiang, Y. Application of ultrasonication or high-pressure extraction of flavonoids from litchi fruit pericarp. *J. Food Process. Eng.* **2009**, *32*, 828–843. [CrossRef]

33. Naczk, M.; Shahidi, F. Extraction and analysis of phenolics in food. *J. Chromatogr. A* **2004**, *1054*, 95–111. [CrossRef]

34. Şahin, S.; Şamli, R. Optimization of olive leaf extract obtained by ultrasound-assisted extraction with response surface methodology. *Ultrason. Sonochem.* **2013**, *20*, 595–602. [CrossRef]

35. Rajha, H.N.; El Darra, N.; Hobaika, Z.; Boussetta, N.; Vorobiev, E.; Maroun, R.G.; Louka, N. Extraction of Total Phenolic Compounds, Flavonoids, Anthocyanins and Tannins from Grape Byproducts by Response Surface Methodology. Influence of Solid-Liquid Ratio, Particle Size, Time, Temperature and Solvent Mixtures on the Optimization Process. *Food Nutr. Sci.* **2014**, *5*, 397–409. [CrossRef]

36. Sheng, Z.L.; Wan, P.F.; Dong, C.L.; Li, Y.H. Optimization of total flavonoids content extracted from Flos Populi using response surface methodology. *Ind. Crop. Prod.* **2013**, *43*, 778–786. [CrossRef]

37. Wang, Y.L.; Xi, G.S.; Zheng, Y.C.; Miao, F.S. Microwave-assisted extraction of flavonoids from Chinese herb Radix puerariae (Ge Gen). *J. Med. Plants Res.* **2010**, *4*, 304–308.

38. Krishnaswamy, K.; Orsat, V.; Gariépy, Y.; Thangavel, K. Optimization of Microwave-Assisted Extraction of Phenolic Antioxidants from Grape Seeds (*Vitis vinifera*). *Food Bioprocess Technol.* **2013**, *6*, 441–455. [CrossRef]

39. Salar, R.K.; Purewal, S.S.; Bhatti, M.S. Optimization of extraction conditions and enhancement of phenolic content and antioxidant activity of pearl millet fermented with *Aspergillus awamori* MTCC-548. *Resour. Effic. Technol.* **2016**, *2*, 148–157. [CrossRef]

40. Herodež, Š.S.; Hadolin, M.; Škerget, M.; Knez, Ž. Solvent extraction study of antioxidants from Balm (*Melissa officinalis* L.) leaves. *Food Chem.* **2003**, *80*, 275–282. [CrossRef]

41. Ghasemzadeh, A.; Baghdadi, A.; Jaafar, H.Z.E.; Swamy, M.K.; Megat Wahab, P.E. Optimization of flavonoid extraction from red and brown rice bran and evaluation of the antioxidant properties. *Molecules* **2018**, *23*, 1863. [CrossRef]

42. Wong, P.Y.Y.; Kitts, D.D. Studies on the dual antioxidant and antibacterial properties of parsley (*Petroselinum crispum*) and cilantro (*Coriandrum sativum*) extracts. *Food Chem.* **2006**, *97*, 505–515. [CrossRef]

43. Zhang, Z.S.; Li, D.; Wang, L.J.; Ozkan, N.; Chen, X.D.; Mao, Z.H.; Yang, H.Z. Optimization of ethanol-water extraction of lignans from flaxseed. *Sep. Purif. Technol.* **2007**, *57*, 17–24. [CrossRef]

44. Karacabey, E.; Mazza, G. Optimisation of antioxidant activity of grape cane extracts using response surface methodology. *Food Chem.* **2010**, *119*, 343–348. [CrossRef]

45. Ghafoor, K.; Choi, Y.H.; Jeon, J.Y.; Jo, I.H. Optimization of ultrasound-assisted extraction of phenolic compounds, antioxidants, and anthocyanins from grape (*Vitis vinifera*) seeds. *J. Agric. Food Chem.* **2009**, *57*, 4988–4994. [CrossRef]

46. Dent, M.; Dragović-Uzelac, V.; Penić, M.; Brñić, M.; Bosiljkov, T.; Levaj, B. The effect of extraction solvents, temperature and time on the composition and mass fraction of polyphenols in dalmatian wild sage (*Salvia officinalis* L.) extracts. *Food Technol. Biotechnol.* **2013**, *51*, 84–91.

47. Jovanović, A.A.; Đorđević, V.B.; Zdunić, G.M.; Pljevljakušić, D.S.; Šavikin, K.P.; Godevac, D.M.; Bugarski, B.M. Optimization of the extraction process of polyphenols from Thymus serpyllum L. herb using maceration, heat- and ultrasound-assisted techniques. *Sep. Purif. Technol.* **2017**, *179*, 369–380. [CrossRef]

48. Fecka, I.; Turek, S. Determination of polyphenolic compounds in commercial herbal drugs and spices from Lamiaceae: Thyme, wild thyme and sweet marjoram by chromatographic techniques. *Food Chem.* **2008**, *108*, 1039–1053. [CrossRef]

49. Mustafa, A.; Turner, C. Pressurized liquid extraction as a green approach in food and herbal plants extraction: A review. *Anal. Chim. Acta* **2011**, *703*, 8–18. [CrossRef]

50. Miron, T.L.; Plaza, M.; Bahrim, G.; Ibáñez, E.; Herrero, M. Chemical composition of bioactive pressurized extracts of Romanian aromatic plants. *J. Chromatogr. A* **2011**, *1218*, 4918–4927. [CrossRef]

51. Vajić, U.J.; Grujić-Milanović, J.; Živković, J.; Šavikin, K.; Godevac, D.; Miloradović, Z.; Bugarski, B.; Mihailović-Stanojević, N. Optimization of extraction of stinging nettle leaf phenolic compounds using response surface methodology. *Ind. Crop. Prod.* **2015**, *74*, 912–917. [CrossRef]

52. Li, H.; Deng, Z.; Wu, T.; Liu, R.; Loewen, S.; Tsao, R. Microwave-assisted extraction of phenolics with maximal antioxidant activities in tomatoes. *Food Chem.* **2012**, *130*, 928–936. [CrossRef]

53. Ahmed, D.; Khan, M.; Saeed, R. Comparative Analysis of Phenolics, Flavonoids, and Antioxidant and Antibacterial Potential of Methanolic, Hexanic and Aqueous Extracts from Adiantum caudatum Leaves. *Antioxidants* **2015**, *4*, 394–409. [CrossRef]

54. Yim, H.S.; Chye, F.Y.; Rao, V.; Low, J.Y.; Matanjun, P.; How, S.E.; Ho, C.W. Optimization of extraction time and temperature on antioxidant activity of Schizophyllum commune aqueous extract using response surface methodology. *J. Food Sci. Technol.* **2013**, *50*, 275–283. [CrossRef]

55. Youn, J.S.; Kim, Y.J.; Na, H.J.; Jung, H.R.; Song, C.K.; Kang, S.Y.; Kim, J.Y. Antioxidant activity and contents of leaf extracts obtained from Dendropanax morbifera LEV are dependent on the collecting season and extraction conditions. *Food Sci. Biotechnol.* **2019**, *28*, 201–207. [CrossRef]

56. Pinelo, M.; Del Fabbro, P.; Manzocco, L.; Nuñez, M.J.; Nicoli, M.C. Optimization of continuous phenol extraction from Vitis vinifera byproducts. *Food Chem.* **2005**, *92*, 109–117. [CrossRef]

57. Md Yusof, A.H.; Abd Gani, S.S.; Zaidan, U.H.; Halmi, M.I.E.; Zainudin, B.H. Optimization of an Ultrasound-Assisted Extraction Condition for Flavonoid Compounds from Cocoa Shells (*Theobroma cacao*) Using Response Surface Methodology. *Molecules* **2019**, *24*, 711. [CrossRef]

58. Liu, Y.; Wei, S.; Liao, M. Optimization of ultrasonic extraction of phenolic compounds from Euryale ferox seed shells using response surface methodology. *Ind. Crop. Prod.* **2013**, *49*, 837–843. [CrossRef]

59. Atanassova, M.; Bagdassarian, V. RUTIN CONTENT IN PLANT PRODUCTS. *J. Univ. Chem. Technol. Metall.* **2009**, *44*, 201–203.

60. Zielińska, D.; Turemko, M.; Kwiatkowski, J.; Zieliński, H. Evaluation of flavonoid contents and antioxidant capacity of the aerial parts of common and tartary buckwheat plants. *Molecules* **2012**, *17*, 9668–9682. [CrossRef]

61. Wells, J.M.; McLuckey, S.A. Collision-induced dissociation (CID) of peptides and proteins. *Methods Enzymol.* **2005**, *402*, 148–185.

62. Wysocki, V.H.; Kenttämaa, H.I.; Cooks, R.G. Internal energy distributions of isolated ions after activation by various methods. *Int. J. Mass Spectrom. Ion Process.* **1987**, *75*, 181–208. [CrossRef]

63. Daji, G.; Steenkamp, P.; Madala, N.; Dlamini, B. Phytochemical Composition of *Solanum retroflexum* Analysed with the Aid of Ultra-Performance Liquid Chromatography Hyphenated to Quadrupole-Time-of-Flight Mass Spectrometry (UPLC-qTOF-MS). *J. Food Qual.* **2018**, *2018*. [CrossRef]

64. Cha, M.N.; Jun, H.I.; Lee, W.J.; Kim, M.J.; Kim, M.K.; Kim, Y.S. Chemical composition and antioxidant activity of Korean cactus (*Opuntia humifusa*) fruit. *Food Sci. Biotechnol.* **2013**, *22*, 523–529. [CrossRef]

65. Tyug, T.S.; Johar, M.H.; Ismail, A. Antioxidant properties of fresh, powder, and fiber products of mango (*Mangifera foetida*) fruit. *Int. J. Food Prop.* **2010**, *13*, 682–691. [CrossRef]

66. Quettier-Deleu, C.; Gressier, B.; Vasseur, J.; Dine, T.; Brunet, C.; Luyckx, M.; Cazin, M.; Cazin, J.C.; Bailleul, F.; Trotin, F. Phenolic compounds and antioxidant activities of buckwheat (*Fagopyrum esculentum* Moench) hulls and flour. *J. Ethnopharmacol.* **2000**, *72*, 35–42. [CrossRef]

67. Re, R.; Pellegrini, N.; Proteggente, A.; Pannala, A.; Yang, M.; Rice-Evans, C. Antioxidant activity applying an improved ABTS radical cation decolorization assay. *Free Radic. Biol. Med.* **1999**, *26*, 1231–1237. [CrossRef]

Article

Antiprotozoal Activity Against *Entamoeba histolytica* of Flavonoids Isolated from *Lippia graveolens* Kunth

Ramiro Quintanilla-Licea [1,*], Javier Vargas-Villarreal [2], María Julia Verde-Star [1], Verónica Mayela Rivas-Galindo [3] and Ángel David Torres-Hernández [1]

[1] Facultad de Ciencias Biológicas, Universidad Autónoma de Nuevo León (UANL), Av. Universidad S/N, Cd. Universitaria, San Nicolás de los Garza, C.P. 66455 Nuevo León, Mexico; maria.verdest@uanl.edu.mx (M.J.V.-S.); angel.torreshr@uanl.edu.mx (Á.D.T.-H.)

[2] Laboratorio de Bioquímica y Biología Celular, Centro de Investigaciones Biomédicas del Noreste (CIBIN), Dos de abril esquina con San Luis Potosí, C.P. 64720 Monterrey, Mexico; jvargas147@yahoo.com.mx

[3] Facultad de Medicina, Universidad Autónoma de Nuevo León (UANL), Madero y Aguirre Pequeño, Mitras Centro, Monterrey, C.P. 64460 Nuevo León, Mexico; veronica.rivasgl@uanl.edu.mx

* Correspondence: ramiro.quintanillalc@uanl.edu.mx; Tel.: +52-81-83763668

Academic Editor: H.P. Vasantha Rupasinghe

Received: 24 April 2020; Accepted: 25 May 2020; Published: 26 May 2020

Abstract: Amebiasis caused by *Entamoeba histolytica* is nowadays a serious public health problem worldwide, especially in developing countries. Annually, up to 100,000 deaths occur across the world. Due to the resistance that pathogenic protozoa exhibit against commercial antiprotozoal drugs, a growing emphasis has been placed on plants used in traditional medicine to discover new antiparasitics. Previously, we reported the in vitro antiamoebic activity of a methanolic extract of *Lippia graveolens* Kunth (Mexican oregano). In this study, we outline the isolation and structure elucidation of antiamoebic compounds occurring in this plant. The subsequent work-up of this methanol extract by bioguided isolation using several chromatographic techniques yielded the flavonoids pinocembrin (**1**), sakuranetin (**2**), cirsimaritin (**3**), and naringenin (**4**). Structural elucidation of the isolated compounds was achieved by spectroscopic/spectrometric analyses and comparing literature data. These compounds revealed significant antiprotozoal activity against *E. histolytica* trophozoites using in vitro tests, showing a 50% inhibitory concentration (IC_{50}) ranging from 28 to 154 µg/mL. Amebicide activity of sakuranetin and cirsimaritin is reported for the first time in this study. These research data may help to corroborate the use of this plant in traditional Mexican medicine for the treatment of dyspepsia.

Keywords: infectious diseases; amoebiasis; Mexican oregano; bioguided isolation; flavonoids; antiprotozoal agents

1. Introduction

Amoebiasis is caused by *Entamoeba histolytica*, which is a protozoan of the family Endomoebidae [1]. It is related to elevated morbidity and mortality worldwide, and has become a serious public health problem in developing countries [2]. Traveling to endemic countries is a risk factor for acquiring an *E. histolytica* infection [3]. After malaria, amoebiasis is the second cause of death due to parasitic diseases [4,5]. The symptoms vary from mild diarrhea to dysentery, but, occasionally, *E. histolytica* can invade the intestinal mucosal barrier and trigger liver abscesses [6]. Asymptomatic infections occur in 90% of individuals, whereas the remaining 10% contract symptomatic infections [7]. Around 50 million people suffer from severe amoebiasis, and 40,000–100,000 deaths occur annually due to this parasitosis [8,9].

Currently, metronidazole is the most used commercial drug for the treatment of amoebiasis, however, since drug resistance by *E. histolytica* is increasing, the use of higher doses to overcome the infection is needed, thus causing unpleasant side effects [10,11]. Considering these undesired side effects as well as the development of resistant strains of *E. histolytica* against metronidazole, more efficient and safer antiamoebic agents are required [12–14].

Natural products occurring in medicinal plants have proved to be an important source of leading compounds for the design of new drugs [15]. Mexican oregano (*Lippia graveolens* Kunth) has been used in traditional Mexican medicine for curing inflammation-related diseases, such as respiratory and digestive disorders, headaches, and rheumatism, among others [16,17]. Oregano's essential oil, regardless of the species, shows a broad range of effects on bacteria, with some of them being resistant to antibiotics of clinical use, as well as on fungi and parasites [18–22].

Recently, we reported the in vitro antiamoebic activity of a methanolic extract of *Lippia graveolens* Kunth and the bioguided isolation of carvacrol, as one of the bioactive compounds with antiprotozoal activity [23]. This work aims to isolate and achieve the structure elucidation of additional antiamoebic compounds present in this plant.

2. Results

2.1. Bioguided Isolation of Flavonoids from Lippia graveolens Kunth

As previously reported, the partition of a methanolic extract of *Lippia graveolens* by extraction with *n*-hexane and fractionation of the hexane phase led to the isolation of carvacrol with excellent antiamoebic activity [23]. The subsequent handling of the remaining methanol (MeOH) by partition between methanol-water and ethyl acetate (EtOAc), carried out in this research, yielded an EtOAc residue with 93.3% growth inhibition of *E. histolytica*. After column chromatography (silica gel, Sephadex), this residue afforded the known flavonoids pinocembrin (**1**), sakuranetin (**2**), cirsimaritin (**3**), and naringenin (**4**), with significant antiamoebic activity (Figure 1).

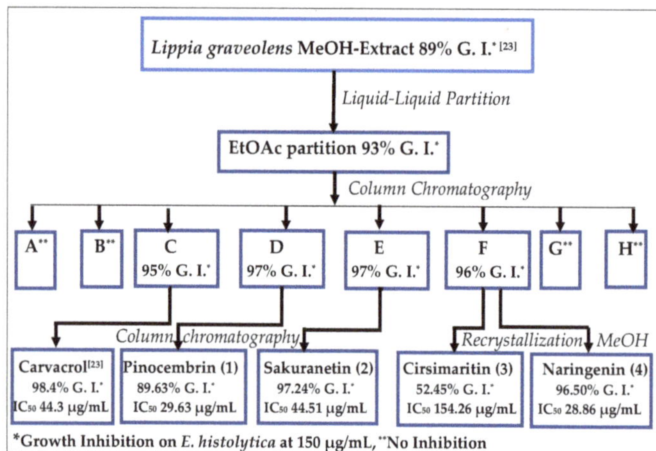

Figure 1. General scheme for the bioguided isolation of compounds with antiamoebic activity from *Lippia graveolens* Kunth (Mexican oregano).

The isolated flavonoids were identified by comparing their physical and spectral data with those reported in the literature.

The electron ionization (EI) mass spectrum showed a molecular ion with $m/z = 256$ for pinocembrin **1** (calcd. for $C_{15}H_{12}O_4$, 256.253), $m/z = 286$ for sakuranetin **2** (calcd. for $C_{16}H_{14}O_5$, 286.283), $m/z = 314$

for cirsimaritin **3** (calcd. for $C_{17}H_{14}O_6$, 314.28), and $m/z = 272$ for naringenin **4** (calcd. for $C_{15}H_{12}O_5$, 272.252).

The infrared (IR) spectrum of all isolated flavonoids (see Supplementary material) contained absorption bands at 1600–1650 cm^{-1} (medium) and 1100–1250 cm^{-1} (strong), consistent with a C=O bond in the molecules [24–27].

One- and two-dimensional nuclear magnetic resonance (NMR) spectra were recorded for the isolated compounds using deuterated dimethyl sulfoxide (DMSO-d_6). ^{1}H- and ^{13}C-NMR chemical shifts (see Supplementary Material) were in accordance with those reported for pinocembrin **1** [28–30], sakuranetin **2** [31–34], cirsimaritin **3** [26,35–37], and naringenin **4** [27,28,38–41]. An unambiguous assignment of the ^{13}C-NMR spectrum of these compounds was deduced from ^{1}H-^{1}H COSY, NOESY, HSQC, and HMBC spectra (see Supplementary Material).

The four isolated flavonoids have the common characteristic that the rotameric hydroxy group at C-5 forms an intramolecular H-bond with the carbonyl group [42]. That explains the shift of this proton absorption to the range of about δ 13.0–12.0 as a sharp singlet in DMSO-d_6 [43].

2.2. Entamoeba Histolytica Growth Parameters

After collecting data from *E. histolytica* growth kinetics experiments, it was estimated that the generation time is 14.76 h and the duplication time is 21.30 h.

The ideal growth time of *E. histolytica* for evaluating the amebicide activity of the compounds was set at 72 h because, at this point, protozoa are still in the exponential and sustained growth stage, thus decreasing the number of false positives.

2.3. In Vitro Assay for Entamoeba histolytica

The pure compounds were dissolved in DMSO to a concentration of 150 μg/mL in a suspension of *E. histolytica* trophozoites in a logarithmic phase in peptone, pancreas, and liver extract plus 10% bovine serum (PEHPS medium). They showed significant growth inhibition of *E. histolytica* at this concentration. The 50% inhibitory concentration (IC$_{50}$) values of these compounds ranged from 28.86 to 154.26 μg/mL (metronidazole IC$_{50}$ 0.205 μg/mL).

Figures 2–5 show the 50% inhibitory concentration of each compound calculated by using a Probit analysis, considering a 95% confidence level.

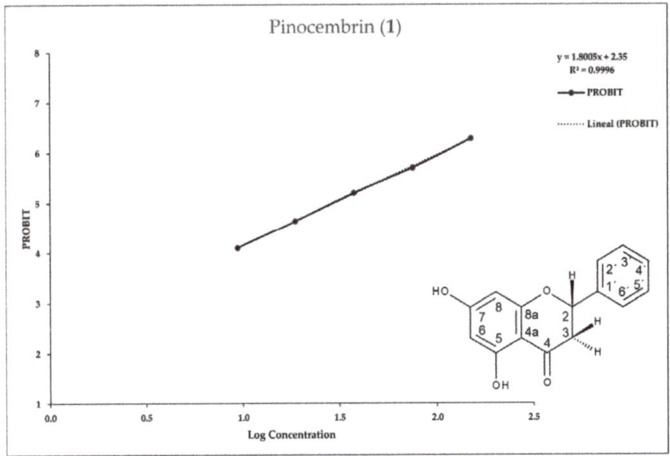

Figure 2. Antiprotozoal activity of Pinocembrin **1** against *Entamoeba histolytica*. Growth inhibition of 89.63% at 150 μg/mL, 50% inhibitory concentration value (IC$_{50}$) = 29.63 μg/mL.

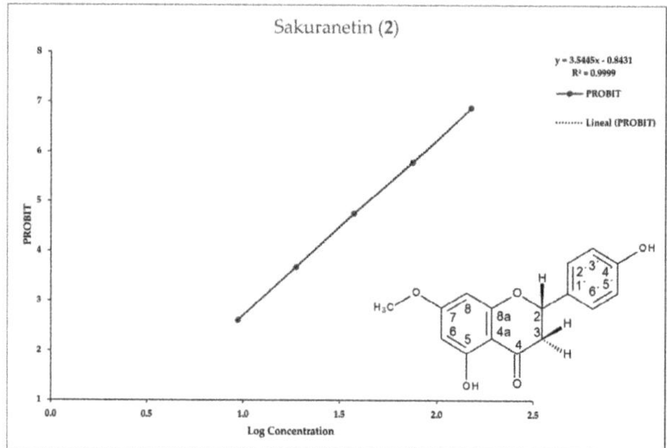

Figure 3. Antiprotozoal activity of Sakuranetin **2** against *Entamoeba histolytica*. Growth inhibition of 97.24% at 150 µg/mL, IC$_{50}$ = 44.51 µg/mL.

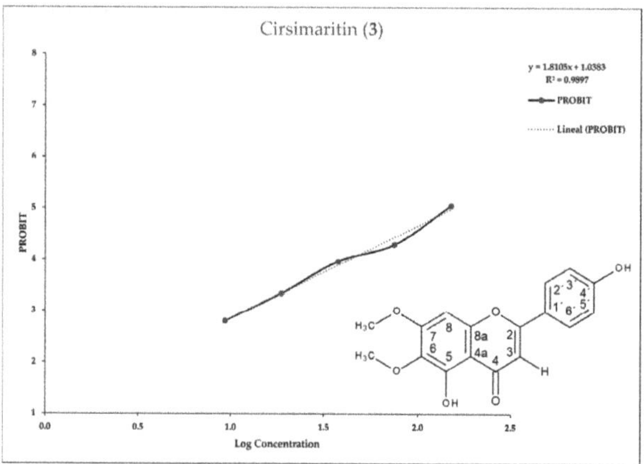

Figure 4. Antiprotozoal activity of Cirsimaritin **3** against *Entamoeba histolytica*. Growth inhibition of 52.45% at 150 µg/mL, IC$_{50}$ = 154.26 µg/mL.

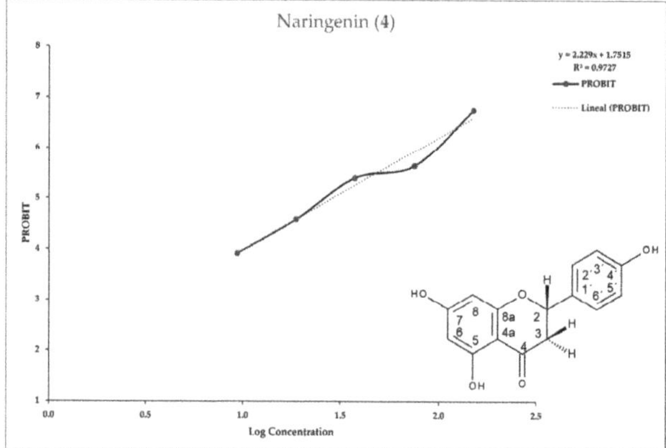

Figure 5. Antiprotozoal activity of Naringenin **4** against *Entamoeba histolytica*. Growth inhibition of 96.50% at 150 µg/mL, IC$_{50}$ = 28.86 µg/mL.

3. Discussion

It is estimated that about 6000 flavonoids are present in different plants worldwide [44,45], and many of them are common ingredients of our daily food. Flavonoids exhibit a variety of biological properties, such as antioxidant, anticancer, antibacterial, antifungal, antiparasitic [46–53], as well as those for treating other kinds of illness [54–57].

More than 20 flavonoids have been identified in the leaves of *Lippia graveolens* by high pressure liquid chromatography (HPLC), according to previous reports [58]. Pinocembrin, sakuranetin, naringenin, and cirsimaritin have already been isolated from this plant [58–62], but, in the respective studies, no antiamoebic activity is reported. Nevertheless, some research groups have reported antiprotozoal activity of these flavonoids isolated from sources other than *Lippia graveolens*.

A weak antiparasitic activity of pinocembrin has been shown against trypomastigotes of *Trypanosoma cruzi*, with inhibition values in the range of 40.13% (at 250 µg/mL) to 43.68% (at 500 µg/mL), according to Grael et al. [63]. Weak inhibition was also observed against *Giardia lamblia* trophozoites, with an IC_{50} of 174.4 µg/mL reported by Alday-Provencio et al. [64] and an IC_{50} of 57.39 µg/mL reported by Calzada et al. [65], respectively.

Sakuranetin presented activity against *Leishmania amazonensis*, *Leishmania brazilians*, *Leishmania major*, and *Leishmania chagasi*, with a range of 43–52 µg/mL, as well as against *T. cruzi* trypomastigotes, with an IC_{50} of 20.17 µg/mL, according to Grecco et al. [39].

Regarding parasitic diseases, cirsimaritin showed potent inhibition against *Plasmodium falciparum* resistant to chloroquine, with an IC_{50} of 16.9 µM [66], and similar activity against *Leishmania donovani* (IC_{50} = 3.9 µg/mL), *Trypanosoma brucei rhodesiense* (IC_{50} = 3.3 µg/mL), and *T. cruzi* (IC_{50} = 19.7 µg/mL), according to Tasdemir et al. [49].

Antiparasitic research has found giardicidal activity in naringenin (**4**), with an IC_{50} of 125.7 µg/mL [64] and 47.84 µg/mL [65], according to Alday-Provencio et al. and Calzada et al., respectively, but no damage against *Trypanosoma cruzi* and *Leishmania* spp. was observed, according to Grecco et al. [39].

Some results of pinocembrin and naringenin tested against the strain *E. histolytica* HM1:IMSS have been published; pinocembrin isolated from *Teloxys graveolens* and naringenin from a commercial source exhibited an IC_{50} of 80.76 and 98.24 µg/mL, respectively [65,67]. This indicates a lower effectiveness compared with our results (IC_{50} of 29.51 µg/mL for pinocembrin and 28.85 µg/mL for naringenin, respectively), which could be explained by the difference between the methodology reported by Calzada [65] and that used by our group. The principal difference was the time that the trophozoites were exposed to the chemical compound. In our methodology, we incubated the trophozoites with the flavonoids for 72 h, while Calzada [65] incubated them for 48 h, which could represent a factor in the certainty of the biological activity.

In our view, this is the first report on the amebicide activity of sakuranetin and cirsimaritin.

The antiprotozoal activity of pinocembrin and naringenin (IC_{50} of 29.63 and 28.86 µg/mL, respectively) was higher compared with sakuranetin (44.51 µg/mL), and the most remarkable comparison was with cirsimaritin (154.26 µg/mL), revealing that a 5,7-dihydroxylated A ring is essential for antiprotozoal activity, as remarked by Calzada [65]. The structure–effect correlations also showed that a 2,3-double bond in ring B (as in cirsimaritin) reduces the antiprotozoal activity.

Currently, there are no available studies on the mechanism of action against *E. histolytica* of the flavonoids isolated from *L. graveolens*, although there are reports on some structurally related flavonoids [68,69]. The ultrastructural changes in the morphology of *Entamoeba histolytica* when it was assayed with (−)-epicatechin, a flavan-3-ol flavonoid, have previously been demonstrated, showing an IC_{50} of 1.9 µg/mL. The results indicated programmed cell death activation with nuclear alterations (small clumps around the nuclear membrane), in addition to cytoplasmatic modifications, such as an increase of glycogen deposits and a reduction of the size and number of vacuoles [70]. Recently, Bolaños et al. [68] have also demonstrated that the flavonoids (−)-epicatechin and kaempferol affect cytoskeleton proteins and functions in *E. histolytica* [68,71], leading to changes in essential cellular mechanisms, such as adhesion, migration, phagocytosis, and cytolysis. These findings lead us to have an idea of the possible targets and mechanisms of action of the flavonoids isolated from *L. graveolens*. None of the flavonoids isolated from *L. graveolens* present a hydroxyl group at position 3, as in the case of epicatechin and kaempferol, so there must be subtle differences in the mechanism of action of pinocembrin, sakuranetin, naringenin, and cirsimaritin, and part of the future work of our group will be devoted to this issue.

The significant inhibitory effect against *E. histolytica* observed for the methanolic extract of *Lippia graveolens* (IC$_{50}$ = 59.15 µg/mL) can be attributed to the presence of the flavonoids isolated from the ethyl acetate partition as well as the carvacrol mainly obtained from the hexane partition [23]. Compounds with a higher polarity occurring in *L. graveolens*, soluble in methanol or water, are not involved in the antiprotozoal activity of this plant against *E. histolytica*.

4. Materials and Methods

4.1. General

An Electrothermal 9100 apparatus (Electrothermal Engineering Ltd., Southend-on-Sea, UK) was used for melting point acquisition. IR spectra were measured on a Frontier Fourier transform infrared (FT-IR) spectrometer (PerkinElmer, Waltham, MA, USA) with an ATR accessory. NMR spectra were recorded on an Avance DPX 400 spectrometer (Bruker, Billerica, MA, USA) running at 400.13 MHz for ^{1}H and 100.61 MHz for ^{13}C. EI-MS were obtained on a MAT 95 spectrometer (70 eV, Finnigan, San Jose, CA, USA). Thin layer chromatography (TLC) was realized on precoated silica gel glass plates (5 × 10 cm, Merck silica gel 60 F$_{254}$, Darmstadt, Germany). Column chromatography was carried out on silica gel (60–200 mesh) purchased from J. T. Baker (Phillipsburg, NJ, USA). Size-exclusion chromatography was performed on Sephadex LH-20 (Lipophilic Sephadex, Amersham Biosciences Ltd., purchased from Sigma-Aldrich Chemie, Steinheim, Germany).

4.2. Plant Material

Aerial parts of *Lippia graveolens* were collected near the town General Cepeda (Mexican State Coahuila) in March 2011 and identified by Maria del Consuelo González. A voucher specimen (No. 025554) was deposited at the Herbario de la Facultad de Ciencias Biológicas (UANL), Nuevo León, México. The plant name has been checked with http://www.theplantlist.org. The vegetal material was dried and ground to powder.

4.3. Plant Extraction and Bioguided Isolation of Antiamoebic Compounds from Lippia graveolens Kunth

In total, 600 g of dried and ground *Lippia graveolens* leaves were extracted in a Soxhlet apparatus for 40 h with MeOH. After filtration, the solvent was removed in a rotatory evaporator to yield 260 g of crude extract. This extract was analyzed for its amebicide activity on trophozoites of *E. histolytica* (HM1:IMSS strain), showing a significant inhibition percentage (89%; IC$_{50}$ 59.15 µg/mL) in terms of the standard concentration of 150 µg/mL. Afterward, the extract was redissolved in 2 L methanol and divided into four portions of 500 mL each for conducting liquid-liquid partition with *n*-hexane. After solvent evaporation, 19.9 g of a residue with high amebicidal activity (90.9% growth inhibition) was obtained. Bioguided fractionation of this hexane partition, using column chromatography on silica gel, provided 2.2 g of carvacrol (98.4% growth inhibition; IC$_{50}$ 44.30 µg/mL), as previously reported [23].

The methanolic phase was concentrated under reduced pressure up to a volume of ca. 500 mL, and 1.5 L distilled water was then added, gradually and under constant stirring. Afterward, the methanol/water mixture was divided into four portions of 500 mL each and submitted to liquid–liquid partition with ethyl acetate to yield, after solvent evaporation, 67.3 g of a combined residue with 93.3% growth inhibition against *E. histolytica*.

The EtOAc partition was suspended in 400 mL of chloroform (CHCl$_3$), and, after stirring, filtration, and solvent evaporation, 25.8 g of CHCl$_3$-soluble residue was obtained. The material recovered from the filter was then suspended in 400 mL of EtOAc, and, after stirring, filtration, and solvent evaporation, 8.8 g of EtOAc-soluble residue was obtained. This second material recovered from the filter was then suspended in 100 mL MeOH, and the same procedure was applied to yield 8.8 g of MeOH-soluble residue and ca. 24 g of an insoluble powder from the final filtration. Each residue was analyzed for its anti-*Entamoeba histolytica* activity, and remarkable results were obtained, mainly for the $CHCl_3$-residue, with 90.9% growth inhibition, followed by the EtOAc residue, with 61.6% growth inhibition. The MeOH residue and insoluble powder did not show any activity.

The $CHCl_3$ fraction was divided into five portions of ca. 5 g, and each of them was chromatographed on a silica gel (100 g) column (60 × 2.6 cm) and eluted with stepwise gradients of chloroform-ethyl acetate, and finally with methanol. For each column, a total of 110 subfractions (50 mL) were obtained and collected, and considering their TLC ($CHCl_3$–EtOAc, 9:1) profiles, split into eight main fractions (A-H). These main fractions, containing the nonpolar to the more polar compounds, were used for amebicide assays. Out of the eight fractions, only fractions C, D, E, and F showed amebicide activity, with 95.34%, 96.89%, 97.24%, and 95.85% growth inhibition, respectively.

Fraction C (showing the main compound, according to TLC, with R_f = 0.77; $CHCl_3$-Ethyl acetate, 9:1) was divided into five portions of approx. 1 g, and each of them was chromatographed on a silica gel (20 g) column (39 × 2 cm) and eluted with a stepwise gradient solvent system of chloroform and ethyl acetate, and finally with methanol. Sixty subfractions (10 mL) were collected for each column and combined, based on their TLC ($CHCl_3$-Ethyl acetate, 9:1) profiles, into five main fractions (CA-CE). Fraction CB presented the main compound, with R_f = 0.77, so it was subjected to subsequent purification with three columns packed with silica gel as stationary phase (data not shown). Afterward, 315 mg of a viscous liquid with a high amebicide activity (96.76% growth inhibition; IC_{50} 44.1 µg/mL) was recovered. The spectroscopic results indicated that this compound is carvacrol, previously isolated from hexane partition [23].

Fraction D (showing the main compound, according to TLC, with R_f = 0.51; $CHCl_3$-Ethyl acetate, 9:1) was divided into four portions of ca. 1 g and each of them was submitted to chromatography on a silica gel (20 g) column (39 × 2 cm) and eluted with a stepwise gradient solvent system of chloroform and ethyl acetate, and finally with methanol. Sixty subfractions (10 mL) were collected for each column and combined, based on their TLC ($CHCl_3$-Ethyl acetate, 9:1) profiles, into five main fractions (DA-DE). Fraction DB presented the main compound, with R_f = 0.51, so it was subjected to subsequent purification with several columns packed with Sephadex as stationary phase (data not shown). Afterward, 32 mg of a solid with a high amebicide activity (89.63% growth inhibition; IC_{50} 29.63 µg/mL) was recovered. The spectroscopic results indicated that this compound is pinocembrin **1** ($C_{15}H_{12}O_4$; M.p. 210 °C: Lit. 223–236 °C [72], 191–193 °C [24]).

Fraction E (showing the main compound, according to TLC, with R_f = 0.33; $CHCl_3$-Ethyl acetate, 9:1) was divided into three portions of approx. 1 g and each of them was chromatographed on a silica gel (20 g) column (39 × 2 cm) and eluted with a stepwise gradient solvent system of chloroform and ethyl acetate, and finally with methanol. Sixty subfractions (10 mL) were collected for each column and combined, based on their TLC ($CHCl_3$-Ethyl acetate, 9:1) profiles, into five main fractions (EA-EE). Fractions EB and EC presented the main compound, with R_f = 0.33, so they were subjected to subsequent purification with several columns packed with Sephadex as stationary phase (data not shown). Afterward, 102 mg of a solid with a high amebicide activity (97.24% growth inhibition; IC_{50} 44.51 µg/mL) was recovered. The spectroscopic results indicated that this compound is sakuranetin **2** ($C_{16}H_{14}O_5$; M.p. 160 °C: Lit. 151–153 °C [73], 143–144 °C [74]).

Fraction F showed only a main compound, according to TLC, with R_f = 0.10 ($CHCl_3$-Ethyl acetate, 9:1). Using an eluent of a higher polarity ($CHCl_3$-Ethyl acetate, 1:1), this compound was revealed to be a mixture of two compounds, with an R_f of 0.70 and 0.53, respectively. This mixture was no longer soluble in chloroform and it was therefore not possible to use column chromatography with silica gel of a normal phase to try to separate the two compounds. Then, this fraction was suspended in 30 mL of chloroform and heated to reflux for 15 min, noting that a large quantity of the material remained insoluble. After cooling and filtering, 645 mg of an insoluble solid consisting of the two compounds of the mixture (TLC) was recovered, while in the chloroform solution, a mixture of the same compounds was also observed, but with additional impurities, mainly of a lower polarity. The insoluble solid (645 mg) recovered from the filter was suspended in 30 mL methanol and heated to reflux until complete solubility was observed. After cooling, the MeOH solution was kept refrigerated for 72 h with hermetic closure, and during that time, 50.6 mg of a greenish powder containing only the compound with R_f = 0.53 (CHCl$_3$-Ethyl acetate, 1:1) was separated. This solid presented no significant amebicide activity (52.45% growth inhibition; IC$_{50}$ 154.26 µg/mL). The spectroscopic results indicated that this compound is cirsimaritin **3** (C$_{17}$H$_{14}$O$_6$; M.p. 268 °C: Lit. 256–258 °C [26], 267–268 °C [75]).

The filtered MeOH solution was concentrated and chromatographed on a Sephadex (50 g) column (160 × 1.5 cm) and eluted with methanol (300 mL). A total of 60 subfractions (5 mL) were collected and combined, based on their TLC (CHCl$_3$-Ethyl acetate, 1:1) profiles, into seven main fractions (FA-FG). From fraction FC, additional cirsimaritin with R_f = 0.53 was recovered. Fraction FF showed only the pure compound, with R_f = 0.70. This solid also presented high amebicide activity (96.50% growth inhibition; IC$_{50}$ 28.86 µg/mL). The spectroscopic results indicated that this compound is naringenin **4** (C$_{15}$H$_{12}$O$_5$; M.p. 254 °C: Lit. 250–252 °C [27], 248–250 °C [73]).

4.4. Antiprotozoal Assay

4.4.1. Test Microorganisms

Strain HM-1:IMSS of *Entamoeba histolytica* was obtained from the microorganism culture collection of the Centro de Investigación Biomédica del Noreste (CIBIN) in Monterrey, Mexico. The trophozoites were grown axenically and maintained in peptone, pancreas, and liver extract plus bovine serum and employed at the log phase of growth (2×10^4 cells/mL) by all of the bioassays performed [76,77].

The procedure for determining the growth curve for *E. histolytica* was performed in 13 × 100 mm screw cap test tubes, by inoculating 20,000 trophozoites of *E. histolytica* in 5 mL of PEHPS medium, to which 10% bovine serum was added. Subsequently, they were incubated at 36.5 °C for 120 h, and every 24 h, the number of trophozoites was determined and the growth parameters in the medium were evaluated. The process was conducted in three separate experiments per triplicate.

4.4.2. In Vitro Assay for Entamoeba histolytica

Each compound was dissolved in DMSO and adjusted to a concentration of 150 µg/mL in a suspension of *E. histolytica* trophozoites at a logarithmic phase in PEHPS medium with 10% bovine serum. Vials were incubated for 72 h, and then chilled in iced water for 20 min, and, by using a hemocytometer, the number of dead trophozoites per milliliter was calculated. Each extract assay was carried out in triplicate. Metronidazole was used as a positive control, and as a negative control, an *E. histolytica* suspension in PEHPS medium with no extract added was used. The percentage of inhibition was estimated as the number of dead trophozoites compared to the negative controls.

4.4.3. In Vitro IC$_{50}$ Determination

Each compound was dissolved in dimethyl sulfoxide and adjusted to 150, 75, 37.5, 18.75, and 9.375 µg/mL by adding a suspension of *E. histolytica* trophozoites at a logarithmic phase in PEHPS medium with 10% bovine serum. Vials were incubated for 72 h, and then chilled in cold water for 20 min, and the number of dead trophozoites per milliliter was evaluated using a hemocytometer. All assays were performed in triplicate. Metronidazole was used as a positive control, and as a negative control, an *E. histolytica* suspension in PEHPS medium with no extract added was used. The percentage of inhibition was estimated as the number of dead trophozoites compared to the negative controls. The 50% inhibitory concentration of each compound was calculated by using a Probit analysis, considering a 95% confidence level.

5. Conclusions

The isolated and pure flavonoids from *L. graveolens* showed significant growth inhibition against *E. histolytica* (52% to 97% at a concentration of 150 µg/mL). The IC_{50} values of these compounds ranged from 28.86 to 154.26 µg/mL, so they were not as effective as metronidazole (IC_{50} 0.205 µg/mL), but these IC_{50} values can be used as a guideline for further research on this plant as a source of potential antiamoebic agents.

The main contribution of this research work lies in the fact that it has shown that the presence of the flavonoids described herein in *Lippia graveolens* has a direct relationship with the antiprotozoal activity of extracts of this plant against *Entamoeba histolytica*. These flavonoids could be used as biomarkers [78] for the quality control of phytotherapeutics developed based on this work.

The results of our research may also form the basis for directly incorporating the use of *Lippia graveolens* extracts into conventional and complementary medicine for the treatment of amebiasis, as well as other infectious diseases.

Supplementary Materials: The following are available online, Figures S1–S30: supplementary spectroscopic data.

Author Contributions: Conceptualization, R.Q.-L.; investigation, R.Q.-L., J.V.-V., M.J.V.-S., and V.M.R.-G.; writing—review and editing, R.Q.-L. and Á.D.T.-H.; R.Q.-L. conceived and designed the experiments; R.Q.-L. and M.J.V.-S. performed the extraction and isolation; V.M.R.-G. performed the spectral analysis and structure determination; J.V.-V. conceived, designed, and performed the antiprotozoal assay; R.Q.-L. wrote the paper. All authors read and approved the final manuscript.

Funding: We thank the Universidad Autónoma de Nuevo León (Mexico) for PAICYT grants CN-422-10, and CN-662-11.

Acknowledgments: For the spectroscopic measurements, we thank Noemi Waksman-Minsky of the College of Medicine from Universidad Autónoma de Nuevo León.

Conflicts of Interest: The authors declare no conflicts of interest.

References

1. Ávila-García, R.; Valdés, J.; Jáuregui-Wade, J.M.; Ayala-Sumuano, J.T.; Cerbón-Solórzano, J. The metabolic pathway of sphingolipids biosynthesis and signaling in Entamoeba histolytica. *Biochem. Biophys. Res. Commun.* **2020**, *522*, 574–579. [CrossRef] [PubMed]

2. Carrero, J.C.; Reyes-López, M.; Serrano-Luna, J.; Shibayama, M.; Unzueta, J.; León-Sicairos, N.; De La Garza, M. Intestinal amoebiasis: 160 years of its first detection and still remains as a health problem in developing countries. *Int. J. Med. Microbiol.* **2020**, *310*, 151358. [CrossRef] [PubMed]

3. Broucke, S.V.D.; Verschueren, J.; Van Esbroeck, M.; Bottieau, E.; Ende, J.V.D. Clinical and microscopic predictors of Entamoeba histolytica intestinal infection in travelers and migrants diagnosed with Entamoeba histolytica/dispar infection. *PLoS Negl. Trop. Dis.* **2018**, *12*, e0006892. [CrossRef]

4. Aguilar-Rojas, A.; Olivo-Marin, J.-C.; Guillen, N. The motility of Entamoeba histolytica: Finding ways to understand intestinal amoebiasis. *Curr. Opin. Microbiol.* **2016**, *34*, 24–30. [CrossRef] [PubMed]

5. Muñoz, P.L.A.; Minchaca, A.Z.; Mares, R.E.; Ramos-Ibarra, M.A. Activity, stability and folding analysis of the chitinase from Entamoeba histolytica. *Parasitol. Int.* **2016**, *65*, 70–77. [CrossRef]

6. Guo, F.; Forde, M.; Werre, S.R.; Krecek, R.C.; Zhu, G. Seroprevalence of five parasitic pathogens in pregnant women in ten Caribbean countries. *Parasitol. Res.* **2016**, *116*, 347–358. [CrossRef]

7. Singh, R.S.; Walia, A.K.; Kanwar, J.R.; Kennedy, J.F. Amoebiasis vaccine development: A snapshot on E. histolytica with emphasis on perspectives of Gal/GalNAc lectin. *Int. J. Boil. Macromol.* **2016**, *91*, 258–268. [CrossRef]

8. Iyer, L.R.; Verma, A.K.; Paul, J.; Bhattacharya, A. Phagocytosis of Gut Bacteria by Entamoeba histolytica. *Front. Microbiol.* **2019**, *9*, 34. [CrossRef]

9. Institut-Pasteur Amoebiasis. Available online: https://www.pasteur.fr/en/medical-center/disease-sheets/amoebiasis-0 (accessed on 19 April 2020).

10. Gómez-García, C.; Marchat, L.A.; López-Cánovas, L.; Pérez-Ishiwara, D.G.; Rodriguez, M.A.; Orozco, E. Drug Resistance Mechanisms in Entamoeba histolytica, Giardia lamblia, Trichomonas vaginalis, and Opportunistic Anaerobic Protozoa. In *Antimicrobial Drug Resistance. Mechanisms of Drug Resistance*; Mayers, D.L., Sobel, J.D., Oullette, M., Kaye, K.S., Marchaim, D., Eds.; Springer International Publishing AG: Cham, Switzerland, 2017; pp. 613–628. ISBN 978-3-319-47264-5.

11. Luevano, J.H.E.; Castro-Ríos, R.; Sánchez-García, E.; Hernández-García, M.E.; Villarreal, J.V.; Rodríguez-Luis, O.E.; Chávez-Montes, A. In Vitro Study of Antiamoebic Activity of Methanol Extracts of Argemone mexicana on Trophozoites of Entamoeba histolytica HM1-IMSS. *Can. J. Infect. Dis. Med. Microbiol.* **2018**, *2018*, 1–8. [CrossRef]

12. Singh, B.; Singh, J.P.; Kaur, A.; Singh, N. Bioactive compounds in banana and their associated health benefits—A review. *Food Chem.* **2016**, *206*, 1–11. [CrossRef]

13. Hayat, F.; Azam, A.; Shin, D. Recent progress on the discovery of antiamoebic agents. *Bioorganic Med. Chem. Lett.* **2016**, *26*, 5149–5159. [CrossRef] [PubMed]

14. Varghese, S.S.; Ghosh, S.K. Stress-responsive Entamoeba topoisomerase II: A potential antiamoebic target. *FEBS Lett.* **2020**, *594*, 1005–1020. [CrossRef] [PubMed]

15. Newman, D.J.; Cragg, G.M. Natural Products as Sources of New Drugs from 1981 to 2014. *J. Nat. Prod.* **2016**, *79*, 629–661. [CrossRef]

16. Leyva-López, N.; Nair, V.; Bang, W.Y.; Cisneros-Zevallos, L.; Heredia, J.B. Protective role of terpenes and polyphenols from three species of Oregano (Lippia graveolens, Lippia palmeri and Hedeoma patens) on the suppression of lipopolysaccharide-induced inflammation in RAW 264.7 macrophage cells. *J. Ethnopharmacol.* **2016**, *187*, 302–312. [CrossRef]

17. Ramirez-Moreno, E.; Soto-Sanchez, J.; Rivera, G.; Marchat, L.A. Mexican Medicinal Plants as an Alternative for the Development of New Compounds Against Protozoan Parasites. In *Natural Remedies in the Fight Against Parasites*; Hanem, K., Govindarajan, M., Benelli, G., Eds.; INTECH: London, UK, 2017; pp. 61–91. ISBN 978-953-51-3290-5.

18. Martínez-Rocha, A.; Puga, R.; Hernández-Sandoval, L.; Loarca-Piña, G.; Mendoza, S. Antioxidant and Antimutagenic Activities of Mexican Oregano (Lippia graveolens Kunth). *Plant Foods Hum. Nutr.* **2007**, *63*, 1–5. [CrossRef] [PubMed]

19. Arana-Sánchez, A.; Estarrón-Espinosa, M.; Obledo-Vázquez, E.; Padilla-Camberos, E.; Silva-Vázquez, R.; Lugo-Cervantes, E. Antimicrobial and antioxidant activities of Mexican oregano essential oils (Lippia graveolens H. B. K.) with different composition when microencapsulated inβ-cyclodextrin. *Lett. Appl. Microbiol.* **2010**, *50*, 585–590. [CrossRef] [PubMed]

20. Rivero-Cruz, I.; Duarte, G.; Navarrete, A.; Bye, R.; Linares, E.; Mata, R. Chemical composition and antimicrobial and spasmolytic properties of Poliomintha longiflora and Lippia graveolens essential oils. *J. Food Sci.* **2011**, *76*, C309–C317. [CrossRef]

21. Adame-Gallegos, J.; Ochoa, S.A.; Nevárez-Moorillón, G.V. Potential Use of Mexican Oregano Essential Oil against Parasite, Fungal and Bacterial Pathogens. *J. Essent. Oil Bear. Plants* **2016**, *19*, 553–567. [CrossRef]

22. García-Heredia, A.; Garcia, S.; Merino-Mascorro, J.Á.; Feng, P.; Heredia, N. Natural plant products inhibits growth and alters the swarming motility, biofilm formation, and expression of virulence genes in enteroaggregative and enterohemorrhagic Escherichia coli. *Food Microbiol.* **2016**, *59*, 124–132. [CrossRef]

23. Quintanilla-Licea, R.; Mata-Cárdenas, B.D.; Villarreal, J.V.; Bazaldúa-Rodríguez, A.F.; Kavimngeles-Hernández, I.; Garza-González, J.N.; Hernández-García, M.E.; Ángeles-Hernández, I.K. Antiprotozoal Activity against Entamoeba histolytica of Plants Used in Northeast Mexican Traditional Medicine. Bioactive Compounds from Lippia graveolens and Ruta chalepensis. *Molecules* **2014**, *19*, 21044–21065. [CrossRef]

24. Ching, A.Y.L.; Wah, T.S.; Sukari, M.A.; Lian, G.E.C.; Rahmani, M.; Khalid, K. Characterization of Flavonoid Derivatives From Boesenbergia Rotunda (L.). *Malays. J. Anal. Sci.* **2007**, *11*, 154–159.

25. Jerz, G.; Waibel, R.; Achenbach, H. Cyclohexanoid protoflavanones from the stem-bark and roots of Ongokea gore. *Phytochemistry* **2005**, *66*, 1698–1706. [CrossRef] [PubMed]

26. Suleimen, Y.; Raldugin, V.A.; Adekenov, S.M. Cirsimaritin from Stizolophus balsamita. *Chem. Nat. Compd.* **2008**, *44*, 398. [CrossRef]

27. Jain, R.; Mittal, M. Naringenin, a flavonone from the stem of Nyctanthes arbortristis Linn. *Int. J. Biol. Pharm. Allied Sci.* **2012**, *1*, 964–972.

28. Bertelli, D.; Papotti, G.; Bortolotti, L.; Marcazzan, G.L.; Plessi, M. 1H-NMR Simultaneous Identification of Health-Relevant Compounds in Propolis Extracts. *Phytochem. Anal.* **2011**, *23*, 260–266. [CrossRef]

29. Katerere, D.R.; Gray, A.I.; Nash, R.J.; Waigh, R.D. Phytochemical and antimicrobial investigations of stilbenoids and flavonoids isolated from three species of Combretaceae. *Fitoterapia* **2012**, *83*, 932–940. [CrossRef]

30. Ramírez, J.; Cartuche, L.; Morocho, V.; Aguilar, S.; Malagón, O. Antifungal activity of raw extract and flavanons isolated from Piper ecuadorense from Ecuador. *Rev. Bras. Farm.* **2013**, *23*, 370–373. [CrossRef]

31. Guzmán-Avendaño, A.J.; Torrenegra-Guerrero, R.D.; Rodríguez-Aguirre, O.E. Flavonoids from Chromolaena subscandens (Hieron.) R.M. King & H. Rob. *Rev. Prod. Nat.* **2008**, *2*, 1–5.

32. McNulty, J.; Nair, J.J.; Bollareddy, E.; Keskar, K.; Thorat, A.; Crankshaw, D.; Holloway, A.C.; Khan, G.; Wright, G.D.; Ejim, L. Isolation of flavonoids from the heartwood and resin of Prunus avium and some preliminary biological investigations. *Phytochemistry* **2009**, *70*, 2040–2046. [CrossRef]

33. Grecco, S.D.S.; Dorigueto, A.C.; Landre, I.M.; Soares, M.G.; Martho, K.; Lima, R.; Pascon, R.C.; Vallim, M.A.; Capello, T.M.; Romoff, P.; et al. Structural Crystalline Characterization of Sakuranetin—An Antimicrobial Flavanone from Twigs of Baccharis retusa (Asteraceae). *Molecules* **2014**, *19*, 7528–7542. [CrossRef]

34. Yamashita, Y.; Hanaya, K.; Shoji, M.; Sugai, T. Simple Synthesis of Sakuranetin and Selinone via a Common Intermediate, Utilizing Complementary Regioselectivity in the Deacetylation of Naringenin Triacetate. *Chem. Pharm. Bull.* **2016**, *64*, 961–965. [CrossRef]

35. Alwahsh, M.A.A.; Khairuddean, M.; Chong, W.K. Chemical constituents and antioxidant activity of Teucrium barbeyanum Aschers. *Rec. Nat. Prod.* **2015**, *9*, 159–163.

36. Exarchou, V.; Kanetis, L.; Charalambous, Z.; Apers, S.; Pieters, L.; Gekas, V.; Goulas, V. HPLC-SPE-NMR Characterization of Major Metabolites in Salvia fruticosa Mill. Extract with Antifungal Potential: Relevance of Carnosic Acid, Carnosol, and Hispidulin. *J. Agric. Food Chem.* **2015**, *63*, 457–463. [CrossRef] [PubMed]

37. Lee, S.-J.; Jang, H.-J.; Kim, Y.; Oh, H.-M.; Lee, S.; Jung, K.; Kim, Y.-H.; Lee, W.-S.; Lee, S.W.; Rho, M.-C. Inhibitory effects of IL-6-induced STAT3 activation of bio-active compounds derived from Salvia plebeia R.Br. *Process. Biochem.* **2016**, *51*, 2222–2229. [CrossRef]

38. Gaggeri, R.; Rossi, D.; Christodoulou, M.S.; Passarella, D.; Leoni, F.; Azzolina, O.; Collina, S. Chiral Flavanones from Amygdalus lycioides Spach: Structural Elucidation and Identification of TNFalpha Inhibitors by Bioactivity-guided Fractionation. *Molecules* **2012**, *17*, 1665–1674. [CrossRef]

39. Grecco, S.D.S.; Reimão, J.Q.; Tempone, A.G.; Sartorelli, P.; Cunha, R.L.O.R.; Romoff, P.; Ferreira, M.J.P.; Fávero, O.A.; Lago, J. In vitro antileishmanial and antitrypanosomal activities of flavanones from Baccharis retusa DC. (Asteraceae). *Exp. Parasitol.* **2012**, *130*, 141–145. [CrossRef]

40. Kyriakou, E.; Primikyri, A.; Charisiadis, P.; Katsoura, M.; Stamatis, H.; Gerothanassis, I.P.; Tzakos, A. Unexpected enzyme-catalyzed regioselective acylation of flavonoid aglycones and rapid product screening. *Org. Biomol. Chem.* **2012**, *10*, 1739. [CrossRef]

41. Ashour, M.A.-G. New diacyl flavonoid derivatives from the Egyptian plant Blepharis edulis (Forssk.) Pers. *Bull. Fac. Pharm. Cairo Univ.* **2015**, *53*, 11–17. [CrossRef]

42. Joseph-Nathan, P.; Gordillo-Román, B. Vibrational Circular Dichroism Absolute Configuration Determination of Natural Products. In *Progress in the Chemistry of Organic Natural Products*; Kinghorn, A.D., Falk, H., Kobayashi, J., Eds.; Springer International Publishing Switzerland: Cham, Switzerland, 2015; Volume 100, pp. 311–452.

43. Silverstein, R.M.; Bassler, G.C. Spectrometric identification of organic compounds. *J. Chem. Educ.* **1962**, *39*, 546. [CrossRef]

44. Grotewold, E. *The Science of Flavonoids*; Springer Science+Business Media, Inc.: New York, NY, USA, 2006; ISBN 9780387288222.

45. Pinheiro, P.F.; Justino, G.C. Structural analysis of flavonoids and related compounds—A review of spectroscopic applications. In *Phytochemicals–A Global Perspective of Their Role in Nutrition and Health*; Rao, V., Ed.; InTech Europe: Rijeka, Croatia, 2012; pp. 33–56. ISBN 978-953-51-0296-0.

46. Ren, W.; Qiao, Z.; Wang, H.; Zhu, L.; Zhang, L. Flavonoids: Promising anticancer agents. *Med. Res. Rev.* **2003**, *23*, 519–534. [CrossRef]

47. Cai, Y.; Luo, Q.; Sun, M.; Corke, H. Antioxidant activity and phenolic compounds of 112 traditional Chinese medicinal plants associated with anticancer. *Life Sci.* **2004**, *74*, 2157–2184. [CrossRef] [PubMed]

48. Cushnie, T.T.; Lamb, A.J. Antimicrobial activity of flavonoids. *Int. J. Antimicrob. Agents* **2005**, *26*, 343–356. [CrossRef] [PubMed]

49. Tasdemir, D.; Kaiser, M.; Brun, R.; Yardley, V.; Schmidt, T.J.; Tosun, F.; Ruedi, P. Antitrypanosomal and Antileishmanial Activities of Flavonoids and Their Analogues: In Vitro, In Vivo, Structure-Activity Relationship, and Quantitative Structure-Activity Relationship Studies. *Antimicrob. Agents Chemother.* **2006**, *50*, 1352–1364. [CrossRef] [PubMed]

50. Plochmann, K.; Korte, G.; Koutsilieri, E.; Richling, E.; Riederer, P.; Rethwilm, A.; Schreier, P.; Scheller, C. Structure–activity relationships of flavonoid-induced cytotoxicity on human leukemia cells. *Arch. Biochem. Biophys.* **2007**, *460*, 1–9. [CrossRef] [PubMed]

51. Delgado, L.; Fernandes, I.; González-Manzano, S.; De Freitas, V.; Mateus, N.; Santos-Buelga, C. Anti-proliferative effects of quercetin and catechin metabolites. *Food Funct.* **2014**, *5*, 797. [CrossRef] [PubMed]

52. Resende, F.A.; Almeida, C.P.D.S.; Vilegas, W.; Varanda, E.A. Differences in the hydroxylation pattern of flavonoids alter their chemoprotective effect against direct- and indirect-acting mutagens. *Food Chem.* **2014**, *155*, 251–255. [CrossRef]

53. Martínez-Castillo, M.; Pacheco-Yépez, J.; Flores, N.; Guzman, P.G.; Jarillo-Luna, R.A.; Cárdenas-Jaramillo, L.M.; Campos-Rodríguez, R.; Shibayama, M. Flavonoids as a Natural Treatment Against Entamoeba histolytica. *Front. Microbiol.* **2018**, *8*, 1–14. [CrossRef]

54. Vaya, J.; Tamir, S. The relation between the chemical structure of flavonoids and their estrogen-like activities. *Curr. Med. Chem.* **2004**, *11*, 1333–1343. [CrossRef]

55. Agrawal, A.D. Pharmacological activities of flavonoids: A review. *Int. J. Pharm. Sci. Nanotechnol.* **2011**, *4*, 1394–1398.

56. Beekmann, K.; Actis-Goretta, L.; Van Bladeren, P.J.; Dionisi, F.; Destaillats, F.; Rietjens, I.M.C.M. A state-of-the-art overview of the effect of metabolic conjugation on the biological activity of flavonoids. *Food Funct.* **2012**, *3*, 1008. [CrossRef]

57. Kumar, S.; Pandey, A.K. Chemistry and Biological Activities of Flavonoids: An Overview. *Sci. World J.* **2013**, *2013*, 1–16. [CrossRef] [PubMed]

58. Lin, L.-Z.; Mukhopadhyay, S.; Robbins, R.J.; Harnly, J.M. Identification and quantification of flavonoids of Mexican oregano (Lippia graveolens) by LC-DAD-ESI/MS analysis. *J. Food Compos. Anal.* **2007**, *20*, 361–369. [CrossRef] [PubMed]

59. Dominguez, S.; Sánchez, H.; Suárez, M.; Baldas, J.; González, M.D.R. Chemical Constituents of Lippia graveolens. *Planta Med.* **1989**, *55*, 208–209. [CrossRef]

60. González-Güereca, M.C.; Soto-Hernández, M.; Kite, G.; Martínez-VÁzquez, M. Antioxidant activity of flavonoids from the stem of the Mexican oregano (Lippia graveolens HBK var. berlandieri Schauer). *Rev. Fitotec. Mex.* **2007**, *30*, 43–49.

61. Rufino-González, Y.; Ponce-Macotela, M.; González-Maciel, A.; Reynoso-Robles, R.; Jiménez-Estrada, M.; Sanchez-Contreras, A.; Martínez-Gordillo, M.N. In vitro activity of the F-6 fraction of oregano against Giardia intestinalis. *Parasitology* **2012**, *139*, 434–440. [CrossRef] [PubMed]

62. Moreno-Rodríguez, A.; Vázquez-Medrano, J.; Hernández-Portilla, L.B.; Peñalosa-Castro, I.; Canales-Martínez, M.; Orozco-Segovia, A.; Jiménez-Estrada, M.; Colville, L.; Pritchard, H.W.; Flores-Ortiz, C.M. The effect of light and soil moisture on the accumulation of three flavonoids in the leaves of Mexican oregano (Lippia graveolens Kunth). *J. Food, Agric. Environ.* **2014**, *12*, 1272–1279.

63. Grael, C.; De Albuquerque, S.; Lopes, J.L. Chemical constituents of Lychnophora pohlii and trypanocidal activity of crude plant extracts and of isolated compounds. *Fitoterapia* **2005**, *76*, 73–82. [CrossRef]

64. Alday-Provencio, S.; Diaz, G.; Rascon, L.; Quintero, J.; Alday, E.; Robles-Zepeda, R.; Garibay-Escobar, A.; Astiazaran, H.; Hernández, J.; Velázquez, C. Sonoran Propolis and Some of its Chemical Constituents Inhibit in vitro Growth of Giardia lamblia Trophozoites. *Planta Med.* **2015**, *81*, 742–747. [CrossRef]

65. Calzada, F.; Meckes, M.; Cedillo-Rivera, R. Antiamoebic and Antigiardial Activity of Plant Flavonoids. *Planta Med.* **1999**, *65*, 78–80. [CrossRef]

66. Tasdemir, D.; Lack, G.; Brun, R.; Rüedi, P.; Scapozza, L.; Perozzo, R. Inhibition of Plasmodium falciparum fatty acid biosynthesis: Evaluation of FabG, FabZ, and FabI as drug targets for flavonoids. *J. Med. Chem.* **2006**, *49*, 3345–3353. [CrossRef]

67. Calzada, F.; Velazquez, C.; Cedillo-Rivera, R.; Esquivel, B. Antiprotozoal activity of the constituents of Teloxys graveolens. *Phytother. Res.* **2003**, *17*, 731–732. [CrossRef] [PubMed]

68. Bolaños, V.; Díaz-Martínez, A.; Soto, J.; Marchat, L.A.; Sánchez-Monroy, V.; Ramírez-Moreno, E. Kaempferol inhibits Entamoeba histolytica growth by altering cytoskeletal functions. *Mol. Biochem. Parasitol.* **2015**, *204*, 16–25. [CrossRef] [PubMed]

69. Fatima, S.; Gupta, P.; Agarwal, S.M. Insight into structural requirements of antiamoebic flavonoids: 3D-QSAR and G-QSAR studies. *Chem. Boil. Drug Des.* **2018**, *92*, 1743–1749. [CrossRef] [PubMed]

70. Soto, J.; Gómez, C.; Calzada, F.; Ramírez, M.E. Ultrastructural Changes on Entamoeba histolytica HM1-IMSS Caused by the Flavan-3-Ol, (-)-Epicatechin. *Planta Med.* **2009**, *76*, 611–612. [CrossRef] [PubMed]

71. Bolaños, V.; Díaz-Martínez, A.; Soto, J.; Rodríguez, M.A.; López-Camarillo, C.; Marchat, L.A.; Ramírez-Moreno, E. The flavonoid (-)-epicatechin affects cytoskeleton proteins and functions in Entamoeba histolytica. *J. Proteom.* **2014**, *111*, 74–85. [CrossRef]

72. Yi, W.; Wei, Q.; Di, G.; Jing-Yu, L.; Luo, Y.-L. Phenols and flavonoids from the aerial part of Euphorbia hirta. *Chin. J. Nat. Med.* **2012**, *10*, 40–42.

73. Vasconcelos, J.M.; Silva, A.M.S.; Cavaleiro, J.A.S. Chromones and flavanones from artemisia campestris subsp. maritima. *Phytochemistry* **1998**, *49*, 1421–1424. [CrossRef]

74. Oyama, K.-I.; Kondo, T. Total Synthesis of Flavocommelin, a Component of the Blue Supramolecular Pigment from Commelinacommunis, on the Basis of Direct 6-C-Glycosylation of Flavan. *J. Org. Chem.* **2004**, *69*, 5240–5246. [CrossRef]

75. Isobe, T.; Doe, M.; Morimoto, Y.; Nagata, K.; Ohsaki, A. The anti-Helicobacter pylori flavones in a Brazilian plant, Hyptis fasciculata, and the activity of methoxyflavones. *Boil. Pharm. Bull.* **2006**, *29*, 1039–1041. [CrossRef]

76. Mata-Cárdenas, B.D.; Vargas-Villarreal, J.; González-Salazar, F.; Palacios-Corona, R.; Salvador, S.-F. A new vial microassay to screen antiprotozoal drugs. *Pharmacologyonline* **2008**, *1*, 529–537.

77. González-Salazar, F.; Mata-Cárdenas, B.D.; Vargas-Villarreal, J. Sensibilidad de trofozoitos de Entamoeba histolytica a Ivermectina. *Medicina (Buenos Aires)* **2009**, *69*, 318–320.

78. Hirudkar, J.R.; Parmar, K.M.; Prasad, R.S.; Sinha, S.K.; Jogi, M.S.; Itankar, P.R.; Prasad, S.K. Quercetin a major biomarker of Psidium guajava L. inhibits SepA protease activity of Shigella flexneri in treatment of infectious diarrhoea. *Microb. Pathog.* **2019**, *138*, 103807. [CrossRef] [PubMed]

Sample Availability: Not available.

MDPI
St. Alban-Anlage 66
4052 Basel
Switzerland
Tel. +41 61 683 77 34
Fax +41 61 302 89 18
www.mdpi.com

Molecules Editorial Office
E-mail: molecules@mdpi.com
www.mdpi.com/journal/molecules

www.ingramcontent.com/pod-product-compliance
Lightning Source LLC
LaVergne TN
LVHW071548170726
843515LV00008B/2235